Introductory Digital Wireless Communications

Gordon Stuber

Introductory Digital Wireless Communications

Springer

Gordon Stuber
School of Electrical and Computer Engineering
Georgia Institute of Technology
Atlanta, GA, USA

Request lecturer material: sn.pub/lecturer-material in the below link https://springernature.sharepoint.com/:u:/r/sites/Research_IT_SGW-ESM/Shared%20Documents/Lecturer%20Materials/978-3-032-01558-7/solution-manual.zip?csf=1&web=1&e=J8wHby

ISBN 978-3-032-01558-7 ISBN 978-3-032-01559-4 (eBook)
https://doi.org/10.1007/978-3-032-01559-4

This Springer imprint is published by the registered company Springer Nature Switzerland AG
The registered company address is: Gewerbestrasse 11, 6330 Cham, Switzerland

To the alumni of the Wireless Systems Laboratory

Preface

Introductory Digital Wireless Communications is a senior-level or fourth year undergraduate textbook for courses in digital communications, information theory and coding, or wireless communications. The textbook draws some basic material from my graduate-level textbook *Principles of Mobile Communication*; however, the presentation is more tutorial in nature and less dense. As such, the textbook is also useful reference for graduate students that have no prior course work on these topics. *Introductory Digital Wireless Communications* assumes a background in introductory signal processing, signals and systems, Fourier transforms, and probability. It takes a mathematical modeling approach to physical layer communication systems, and provides detailed step-by-step derivations of key results, along with examples, to guide the student in the subject matter.

Most senior undergraduate students in Electrical and Computer Engineering will have taken a course on probability and statistics, but lack exposure to random processes. Since these topics are critical for the modeling and analysis of physical layer communications, the text includes a review of probability in Chap. 2 and an introduction to random processes that is tailored to communication systems in Chap. 3. Concepts from these two foundation chapters are used throughout the remainder of the textbook.

Chapter 4 introduces bandpass communication signals and channels, including their complex envelope representations and bandpass random signals. Chapter 5 introduces communication waveforms and their vector representation, basic linear modulation schemes, and pulse shaping and power spectrum. Chapter 6 is concerned with the detection of bandpass modulated waveforms, both coherent and noncoherent detection, and the derivation of error probability for the linear modulation schemes introduced in Chap. 5.

The solution manual available in individual chapters needs to be removed, and it will be made available only upon request, not to students. The original version of the book has been revised. A correction to this book can be found at https://doi.org/10.1007/978-3-032-01559-4_14

Chapter 7 introduces information theory, including the concepts of entropy, mutual information, and capacity through a series of theorems. Both discrete and continuous communication channels are considered. Chapter 8 treats error control methods, including forward error correction techniques such as binary block coding, convolutional coding, and trellis-coded modulation. It also covers automatic repeat request techniques that use a feedback channel from the receiver to the transmitter.

Chapters 9 and 10 deal with inter-symbol interference (ISI). Chapter 9 considers front-end processing for ISI channels, and then explores a time-domain approach to equalization, studying the classical zero-forcing equalizer, its tap solution, and its adaptive version. The chapter also considers partial response signals that intentionally induce ISI. Chapter 10 introduces orthogonal frequency division multiplexing (OFDM) and its DFT implementation, where equalization is done efficiently in the frequency domain.

Chapters 11 to 13 are relevant to wireless communication systems. Chapter 11 introduces the land mobile radio channel, including the modeling, statistical characterization and simulation of multipath-fading channels, and the modeling of shadow fading and path loss. Chapter 12 is concerned with multi-antenna techniques that provide for receiver diversity, transmitter diversity, and multi-input multi-output (MIMO) communications, and includes basic space-time coding and spatial multiplexing concepts. Chapter 13 is devoted to cellular land mobile radio deployment and can be largely understood independent of the other chapters.

A course in digital communications may focus on Chaps. 1 through 6, and optional material from Chaps. 7, 8, 9, and 10. A course in physical wireless communications may start with Chap. 13, followed by Chaps. 2 through 4, parts of Chaps. 5 linkdes and 6, and Chaps. 10 through 12. Problems are provided at the end of each chapter.

Atlanta, GA, USA
June 2025

Gordon Stüber

Contents

Acronyms

ACK	Acknowledgment
A/D	Analog to digital converter
AM	Amplitude modulation
AoA	Angle-of-arrival
AoD	Angle-of-departure
ARQ	Automatic repeat request
AWGN	Additive white Gaussian noise
BCH	Bose-Chaudhuri-Hocquenghem
BER	Bit error rate
BFSK	Binary frequency shift keying
BPSK	Binary phase shift keying
BS	Base station
BSC	Binary symmetric channel
BEC	Binary erasure channel
BU	Bad urban
CCI	Co-channel interference
cdf	Cumulative distribution function
cdfc	Complementary distribution function
CDMA	Code division multiple access
D/A	Digital-to-analog converter
DFT	Discrete Fourier transform
DMC	Discrete memoryless channel
DMS	Discrete memoryless source
FDD	Frequency division duplexing
FEC	Forward error correction
FEPL	Flat Earth path loss
FFT	Fast Fourier transform
FIR	Finite impulse response
FM	Frequency modulation
FSK	Frequency shift keying
FSM	Finite-state machine

FSPL	Free-space path loss
GBN	Go-back-N
GSM	Global system for mobile communication
HT	Hilly terrain
IDFT	Inverse discrete Fourier transform
IFFT	Inverse fast Fourier transform
i.i.d.	Independent and identically distributed
ISI	Intersymbol interference
LoS	Line-of-sight
MAHO	Mobile-assisted handoff
MAP	Maximum aposteriori probability
MDS	Maximum distance separable
MFSK	M-ary frequency shift keying
MIMO	Multiple-input multiple-output
ML	Maximum likelihood
MLSE	Maximum likelihood sequence estimation
MPSK	M-ary phase shift keying
MRC	Maximal ratio combining
MS	Mobile station
NACK	Negative acknowledgement
NLoS	Non-line-of-sight
NXT	New transmission
OFDM	Orthogonal frequency division multiplexing
OFDMA	Orthogonal frequency division multiple access
OOK	On-off keying
PAM	Pulse amplitude modulation
PAPR	Peak-to-average power ratio
pdf	Probability density function
psd	Power spectral density
PSK	Phase shift keying
QAM	Quadrature amplitude modulation
QPSK	Quadrature phase shift keying
QoS	Quality of service
RA	Rural area
RF	Radio frequency
rms	Root mean square
RBU	Reduced bad urban
RHT	Reduced hilly terrain
RTX	Retransmission
RTU	Reduced typical urban
SAW	Stop-and-wait
SC	Selective combining
SNR	Signal-to-noise ratio
SoS	Sum-of-sinusoids
SR	Selective repeat

SSK	Space shift keying
STD	Selective transmit diversity
SVD	Singular value decomposition
TCM	Trellis coded modulation
TDD	Time division duplexing
TDTD	Time division transmit diversity
TSTD	Time switched transmit diversity
TU	Typical urban
UWB	Ultra wide-band
WSS	Wide sense stationary
WSSUS	Wide sense stationary uncorrelated scattering
ZF	Zero-forcing

Chapter 1
Introduction

Legacy *analog communication* systems transmit information using analog waveforms that are not constrained to a finite set of signal waveforms. Only a few analog communication systems currently remain, such as broadcast amplitude modulation/frequency modulation (AM/FM) radio. All modern communication systems use *digital communication*, where information is exchanged between a source and destination by using a *finite* set of finite-energy continuous-time waveforms. Digital communication is preferred over analog communication for a variety of reasons, including the following:

- Digital communications is a natural choice for digital sources, e.g., computer communications.
- Source encoding or data compression techniques can reduce the required transmission bandwidth with either no loss of information or a controlled amount of message distortion.
- Digital communication is more robust to channel impairments such as:
  - internal receiver thermal noise
  - external co-channel and adjacent channel interference
  - multipath-fading found in wireless channels
  - surface defects found in recording media such as optical and magnetic disks.
- Digital communication has higher bandwidth efficiency and spectral efficiency than analog communications.
- Data encryption and multiplexing is much easier.
- Digital communication systems benefit from using powerful digital signal processing techniques.

Consider the traditional partitioned communication system shown in Fig. 1.1 showing a communication link between an information source and sink. The source encoder is intended to represent the source message such as voice or video in a compact fashion. For example, a vocoder or voice coder is a type of source

G. Stuber, *Introductory Digital Wireless Communications*,
https://doi.org/10.1007/978-3-032-01559-4_1

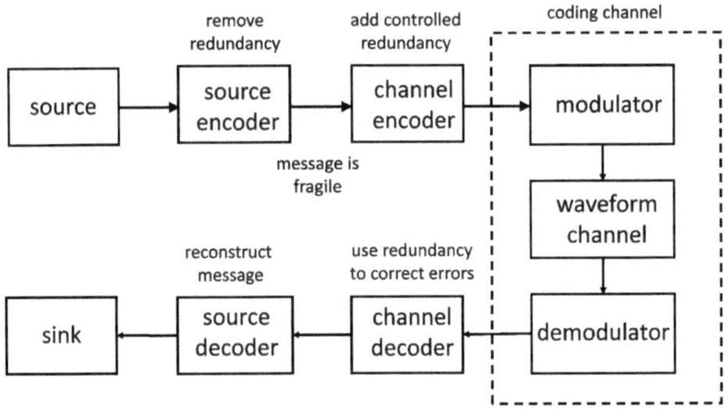

Fig. 1.1 Partitioned communication system

coder that compresses a speech waveform into a sequence of binary data bits. The binary data bit rate at the vocoder output, known as the vocoder rate, is a key parameter. The vocoder rate can be a low as 2 kb/s to as high as 9.6 kb/s or more. A vocoder is not lossless as it reduces the information content of the voice message by introducing distortion into the reconstructed voice waveform at the destination. A greater amount of distortion will generally be introduced as the vocoder rate is reduced.

The compressed message bits at the output of the source coder are fragile, such that any errors in their transmission between source and destination will lead to noticeable artifacts in the reconstructed source message, for example, distortion in a reconstructed speech waveform. This problem becomes more acute as the degree of source compression increases. To remedy this problem, channel coding is used, where controlled redundancy is introduced into the source encoder output that can be exploited at the receiver to correct for transmission errors. Since the source encoder removes redundancy while the channel coder adds redundancy, there is a complicated trade-off between the two to achieve a given end-to-end quality of service (QoS).

At the transmitter, the output of the channel coder is applied to a digital modulator that maps the binary code bits onto a finite set of communication waveforms. Exactly how this mapping is done is critical to the bit error rate performance of the communication link. These baseband waveforms may then be radio frequency (RF) carrier modulated and transmitted over a waveform channel as a bandpass waveform whose bandwidth is typically much less than the RF carrier frequency. Modulation schemes are also possible that do not use RF carrier modulation but instead communicate over a large bandwidth by using very short duration pulses, typically on the order of a nanosecond, commonly known as ultra-wide-band (UWB) signaling. Since their bandwidth is large and may overlap and cause interference to other spectrum users, UWB systems typically have strict emission constraints that limit their use to communication over short distances.

The waveform channel is the physical medium used to transmit information and may assume a wide variety of forms. These include, among other things, the following:

- wireline channels, e.g., twisted copper pair, coaxial cable, residential power lines
- fiber-optic cable
- wireless channels, e.g., microwave and mm-wave channels
  - line-of-sight, e.g., satellite, point-to-point microwave radio
  - non-line-of-sight, e.g., cellular, wireless local area networks
- underwater acoustic channels that transmit information using sound
- storage channels, e.g., optical and magnetic disks.
  - communication with storage channels occurs from the present into the future.

At the receiver, the received bandpass waveform is processed to perform RF demodulation followed by baseband signal detection. The output of the detector is input to the channel decoder that corrects and/or detects any transmission errors. If uncorrectable transmission errors are detected, the receiver may implement an automatic repeat request (ARQ) protocol that asks the transmitter to resend the message via a return channel. Finally, the output of the channel decoder is input to the source decoder that reconstructs the original source message, for example, a speech waveform.

The definition of what constitutes the communication channel depends upon how it is defined exactly in the communication chain in terms of its input and output. For example, a communication channel may have as its input a bandpass communication waveform that is radiated from the antenna of a communication transmitter, while the output of the communication channel is the bandpass waveform that is incident upon the antenna of a communication receiver. Alternatively, the transmitter and receiver antenna subsystems may be incorporated into the communication channel. In this case, the communication channel input is the bandpass waveform input to the transmitter antenna, while the communication channel output is the bandpass waveform at the output of the receiver antenna.

To design effective digital communication systems and to analyze their theoretical performance, mathematical models are needed for the communication channels. There are several commonly used mathematical channel models.

Additive White Gaussian Noise (AWGN) Channel

The simplest of all communication channel models is the additive white Gaussian noise (AWGN) channel as depicted in Fig. 1.2.

Considering a wireless communication link, the signal at the output of a communication receiver antenna typically has a very low power level, so it must be amplified before further processing. The amplifier electronics will introduce thermal noise, present in all communication receivers that is modeled as AWGN.

The AWGN waveform, $w(t)$, has a spectrally flat or "white" power spectral density of $S_{ww}(f) = N_o/2$ Watts/Hz, where the parameter $N_o = kT$ is directly

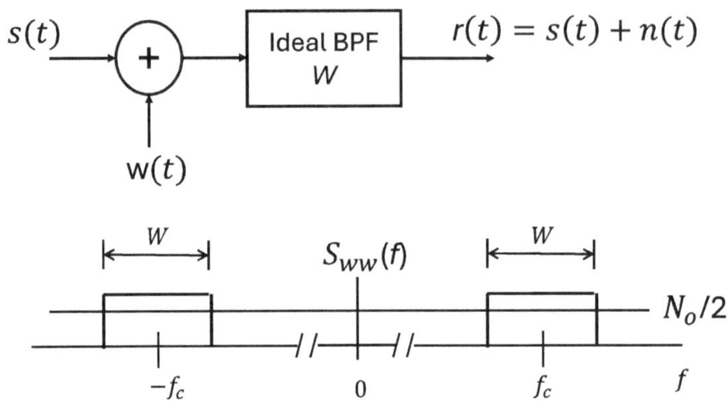

Fig. 1.2 Additive white Gaussian noise (AWGN) channel model

proportional to the operating temperature of the receiver, T^o Kelvin, where the constant of proportionality $k = 1.38 \times 10^{-23}$ Joules/degree is Boltzmann's constant. A white noise model implies that all frequency components are present and with equal strength in the white noise waveform $w(t)$ similar to how all colors are present with equal intensity in white light.

In order to limit the effects of receiver thermal noise and other externally received emissions falling outside of the signal bandwidth, W, the received signal plus noise, $s(t) + w(t)$, is bandpass filtered to yield the output $r(t) = s(t) + n(t)$. Assuming an ideal bandpass filter with passband W as in Fig. 1.2, the thermal noise power at the filter output is

$$\frac{N_o}{2} \cdot 2 \cdot W = N_o W \quad \text{Watts} \tag{1.1}$$

where both the positive and negative frequencies are included in the calculation. The bandpass-filtered AWGN, $n(t)$, sampled at any time t_1, $n = n(t_1)$, is a zero-mean Gaussian random variable having variance $N_o W$, as denoted by $f(n) \sim \mathcal{N}(0, N_o W)$. Given the signal $s(t)$ at any time t_1, $s = s(t_1)$, the filter output $r(t)$ at time t_1, $r = r(t_1)$ has the conditional Gaussian distribution $f(r|s) \sim \mathcal{N}(s, N_o W)$.

Linear Filter Channel Model

The linear filter channel model goes a step beyond the AWGN channel model, by passing the message waveform $s(t)$ through a linear filter having an impulse response $h(t)$, prior to the addition of the AWGN as shown in Fig. 1.3.

The channel output is given by

$$r(t) = \int_{-\infty}^{\infty} h(\tau)s(t-\tau)d\tau + n(t) = s(t) * h(t) + n(t) \tag{1.2}$$

where the first term on the R.H.S. is the continuous-time convolution of $s(t)$ and $h(t)$ and the second term is the AWGN $n(t) = w(t) * h(t)$. By taking Fourier

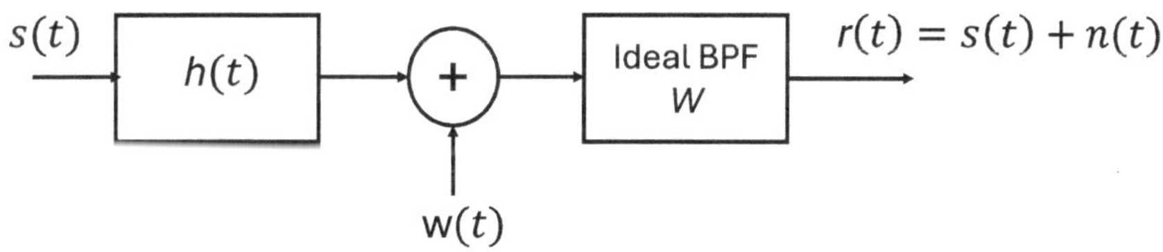

Fig. 1.3 Linear filter channel with AWGN

transforms of both sides of the above equation, convolution in the time domain becomes multiplication in the frequency domain, so

$$R(f) = H(f)S(f) + N(f). \tag{1.3}$$

An *ideal channel* is one having an impulse response of the form:

$$h(t) = \alpha\delta(t - t_0) \tag{1.4}$$

and, therefore,

$$r(t) = \alpha s(t - t_o) + n(t). \tag{1.5}$$

The ideal channel simply delays the input signal by t_o seconds and attenuates it by a scalar α, but otherwise leaves the input signal undistorted. The transfer function of the ideal channel in (1.4) is the Fourier transform of $h(t)$:

$$H(f) = \mathcal{F}\{h(t)\} = \alpha e^{-j2\pi f t_o}. \tag{1.6}$$

An ideal channel has two characteristics:

- the *magnitude response* $|H(f)| = \alpha$ is flat or constant in frequency f.
- the *phase response* $\angle H(f) = -2\pi f t_o$ is linear in frequency f.

An ideal channel will always have a single impulse function. By property of the Fourier transform, anything other than a single impulse function, e.g., a pair of impulses having different delays, will result in a frequency response that varies with frequency f.

Coding Channel
As shown in Fig. 1.1, the coding channel is that portion of the communication system "seen" by the coding system. The coding channel accepts as inputs the outputs of the channel encoder, and the coding channel outputs are input the channel decoder. The coding channel itself will depend on the type of modulation and detection that is used and the waveform channel. In a digital communication system, the coding channel inputs are chosen from a finite alphabet. The coding channel outputs will depend on the type of detection and decoding that is used.

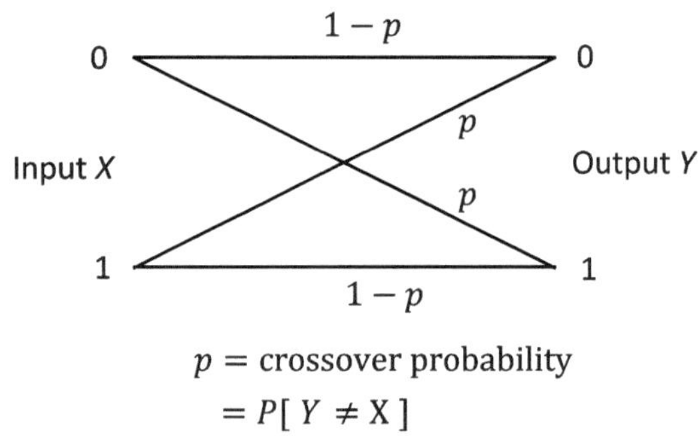

Fig. 1.4 Binary symmetric channel (BSC)

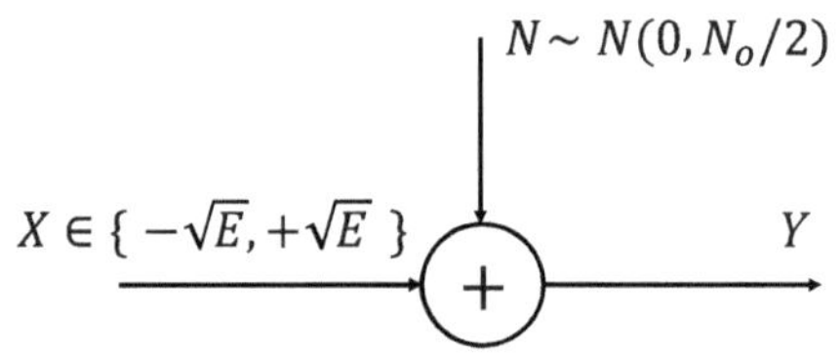

Fig. 1.5 Discrete-input AWGN coding channel

Suppose that a binary error control code is used, and the binary-valued code bits $X \in \{0, 1\}$ are sent over a waveform channel by using a pair of distinct waveforms. Due to noise and other impairments, errors are introduced into the coding channel outputs $Y \in \{0, 1\}$ that are input to the decoder. Such a coding channel may be described by the four channel transition probabilities $P[Y = y_j | X = x_i], i, j \in \{0, 1\}$. In the case of a binary symmetric channel (BSC), $P[Y = 1|X = 0] = P[Y = 0|X = 1] = p$, such that the coding channel has a "symmetry" as depicted in Fig. 1.4. Chapter 6 shows that the crossover or code bit error probability p will depend on the type of modulation and detection being used, along with the received bit energy-to-noise ratio, E_b/N_o.

The convolutional and trellis codes described in Chap. 8 benefit from using a *soft decision decoder* that does not make hard decisions on the received code symbols or code bits before they are input to the channel decoder. Rather the soft values at the output of the demodulator/detector are input directly to the channel decoder that makes a sequence decision on the entire codeword. Generally, soft decision decoding results in better performance than hard decision decoding. In the case of binary modulation and an AWGN channel, the inputs to the coding channel are the binary modulation symbols, and the outputs of the coding channel are the noise-corrupted modulation symbols. As shown in Fig. 1.5, $X \in \{\sqrt{E}, -\sqrt{E}\}$ is the binary code symbol having energy E that is input to the coding channel, N is the AWGN such that $N \sim \mathcal{N}(0, N_o/2)$, where $N_o/2$ is the AWGN power spectral density, and $Y = X + N$ is the output of the coding channel.

Signal-to-Noise Ratio

The term *signal-to-noise ratio* (SNR) refers to a ratio of signal power P to noise power N, given by P/N. For AWGN channels, the SNR may be related to the modulation symbol-energy-to-noise ratio E_s/N_o, where E_s is the energy per modulation symbol and $N_o/2$ is the two-sided power spectral density of the AWGN. Doing so requires two things, one is the modulation symbol period T or baud rate $R = 1/T$ (symbols/second) and the other is the noise equivalent bandwidth B_w that is explained further in Sect 4.4.4. Using these quantities

$$\frac{E_s}{N_o} = \frac{P}{N} \cdot \frac{B_w}{R} \tag{1.7}$$

where $E_s = PT = P/R$ and $N_o = N/B_w$.

As shown in Fig. 1.6, the SNR may be related to the information bit-SNR, E_b/N_o, codebit-SNR, E_r/N_o, modulation symbol-SNR, E_s/N_o, or chip-SNR, E_c/N_o. They are all interrelated in the communication chain consisting of the channel encoder, bit mapping, symbol modulator, and time/frequency spreading. Performance comparisons of different modulation, coding, and spreading options should always be made on the basis of equal information bit energy-to-noise ratio or bit-SNR, E_b/N_o, to ensure a fair comparison.

Channel Capacity

Claude Shannon in his pioneering paper "A Mathematical Theory of Communication" [27] proved that every coding channel has a *channel capacity*, C, defined as the maximum possible rate that information can be transmitted over the coding channel with *arbitrary reliability*. Arbitrary reliability means that as long as the data rate R is less the coding channel capacity C, the information bit error rate (BER) can theoretically be made as small as desired. Conversely, information cannot be transmitted reliably over a coding channel at any data rate R greater than the coding

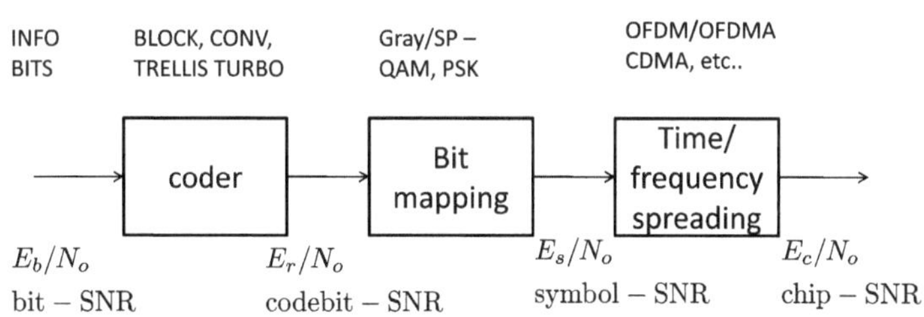

$$\begin{aligned} E_b &= \text{energy per information bit} \\ E_r &= \text{energy per code bit} \\ E_s &= \text{energy per modulated symbol} \\ E_c &= \text{energy per spreading chip} \end{aligned}$$

Fig. 1.6 Various definitions of signal-to-noise ratio

channel capacity, C, as the information BER is 1/2 meaning that the coding channel is useless. Arbitrary reliability is realized in practice not by increasing the signal power but by using increasingly more powerful error control coding techniques.

Consider the band-limited AWGN channel shown in Fig. 1.7. The ideal low-pass filter has transfer function:

$$H(f) = \begin{cases} 1 \ , & -W \leq f \leq W \\ 0 \ , & \text{elsewhere} \end{cases}$$

The AGWN $w(t)$ has a two-sided power spectral density of $N_o/2$ watts/Hz. The input to the coding channel, $s(t)$, is assumed to be unconstrained except that the average input power is P watts.

The capacity of the bandlimited power-constrained AWGN channel is derived in Sect. 7.5.2 as

$$C = W\log_2\left(1 + \frac{P}{N_o W}\right) \quad \text{bits/s} \tag{1.8}$$

where

$$\begin{aligned} W &= \text{channel bandwidth (Hz)} \\ P &= \text{average signal power (watts)} \\ N_o/2 &= \text{noise power spectral density (watts/Hz)} \end{aligned}$$

Dividing both sides of the above equation by W yields:

$$\frac{C}{W} = \log_2\left(1 + \frac{P}{N_o W}\right) = \log_2\left(1 + \frac{E_b}{N_o} \cdot \frac{R}{W}\right) \quad \text{bits/s/Hz}$$

$$\begin{aligned} R = 1/T &= \text{data rate (bits/second)} \\ E_b &= \text{energy per information data bit (Joules)} = PT \\ E_b/N_o &= \text{bit energy-to-noise spectral density ratio (dimensionless)} \\ R/W &= \text{bandwidth efficiency (bits/s/Hz)} \end{aligned}$$

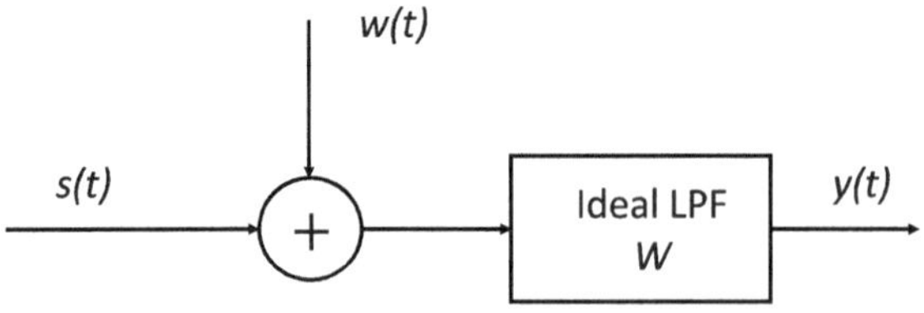

Fig. 1.7 Bandlimited AWGN channel

If information is transmitted at a rate R equal to the channel capacity C, i.e., $R = C$, then

$$\frac{C}{W} = \log_2\left(1 + \frac{E_b}{N_o} \cdot \frac{C}{W}\right) \text{ bits/s/Hz}$$

Inverting this equation gives E_b/N_o as a function of C/W

$$\frac{E_b}{N_o} = \frac{2^{C/W} - 1}{C/W} \tag{1.9}$$

The above equation may be used to plot C/W vs. E_b/N_o as shown in Fig. 1.8.

Example 1.1 Suppose that $W = 6$ MHz (TV channel bandwidth) and the signal-to-noise power ratio is $P/N = P/(N_o W) = 20$ dB. What is the channel capacity?

In this case, the channel capacity is $C = (6 \times 10^6)\log_2(1 + 100) =$ 40 Mbps.

This means that it is theoretically possible to transmit information with arbitrary reliably through this channel for all data rates up to 40 Mbps. However, any data rate greater than 40 Mbps is theoretically infeasible.

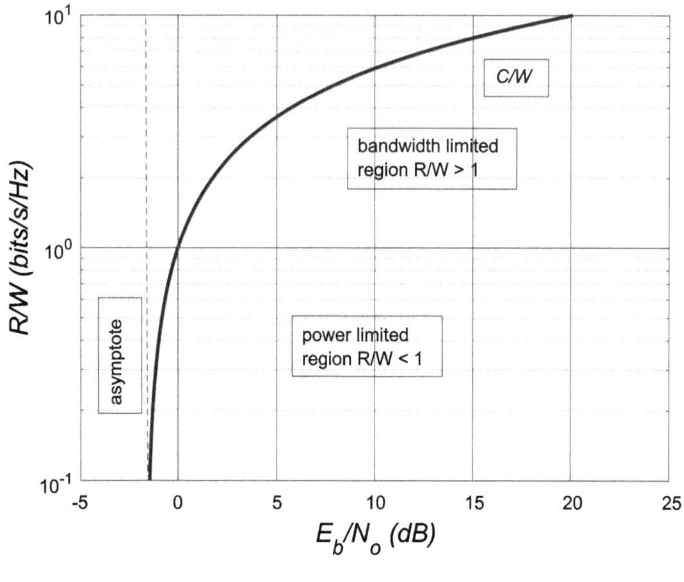

Fig. 1.8 Capacity of bandlimited Gaussian channel

Asymptotic Behavior An asymptotic behavior occurs as the bandwidth efficiency $C/W \to 0$. Using L'Hôpital's rule

$$\lim_{C/W \to 0} \frac{E_b}{N_o} = \lim_{C/W \to 0} 2^{C/W} \ln 2 = \ln 2 = 0.693 = -1.6\,\text{dB}$$

The above asymptote means that if $E_b/N_o < -1.6$ dB, it is theoretically impossible to reliably communicate on a bandlimited AWGN channel at any data rate R no matter how much bandwidth W is used.

Regions of Operation From Fig. 1.8, two regions of feasible operation may be identified depending on the bandwidth efficiency:

Power Efficient Region $R/W < 1$ bits/s/Hz. In this region the bandwidth resources are plentiful, but power is limited. The objective in this region of operation is to enable operation at a low received power level, e.g., deep space communications.

Bandwidth Efficient Region $R/W > 1$ bits/s/Hz. In this region power resources are plentiful, but bandwidth is limited. The objective in this region of operation is to use the available bandwidth as efficiently as possible, e.g., commercial wireless communications.

Most uncoded digital communication systems operate roughly 10 dB or more from the Shannon capacity limit for a message error rate of 10^{-5}. Powerful error control coding schemes can close this gap to less than 1 dB. However, they do so at the expense of processing complexity and communication delay. Generally, in any point-to-point communication link, it is desirable to optimize *power*, *bandwidth*, *processing complexity*, and *delay*.

Problems

1.1 A customer desires a communication system that is capable of conveying 60 kb/s through a channel bandwidth of $W = 10$ kHz. The customer can achieve a received signal power of $P = 1$ picowatt. The two-sided channel noise power spectral density due to internal receiver thermal noise is $N_o/2 = 10^{-19}$ watts per hertz of bandwidth. Is it possible to build the communication system?

1.2 Consider a $W = 6$ MHz television channel and assume that the received signal-to-noise power ratio $P/(N_o W)$ is equal to 10 dB:

(a) What is the maximum possible date rate that can be transmitted over this channel?

(b) Suppose the received bit energy-to-noise ratio is $E_b/N_o = 10$ dB. What is the minimum possible bandwidth W needed to support a data rate of 100 Mbps (100×10^6 bits per second)?

Chapter 2
Probability

Probability and random processes find application in the characterization, design, and performance analysis of digital communication systems. Due to the presence of noise, interference, and other transmission impairments, the behavior and performance of digital communication systems must be described in a probabilistic manner. Moreover, the transmitted waveforms, waveform channels, and additive impairments like interference and noise are all random processes or stochastic processes. This chapter presents a review of the basic concepts of probability, while Chap. 3 treats random processes.

Section 2.1 begins this chapter with the basic notions and concepts of probability. Section 2.2 introduces discrete and continuous random variables, while Sect. 2.3 considers their various moments or expected values. Section 2.4 considers bivariate random variables, while Section 2.5 treats functions of single random variable and bivariate transformations on a pair of random variables. Finally, Sects. 2.6 and 2.7 present a variety of discrete and continuous probability density functions, respectively, which are useful for describing and analyzing communication channels and the digital communication systems that operate on them.

2.1 Basic Definitions and Concepts

Consider an experiment where the outcome of the experiment is random or uncertain, but can be repeated under the same conditions, e.g., rolling dice. The *sample space* is defined as the set of all possible outcomes of the experiment, and the size of the sample space can be either finite or infinite. An *event* can be any element or subset of the sample space to which a probability can be assigned. The set of all possible events is called the *event space*.

G. Stuber, *Introductory Digital Wireless Communications*,
https://doi.org/10.1007/978-3-032-01559-4_2

Example 2.1 Consider an experiment consisting of rolling a fair die. The sample space $\mathcal{S}$ is the set of die faces $\mathcal{S} = \{1, 2, 3, 4, 5, 6\}$. The event space is the set of all $2^6 = 64$ subsets of the sample space, including the null set $\emptyset$ and the sample space $\mathcal{S}$.

Suppose that events A and B are defined as the outcome of rolling an even die number $\{2, 4, 6\}$ and odd die number $\{1, 3, 5\}$, respectively. Then, the probability of events A and B, denoted as $\mathrm{P}[A]$ and $\mathrm{P}[B]$ are, respectively

$$\mathrm{P}[A] = \tfrac{1}{2}$$
$$\mathrm{P}[B] = \tfrac{1}{2}$$

Any experiment with a random output has the properties:

1. For any event, A, $0 \leq \mathrm{P}[A] \leq 1$.
2. For an impossible event, or null set $\emptyset$, $\mathrm{P}[\emptyset] = 0$.
3. For a sure event consisting of the sample space S, $\mathrm{P}[S] = 1$.
4. For the events A and B with $A \bigcap B = \emptyset$, $\mathrm{P}[A \bigcup B] = \mathrm{P}[A] + \mathrm{P}[B]$, where $A \bigcap B$ is the intersection of events A and B, while $A \bigcup B$ is their union.

2.1.1 Conditional Probability, Independence, and Bayes' Theorem

The *conditional probability* of event A given the occurrence of event B is

$$\mathrm{P}[A|B] = \frac{\mathrm{P}[A \bigcap B]}{\mathrm{P}[B]} \tag{2.1}$$

where $\mathrm{P}[A \bigcap B]$ is the probability that events A and B both occur and $\mathrm{P}[B] \neq 0$. If $\mathrm{P}[B] = 0$, then $\mathrm{P}[A|B]$ is undefined. There are several special cases:

- If $A \bigcap B = \emptyset$, then the events A and B are *mutually exclusive*, meaning that if event B occurs, then event A could not have occurred and $\mathrm{P}[A|B] = 0$. Likewise, if event A occurs, then event B could not have occurred and $\mathrm{P}[B|A] = 0$.
- If $B \subset A$, then the knowledge that event B has occurred implies that the event A has occurred and so $\mathrm{P}[A|B] = 1$.
- If A and B are *statistically independent*, then $\mathrm{P}[A \bigcap B] = \mathrm{P}[A]\mathrm{P}[B]$ and so $\mathrm{P}[A|B] = \mathrm{P}[A]$. This means that the occurrence of event B has no effect on the probability of event A occurring. Likewise, for statistically independent events, $\mathrm{P}[B|A] = \mathrm{P}[B]$.

There is a strong relationship between mutually exclusive events and independent events. While it may seem that mutually exclusive events are independent events, the exact opposite is true. Consider two events A and B with non-zero probabilities $\mathrm{P}[A] > 0$ and $\mathrm{P}[B] > 0$. If A and B are mutually exclusive, then $A \bigcap B = \emptyset$ and $\mathrm{P}[A \bigcap B] = 0 \neq \mathrm{P}[A]\mathrm{P}[B]$. Therefore, mutually exclusive events that each occur with non-zero probability cannot be independent. Thus, the disjointness of events is a property of the events themselves, while independence is a property of their probabilities.

In general, the events $A_i, i = 1, \ldots, n$, are independent if and only if for all collections of k distinct integers $(i_1, i_2, \ldots, i_k)$ chosen from the set $(1, 2, \ldots, n)$,

$$\mathrm{P}\left[A_{i_1} \bigcap A_{i_2} \bigcap \cdots \bigcap A_{i_k}\right] = \mathrm{P}[A_{i_1}]\mathrm{P}[A_{i_2}] \cdots \mathrm{P}[A_{i_k}]$$

for $2 \leq k \leq n$. In summary,

- If $A_i, i = 1, \ldots, n$ is a collection of mutually exclusive events, then

$$\mathrm{P}\left[\bigcup_{i=1}^{n} A_i\right] = \sum_{i=1}^{n} \mathrm{P}[A_i]. \tag{2.2}$$

- If $A_i, i = 1, \ldots, n$ is a collection of independent events, then

$$\mathrm{P}\left[\bigcap_{i=1}^{n} A_i\right] = \prod_{i=1}^{n} \mathrm{P}[A_i]. \tag{2.3}$$

Total Probability

The collection of sets $\{B_i\}$, $i = 1, \ldots, n$ forms a *partition* of the sample space S if $B_i \bigcap B_j = \emptyset$, $i \neq j$ and $\bigcup_{i=1}^{n} B_i = S$. For any event $A \subset S$

$$A = \bigcup_{i=1}^{n} (A \bigcap B_i). \tag{2.4}$$

That is, every element of A is contained in one and only one B_i. Since $(A \bigcap B_i) \bigcap (A \bigcap B_j) = \emptyset$, $i \neq j$, the sets $A \bigcap B_i$ are mutually exclusive. Therefore,

$$\begin{aligned} \mathrm{P}[A] &= \sum_{i=1}^{n} \mathrm{P}[A \bigcap B_i] \\ &= \sum_{i=1}^{n} \mathrm{P}[A|B_i]\mathrm{P}[B_i]. \end{aligned} \tag{2.5}$$

This last equation is often referred to as the *theorem of total probability*.

Bayes' Theorem

Let the events B_i, $i = 1, \ldots, n$ be mutually exclusive such that $\bigcup_{i=1}^{n} B_i = S$, where S is the sample space. Let A be an event occurring with non-zero probability. Then, as a result of conditional probability and total probability

$$\begin{aligned} \mathrm{P}[B_i|A] &= \frac{\mathrm{P}[B_i \bigcap A]}{\mathrm{P}[A]} \\ &= \frac{\mathrm{P}[A|B_i]\mathrm{P}[B_i]}{\sum_{i=1}^{n} \mathrm{P}[A|B_i]\mathrm{P}[B_i]}. \end{aligned} \tag{2.6}$$

a result known as Bayes' theorem. Bayes' theorem allows the conditional probabilities $\mathrm{P}[B_i|A]$ to be determined when the conditional probabilities $\mathrm{P}[A|B_i]$ and event probabilities $\mathrm{P}[B_i]$ are known. This reversal of conditional probabilities is useful in problems relating to detection. The example below is a typical application of Bayes' theorem to a communication problem.

Example 2.2 Consider the binary symmetric channel (BSC) shown in Fig. 2.1. The BSC accepts as inputs 0's and 1's and outputs 0's and 1's. The branches are labeled with the transition probabilities $\mathrm{P}[j|k]$, such that

$$\mathrm{P}[j = 0|k = 0] = \mathrm{P}[j = 1|k = 1] = 1 - p$$

$$\mathrm{P}[j = 1|k = 0] = \mathrm{P}[j = 0|k = 1] = p$$

where p is the *crossover probability* or *bit error probability*.

Suppose the input distribution is

$$\mathrm{Q}[k = 0] = q$$

$$\mathrm{Q}[k = 1] = 1 - q$$

If the output is $j = 1$ what is the probability that the input $k = 0$?

Ans: This seemingly difficult problem can be solved quite easily by using Bayes' theorem to write:

$$\begin{aligned} \mathrm{P}[k = 0|j = 1] &= \frac{\mathrm{P}[j = 1|k = 0]\mathrm{Q}[k = 0]}{\mathrm{P}[j = 1|k = 0]\mathrm{Q}[k = 0] + \mathrm{P}[j = 1|k = 1]\mathrm{Q}[k = 1]} \\ &= \frac{pq}{pq + (1 - p)(1 - q)}. \end{aligned}$$

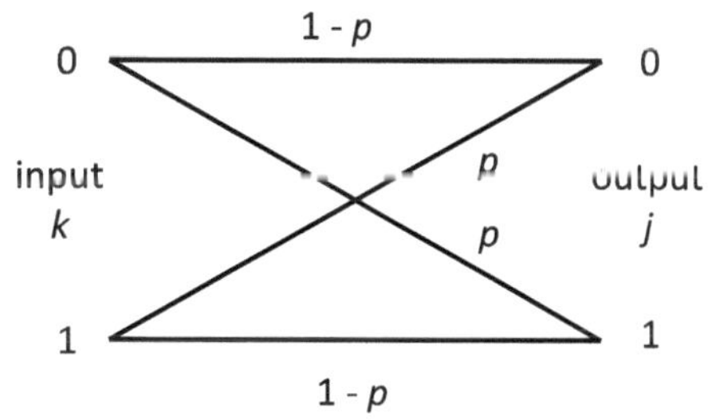

Fig. 2.1 Binary symmetric channel with crossover probability p

2.2 Discrete and Continuous Random Variables

For any random variable X, the *cumulative distribution function* (cdf) of $F_X(x)$ can be defined as follows:

$$F_X(x) \triangleq \mathrm{P}[s \in \mathcal{S} : X(s) \leq x] \equiv \mathrm{P}[X \leq x] \tag{2.7}$$

Note that $F_X(-\infty) = 0$ and $F_X(\infty) = 1$, and at any point x, $F_X(x)$ represents the probability that $X(s) \leq x$. The *complementary distribution function* (cdfc) is defined as

$$F_X^c(x) \triangleq \mathrm{P}[s \in \mathcal{S} : X(s) > x] \equiv \mathrm{P}[X > x] = 1 - F_X(x). \tag{2.8}$$

Note that the sample point s is mapped to the number $X(s)$. For notational simplicity in what follows this dependency of $X(s)$ on s will be dropped, as illustrated by the above equivalences.

There are two types of random variables, discrete and continuous. A *discrete random variable* X is one that takes on only a countable number of distinct values, each with an associated probability. An example is the outcome from rolling a die, taking on values from the set $\{1, 2, 3, 4, 5, 6\}$. A *continuous random variable* X is one that can assume an infinite number of possible values. An example is the lifetime of a light bulb. For continuous random variables, the probability of observing any single value is equal to 0, since the number of values that the random variable can assume is infinite.

Continuous random variables are not defined at specific values. Instead, a continuous random variable X is defined over an interval of values and is described by a *probability density function* (pdf) $f_X(x)$ whose integral across an interval gives the probability that the random variable X lies within the interval. The probability density function of a continuous random variable must satisfy the following: i) $f_X(x) \geq 0$, and ii) $\int_{-\infty}^{\infty} f_X(x)\mathrm{d}x = 1$. For a continuous random variable, the cdf of X is given by

$$F_X(x) = \int_{-\infty}^{x} f_X(x)\mathrm{d}x. \tag{2.9}$$

For a continuous random variable, the cdf is a continuous function. For a continuous random variable, the cdf is differentiable and its derivative gives the pdf:

$$f_X(x) = \frac{d}{dx} F_X(x). \tag{2.10}$$

The *probability distribution* of a discrete random variable is a list of the probabilities associated with each of its possible values. The discrete random variable X has the n-element sample space, $\mathcal{S} = \{x_1, x_2, \ldots, x_n\}$, where the probability that $X = x_i$ is defined to be $\mathrm{P}[X = x_i] = p_X(x_i)$, $i = 1, \ldots, n$. The probabilities $p_X(x_i)$ must satisfy the following: i) $0 \leq p_X(x_i) \leq 1$ for $1 \leq i \leq n$, and ii) $\sum_{i=1}^{n} p_X(x_i) = 1$. For a discrete random variable, the cdf of X is given by

$$F_X(x) = \sum_{x_i \leq x} p_X(x_i). \tag{2.11}$$

For discrete random variables, the cdf is a staircase function. As a result of the discontinuities, the cumulative distribution function is not differentiable. However, by using dirac impulse functions, a discrete random variable can be treated as a continuous random variable having the pdf:

$$f_X(x) = \sum_{i=1}^{n} p_X(x_i)\delta(x - x_i), \tag{2.12}$$

which is differentiable.

Example 2.3 Consider a fair die with faces $\mathcal{S} = \{1, 2, 3, 4, 5, 6\}$. The probability distribution of rolling the die, X, is

$$p_X(k) = \frac{1}{6}, \quad k = 1, 2, \ldots, 6$$

which is depicted in the stem plot of Fig. 2.2a. By using dirac impulse functions, the probability distribution of X can be expressed as the pdf:

$$f_X(x) = \frac{1}{6} \sum_{i=1}^{n} \delta(x - k)$$

which is plotted in Fig. 2.2b.

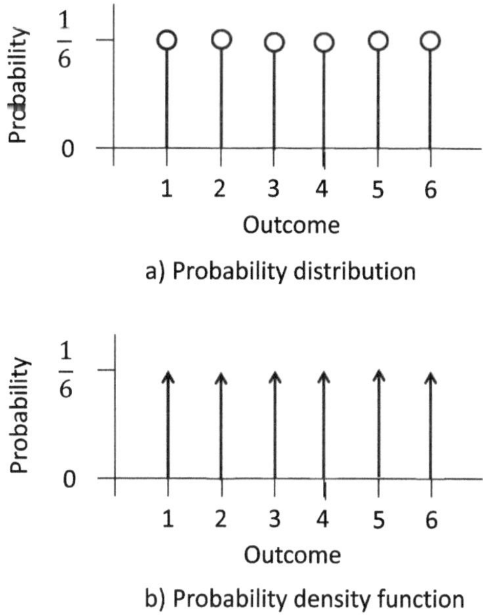

Fig. 2.2 Fair die (**a**) probability distribution, (**b**) probability density function

2.3 Means, Moments, and Moment Generating Functions

The kth moment of a random variable X, $\mathrm{E}[X^k]$, is defined as

$$\mathrm{E}[X^k] \triangleq \begin{cases} \sum_{x_i \in \mathcal{S}} x_i^k p_X(x_i) & \text{if } X \text{ is discrete} \\ \int_{-\infty}^{\infty} x^k f_X(x)\mathrm{d}x & \text{if } X \text{ is continuous} \end{cases}. \tag{2.13}$$

The kth central moment of the random variable X is $\mathrm{E}[(X - \mathrm{E}[X])^k]$. The mean of X is the first moment:

$$\mu_X = \mathrm{E}[X] \tag{2.14}$$

and the variance of X is the second central moment:

$$\sigma_X^2 = \mathrm{E}[(X - \mu_X)^2] = \mathrm{E}[X^2] - \mu_X^2. \tag{2.15}$$

The moment-generating function or characteristic function of a random variable X is

$$\triangleq \mathrm{E}[\mathrm{e}^{j\upsilon X}] = \begin{cases} \sum_{x_i \in \mathcal{S}} \mathrm{e}^{j\upsilon x_i} p_X(x_i) & \text{if } X \text{ is discrete} \\ \int_{-\infty}^{\infty} \mathrm{e}^{j\upsilon x} p_X(x)\mathrm{d}x & \text{if } X \text{ is continuous} \end{cases}, \tag{2.16}$$

where $j = \sqrt{-1}$. Note that the characteristic function of a continuous random variable is a Fourier transform, except for the sign in the exponent. Likewise, the characteristic function for a discrete random variable is a z-transform, except for the sign in the exponent. The probability distributions and pdfs of discrete and continuous random variables, respectively, can be obtained by taking the inverse transforms of the characteristic functions, i.e.,

$$f_X(x) = \frac{1}{2\pi} \int_{-\infty}^{\infty} \psi_X(jv) \mathrm{e}^{-jvx} \mathrm{d}v \tag{2.17}$$

and

$$p_X(x_k) = \frac{1}{2\pi} \oint_C \psi_X(jv) \mathrm{e}^{-jvx_k} \mathrm{d}v. \tag{2.18}$$

The characteristic function can be used to generate moments of a random variable X. Provided that the nth moment exists, it can be obtained by differentiating the characteristic function n times:

$$\frac{\mathrm{d}^n}{\mathrm{d}v^n} \psi_X(jv) = \mathrm{E}[\mathrm{e}^{jvX}(jX)^n], \tag{2.19}$$

then evaluate the result at $v = 0$ to give

$$\mathrm{E}[X^n] = j^{-n} \left[\frac{\mathrm{d}^n}{\mathrm{d}v^n} \psi_X(jv) \right]_{v=0}. \tag{2.20}$$

2.4 Bivariate Random Variables

Consider the pair of continuous real-valued random variables X and Y.

Joint cdf and pdf

The *joint cdf* of X and Y is

$$F_{XY}(x, y) = \mathrm{P}[X \leq x, Y \leq y], \quad 0 \leq F_{XY}(x, y) \leq 1, \tag{2.21}$$

and the *joint cdfc* of X and Y is

$$F^c_{XY}(x, y) = \mathrm{P}[X > x, Y > y] = 1 - F_{XY}(x, y), \quad 0 \leq F^c_{XY}(x, y) \leq 1. \tag{2.22}$$

The joint pdf $f_{XY}(x, y)$ and joint cdf $F_{XY}(x, y)$ of X and Y are related as

$$f_{XY}(x, y) = \frac{\partial^2 F_{XY}(x, y)}{\partial x \partial y}, \qquad F_{XY}(x) = \int_{-\infty}^{x} \int_{-\infty}^{y} f_{XY}(x, y) \mathrm{d}x \mathrm{d}y. \tag{2.23}$$

Marginal and Conditional pdfs
The *marginal pdfs* of X and Y are obtained from the joint pdf $f_{XY}(x, y)$ as

$$f_X(x) = \int_{-\infty}^{\infty} f_{XY}(x, y)\mathrm{d}y \qquad\qquad f_Y(x) = \int_{-\infty}^{\infty} f_{XY}(x, y)\mathrm{d}x. \tag{2.24}$$

The *conditional pdfs* of X given Y, and Y given X, are

$$f_{X|Y}(x|y) = \frac{f_{XY}(x, y)}{f_Y(y)} \qquad\qquad f_{Y|X}(y|x) = \frac{f_{XY}(x, y)}{f_X(x)}. \tag{2.25}$$

Joint Moments, Correlations, and Independence
The joint moments of X and Y are

$$\mathrm{E}[X^i Y^j] = \int_{-\infty}^{\infty} x^i y^j f_{XY}(x, y)\mathrm{d}x\mathrm{d}y. \tag{2.26}$$

The *covariance* of X and Y is

$$\begin{aligned}\lambda_{XY} &= \mathrm{E}[(X - \mu_X)(Y - \mu_Y)] \\ &= \mathrm{E}[XY - X\mu_Y - Y\mu_X + \mu_X\mu_Y] \\ &= \mathrm{E}[XY] - \mu_X\mu_Y \end{aligned} \tag{2.27}$$

The *correlation coefficient* of X and Y is

$$\rho_{XY} = \frac{\lambda_{XY}}{\sigma_X\sigma_Y}, \tag{2.28}$$

with the property $0 \leq \rho \leq 1$. Two random variables X and Y are said to be *uncorrelated* if and only if $\lambda_{X,Y} = 0$. Two random variables X and Y are said to be *orthogonal* if and only if $\mathrm{E}[XY] = 0$. When $\rho = \pm 1$, the random variables X and Y are linearly dependent such that $Y = aX + b$.

If two random variables X and Y are *independent*, then their joint pdf has the product form:

$$f_{XY}(x, y) = f_X(x)f_Y(x). \tag{2.29}$$

For independent random variables, $\mathrm{E}[XY] = \mathrm{E}[X]\mathrm{E}[Y]$, and it follows from (2.27) that X and Y are uncorrelated. However, uncorrelated random variables are not necessarily independent.

Characteristic Function

The joint characteristic or moment generating function of the random variables X and Y is

$$\Phi_{XY}(v_1, v_2) = \mathrm{E}[\mathrm{e}^{jv_1X+jv_2Y}] = \int_{-\infty}^{\infty}\int_{-\infty}^{\infty} f_{XY}(x, y)\mathrm{e}^{jv_1x+jv_2y}\mathrm{d}x\mathrm{d}y. \tag{2.30}$$

If X and Y are independent, then

$$\begin{aligned}\Phi_{XY}(v_1, v_2) &= \mathrm{E}[\mathrm{e}^{jv_1X+jv_2Y}] \\ &= \int_{-\infty}^{\infty} f_X(x)\mathrm{e}^{jv_1x}\mathrm{d}x \int_{-\infty}^{\infty} f_Y(y)\mathrm{e}^{jv_2y}\mathrm{d}y \\ &= \Phi_X(v_1)\Phi_Y(v_2).\end{aligned} \tag{2.31}$$

Joint moments can be generated according to

$$\mathrm{E}[X^nY^m] = -\frac{\partial^{n+m}\Phi_{XY}(v_1, v_2)}{\partial^n v_1 \partial^m v_2}\Big|_{v_1=v_2=0}\,. \tag{2.32}$$

2.5 Functions of Random Variables

A function $Y = G(X)$ on the random variable X will produce a new random variable Y. For example, a random variable X that is input to a square-law device will produce the output random variable $Y = X^2$. Having knowledge of the pdf of X and the function $Y = G(X)$, the pdf of Y is of interest.

2.5.1 *Functions of Single Random Variable*

Consider the continuous random variable X with pdf $f_X(x)$ and cdf $F_X(x)$. Construct the new random variable $Y = G(X)$, where $G(X)$ is a continuous-valued monotonically increasing or decreasing function. Our objective is to determine the pdf and cdf of Y, given the pdf and cdf of X.

Suppose that $Y = G(X)$ is a monotonically increasing function. The function $G(X)$ has an inverse, such that $X = G^{-1}(Y)$, so that

$$\begin{aligned}F_Y(y) &= \mathrm{P}[Y \le y] = \mathrm{P}[X \le G^{-1}(y)] \\ &= F_X(G^{-1}(y))\end{aligned}$$

Differentiating the above expression with respect to y gives the pdf of Y:

$$f_Y(y) = \frac{\mathrm{d}}{\mathrm{d}y} F_Y(y) = \frac{\mathrm{d}}{\mathrm{d}y} F_X(G^{-1}(y))$$

Applying the chain rule of derivatives gives:

$$\begin{aligned} f_Y(y) &= \frac{\mathrm{d}}{\mathrm{d}y} F_X(G^{-1}(y)) = f_X(G^{-1}(y)) \frac{\mathrm{d}}{\mathrm{d}y} G^{-1}(y) \\ &= f_X(x) \frac{\mathrm{d}x}{\mathrm{d}y}, \quad x = G^{-1}(y) \end{aligned} \tag{2.33}$$

where the second line follows with the understanding that wherever x appears on the right-hand side it is replaced by $x = G^{-1}(y)$.

Suppose now that the function $Y = G(X)$ is monotonically decreasing. In this case the event $Y \leq y$ corresponds to $X \geq G^{-1}(Y)$, implying that

$$\begin{aligned} F_Y(y) &= \mathrm{P}[Y \leq y] = \mathrm{P}[X \geq G^{-1}(y)] \\ &= 1 - F_X(G^{-1}(y)) \end{aligned}$$

Once again, differentiating with respect y gives:

$$f_Y(y) = -f_X(x) \frac{\mathrm{d}x}{\mathrm{d}y}, \quad x = G^{-1}(y) \tag{2.34}$$

However, since $Y = G(X)$ is monotonically decreasing, the derivative $\mathrm{d}y/\mathrm{d}x$ is negative in this case. It follows that (2.33) and (2.34) can be combined as

$$f_Y(y) = f_X(x) \left| \frac{\mathrm{d}x}{\mathrm{d}y} \right|, \quad x = G^{-1}(y) \tag{2.35}$$

Example 2.4 Consider the transformation $Y = X^2$. Although this is not a one-to-one function since two values of X are mapped onto the same value of Y, it is possible to modify the above results to obtain the density of $Y = X^2$.

The probability $\mathrm{P}[Y \leq y]$ is equivalent to $\mathrm{P}[-\sqrt{y} \leq X \leq \sqrt{y}]$. Hence,

$$\begin{aligned} F_Y(y) &= \mathrm{P}[-\sqrt{y} \leq X \leq \sqrt{y}] \\ &= F_X(\sqrt{y}) - F_X(-\sqrt{y}) \end{aligned}$$

(continued)

Example 2.4 (continued)
Using the chain rule of derivatives to differentiate this expression with respect to y gives its pdf:

$$\begin{aligned} f_Y(y) &= \frac{\mathrm{d}}{\mathrm{d}y} F_Y(y) \\ &= \frac{\mathrm{d}}{\mathrm{d}y}\left(F_X(\sqrt{y}) - F_X(-\sqrt{y})\right) \\ &= f_X(\sqrt{y})\frac{1}{2\sqrt{y}} + f_X(-\sqrt{y})\frac{1}{2\sqrt{y}} \\ &= \frac{f_X(\sqrt{y}) + f_X(-\sqrt{y})}{2\sqrt{y}} \end{aligned} \tag{2.36}$$

2.5.2 *Functions of Two Random Variables*

Consider the pair of random variables X and Y with a known joint pdf $f_{XY}(x, y)$. Let

$$U = G_1(X, Y) \quad \text{and} \quad V = G_2(X, Y) \tag{2.37}$$

be a *bivariate transformation* of the random variables X and Y into the two new random variables U and V. Often only the function $U = G_1(X, Y)$ and its pdf $f_U(u)$ is of interest. However, to obtain $f_U(u)$, it is necessary to define a suitable auxiliary function $V = G_2(X, Y)$.

It is assumed that the transformations are one-to-one, that is, each point (x, y) is mapped onto one and only one point (u, v), and the inverse transforms exist such that

$$X = H_1(U, V) \quad \text{and} \quad Y = H_2(U, V) \tag{2.38}$$

Our approach is to determine the joint pdf of U and V, $f_{UV}(u, v)$, from which the marginal density functions $f_U(u)$ and $f_V(v)$ can be obtained.

There are several methods for solving this problem that are described in various introductory probability texts. Our approach follows an intuitive approach. Suppose that a very small rectangle A having area A in the $x - y$ plane is mapped as shown in Fig. 2.3 onto a similarly very small area B in the $u - v$ plane according to the transformation $U = G_1(X, Y)$ and $V = G_2(X, Y)$.

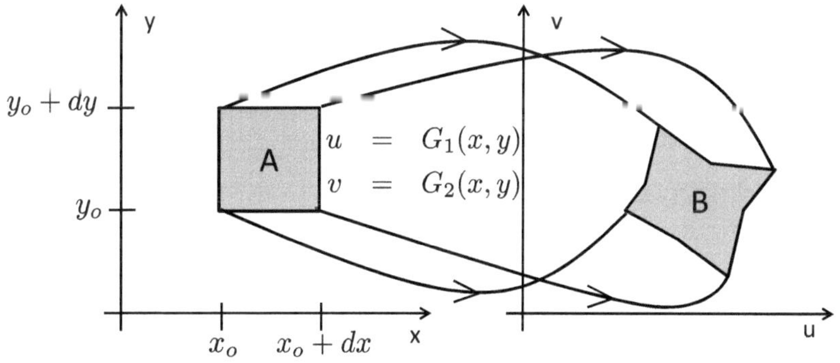

Fig. 2.3 Mapping from $x - y$ to $u - v$ plane

The two areas A and B represent equivalent events, due to the assumption of one-to-one transformations. Hence,

$$P((X, Y) \in A) = P((U, V) \in B) \tag{2.39}$$

This equality can be expressed as

$$f_{XY}(x_o, y_o)\mathrm{d}x\mathrm{d}y = f_{UV}(u_o, v_o) \times \text{area of } B \tag{2.40}$$

where $u_o = G_1(x_o, y_o)$ and $v_o = G_2(x_o, y_o)$. The remaining problem is to determine the area B.

The area B is redrawn in Fig. 2.4 to show more detail. For very small $\mathrm{d}x$ and $\mathrm{d}y$ the following approximations can be made:

$$G_1(x_o + \mathrm{d}x, y_o) \approx G_1(x_o, y_o) + \frac{\partial}{\partial x} G_1(x_o, y_o)\mathrm{d}x \tag{2.41a}$$

$$G_1(x_o, y_o + \mathrm{d}y) \approx G_1(x_o, y_o) + \frac{\partial}{\partial y} G_1(x_o, y_o)\mathrm{d}y \tag{2.41b}$$

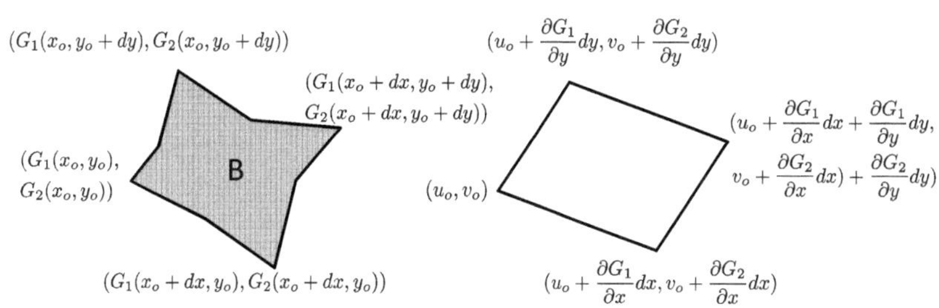

Fig. 2.4 Detailed mapping from $x - y$ to $u - v$ plane

The same can be done for the second function $G_2(X, Y)$. Based on the above, the linearized version of the area B is shown in Fig. 2.4 using the approximate coordinates. The approximate area has the shape of a parallelogram, and its area can be derived as

$$\left| \frac{\partial G_1(x_o, y_o)}{\partial x} \cdot \frac{\partial G_2(x_o, y_o)}{\partial y} - \frac{\partial G_2(x_o, y_o)}{\partial x} \cdot \frac{\partial G_1(x_o, y_o)}{\partial y} \right| \mathrm{d}x\mathrm{d}y \tag{2.42}$$

The quantity inside the absolute value sign is called the *Jacobian* of the bivariate transformation and can be expressed as the matrix determinant:

$$\begin{aligned} J(x, y) &= \det \begin{bmatrix} \frac{\partial G_1(x,y)}{\partial x} & \frac{\partial G_1(x,y)}{\partial y} \\ \frac{\partial G_2(x,y)}{\partial x} & \frac{\partial G_2(x,y)}{\partial y} \end{bmatrix} \\ &= \det \begin{bmatrix} \frac{\partial u}{\partial x} & \frac{\partial u}{\partial y} \\ \frac{\partial v}{\partial x} & \frac{\partial v}{\partial y} \end{bmatrix} \end{aligned} \tag{2.43}$$

It can be similarly shown that

$$\begin{aligned} J(u, v) &= \det \begin{bmatrix} \frac{\partial H_1(u,v)}{\partial u} & \frac{\partial H_1(u,v)}{\partial v} \\ \frac{\partial H_2(u,v)}{\partial u} & \frac{\partial H_2(u,v)}{\partial v} \end{bmatrix} \\ &= \det \begin{bmatrix} \frac{\partial x}{\partial u} & \frac{\partial x}{\partial v} \\ \frac{\partial y}{\partial u} & \frac{\partial y}{\partial v} \end{bmatrix} \end{aligned} \tag{2.44}$$

and as a consequence

$$|J(u, v)| = \frac{1}{|J(x, y)|}, \quad \text{note the absolute values of the determinants}$$

Thus (2.40) has the form:

$$f_{XY}(x_o, y_o)\mathrm{d}x\mathrm{d}y = f_{UV}(G_1(x_o, y_o), G_2(x_o, y_o))|J(x, y)|\mathrm{d}x\mathrm{d}y$$

and from this the inverse form is obtained:

$$f_{UV}(u_o, v_o)\mathrm{d}x\mathrm{d}y = f_{XY}(H_1(u_o, v_o), H_2(u_o, v_o))\frac{1}{|J(x, y)|}\mathrm{d}x\mathrm{d}y$$

This yields the final result:

$$f_{UV}(u, v) = f_{XY}(H_1(u, v), H_2(u, v))|J(u, v)|, \quad u, v \in R_{UV} \tag{2.45}$$

To determine the joint pdf $f_{UV}(u, v)$, the joint pdf $f_{XY}(x, y)$ is used with $x = H_1(u, v)$ and $y = H_2(u, v)$. Then, multiply the result by the magnitude of the

Jacobian $|J(u, v)|$, which may be calculated as $1/|J(x, y)|$ again with $x = H_1(u, v)$ and $y = H_2(u, v)$. The main tasks with this approach are to identify the inverse functions $X = H_1(U, V)$ and $Y = H_2(U, V)$, a procedure which proceeds more by intuition for each individual case than by any set process. One proceeds similarly to find the range R_{UV} for each individual case, again by intuition and without set process.

In the special cases that follow, the joint pdf $f_{XY}(x, y)$ is not specified so the ranges R_{UV} are not determined.

pdf of U = X + Y
The bivariate transformation of random variables X and Y

$$U = X + Y \qquad \text{and} \qquad V = X$$

has the inverse functions

$$X = V \qquad \text{and} \qquad Y = U - V$$

The Jacobian is

$$J(u, v) = \det \begin{bmatrix} \frac{\partial x}{\partial u} & \frac{\partial x}{\partial v} \\ \frac{\partial y}{\partial u} & \frac{\partial y}{\partial v} \end{bmatrix} = \det \begin{bmatrix} 0 & 1 \\ 1 & -1 \end{bmatrix} = -1$$

and $|J(u, v)| = 1$. Hence, the joint pdf is

$$f_{UV}(u, v) = f_{XY}(v, u - v)$$

The marginal density of U is

$$f_U(u) = \int f_{XY}(v, u - v)dv \tag{2.46}$$

pdf of U = XY
The bivariate transformation

$$U = XY \qquad \text{and} \qquad V = X$$

has the inverse functions

$$X = V \qquad \text{and} \qquad Y = U/V$$

The Jacobian is

$$J(u, v) = \det \begin{bmatrix} \frac{\partial x}{\partial u} & \frac{\partial x}{\partial v} \\ \frac{\partial y}{\partial u} & \frac{\partial y}{\partial v} \end{bmatrix} = \det \begin{bmatrix} 0 & 1 \\ 1/v & -u/v^2 \end{bmatrix} = -\frac{1}{v}$$

and $|J(u, v)| = 1$. Hence, the joint pdf is

$$f_{UV}(u, v) = f_{XY}(v, u/v)|1/v|$$

The marginal density of U is

$$f_U(u) = \int f_{XY}(v, u/v)|1/v|\mathrm{d}v \tag{2.47}$$

pdf of U = X/Y
The bivariate transformation

$$U = X/Y \qquad \text{and} \qquad V = X$$

has the inverse functions

$$X = V \qquad \text{and} \qquad Y = V/U$$

The Jacobian is

$$J(u, v) = \det \begin{bmatrix} \frac{\partial x}{\partial u} & \frac{\partial x}{\partial v} \\ \\ \frac{\partial y}{\partial u} & \frac{\partial y}{\partial v} \end{bmatrix} = \det \begin{bmatrix} 0 & 1 \\ -v/u^2 & 1/u \end{bmatrix} = \frac{v}{u^2}$$

and $|J(u, v)| = 1$. Hence, the joint pdf is

$$f_{UV}(u, v) = f_{XY}(v, v/u)|v/u^2|$$

The marginal density of U is

$$f_U(u) = \int f_{XY}(v, v/u)|v/u^2|\mathrm{d}v \tag{2.48}$$

2.6 Discrete Distributions

Digital communication systems often involve discrete random variables. For example, the inputs and outputs of a BSC are discrete random variables as are the channel transition probabilities. This section presents some useful discrete distributions.

2.6.1 Bernoulli Distribution

A Bernoulli random variable X takes on the values 0 and 1, and has the probability distribution:

$$p_X(x_i) = \begin{cases} 1 \ , & p \\ 0 \ , & 1-p \end{cases} \tag{2.49}$$

By using dirac impulse functions, X may be written as a continuous random variable having the pdf:

$$f_X(x) = (1-p)\delta(x) + p\delta(x-1). \tag{2.50}$$

The Bernoulli random variable has mean

$$\mu_X = \mathrm{E}[X] = p \tag{2.51}$$

and variance

$$\sigma_X^2 = \mathrm{E}[X^2] - (\mathrm{E}[X])^2 = p(1-p) \,. \tag{2.52}$$

2.6.2 Binomial Distribution

Consider a random experiment that generates a vector of n independent and identically distributed (i.i.d.) Bernoulli random variables $\mathbf{X} = \{X_1, X_2, \ldots, X_n\}$. The sum $Y = \sum_{i=1}^{n} X_i$ is a random variable spanning the set of integers $\{0, 1, \ldots, n\}$, and its probability distribution $p_Y(k)$ is of interest.

Any outcome having k 1's and $n-k$ 0's in the vector $\mathbf{X}$ occurs with probability $p^k(1-p)^{n-k}$ and yields $Y = k$. The number of distinct ways of placing k 1's in the n coordinates of the vector $\mathbf{X}$ is given by the binomial coefficient:

$$\binom{n}{k} = \frac{n!}{k!(n-k)!} \tag{2.53}$$

Hence,

$$p_Y(k) = \binom{n}{k} p^k (1-p)^{n-k} \ , \quad k = 0, 1, \ldots, n \tag{2.54}$$

known as the binomial distribution. By using dirac delta functions, the binomial random variable Y has the pdf:

$$f_Y(y) = \sum_{k=0}^{n} \binom{n}{k} p^k (1-p)^{n-k} \delta(y-k). \tag{2.55}$$

The mean of the binomial random variable Y is

$$\mu_Y = \sum_{i=1}^{n} \mathrm{E}[X_i] = np \tag{2.56}$$

and its variance is

$$\sigma_Y^2 = \sum_{i=1}^{n} \sigma_{X_i}^2 = np(1-p) \tag{2.57}$$

Example 2.5 Suppose that binary data is to be transmitted over a BSC channel with crossover probability p as shown in Fig. 2.1. To make the data transmission more reliable, a simple length-n binary repetition code is used where each data bit is repeated and transmitted n times as follows:

$$0 \to (0, 0, \ldots, 0)_n$$
$$1 \to (1, 1, \ldots, 1)_n$$

At the receiver a majority logic decision is made, such that if more 1's than 0's are received, the receiver decides that a 1 was sent and vice versa. If an equal number of 0's and 1's are received, the tie is resolved randomly.

Assuming that data bit 0's and 1's are transmitted with equal probability, the probability of bit error, P_b, can be obtained by using the binomial distribution.

For odd values of n, there is never a tie and the probability of data bit error is

$$P_b = \sum_{k=(n+1)/2}^{n} \binom{n}{k} p^k (1-p)^{n-k}$$

For example, if $p = 0.1$ and $n = 5$, then

$$P_b = \binom{5}{3}(0.1)^3(1-0.1)^2 + \binom{5}{4}(0.1)^4(1-0.1) + \binom{5}{5}(0.1)^5$$

(continued)

Example 2.5 (continued)

$$= 8.56 \times 10^{-3}$$

Note that with $p = 0.1$, a simple length-5 repetition code reduces the probability of bit error by more than an order of magnitude. The cost of this increased reliability is data rate and power, since each data bit consumes 5 channel uses instead of one.

For even values of n, a tie occurs when $k = n/2$ that is resolved randomly, so that the probability of data bit error is

$$P_b = \sum_{k=n/2+1}^{n} \binom{n}{k} p^k (1-p)^{n-k} + \frac{1}{2}\binom{n}{n/2} p^n/2(1-p)^{n/2}$$

2.6.3 Geometric Distribution

The geometric distribution describes the number of failures before one success in a series of independent trials, where each trial results in either a success or failure, and the probability of success in any individual trial is equal to a constant p. The probability of success on the kth trial is described by the geometric distribution:

$$p_X(k) = (1-p)^{k-1} p\,, \quad k = 1, 2, \ldots, \infty. \tag{2.58}$$

The mean and variance of the geometric distribution are $\mu_X = 1/p$ and $\sigma_X^2 = (1-p)/p^2$, respectively.

Example 2.7 Consider a packet data communication system, where packets are transmitted and if necessary retransmitted until they are correctly received. Suppose that the probability of any given packet transmission or retransmission being received in error is equal to ε. What is the probability that the packet will be received correctly in three or fewer attempts?

For this problem the probability of success (correct packet reception) for any trial is $1-\varepsilon$. Therefore, the probability of correct packet reception in three or fewer attempts is

$$P_C = (1-\varepsilon) + \varepsilon(1-\varepsilon) + \varepsilon^2(1-\varepsilon)$$

2.6.4 Poisson Distribution

The Poisson distribution describes the probability that a given number of events, X, will occur in a fixed interval of time and/or space if these events occur with a known average rate and independently of the time and/or location since the last event. If the average number of events occurring in a given temporal or spatial interval is λ, then the probability of k events in the same temporal or spatial interval has the Poisson probability distribution:

$$p_X(k) = \frac{\lambda^k \mathrm{e}^{-\lambda}}{k!}, \quad k = 0, 1, \ldots, \infty \tag{2.59}$$

The mean and variance of the Poisson distribution are $\mu_X = \lambda$ and $\sigma_X^2 = \lambda$.

Example 2.6 In a telephone network, the term busy hour call attempts (BHCA) refers to the number of telephone calls attempted at the busiest hour of the day (peak hour). Suppose the mean BHCA is equal to 100, as observed over many days. If the BHCA on any given day is modeled as a Poisson random variable, what is the probability that on any given day BCHA is equal to 120?

For this problem $\lambda = 100$, so that

$$\mathrm{P[BCHA = 120]} = \frac{100^{120}\mathrm{e}^{-100}}{120!} = 0.00556$$

2.7 Continuous Distributions

In digital communication systems, thermal noise which is present in all communication receivers can be modeled as a Gaussian random processes. As will be detailed in Chap. 11, multipath-fading channels can be modeled as complex Gaussian random processes. Hence, Gaussian or normal random variables and functions of Gaussian random variables play a central role in communication systems.

2.7.1 Gaussian Distribution

A real-valued Gaussian, or normal random variable, X has the pdf:

$$f_X(x) = \frac{1}{\sqrt{2\pi}\sigma_X} \exp\left\{-\frac{(x-\mu_X)^2}{2\sigma_X^2}\right\}, \tag{2.60}$$

where $\mu_X = \text{E}[X]$ is the mean of X and $\sigma_X^2 = \text{E}[(X - \mu)^2]$ is the variance of X. Sometimes, the shorthand notation $X \sim \mathcal{N}(\mu_X, \sigma_X^2)$ is used to describe a Gaussian random variable. The random variable X is said to have a standard normal distribution if $X \sim \mathcal{N}(0, 1)$. The cumulative distribution function (cdf) of a Gaussian random variable X is

$$F_X(x) = \int_{-\infty}^{x} \frac{1}{\sqrt{2\pi}\sigma_X} \exp\left\{-\frac{(y - \mu_X)^2}{2\sigma_X^2}\right\} \text{d}y. \tag{2.61}$$

The cdf of a standard normal distribution defines the *Gaussian*: Φ *function*

$$\Phi(\alpha) \triangleq \int_{-\infty}^{\alpha} \frac{1}{\sqrt{2\pi}} \text{e}^{-x^2/2} \text{d}x \tag{2.62}$$

while its cdfc defines the *Gaussian Q function*

$$Q(\alpha) \triangleq \int_{\alpha}^{\infty} \frac{1}{\sqrt{2\pi}} \text{e}^{-x^2/2} \text{d}x \tag{2.63}$$

and $Q(\alpha) + \Phi(\alpha) = 1$. The Gaussian Q function gives the tail area under a standard normal distribution as shown in Fig. 2.5. The cdf and cdfc for the nonstandard normal random variable $X \sim \mathcal{N}(\mu_X, \sigma_X^2)$ are, respectively,

$$F_X(\alpha) = \Phi\left(\frac{\alpha - \mu_X}{\sigma_X}\right) \tag{2.64}$$

$$F_X^c(\alpha) = Q\left(\frac{\alpha - \mu_X}{\sigma_X}\right). \tag{2.65}$$

The cdfc for a nonstandard normal random variable gives the tail area as shown in Fig. 2.6.

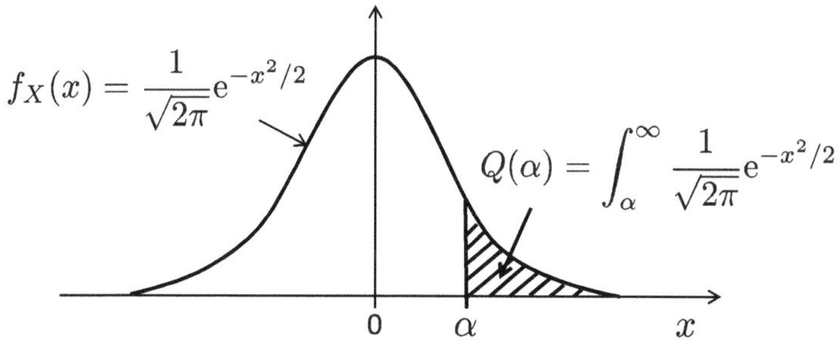

Fig. 2.5 The Gaussian Q function is the tail area under a standard normal distribution

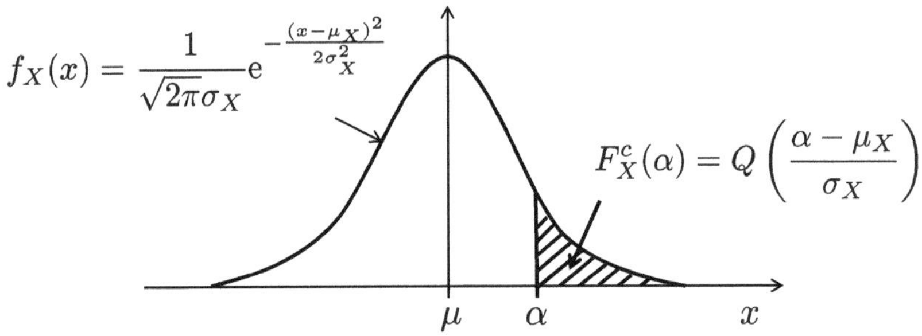

Fig. 2.6 The cdfc of a nonstandard Gaussian random variable expressed in terms of the Gaussian Q function

Sometimes the tail area under Gaussian pdf is described by the *complementary error function* $\text{erfc}(x)$, defined as

$$\text{erfc}(x) \triangleq \frac{2}{\sqrt{\pi}} \int_x^{\infty} \text{e}^{-y^2} \text{d}y. \tag{2.66}$$

The complementary error function and the Gaussian Q function are related as follows:

$$\text{erfc}(x) = 2Q(\sqrt{2}x) \tag{2.67}$$

$$Q(x) = \frac{1}{2}\text{erfc}\left(\frac{x}{\sqrt{2}}\right). \tag{2.68}$$

These identities can be derived by starting with the definition of the Gaussian Q function in (2.63) and changing the variable of integration.

The *error function* of a Gaussian random variable is defined as

$$\text{erf}(x) \triangleq \frac{2}{\sqrt{\pi}} \int_0^{x} \text{e}^{-y^2} \text{d}y. \tag{2.69}$$

Note that $\text{erfc}(x) + \text{erf}(x) \neq 1$. Also,

$$Q(x) = \frac{1}{2} - \frac{1}{2}\text{erf}\left(\frac{x}{\sqrt{2}}\right), \quad x \geq 0. \tag{2.70}$$

The functions $Q(x)$, $\Phi(x)$, $\text{erfc}(x)$, and $\text{erf}(x)$ do not exist in closed form. However, many scientific calculators will have a built-in function to calculate $Q(x)$. MATLAB has built-in functions to calculate $\text{erfc}(x)$ and $\text{erf}(x)$:

Example 2.8 Suppose that $X \sim N(\mu, \sigma^2)$, where μ is positive. What is the probability that $X < 0$?
Answer: From the symmetry of a Gaussian random variable around its mean μ

$$\begin{aligned} P[X < 0] &= P[X > 2\mu] \\ &= Q\left(\frac{2\mu - \mu}{\sigma}\right) \\ &= Q\left(\frac{\mu}{\sigma}\right) \end{aligned}$$

2.7.2 Multivariate Gaussian Distribution

Let $X_i \sim \mathcal{N}(\mu_i, \sigma_i^2), i = 1, \ldots, n$, be a collection of n real-valued Gaussian random variables having means $\mu_i = \mathrm{E}[X_i]$ and covariances:

$$\begin{aligned} \lambda_{X_i X_j} &= \mathrm{E}\big[(X_i - \mu_i)(X_j - \mu_j)\big] \\ &= \mathrm{E}\big[X_i X_j\big] - \mu_i \mu_j, \quad 1 \le i, j \le n. \end{aligned} \tag{2.71}$$

To express the joint pdf of $X_i, i = 1, \ldots, n$ in a compact form, define the following vectors:

$$\mathbf{X} = (X_1, X_2, \ldots, X_n) \tag{2.72a}$$

$$\mathbf{x} = (x_1, x_2, \ldots, x_n) \tag{2.72b}$$

$$\boldsymbol{\mu}_X = (\mu_1, \mu_2, \ldots, \mu_n). \tag{2.72c}$$

The $n \times n$ covariance matrix $\mathbf{\Lambda}$ of the Gaussian vector $\mathbf{X}$ is

$$\mathbf{\Lambda} = \mathrm{E}\big[(\mathbf{X} - \boldsymbol{\mu})^T (\mathbf{X} - \boldsymbol{\mu})\big] = \begin{bmatrix} \lambda_{X_1 X_1} & \cdots & \lambda_{X_1 X_n} \\ \vdots & & \vdots \\ \lambda_{X_n X_1} & \cdots & \lambda_{X_n X_n} \end{bmatrix}. \tag{2.73}$$

where $\mathbf{x}^T$ is the transpose of the vector $\mathbf{x}$. The diagonal elements of $\boldsymbol{\Lambda}$ are the variances, $\lambda_{X_i X_i} = \sigma_{X_i}^2, i = 1, \ldots, n$.

By using the above definitions, the Gaussian random vector $\mathbf{X}$ has the multivariate Gaussian distribution:

$$f_{\mathbf{X}}(\mathbf{x}) = \frac{1}{(2\pi)^{n/2}|\boldsymbol{\Lambda}|^{1/2}} \exp\left\{-\frac{1}{2}(\mathbf{x}-\boldsymbol{\mu}_X)\boldsymbol{\Lambda}^{-1}(\mathbf{x}-\boldsymbol{\mu}_X)^T\right\}, \tag{2.74}$$

where $|\boldsymbol{\Lambda}|$ is the determinant of $\boldsymbol{\Lambda}$ and $\boldsymbol{\Lambda}^{-1}$ is the inverse of $\boldsymbol{\Lambda}$.

If the Gaussian random variables are all uncorrelated, then the covariance matrix becomes the $n \times n$ diagonal matrix:

$$\boldsymbol{\Lambda} = \begin{bmatrix} \sigma_{X_1}^2 & 0 & \cdot\;\cdot & \cdot & 0 \\ 0 & \sigma_{X_2}^2 & 0\;\cdot & \cdot & 0 \\ \vdots & & & & \vdots \\ 0 & \cdot & \cdot\;0 & \sigma_{X_{n-1}}^2 & 0 \\ 0 & \cdot & \cdot\;\cdot & 0 & \sigma_{X_n}^2 \end{bmatrix}.$$

and the expression in (2.74) simplifies. In particular,

$$|\boldsymbol{\Lambda}|^{1/2} = \left(\prod_{i=1}^{n} \sigma_{X_i}^2\right)^{1/2} = \prod_{i=1}^{n} \sigma_{X_i}$$

and

$$\boldsymbol{\Lambda}^{-1} = \begin{bmatrix} 1/\sigma_{X_1}^2 & 0 & \cdot\;\cdot & \cdot & 0 \\ 0 & 1/\sigma_{X_2}^2 & 0\;\cdot & \cdot & 0 \\ \vdots & & & & \vdots \\ 0 & \cdot & \cdot\;0 & 1/\sigma_{X_{n-1}}^2 & 0 \\ 0 & \cdot & \cdot\;\cdot & 0 & 1/\sigma_{X_n}^2 \end{bmatrix}.$$

so that the joint pdf in (2.74) reduces to the simple product form

$$f_{\mathbf{X}}(\mathbf{x}) = \prod_{i=1}^{n} \frac{1}{\sqrt{2\pi}\sigma_{X_i}} \exp\left\{-\frac{(x_i-\mu_{X_i})^2}{2\sigma_{X_i}^2}\right\} \tag{2.75}$$

$$= \prod_{i=1}^{n} f_{X_i}(x_i). \tag{2.76}$$

Notice that uncorrelated Gaussian random variables are statistically independent, a property that does not generally apply to other types of random variables.

Bivariate Gaussian Distribution

Consider the case of two correlated jointly Gaussian random variables $X_1 \sim \mathcal{N}(\mu_{X_1}, \sigma^2_{X_1})$ $X_2 \sim \mathcal{N}(\mu_{X_2}, \sigma^2_{X_2})$. From the multivariate Gaussian distribution in (2.74)

$$\boldsymbol{\mu}_X = (\mu_{X_1}, \mu_{X_2})$$

$$\boldsymbol{\Lambda} = \begin{bmatrix} \sigma^2_{X_1} & \lambda_{X_1X_2} \\ \lambda_{X_1X_2} & \sigma^2_{X_2} \end{bmatrix}$$

Then

$$|\boldsymbol{\Lambda}| = \sigma^2_{X_1}\sigma^2_{X_2}(1-\rho^2)$$

where

$$\rho = \frac{\lambda_{X_1X_2}}{\sigma_{X_1}\sigma_{X_2}}$$

and

$$\begin{aligned}\boldsymbol{\Lambda}^{-1} &= \frac{1}{\sigma^2_{X_1}\sigma^2_{X_2} - \lambda^2_{X_1X_2}} \begin{bmatrix} \sigma^2_{X_2} & -\lambda_{X_1X_2} \\ -\lambda_{X_1X_2} & \sigma^2_{X_1} \end{bmatrix} \\ &= \frac{1}{\sigma^2_{X_1}\sigma^2_{X_2}(1-\rho^2)} \begin{bmatrix} \sigma^2_{X_2} & -\rho\sigma_{X_1}\sigma_{X_2} \\ -\rho\sigma_{X_1}\sigma_{X_2} & \sigma^2_{X_1} \end{bmatrix}\end{aligned}$$

From (2.74), the bivariate Gaussian pdf is

$$\begin{aligned} f_{X_1,X_2}(x_1,x_2) &= \frac{1}{2\pi\sigma_{X_1}\sigma X_2\sqrt{1-\rho^2}} \\ &\times \exp\left\{\frac{-1}{2(1-\rho^2)}\left[\frac{(x_1-\mu_{X_1})^2}{\sigma^2_{X_1}} - \frac{2\rho(x_1-\mu_{X_1})(x_2-\mu_{X_2})}{\sigma_{X_1}\cdot\sigma_{X_2}} \right.\right. \\ &\qquad \left.\left. + \frac{(x_2-\mu_{X_2})^2}{\sigma^2_{X_2}}\right]\right\} \end{aligned} \tag{2.77}$$

For the case of two correlated Gaussian random variables that are identically distributed with zero mean and variance σ^2

$$f_{X_1,X_2}(x_1,x_2) = \frac{1}{2\pi\sigma^2\sqrt{1-\rho^2}} \exp\left[-\frac{x_1^2 - 2\rho x_1 x_2 + x_2^2}{2\sigma^2(1-\rho^2)}\right]. \tag{2.78}$$

The case $\rho = 0$ corresponds to independent Gaussian random variables. In this case the 2-D joint pdf has circular contours and is said to be *circularly symmetric*.

As $\rho \to 1$, the bivariate Gaussian density becomes increasingly peaked along the line $x_2 = x_1$. When $\rho = \pm 1$, $x_2 = \pm x_1$ and the random variables are perfectly correlated. In this case, the joint density function can be derived using conditional probability as

$$\begin{aligned} f_{X_1,X_2}(x_1, x_2) &= f_{X_2|X_1}(x_2|x_1) f_{X_1}(x_1) \\ &= \frac{1}{\sqrt{2\pi}\sigma} \exp\left\{-\frac{x_1^2}{2\sigma^2}\right\} \delta(x_2 \mp x_1). \end{aligned}$$

where $f_{X_2|X_1}(x_2|x_1) = \delta(x_2 \mp x_1)$ due to the linear dependency $x_2 = \pm x_1$.

Example 2.9 Suppose that X and Y are i.i.d. Gaussian random variables each with mean μ and variance σ^2. What is the probability of the joint event $X < 0$ and $Y < 0$?

Answer: Exploiting the circular symmetry of the joint pdf and independence of X and Y

$$\begin{aligned} P[X < 0, Y < 0] &= P[X > 2\mu, Y > 2\mu] \\ &= P[X > 2\mu] P[Y > 2\mu] \\ &= Q^2\left(\frac{\mu}{\sigma}\right) \end{aligned}$$

Example 2.10 Suppose that X and Y are i.i.d. Gaussian random variables each with mean μ and variance σ^2. What is the mean and variance of the random variable XY?

Answer: Theoretically, a bivariate transformation of random variables may be used to obtain the pdf of $Z = XY$ and then solve the integrals:

$$\mu_Z = \int_{-\infty}^{\infty} z f_Z(z) dz$$

$$\sigma_Z^2 = \int_{-\infty}^{\infty} (z - \mu_Z)^2 f_Z(z) dz$$

(continued)

Example 2.10 (continued)
There is a much simpler approach that does not require the pdf $f_Z(z)$. The mean is given by

$$\mu_{XY} = \mathrm{E}[XY] = \mathrm{E}[X]\mathrm{E}[Y] = \mu^2$$

and the variance given by

$$\begin{aligned}\sigma_{XY}^2 &= \mathrm{E}[(XY - \mu_{XY})^2] \\ &= \mathrm{E}[(XY)^2 - 2\mathrm{E}[XY]\mu_{XY} + \mu_{XY}^2 \\ &= \mathrm{E}[X^2]\mathrm{E}[Y^2] - \mu_{XY}^2 \\ &= (\sigma^2 + \mu^2)^2 - \mu^4 \\ &= \sigma^4 + 2\mu^2\sigma^2\end{aligned}$$

2.7.3 Complex Gaussian Distribution

A complex-valued random variable has the form $Z = X + jY$, and its statistics are described by the joint pdf of its real and imaginary parts:

$$f_Z(z) = f_{XY}(x, y) \tag{2.79}$$

A complex Gaussian random variable $Z = X + jY$ is denoted as $\mathcal{CN}(\mu_z, \sigma_Z^2)$, where $X \sim \mathcal{N}(\mu_X, \sigma_X^2)$ and $Y \sim \mathcal{N}(\mu_Y, \sigma_Y^2)$ are real-valued Gaussian random variables that are generally correlated.

Consider the important case where X and Y are i.i.d. zero-mean Gaussian random variables with variance σ^2, i.e., $X, Y \sim \mathcal{N}(0, \sigma^2)$. In this case, the complex Gaussian random variable $Z = X + jY$ has the zero-mean *circularly symmetric* complex Gaussian pdf:

$$\begin{aligned}f_Z(z) = f_X(x)f_Y(y) &= \frac{1}{\sqrt{2\pi}\sigma}\exp\left\{-\frac{x^2}{2\sigma^2}\right\} \cdot \frac{1}{\sqrt{2\pi}\sigma}\exp\left\{-\frac{y^2}{2\sigma^2}\right\} \\ &= \frac{1}{2\pi\sigma^2}\exp\left\{-\frac{(x^2 + y^2)}{2\sigma^2}\right\} \\ &= \frac{1}{2\pi\sigma_Z^2}\exp\left\{-\frac{|z|^2}{2\sigma_Z^2}\right\},\end{aligned} \tag{2.80}$$

where the variance of Z is defined as

$$\sigma_Z^2 \triangleq \frac{1}{2}\mathrm{E}[|Z|^2] = \frac{1}{2}\mathrm{E}[X^2 + Y^2] = \frac{1}{2}\mathrm{E}[X^2] + \frac{1}{2}\mathrm{E}[Y^2] = \sigma^2. \tag{2.81}$$

Note that the factor of $\frac{1}{2}$ in definition of the variance of a complex random variable, σ_Z^2 is necessary to maintain the definition of the pdf in (2.79).

In the case that X and Y are independent and with non-zero means and identical variance σ^2, the pdf of $Z = X + jY$ is circularly symmetric about the complex-valued mean: $\mu_Z = \mu_X + j\mu_Y$

$$p_Z(z) = \frac{1}{2\pi\sigma_Z^2}\exp\left\{-\frac{|z-\mu_Z|^2}{2\sigma_Z^2}\right\}, \tag{2.82}$$

where, again, $\sigma_Z^2 = \sigma^2$.

2.7.4 Rayleigh Distribution

Let $X \sim \mathcal{N}(0, \sigma^2)$ and $Y \sim \mathcal{N}(0, \sigma^2)$ be a pair of i.i.d. zero-mean Gaussian random variables. Consider the random variable $R = \sqrt{X^2 + Y^2}$. Note that for the complex Gaussian random variable, $Z = X + jY$, $R = |Z|$ is its magnitude.

The function $R = \sqrt{X^2 + Y^2}$ is a bivariate transformation of the random variables X and Y as discussed in Sect. 2.5.2. The pdf of R may be obtained by first defining an auxiliary variable V such that X and Y can be expressed as inverse functions of R and V. For the problem at hand, the auxiliary variable can be defined as the angle:

$$V = \mathrm{Tan}^{-1}(Y/X).$$

The inverse functions are obtained from basic trigonometry as

$$X = R\cos V$$
$$Y = R\sin V.$$

By using the bivariate transformation of random variables in (2.45), the joint density function $f_{RV}(r, v)$ may be obtained as

$$f_{RV}(r, v) = f_{XY}(r\cos v, r\sin v)\,|J(r, v)|$$

where $J(r, v)$ is the Jacobian

$$J(r, v) = \det \begin{bmatrix} \frac{\partial x}{\partial r} & \frac{\partial x}{\partial v} \\ \frac{\partial y}{\partial r} & \frac{\partial y}{\partial v} \end{bmatrix} = \det \begin{bmatrix} \cos v & r \sin v \\ \sin v & r \cos v \end{bmatrix} = r(\cos^2 v + \sin^2 v) = r$$

Because X and Y are independent, their joint distribution has the product form:

$$f_{XY}(x, y) = f_X(x) f_Y(y) = \frac{1}{2\pi\sigma^2} \exp\left\{-\frac{x^2 + y^2}{2\sigma^2}\right\}. \tag{2.83}$$

Substituting $x = r \cos v$, $y = r \sin v$, using the Jacobian $J(r, v) = r$, and recognizing that $|r| = r$, yields the joint pdf:

$$f_{RV}(r, v) = \frac{r}{2\pi\sigma^2} \exp\left\{-\frac{r^2}{2\sigma^2}\right\}, \quad r \geq 0, \ -\pi \leq v \leq \pi. \tag{2.84}$$

The marginal pdf $f_R(r)$ is obtained by integrating the joint density over the range of the random variable V. Hence, the marginal pdf of R is

$$\begin{aligned} f_R(r) &= \int_{-\pi}^{\pi} f_{RV}(r, v) \mathrm{d}v \\ &= \frac{r}{\sigma^2} \exp\left\{-\frac{r^2}{2\sigma^2}\right\}, \quad r \geq 0. \end{aligned} \tag{2.85}$$

The pdf in (2.85) is called a *Rayleigh distribution*. As discussed in Sect. 11.2.1, the Rayleigh distribution arises in the context of Rayleigh fading, where the magnitude of the complex Gaussian channel gain has a Rayleigh distribution. The Rayleigh distribution also describes the pdf of the complex envelope of bandpass additive white Gaussian noise in Sect. 4.4.3.

The cdf of R is

$$F_R(r) = 1 - \exp\left\{-\frac{r^2}{2\sigma^2}\right\}, \quad r \geq 0. \tag{2.86}$$

The mean and variance of the Rayleigh distribution are, respectively,

$$\mu_R = \sqrt{\frac{\pi}{2}}\sigma \tag{2.87a}$$

$$\sigma_R^2 = \left(2 - \frac{\pi}{2}\right)\sigma^2 \tag{2.87b}$$

It is sometimes useful to express the pdf and cdf of a Rayleigh random variable in terms of its second moment:

$$\Omega = \mathrm{E}[R^2] = 2\sigma^2 \tag{2.88}$$

as, respectively,

$$f_R(r) = \frac{2r}{\Omega}\exp\left\{-\frac{r^2}{\Omega}\right\}, \quad r \geq 0 \tag{2.89a}$$

$$F_R(r) = 1 - \exp\left\{-\frac{r^2}{\Omega}\right\}, \quad r \geq 0. \tag{2.89b}$$

Finally, the marginal pdf of the angle V is

$$\begin{aligned} f_V(v) &= \int_0^\infty f_{RV}(r, v)\mathrm{d}r \\ &= \frac{1}{2\pi}, \quad \pi \leq v \leq \pi. \end{aligned} \tag{2.90}$$

which is a uniformly distributed random variable on the interval $[-\pi, \pi)$.

2.7.5 Exponential Distribution

Let $X \sim \mathcal{N}(0, \sigma^2)$ and $Y \sim \mathcal{N}(0, \sigma^2)$ be two i.i.d. zero-mean Gaussian random variables. Consider the random variable $U = X^2 + Y^2$. Note that for the complex random variable, $Z = X + jY$, $U = |Z|^2$ is its squared magnitude.

While $f_U(u)$ may be obtained using a bivariate transformation of random variables, an easier approach uses the fact that $U = R^2$, where $R = \sqrt{x^2 + Y^2}$ has the Rayleigh pdf in (2.85). Using (2.36), the pdf of U is

$$\begin{aligned} f_U(u) &= \frac{f_R(\sqrt{u}) + f_R(-\sqrt{u})}{2\sqrt{u}} \\ &= \frac{1}{2\sqrt{u}} \cdot \frac{\sqrt{u}}{\sigma^2}\exp\left\{-\frac{u}{2\sigma^2}\right\} \\ &= \frac{1}{2\sigma^2}\exp\left\{-\frac{u}{2\sigma^2}\right\}, \quad u \geq 0. \end{aligned} \tag{2.91}$$

where the second line follows because the pdf for R is nonnegative, meaning that $f_R(-\sqrt{u}) = 0$ for $u > 0$. The pdf in (2.91) is said to be an *exponential distribution.* The cumulative distribution function of U is

$$F_U(u) = 1 - \exp\left\{-\frac{u}{2\sigma^2}\right\}, \quad u \geq 0. \tag{2.92}$$

The mean and variance of the exponential distribution are given by $\mu_U = 2\sigma^2$ and $\sigma_U^2 = 4\sigma^4$.

The pdf and cdf of an exponential random variable is sometimes expressed in terms of its mean value:

$$\Omega = \mathrm{E}[U] = \mathrm{E}[R^2] = 2\sigma^2 \tag{2.93}$$

as, respectively,

$$f_U(u) = \frac{1}{\Omega}\exp\left\{-\frac{u}{\Omega}\right\}, \quad u \geq 0 \tag{2.94a}$$

$$F_U(u) = 1 - \exp\left\{-\frac{u}{\Omega}\right\}, \quad u \geq 0. \tag{2.94b}$$

2.7.6 Rice Distribution

Let $X \sim \mathcal{N}(\mu_1, \sigma^2)$ and $Y \sim \mathcal{N}(\mu_2, \sigma^2)$ be pair of independent Gaussian random variables with non-zero mean values μ_1 and μ_2, respectively. Consider again the random variable $R = \sqrt{X^2 + Y^2}$. As with the derivation of the Rayleigh distribution, define the auxiliary variable $V = \mathrm{Tan}^{-1}(Y/X)$. This leads to the Jacobian $J(r, v) = r$ and the bivariate transformation:

$$f_{RV}(r, v) = r \cdot f_{XY}(r\cos v, r\sin v). \tag{2.95}$$

The joint pdf of X and Y is

$$\begin{aligned} f_{XY}(x, y) &= \frac{1}{2\pi\sigma^2}\exp\left\{-\frac{(x-\mu_1)^2 + (y-\mu_2)^2}{2\sigma^2}\right\} \\ &= \frac{1}{2\pi\sigma^2}\exp\left\{-\frac{x^2 + y^2 + \mu_1^2 + \mu_2^2 - 2(x\mu_1 + y\mu_2)}{2\sigma^2}\right\}. \end{aligned} \tag{2.96}$$

Hence,

$$f_{RV}(r, v) = \frac{r}{2\pi\sigma^2} \exp\left\{-\frac{r^2 + \mu_1^2 + \mu_2^2 - 2r(\mu_1 \cos v + \mu_2 \sin v)}{2\sigma^2}\right\}. \tag{2.97}$$

Now define:

$$s \triangleq \sqrt{\mu_1^2 + \mu_2^2},\ s \geq 0 \tag{2.98a}$$

$$t \triangleq \mathrm{Tan}^{-1}\mu_2/\mu_1,\ -\pi \leq t \leq \pi \tag{2.98b}$$

so that $\mu_1 = s \cos t$ and $\mu_2 = s \sin t$. Then

$$\begin{aligned} f_{RV}(r, v) &= \frac{r}{2\pi\sigma^2} \exp\left\{-\frac{r^2 + s^2 - 2rs(\cos t \cos v + \sin t \sin v)}{2\sigma^2}\right\} \\ &= \frac{r}{2\pi\sigma^2} \exp\left\{-\frac{r^2 + s^2 - 2rs\cos(v - t)}{2\sigma^2}\right\},\ r \geq 0,\ -\pi \leq v \leq \pi. \end{aligned} \tag{2.99}$$

The marginal pdf of R is obtained by integrating over the range of v, which may be taken as $[0, 2\pi)$:

$$f_R(r) = \frac{r}{\sigma^2}\exp\left\{-\frac{r^2 + s^2}{2\sigma^2}\right\} \frac{1}{2\pi} \int_0^{2\pi} \exp\left\{\frac{rs}{\sigma^2}\cos(v - t)\right\} \mathrm{d}v. \tag{2.100}$$

Observe that the value of t is irrelevant in (2.100) due to integration over the entire 2π range of V and t may be set to zero.

The zero-order modified Bessel function of the first kind is defined as

$$I_0(x) \triangleq \frac{1}{2\pi} \int_0^{2\pi} \mathrm{e}^{x\cos\theta} \mathrm{d}\theta. \tag{2.101}$$

Hence,

$$f_R(r) = \frac{r}{\sigma^2}\exp\left\{-\frac{r^2 + s^2}{2\sigma^2}\right\} I_0\left(\frac{rs}{\sigma^2}\right),\ r \geq 0. \tag{2.102}$$

The pdf in (2.102) is called a *Rice distribution*. The cdf of R does not exist in closed form, but can be expressed as

$$F_R(r) = \int_0^r p_R(r)\mathrm{d}r = 1 - Q\left(\frac{s}{\sigma}, \frac{r}{\sigma}\right), \tag{2.103}$$

where $Q(a, b)$ is called a Marcum Q-function. The first two moments of the Rice distribution are, respectively,

$$\mu_R = \sqrt{\frac{\pi}{2}}\sigma\, {}_1F_1\left(-\frac{1}{2}, 1; -\frac{s^2}{2\sigma^2}\right) \tag{2.104a}$$

$$\mathrm{E}[R^2] = s^2 + 2\sigma^2 \tag{2.104b}$$

where ${}_1F_1(a, b; x)$ is the *confluent hypergeometric function* given by

$${}_1F_1(a, b; x) = \frac{\Gamma(b)}{\Gamma(b-a)\Gamma(a)} \int_0^1 \mathrm{e}^{xt} t^{a-1}(1-t)^{b-a-1}\mathrm{d}t \tag{2.104c}$$

and $\Gamma(a)$ is the *Gamma function*

$$\Gamma(a) = \int_0^\infty t^{a-1}\mathrm{e}^{-t}\mathrm{d}t \tag{2.104d}$$

Sometimes it is useful to express the Rice distribution in terms of its *Rice factor* $K = s^2/2\sigma^2$ and second moment $\Omega = \mathrm{E}[R^2] = s^2 + 2\sigma^2$ by first noting that

$$s^2 = \frac{K\Omega}{K+1}, \qquad \sigma^2 = \frac{\Omega}{2(K+1)}. \tag{2.104e}$$

Substituting s^2 and σ^2 into (2.102) yields:

$$p_\alpha(x) = \frac{2(K+1)x}{\Omega} \exp\left\{-K - \frac{(K+1)x^2}{\Omega}\right\} I_o\left(2x\sqrt{\frac{K(K+1)}{\Omega}}\right), \quad x \geq 0. \tag{2.104f}$$

Figure 2.7 plots the Rice pdf for several values of K and $\Omega = 1$. The curve for $K = 0$ is the Rayleigh pdf in (11.18). As K increases, the Rice pdf becomes more peaked around its second moment $\Omega = 1$, and in the limit $K \to \infty$ becomes a unit impulse at $\Omega = 1$.

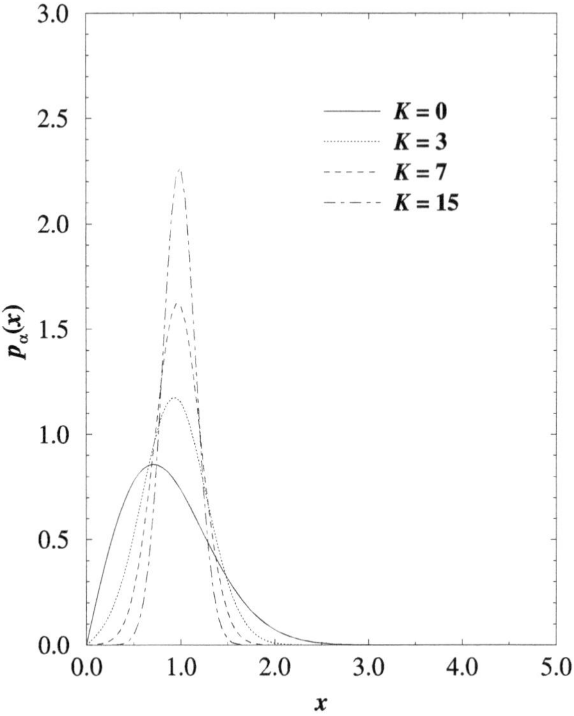

Fig. 2.7 The Rice pdf for several values of K with $\Omega = 1$. The case $K = 0$ is the Rayleigh pdf

Problems

2.1 Determine the parameters L and M so that each of the following is a valid probability distribution function:

(i) $p_X(k) = L^{k+M}, \quad k = 0, 1, 2, \ldots$
(ii) $p_Y(k) = ML^k, \quad k = 0, 1, 2, \ldots$
(iii) $p_Z(k) = L^k + M, \quad k = 0, 1, 2.$

2.2 Determine the parameters K and N so that each of the following is a valid pdf:

(i) $f_X(x) = \frac{K}{(x+N)^2} u(x)$
(ii) $f_Y(y) = K\mathrm{rect}(x) + N\delta(x)$
(iii) $f_Z(x) = e^{Kx^2} + N$

Note:

$$\mathrm{rect}(x) = \begin{cases} 1 \text{ for } |x| \le 1/2 \\ 0 \text{ otherwise} \end{cases}$$

2.3 Let a pdf be given by $f(x) = Ke^{-b|x|}$, $-\infty \leq x \leq \infty$, where K and b are positive constants:

(a) Find the value required for K in terms of b.
(b) Find the variance σ_X^2 in terms of b.
(c) Find the cdf $F_X(x)$.

2.4 A zero-mean Gaussian random variable X has variance $\sigma_X^2 = 2$. Let A be the event such that $|X| < 3$:

(a) Find an expression for the conditional pdf $f_{x|A}(x)$.
(b) Plot $f_{x|A}(x)$ over the range $|x| < 5$.
(c) Plot $f_X(x)$ over the range $|x| < 5$ and compare the plot in part b).

2.5 Consider a binary discrete memoryless channel (DMC) shown in Fig. 2.8, having the inputs $X \in \{0, 1\}$ and outputs $Y \in \{0, 1\}$, such that

$$\mathrm{P}[Y = 0|X = 0] = 0.75 \qquad \mathrm{P}[Y = 0|X = 1] = 0.1$$
$$\mathrm{P}[Y = 1|X = 0] = 0.25 \qquad \mathrm{P}[Y = 1|X = 1] = 0.9$$
$$\mathrm{P}[X = 0] = 0.4 \qquad \mathrm{P}[X = 1] = 0.6$$

(a) Find the probabilities $\mathrm{P}[Y = 0]$ and $\mathrm{P}[Y = 1]$ at the channel output.
(b) By using Bayes' rule, find the probability $\mathrm{P}[X = 1|Y = 1]$.
(c) Find the probability of error, i.e., $\mathrm{P}[Y \neq X]$.
(d) Repeat parts a), b), and c) if two such DMCs are concatenated together in a serial fashion, i.e., the output of the first DMC is input to the second DMC.

2.6 Consider the serial concatenation of three binary asymmetric "Z-channels" in Fig. 2.9, where the output from one channel is input to the next channel. Assume that the inputs to the first channel are used with equal probability, i.e., $P[X = 0] = P[X = 1] = 1/2$.

(a) The overall channel from input to output can be redrawn as the equivalent binary asymmetric channel shown in Fig. 2.10. Determine the values of α and β in terms of ε.
(b) What is the probability of bit error?

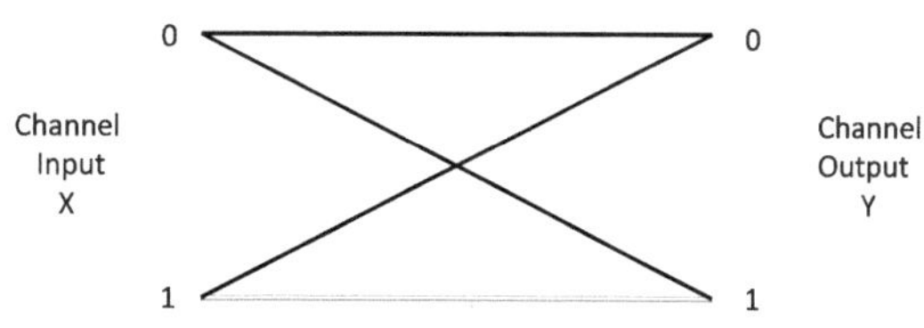

Fig. 2.8 Binary asymmetric channel for Problem 2.5

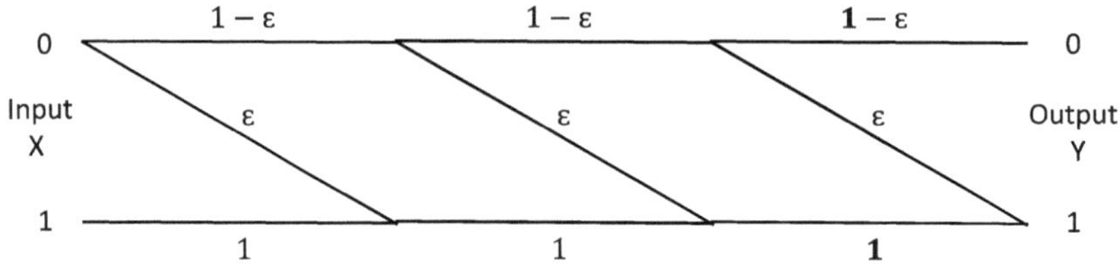

Fig. 2.9 Overall binary asymmetric channel

Fig. 2.10 Serial concatenation of Z channels

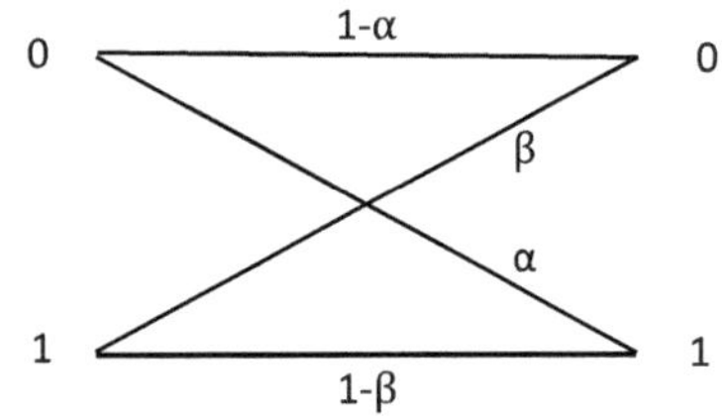

(c) What is the probability $P[X = 1|Y = 1]$? Evaluate your result when $\varepsilon = 0.1$.

2.7 Consider the random variables $X_1 = a_1 + n_1$ and $X_2 = a_2 + n_2$, where a_1 and a_2 are real-valued constants and n_1 and n_2 are independent zero-mean Gaussian random variables with variance σ^2:

(a) Write down an expression for the joint pdf $f_{X_1,X_2}(x_1, x_2)$.
(b) Find the mean and variance of the random variable $Y = X_1 X_2$.

2.8 Let

$$Y = \sum_{k=1}^{N} a_k X_k$$

be the weighted sum of N independent random variables, $X_k, k = 1, \ldots, N$, each having mean μ_{X_k} and variance $\sigma_{X_k}^2$. The weights a_k are real-valued constants:

(a) Find the mean μ_Y.
(b) Find the variance σ_Y^2.

2.9 Consider a zero-mean Gaussian random variable X with variance σ_X^2. This random variable is passed through a system whose input-output relation is given by $Y = g(X)$. Find the pdf of the random variable at the output, Y, in each of the following cases:

(a) full-wave rectifier: $g(X) = |X|$
(b) square-law device: $g(X) = X^2$

(c) limiting amplifier:

$$g(X) = \begin{cases} -b, & X < -b \\ b, & X \geq b \\ X, & |X| \leq b \end{cases}$$

2.10 Suppose the random variables X and Y satisfy $E[X^2] = E[Y^2]$. Define the random variables U and V by $U = X + Y$ and $V = X - Y$:

(a) Show that U and V are orthogonal.
(b) If, in addition, X and Y have the same mean, are U and V uncorrelated? Prove your answer.

2.11 Let Θ be uniformly distributed on $[0, \pi]$ and let the random variables X and Y be defined by $X = \cos\Theta$ and $Y = \sin\Theta$. Show that X and Y are uncorrelated but they are not independent.

2.12 Let X and Y be independent random variables with

$$f_X(x) = \begin{cases} \alpha e^{-\alpha x}, & x > 0 \\ 0, & \text{otherwise} \end{cases}$$

and

$$f_Y(y) = \begin{cases} \beta e^{-\beta y}, & y > 0 \\ 0, & \text{otherwise} \end{cases}$$

where α and β are assumed to be positive constants. Find the pdf of $X + Y$. Treat the case of $\alpha = \beta$ separately.

2.13 If X and Y are independent standard normal random variables, find the pdf of their quotient $Z = X/Y$.

2.14 Let $\mathbf{X} = (X_1, X_2)$ be a Gaussian random vector with mean vector and covariance matrix:

$$\boldsymbol{\mu}_X = (2, -1), \qquad \boldsymbol{\Lambda}_X = \begin{bmatrix} 1 & 0 \\ 0 & 3 \end{bmatrix}$$

and let $Z = X_1 + X_2$:

(a) Are X_1 and X_2 statistically independent?
(b) Find the pdf of Z.

2.15 Suppose the random variables X and Y are jointly Gaussian with mean vector and covariance matrix:

$$\boldsymbol{\mu} = (1\ \ 2), \qquad \boldsymbol{\Lambda} = \begin{bmatrix} 4 & -4 \\ -4 & 9 \end{bmatrix}$$

(a) Find the correlation coefficient between X and Y.
(b) If $Z = 2X + Y$ and $W = X - Y$, find the covariance Z and W.
(c) Find the pdf of Z.

2.16 Consider a collection of N independent Gaussian random variables, $X_1, \ldots, X_N$, such that

$$X_1 \sim \mathcal{N}(\mu, \sigma^2)$$
$$X_i \sim \mathcal{N}(0, \sigma^2), \quad i = 2, \ldots, N$$

Derive an integral expression for the probability of the event A, where

$$A = \bigcap_{i=2}^{N} \{X_i \leq X_1\}$$

2.17 Let $\mathbf{X} = (X_1, X_2, \ldots, X_n)$ be a vector of Gaussian random variables with mean vector $\boldsymbol{\mu} = (\mu_{X_1}, \mu_{X_2}, \ldots, \mu_{X_n})$ and covariance matrix $\boldsymbol{\Lambda}$. Define a new random vector $\mathbf{Y} = (Y_1, Y_2, \ldots, Y_n)$, such that

$$\mathbf{Y} = \mathbf{A}\mathbf{X} + \mathbf{b}$$

where $\mathbf{A}$ is an $n \times n$ constant matrix, and $\mathbf{b} = (b_1, b_2, \ldots, b_n)$ is a constant vector. Find the mean vector and covariance matrix of $\mathbf{Y}$.

Chapter 3
Random Processes

A random process, or stochastic process, $X(t)$, is an ensemble of *sample functions* $\{X_1(t), X_2(t), \ldots, X_\xi(t)\}$ that are a function of the time variable t, together with a probability rule that assigns a probability to any meaningful event associated with the observation of these sample functions. Consider the set of sample functions shown in Fig. 3.1.

The sample function $X_i(t)$ corresponds to the sample point s_i in the sample space and occurs with probability $\mathrm{P}[s_i]$. The number of sample functions, ξ, in the ensemble may be finite or infinite. The function $X_i(t)$ is a deterministic function of time t once the index i is known. The sample functions may be defined at discrete or continuous instants in time, which define discrete- or continuous-time random processes, respectively. Furthermore, their values (or parameters) at these time instants may be either discrete or continuous valued, which defines a discrete- or continuous-parameter random process, respectively.

Suppose that the random process $X(t)$ is observed at time instant t_1. The observed value is randomly drawn from the set of numbers $\{X_i(t_1)\}, i = 1, 2, \ldots, \xi$. Since $X_i(t_1)$ occurs with probability $\mathrm{P}[s_i]$, the collection of numbers $\{X_i(t_1)\}, i = 1, 2, \ldots, \xi$, constitutes a random variable, denoted by $X(t_1)$. By observing the set of waveforms at another time instant t_2, a different random variable $X(t_2)$ is obtained. The collection of n such random variables, $X(t_1),\ X(t_2), \ldots,\ X(t_n)$, obtained by observing $X(t)$ at time instants $t_1, t_2, \ldots, t_n$ has the joint cdf:

$$F_{X(t_1),\ldots,\ X(t_n)}(x_1, \ldots,\ x_n) = \mathrm{P}[X(t_1) < x_1, \ldots, X(t_n) < x_n]. \tag{3.1}$$

A more compact notation can be obtained by defining the vectors:

$$\mathbf{x} \triangleq (x_1, x_2, \ldots, x_n) \tag{3.2a}$$

$$\mathbf{X}(t) \triangleq (X(t_1), X(t_2), \ldots, X(t_n)). \tag{3.2b}$$

G. Stuber, *Introductory Digital Wireless Communications*,
https://doi.org/10.1007/978-3-032-01559-4_3

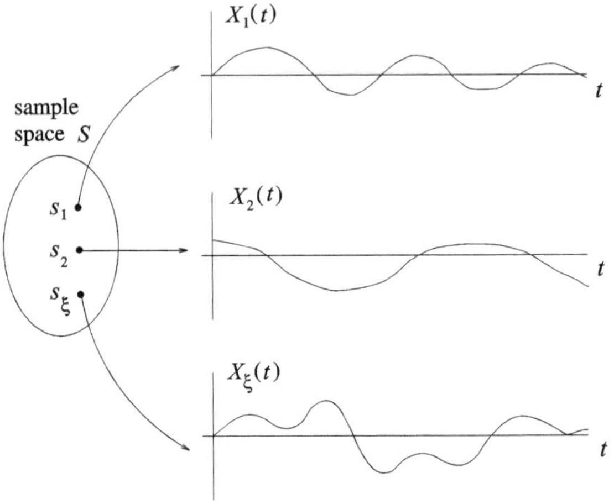

Fig. 3.1 Ensemble of sample functions for a random process

Then, the joint cdf and joint pdf of the random vector $\mathbf{X}(t)$ are, respectively,

$$F_{\mathbf{X}(t)}(\mathbf{x}) = \mathrm{P}(\mathbf{X}(t) \leq \mathbf{x}) \tag{3.3}$$

$$f_{\mathbf{X}(t)}(\mathbf{x}) = \frac{\partial^n F_{\mathbf{X}(t)}(\mathbf{x})}{\partial x_1 \partial x_2 \cdots \partial x_n}. \tag{3.4}$$

A random process is said to be *strictly stationary* if and only if the joint pdf $f_{\mathbf{X}(t)}(\mathbf{x})$ is invariant to the time variable t. In this case, the equality

$$f_{\mathbf{X}(t)}(\mathbf{x}) = f_{\mathbf{X}(t+\tau)}(\mathbf{x}) \tag{3.5}$$

holds for all sets of time instants $\{t_1, t_2, \ldots, t_n\}$ and all time shifts τ.

The remainder of this chapter introduces random processes at a level suitable for the remainder of the text. Section 3.1 considers the moments and autocorrelation of a random process. Section 3.2 considers a pair of random processes, and their cross-correlation and crosscovariance. Section 3.3 discusses complex-valued random processes that find application in describing the complex envelope of bandpass random processes. Section 3.4 discusses the power density spectrum of random processes, while Sect. 3.5 considers the filtering of random processes with linear time-invariant filters. Section 3.6 extends the above concepts developed for continuous-time random processes to discrete-time random processes. Section 3.7 wraps up the chapter with the autocorrelation and power density spectrum of continuous- and discrete-time cyclostationary random processes that include digitally modulated waveforms and their sampled versions.

3.1 Moments and Correlation Functions

To describe the moments and correlation functions of a random process, it is useful to define the following two operators:

$$\mathrm{E}[\,\cdot\,] \triangleq \text{ensemble average} \tag{3.6a}$$

$$\langle\,\cdot\,\rangle \triangleq \text{time average} \tag{3.6b}$$

The *ensemble average* of a random process at time t is the mean value of the random variable $X(t)$:

$$\mu_X(t) = \mathrm{E}[X(t)] = \int_{-\infty}^{\infty} x f_{X(t)}(x)\mathrm{d}x. \tag{3.7}$$

Note that the ensemble average is generally a function of time, particularly if the random process is nonstationary.

The *time average* of a random process is

$$\langle X_i(t)\rangle = \lim_{T\to\infty} \frac{1}{2T}\int_{-T}^{T} X_i(t)\mathrm{d}t. \tag{3.8}$$

In general, the time average $\langle X(t)\rangle$ is a random variable, because it depends on the particular sample function, $X_i(t)$, which is selected from the ensemble for time averaging.

Example 3.1 Consider the simple random process shown Fig. 3.2. The *ensemble mean* is computed by averaging across the ensemble of sample functions as

$$\begin{aligned}\mathrm{E}[X(t)] &= X_1(t)P_1 + X_2(t)P_2 + X_3(t)P_3\\ &= a\cdot 1/4 + 0\cdot 1/2 + (-a)\cdot 1/4 = 0\end{aligned}$$

The *time average mean* is computed by selecting a sample function and time averaging as

$$\langle X(t)\rangle = \begin{cases} a & \text{with probability } 1/4\\ 0 & \text{with probability } 1/2\\ -a & \text{with probability } 1/4\end{cases}$$

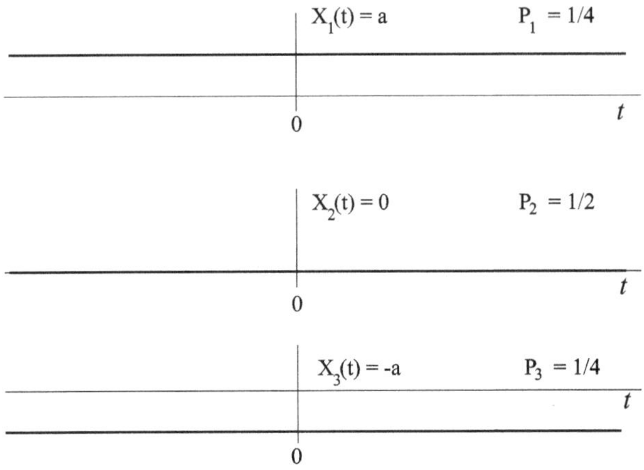

Fig. 3.2 Example of a random process

The *autocorrelation* of a random process $X(t)$ is defined as

$$\phi_{XX}(t_1, t_2) \stackrel{\Delta}{=} \mathrm{E}\left[X(t_1)X(t_2)\right] \tag{3.9}$$

and the *autocovariance* of a random process $X(t)$ is defined as

$$\begin{aligned}\lambda_{XX}(t_1, t_2) &\stackrel{\Delta}{=} \mathrm{E}\left[(X(t_1) - \mu_X(t_1))\,(X(t_2) - \mu_X(t_2))\right] \\ &= \phi_{XX}(t_1, t_2) - \mu_X(t_1)\mu_X(t_2).\end{aligned} \tag{3.10}$$

A random process that is strictly stationary must have

$$\mathrm{E}[X^n(t)] = \mathrm{E}[X^n] \quad \forall\, t, n.$$

Hence, strictly stationary random process must have, among other properties,

$$\begin{aligned}\mu_X(t) &= \mu_X \\ \sigma_X^2(t) &= \sigma_X^2 \\ \phi_{XX}(t_1, t_2) &= \phi_{XX}(t_2 - t_1) \equiv \phi_{XX}(\tau) \\ \lambda_{XX}(t_1, t_2) &= \lambda_{XX}(t_2 - t_1) \equiv \lambda_{XX}(\tau),\end{aligned}$$

where $\tau = t_2 - t_1$.

A *wide-sense stationary* (WSS) random process is defined as a random process that satisfies the following two conditions:

$$\mu_X(t) = \mu_X \tag{3.11a}$$

$$\phi_{XX}(t_1, t_2) = \phi_{XX}(\tau)\,, \quad \tau = t_2 - t_1\,. \tag{3.11b}$$

If a random process is strictly stationary, then it is wide-sense stationary; however, the converse may not be true. A notable exception is the Gaussian random process which is strictly stationary if it is wide-sense stationary, the reason being that the joint pdf of the Gaussian random vector $\mathbf{X}(t) = (X(t_1), X(t_2), \ldots, X(t_n))$ is completely described by the means and covariances of the set of Gaussian random variables $\{X(t_i)\}$. Fortunately, many important random processes are WSS.

3.1.1 Autocorrelation Properties

The autocorrelation $\phi_{XX}(\tau)$ of a WSS random process satisfies the following properties:

1. $\phi_{XX}(0) = \mathrm{E}[X^2(t)]$. This is the total power in the random process.
2. $\phi_{XX}(\tau) = \phi_{XX}(-\tau)$. The autocorrelation must be an even function.
3. $|\phi_{XX}(\tau)| \leq \phi_{XX}(0)$. This is a variant of the Cauchy-Schwartz inequality.
4. $\phi_{XX}(\infty) = \mathrm{E}^2[X(t)] = \mu_X^2$. This holds if $X(t)$ contains no periodic components, and is equal to the d.c. power.

Example 3.2 This example establishes Property 3 above through the following steps:

$$
\begin{aligned}
0 &\leq \mathrm{E}[(X(t) \pm (X(t+\tau))^2] \\
&= \mathrm{E}[X^2(t) + X^2(t+\tau) \pm 2X(t)X(t+\tau)] \\
&= \mathrm{E}[X^2(t)] + \mathrm{E}[X^2(t+\tau)] \pm 2\mathrm{E}[X(t)X(t+\tau)] \\
&= 2\mathrm{E}[X^2(t)] \pm 2\mathrm{E}[X(t)X(t+\tau)] \\
&= 2\phi_{XX}(0) \pm 2\phi_{XX}(\tau).
\end{aligned}
$$

Therefore,

$$
\begin{aligned}
\pm\, \phi_{XX}(\tau) &\leq \phi_{XX}(0) \\
|\phi_{XX}(\tau)| &\leq \phi_{XX}(0).
\end{aligned}
$$

3.1.2 Ergodic Random Processes

A random process $X(t)$ is ergodic in some statistic if that statistic can be estimated by taking the time average of a single and arbitrary sample function of the random process. The random process is said to be ergodic in its mean if

$$\mu_X = \mathrm{E}[X(t)] = \langle X_i(t) \rangle. \tag{3.12}$$

Likewise, a random process is said to be ergodic in its autocorrelation if

$$\phi_{XX}(\tau) = \mathrm{E}[X(t)X(t+\tau)] = \langle X_i(t)X_i(t+\tau) \rangle. \tag{3.13}$$

A random process that is ergodic in its mean and autocorrelation is said to be *wide-sense ergodic*.

Example 3.3 Consider the random process:

$$X(t) = A\cos(2\pi f_c t + \Theta)$$

where A and f_c are constants. The phase Θ is assumed to be a uniformly distributed random variable with pdf

$$f_\Theta(\theta) = \begin{cases} 1/(2\pi)\,, & 0 \le \theta \le 2\pi \\ 0\,, & \text{elsewhere} \end{cases}.$$

The number of sample functions in the ensemble is infinite. The ith sample function is

$$X_i(t) = A\cos(2\pi f_c t + \theta_i)\ , \quad i = 1, \ldots, \infty$$

where θ_i is the ith instance of the random phase Θ.

The ensemble mean of $X(t)$ is obtained by averaging over the pdf of Θ:

$$\begin{aligned}
\mu_X(t) &= \mathrm{E}_\Theta[X(t)] = \mathrm{E}_\Theta[A\cos(2\pi f_c t + \Theta)] \\
&= \int_0^{2\pi} A\cos(2\pi f_c t + \theta) f_\Theta(\theta) d\theta \\
&= \frac{A}{2\pi}\int_0^{2\pi} \cos(2\pi f_c t + \theta) d\theta \\
&= \frac{A}{2\pi}\sin(2\pi f_c t + \theta)\Big|_0^{2\pi} = 0
\end{aligned}$$

(continued)

Example 3.3
The ensemble autocorrelation of $X(t)$ is

$$\begin{aligned}\phi_{XX}(t_1,t_2) &= \mathrm{E}_\Theta[X(t_1)X(t_2)]\\ &= \mathrm{E}_\Theta[A^2\cos(2\pi f_c t_1+\Theta)\cos(2\pi f_c t_2+\Theta)]\\ &= \frac{A^2}{2}\mathrm{E}_\Theta[\cos(2\pi f_c t_1+2\pi f_c t_2+2\Theta)] + \frac{A^2}{2}\mathrm{E}_\Theta[\cos(2\pi f_c(t_2-t_1))]\end{aligned}$$

But

$$\begin{aligned}\mathrm{E}_\Theta[\cos(2\pi f_c t_1+2\pi f_c t_2+2\Theta)] &= \frac{1}{2\pi}\int_0^{2\pi}\cos(2\pi f_c t_1+2\pi f_c t_2+2\theta)d\theta\\ &= \frac{1}{4\pi}\sin(2\pi f_c t_1+2\pi f_c t_2+2\theta)\Big|_0^{2\pi} = 0\\ \mathrm{E}_\Theta[\cos(2\pi f_c(t_2-t_1))] &= \cos(2\pi f_c(t_2-t_1))\end{aligned}$$

Hence,

$$\begin{aligned}\phi_{XX}(t_1,t_2) &= \frac{A^2}{2}\cos(2\pi f_c(t_2-t_1))\\ &= \frac{A^2}{2}\cos(2\pi f_c\tau) \equiv \phi_{XX}(\tau), \quad \tau = t_2-t_1\end{aligned}$$

Since both the mean and autocorrelation are time invariant, this random process is WSS.

The time-average mean of $X(t)$ is

$$\langle X(t)\rangle = \lim_{T\to\infty}\frac{1}{2T}\int_{-T}^{T}A\cos(2\pi f_c t+\theta)\mathrm{d}t = 0.$$

The time-average autocorrelation of $X(t)$ is

$$\begin{aligned}&\langle X(t)X(t+\tau)\rangle\\ &= \lim_{T\to\infty}\frac{A^2}{2T}\int_{-T}^{T}\cos(2\pi f_c t+\theta)\cos(2\pi f_c t+2\pi f_c\tau+\theta)\mathrm{d}t\\ &= \lim_{T\to\infty}\frac{A^2}{4T}\int_{-T}^{T}[\cos(2\pi f_c\tau)+\cos(4\pi f_c t+2\pi f_c\tau+2\theta)]\,\mathrm{d}t\end{aligned}$$

(continued)

Example 3.3 (continued)

$$= \lim_{T\to\infty} \frac{A^2}{4T} \cos(2\pi f_c \tau) \int_{-T}^{T} \mathrm{d}t + \lim_{T\to\infty} \frac{A^2}{4T} \int_{-T}^{T} \cos(4\pi f_c t + 2\pi f_c \tau + 2\theta)\mathrm{d}t$$

$$= \frac{A^2}{2} \cos(2\pi f_c \tau).$$

By comparing the ensemble and time-average mean and autocorrelation, it can be concluded that this random process is wide-sense ergodic.

Example 3.4 Consider again the random process in Example 3.1. The ensemble mean is $\mathrm{E}[X(t)] = 0$. However, the time-averaged mean $\langle X(t) \rangle$ is a random variable, since it depends on the sample function that is chosen for time averaging. Hence, $X(t)$ is not ergodic in its mean.

The ensemble and time averaged autocorrelations are, respectively,

$$\phi_{XX}(\tau) = \mathrm{E}[X(t)X(t+\tau)] = a^2(1/4) + 0(1/2) + (-a)^2(1/4) = a^2/2$$

$$\langle X(t)X(t+\tau) \rangle = \begin{cases} a^2 & \text{with probability } 1/2 \\ 0 & \text{with probability } 1/2 \end{cases}$$

Hence, $X(t)$ is not ergodic in its autocorrelation.

The time averages $\langle X(t) \rangle$ and $\langle X(t)X(t+\tau) \rangle$ are said to provide *unbiased estimates* of μ_X and $\phi_{XX}(\tau)$, respectively, if

$$\mathrm{E}[\langle X(t) \rangle] = \mu_X$$

$$\mathrm{E}[\langle X(t)X(t+\tau) \rangle] = \phi_{XX}(\tau)$$

For this example, the time averages provide unbiased estimates of the ensemble mean and autocorrelation.

Example 3.5 Consider the random process

$$Y(t) = X \cos t, \quad X \sim \mathcal{N}(0, 1).$$

Find the pdf of $Y(0)$, the joint pdf of $Y(0)$, and $Y(\pi)$, and determine whether or not $Y(t)$ is strictly stationary:

(continued)

Example 3.5 (continued)

1. To find the pdf of $Y(0)$

$$Y(0) = X\cos 0 = X.$$

Therefore,

$$f_{Y(0)}(y_0) = \frac{1}{\sqrt{2\pi}}\mathrm{e}^{-y_0^2/2}.$$

2. To find the joint pdf of $Y(0)$ and $Y(\pi)$

$$Y(0) = X = -Y(\pi).$$

Therefore,

$$f_{Y(0)|Y(\pi)}(y_0|y_\pi) = \delta(y_0 + y_\pi)$$

and

$$\begin{aligned} f_{Y(0)Y(\pi)}(y_0, y_\pi) &= f_{Y(0)|Y(\pi)}(y_0|y_\pi)f_{Y(\pi)}(y_\pi) \\ &= \frac{1}{\sqrt{2\pi}}\mathrm{e}^{-y_\pi^2/2}\delta(y_0 + y_\pi). \end{aligned}$$

3. To determine whether or not $Y(t)$ is strictly stationary, note that

$$\begin{aligned} \mathrm{E}[Y(t)] &= \mathrm{E}[X]\cos t = 0 \\ \mathrm{E}[Y^2(t)] &= \mathrm{E}[X^2]\cos^2 t = \cos^2(t). \end{aligned}$$

Since the second moment of this random process varies with time t, it is nonstationary.

3.2 Cross-Correlation and Crosscovariance

Consider a pair of random processes $X(t)$ and $Y(t)$. The cross-correlation of $X(t)$ and $Y(t)$ is

$$\phi_{XY}(t_1, t_2) = \mathrm{E}[X(t_1)Y(t_2)] \tag{3.14a}$$

$$\phi_{YX}(t_1, t_2) = \mathrm{E}[Y(t_1)X(t_2)]. \tag{3.14b}$$

The correlation matrix of $X(t)$ and $Y(t)$ is

$$\mathbf{\Phi}(t_1, t_2) = \begin{bmatrix} \phi_{XX}(t_1, t_2) & \phi_{XY}(t_1, t_2) \\ \phi_{YX}(t_1, t_2) & \phi_{YY}(t_1, t_2) \end{bmatrix}. \tag{3.15}$$

The crosscovariance of $X(t)$ and $Y(t)$ is

$$\begin{aligned} \lambda_{XY}(t_1, t_2) &= \mathrm{E}\left[(X(t_1) - \mu_X(t_1))\,(Y(t_2) - \mu_Y(t_2))\right] \\ &= \phi_{XY}(t_1, t_2) - \mu_X(t_1)\mu_Y(t_2). \end{aligned} \tag{3.16}$$

The covariance matrix of $X(t)$ and $Y(t)$ is

$$\mathbf{\Lambda}(t_1, t_2) = \begin{bmatrix} \lambda_{XX}(t_1, t_2) & \lambda_{XY}(t_1, t_2) \\ \lambda_{YX}(t_1, t_2) & \lambda_{YY}(t_1, t_2) \end{bmatrix}. \tag{3.17}$$

If $X(t)$ and $Y(t)$ are each WSS and jointly WSS, then

$$\mathbf{\Phi}(t_1, t_2) = \mathbf{\Phi}(t_2 - t_1) = \mathbf{\Phi}(\tau) \tag{3.18a}$$

$$\mathbf{\Lambda}(t_1, t_2) = \mathbf{\Lambda}(t_2 - t_1) = \mathbf{\Lambda}(\tau), \tag{3.18b}$$

where $\tau = t_2 - t_1$.

3.2.1 Cross-Correlation Properties

The cross-correlation $\phi_{XY}(\tau)$ has the following properties:

1. $\phi_{XY}(\tau) = \phi_{YX}(-\tau)$
2. $|\phi_{XY}(\tau)| \leq \frac{1}{2}[\phi_{XX}(0) + \phi_{YY}(0)]$
3. $|\phi_{XY}(\tau)|^2 \leq \phi_{XX}(0)\phi_{YY}(0)$ if $X(t)$ and $Y(t)$ have zero mean.

3.2.2 Classifications of Joint Random Processes

The pair of random processes $X(t)$ and $Y(t)$ are said to be:

1. uncorrelated if $\lambda_{XY}(\tau) = 0$.
2. orthogonal if $\phi_{XY}(\tau) = 0$
3. statistically independent if

$$p_{\mathbf{X}(t)\mathbf{Y}(t+\tau)}(\mathbf{x}, \mathbf{y}) = p_{\mathbf{X}(t)}(\mathbf{x})\,p_{\mathbf{Y}(t+\tau)}(\mathbf{y}) \tag{3.19}$$

If $\mu_X = 0$ or $\mu_Y = 0$, then the random processes are orthogonal if they are uncorrelated. Statistically independent random processes are always uncorrelated; however, not all uncorrelated random processes are statistically independent. In the special case of Gaussian random processes, if the processes are uncorrelated, then they are also statistically independent.

Example 3.6 Find the autocorrelation of the random process:

$$Z(t) = X(t) + Y(t)$$

where $X(t)$ and $Y(t)$ are each and jointly WSS random processes.

The autocorrelation of $Z(t)$ is

$$\begin{aligned}\phi_{ZZ}(\tau) &= \mathrm{E}[Z(t)Z(t+\tau)] \\ &= \mathrm{E}\left[(X(t)+Y(t))\,(X(t+\tau)+Y(t+\tau))\right] \\ &= \phi_{XX}(\tau)+\phi_{YX}(\tau)+\phi_{XY}(\tau)+\phi_{YY}(\tau).\end{aligned}$$

If $X(t)$ and $Y(t)$ are uncorrelated, then

$$\phi_{YX}(\tau) = \phi_{XY}(\tau) = \mu_X\mu_Y$$

and

$$\phi_{ZZ}(\tau) = \phi_{XX}(\tau) + \phi_{YY}(\tau) + 2\mu_X\mu_Y.$$

If $X(t)$ and $Y(t)$ are uncorrelated and at least one has zero-mean, then

$$\phi_{ZZ}(\tau) = \phi_{XX}(\tau) + \phi_{YY}(\tau).$$

Example 3.7 Can the following be a correlation matrix for two jointly wide-sense stationary zero-mean random processes?

$$\boldsymbol{\Phi}(\tau) = \begin{bmatrix}\phi_{XX}(\tau) & \phi_{XY}(\tau) \\ \phi_{YX}(\tau) & \phi_{YY}(\tau)\end{bmatrix} = \begin{bmatrix}A^2\cos(\tau) & 2A^2\cos(3\tau/2) \\ 2A^2\cos(3\tau/2) & A^2\sin(2\tau)\end{bmatrix}.$$

The answer is no, because the following three properties are violated:

1. $|\phi_{XY}(\tau)| \leq \frac{1}{2}[\phi_{XX}(0) + \phi_{YY}(0)]$

(continued)

Example 3.7 (continued)
2. $|\phi_{XY}(\tau)|^2 \leq \phi_{XX}(0)\phi_{YY}(0)$ if $X(t)$ and $Y(t)$ have zero mean.
3. $\phi_{YY}(\tau) = \phi_{YY}(-\tau)$

3.3 Complex-Valued Random Processes

A complex-valued random process is given by

$$Z(t) = X(t) \; + \; jY(t) \tag{3.20}$$

where $X(t)$ and $Y(t)$ are real-valued random processes.

3.3.1 *Autocorrelation*

The autocorrelation of a complex-valued random process is

$$\begin{aligned}\phi_{ZZ}(t_1, t_2) &= \tfrac{1}{2}\mathrm{E}[Z^*(t_1)Z(t_2)] \\ &= \tfrac{1}{2}\mathrm{E}\left[(X(t_1) - jY(t_1))\,(X(t_2) + jY(t_2))\right] \\ &= \tfrac{1}{2}\Big(\phi_{XX}(t_1, t_2) + \phi_{YY}(t_1, t_2) + j(\phi_{XY}(t_1, t_2) - \phi_{YX}(t_1, t_2))\Big).\end{aligned} \tag{3.21}$$

Note the factor of $\frac{1}{2}$ is included for a reason similar to variance of a complex Gaussian random variable in (2.7.3). If $Z(t)$ is WSS, then

$$\phi_{ZZ}(t_1, t_2) = \phi_{ZZ}(t_2 - t_1) = \phi_{ZZ}(\tau)\,, \quad \tau = t_2 - t_1. \tag{3.22}$$

3.3.2 *Cross-Correlation*

Consider a pair of complex-valued random processes:

$$\begin{aligned}Z(t) &= X(t) + jY(t) \\ W(t) &= U(t) + jV(t).\end{aligned}$$

The cross-correlation of $Z(t)$ and $W(t)$ is

$$\begin{aligned}\phi_{ZW}(t_1, t_2) &= \tfrac{1}{2}\mathrm{E}[Z^*(t_1)W(t_2)]\\ &= \tfrac{1}{2}\Big(\phi_{XU}(t_1,t_2) + \phi_{YV}(t_1,t_2) + j(\phi_{XV}(t_1,t_2) - \phi_{YU}(t_1,t_2))\Big).\end{aligned} \tag{3.23}$$

If $X(t)$, $Y(t)$, $U(t)$ and $V(t)$ are pairwise WSS random processes, then

$$\phi_{ZW}(t_1, t_2) = \phi_{ZW}(t_2 - t_1) = \phi_{ZW}(\tau). \tag{3.24}$$

The cross-correlation of a pair of complex-valued WSS random process satisfies the following property:

$$\begin{aligned}\phi^*_{ZW}(\tau) &= \tfrac{1}{2}\mathrm{E}[Z(t)W^*(t+\tau)]\\ &= \tfrac{1}{2}\mathrm{E}[Z(\hat{t}-\tau)W^*(\hat{t})]\\ &= \tfrac{1}{2}\mathrm{E}[W^*(\hat{t})Z(\hat{t}-\tau)]\\ &= \phi_{WZ}(-\tau),\end{aligned} \tag{3.25}$$

where the second line uses the change of variable $\hat{t} = t + \tau$. For a complex-valued random process $Z(t)$, it follows that

$$\phi^*_{ZZ}(\tau) = \phi_{ZZ}(-\tau). \tag{3.26}$$

3.4 Power Spectral Density

The *power spectral density* (psd) of a WSS random process $X(t)$ and its autocorrelation are a Fourier transform pair, i.e.,

$$S_{XX}(f) = \int_{-\infty}^{\infty} \phi_{XX}(\tau)\mathrm{e}^{-j2\pi f\tau}\mathrm{d}\tau \tag{3.27a}$$

$$\phi_{XX}(\tau) = \int_{-\infty}^{\infty} S_{XX}(f)\mathrm{e}^{j2\pi f\tau}\mathrm{d}f. \tag{3.27b}$$

If $X(t)$ is a real-valued WSS random process, then its autocorrelation $\phi_{XX}(\tau)$ is real and even. Therefore, $S_{XX}(-f) = S_{XX}(f)$ meaning that the psd $S_{XX}(f)$ is also real and even. If $Z(t)$ is a complex-valued WSS random process, then $\phi_{ZZ}(\tau) = \phi^*_{ZZ}(-\tau)$, and $S^*_{ZZ}(f) = S_{ZZ}(f)$ meaning that the psd $S_{ZZ}(f)$ is real-valued but not necessarily even.

The power P in a WSS random process $X(t)$ is

$$\begin{aligned} P &= \mathrm{E}[X^2(t)] \\ &= \phi_{XX}(0) \\ &= \int_{-\infty}^{\infty} S_{XX}(f)\mathrm{d}f \end{aligned} \tag{3.28}$$

a result known as *Parseval's theorem.*

3.4.1 Properties of Power Spectral Density

The psd, $S_{XX}(\tau)$, of a WSS random process has the following properties:

1. $S_{XX}(0) = \int_{-\infty}^{\infty} \phi_{XX}(\tau)\mathrm{d}\tau$
2. $\int_{0-}^{0+} S_{XX}(f)\mathrm{d}f =$ dc power
3. $\phi_{XX}(0) = \int_{-\infty}^{\infty} S_{XX}(f)\mathrm{d}f =$ total power
4. $S_{XX}(f) \geq 0$ for all f. Power is never negative.
5. $S_{XX}(f) = S_{XX}(-f)$ (even function) if $X(t)$ is a real random process.
6. $S_{XX}(f)$ is always real.

Example 3.8 Consider again the random process in Example 3.3:

$$X(t) = A\cos(2\pi f_c t + \Theta)$$

where A and f_c are constants and

$$p_\Theta(\theta) = \begin{cases} \frac{1}{2\pi}\,, & -\pi \leq \theta \leq \pi \\ 0\,, & \text{elsewhere} \end{cases}$$

Example 3.3 showed that

$$\phi_{XX}(\tau) = \frac{A^2}{2}\cos(2\pi f_c \tau)$$

Taking the Fourier transform of $\phi_{XX}(\tau)$ gives:

$$S_{XX}(f) = \frac{A^2}{2}\mathcal{F}[\cos(2\pi f_c \tau)]$$

(continued)

Example 3.8 (continued)

$$= \frac{A^2}{4}(\delta(f - f_c) + \delta(f + f_c))$$

In this case, the psd consists of a pair of impulses having weight $A^2/4$ at the positive and negative frequencies $\pm f_c$ of the sinusoid as expected. From Parseval's theorem the power in $X(t)$ is

$$P = \phi_{XX}(0) = \int_{-\infty}^{\infty} S_{XX}(f)\mathrm{d}f = A^2/2$$

3.4.2 *Cross Power Spectral Density*

The cross psd between two random processes $X(t)$ and $Y(t)$ is the Fourier transform of their cross-correlation:

$$S_{XY}(f) = \int_{-\infty}^{\infty} \phi_{XY}(\tau)\mathrm{e}^{-j2\pi f\tau}\mathrm{d}\tau. \tag{3.29}$$

If $X(t)$ and $Y(t)$ are both real-valued random processes, then

$$\phi_{XY}(\tau) = \phi_{YX}(-\tau)$$

and

$$S_{XY}(f) = S^*_{YX}(f).$$

If $X(t)$ and $Y(t)$ are complex-valued random processes, then

$$\phi_{XY}(\tau) = \phi^*_{YX}(-\tau)$$

and

$$S_{XY}(f) = S^*_{YX}(f).$$

3.5 Random Processes Filtered by Linear Systems

Consider the linear time-invariant system with impulse response $h(t)$, shown in Fig. 3.3. Assume the input $X(t)$ is a real-valued WSS random process having mean μ_X and autocorrelation $\phi_{XX}(\tau)$. The filter input $X(t)$ and output $Y(t)$ are related through the well-known convolution integral

$$Y(t) = \int_{-\infty}^{\infty} h(\tau)X(t-\tau)\mathrm{d}\tau. \tag{3.30}$$

Taking Fourier transforms of both sides and using the convolution property of the Fourier transform gives:

$$Y(f) = H(f)X(f). \tag{3.31}$$

The output mean is

$$\begin{aligned}\mu_Y &= \mathrm{E}\left[\int_{-\infty}^{\infty} h(\tau)X(t-\tau)\mathrm{d}\tau\right] \\ &= \int_{-\infty}^{\infty} h(\tau)\mathrm{E}[X(t-\tau)]\mathrm{d}\tau \\ &= \mu_X \int_{-\infty}^{\infty} h(\tau)\mathrm{d}\tau \\ &= \mu_X H(0),\end{aligned} \tag{3.32}$$

which is the input mean μ_X multiplied by the d.c. gain $H(0)$ of the filter. Notice how the order of the expectation and integration are exchanged in the second line of (3.32) since both are linear operators. In fact, the expectation operator is just another integral.

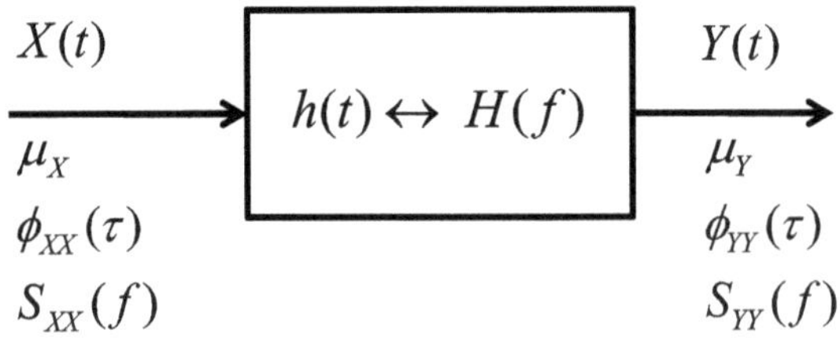

Fig. 3.3 Random process through a linear system

The autocorrelation of the output $Y(t)$ is

$$\begin{aligned}
\phi_{YY}(\tau) &= \mathrm{E}[Y(t)Y(t+\tau)] \\
&= \mathrm{E}\left[\int_{-\infty}^{\infty} h(\beta)X(t-\beta)\mathrm{d}\beta \int_{-\infty}^{\infty} h(\alpha)X(t+\tau-\alpha)\mathrm{d}\alpha\right] \\
&= \int_{-\infty}^{\infty}\int_{-\infty}^{\infty} h(\beta)h(\alpha)\mathrm{E}\left[X(t-\beta)X(t+\tau-\alpha)\right]\mathrm{d}\beta\mathrm{d}\alpha \\
&= \int_{-\infty}^{\infty}\int_{-\infty}^{\infty} h(\beta)h(\alpha)\phi_{XX}(\tau-\alpha+\beta)\mathrm{d}\beta\mathrm{d}\alpha \\
&= \int_{-\infty}^{\infty} h(\alpha)\int_{-\infty}^{\infty} h(\beta)\phi_{XX}(\tau+\beta-\alpha)\mathrm{d}\beta\mathrm{d}\alpha \\
&= \left(\int_{-\infty}^{\infty} h(\beta)\phi_{XX}(\tau+\beta)\mathrm{d}\beta\right) * h(\tau) \\
&= h(-\tau) * \phi_{XX}(\tau) * h(\tau).
\end{aligned} \tag{3.33}$$

Taking the Fourier transform of both sides, the psd of the output process $Y(t)$ is

$$\begin{aligned}
S_{YY}(f) &= H^*(f)S_{XX}(f)H(f) \\
&= |H(f)|^2 S_{XX}(f).
\end{aligned} \tag{3.34}$$

Example 3.9 Consider the linear system shown in Fig. 3.3. Find the cross-correlation between the WSS input and output processes $X(t)$ and $Y(t)$, respectively.

Solution: The cross-correlation $\phi_{XY}(\tau)$ is given by

$$\begin{aligned}
\phi_{XY}(\tau) &= \mathrm{E}[X(t)Y(t+\tau)] \\
&= \mathrm{E}\left[X(t)\int_{-\infty}^{\infty} h(\alpha)X(t+\tau-\alpha)\mathrm{d}\alpha\right] \\
&= \int_{-\infty}^{\infty} h(\alpha)\mathrm{E}[X(t)X(t+\tau-\alpha)]\mathrm{d}\alpha \\
&= \int_{-\infty}^{\infty} h(\alpha)\phi_{XX}(\tau-\alpha)\mathrm{d}\alpha \\
&= h(\tau) * \phi_{XX}(\tau).
\end{aligned}$$

(continued)

Example 3.9 (continued)
Also,

$$S_{XY}(f) = H(f)S_{XX}(f).$$

Example 3.10 Suppose a WSS random process $X(t)$ is input to the first-order low-pass filter shown in Fig. 3.4. If $X(t)$ has autocorrelation $\phi_{XX}(\tau) = e^{-\alpha|\tau|}$, what is the output autocorrelation $\phi_{YY}(\tau)$?

Solution: The output autocorrelation may be obtained by solving the double convolution in (3.33). However, a much simpler approach uses Fourier transforms to solve the problem in the frequency domain.

From basic circuit theory, the transfer function of the RC circuit is

$$H(f) = \frac{1}{1 + j2\pi fRC}$$

From (3.34),

$$\Phi_{YY}(f) = |H(f)|^2\Phi_{XX}(f)$$

where

$$|H(f)|^2 = \frac{1}{1 + (2\pi fRC)^2}$$

$$\Phi_{XX}(f) = \frac{2\alpha}{\alpha^2 + (2\pi f)^2}$$

(continued)

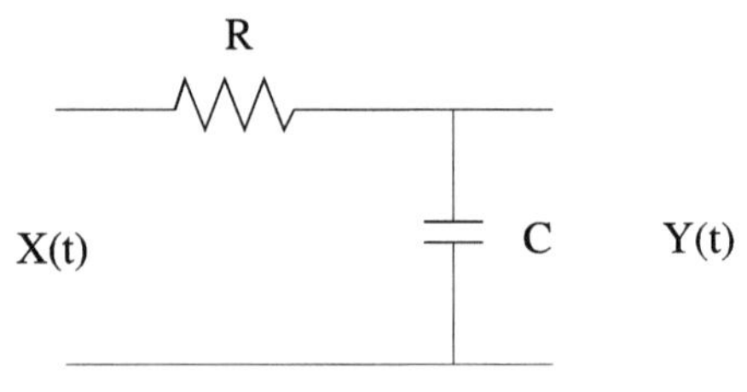

Fig. 3.4 Random process passed through a first-order low-pass filter

Example 3.10 (continued)
Hence,

$$\Phi_{YY}(f) = \frac{1}{1+(2\pi f RC)^2} \cdot \frac{2\alpha}{\alpha^2+(2\pi f)^2}$$

Now use a partial fraction expansion to write:

$$\Phi_{YY}(f) = \frac{A}{\alpha^2+(2\pi f)^2} + \frac{B}{1+(2\pi f RC)^2}$$

Returning to a common denominator, it follows that

$$A(1+(2\pi f RC)^2) + B(\alpha^2+(2\pi f)^2) = 2\alpha$$

Hence,

$$A + B\alpha^2 = 2\alpha$$
$$A(2\pi f RC)^2 + B(2\pi f)^2 = 0$$

giving two equations in two unknowns. From the second equation

$$A = -\frac{B}{(RC)^2} = -B\beta^2$$

where $\beta = 1/(RC)$. Then using the first equation

$$B = \frac{2\alpha}{\alpha^2-\beta^2}$$

Hence,

$$A = -B\beta^2 = -\frac{2\alpha\beta^2}{\alpha^2-\beta^2}$$

Finally, using the above

$$\Phi_{YY}(f) = \frac{\beta^2}{\beta^2-\alpha^2} \cdot \frac{2\alpha}{\alpha^2+(2\pi f)^2} + \frac{\alpha\beta}{\alpha^2-\beta^2} \cdot \frac{2\beta}{\beta^2+(2\pi f)^2}$$

Now take inverse Fourier transforms to get

$$\phi_{YY}(\tau) = \frac{\beta^2}{\beta^2-\alpha^2} \cdot e^{-\alpha|\tau|} + \frac{\alpha\beta}{\alpha^2-\beta^2} \cdot e^{-\beta|\tau|}$$

Example 3.11 Suppose that a real-valued WSS Gaussian random process $X(t)$ with mean μ_X and covariance $\lambda_{XX}(\tau)$ is passed through the linear filter shown in Fig. 3.3. What is the joint pdf of the random variables $X = X(t_1)$ and $Y = Y(t_1 + \tau)$?

Solution: If a Gaussian random process is passed through a linear filter, the output process is also a Gaussian random process. This is due to the fact that a sum of Gaussian random variables is another Gaussian random variable. Hence, X and Y will have the joint Gaussian pdf in (2.74) that is completely described by its 1×2 mean vector and 2×2 covariance matrix.

The crosscovariance of X and Y is

$$\begin{aligned}\lambda_{XY}(\tau) &= \mathrm{E}\left[(X(t) - \mu_X)\,(Y(t+\tau) - \mu_Y)\right] \\ &= \mathrm{E}\left[X(t)Y(t+\tau)\right] - \mu_Y\mu_X.\end{aligned}$$

Now $\mu_Y = H(0)\mu_X$. Also, from Example 3.9

$$\begin{aligned}\mathrm{E}[X(t)Y(t+\tau)] &= \int_{-\infty}^{\infty} h(\alpha)\phi_{XX}(\tau-\alpha)\mathrm{d}\alpha \\ &= \int_{-\infty}^{\infty} h(\alpha)[\lambda_{XX}(\tau-\alpha) + \mu_X^2]\mathrm{d}\alpha \\ &= \int_{-\infty}^{\infty} h(\alpha)\lambda_{XX}(\tau-\alpha)\mathrm{d}\alpha + H(0)\mu_X^2.\end{aligned}$$

Therefore,

$$\lambda_{XY}(\tau) = \int_{-\infty}^{\infty} h(\alpha)\lambda_{XX}(\tau-\alpha)\mathrm{d}\alpha = h(\tau) * \lambda_{XX}(\tau).$$

Also

$$\begin{aligned}\lambda_{YX}(\tau) &= \lambda_{XY}(-\tau) = \lambda_{XY}(\tau) \\ \lambda_{YY}(\tau) &= h(\tau) * h(-\tau) * \lambda_{XX}(\tau),\end{aligned}$$

where the first line follows from the even property of the autocorrelation. Hence, the covariance matrix is

$$\mathbf{\Lambda} = \begin{bmatrix} \lambda_{XX}(0) & \lambda_{XY}(\tau) \\ \lambda_{YX}(\tau) & \lambda_{YY}(0) \end{bmatrix} = \begin{bmatrix} \lambda_{XX}(0) & h(\tau) * \lambda_{XX}(\tau) \\ h(\tau) * \lambda_{XX}(\tau) & h(\tau) * h(-\tau) * \lambda_{XX}(\tau)\,|_{\tau=0} \end{bmatrix}$$

3.6 Discrete-Time Random Processes

This section considers complex-valued discrete-time random processes, and treats real-valued discrete-time random processes as a special case. Let $X[n]$ be a complex-valued discrete-time random process, where n is the integer time index. The mean of $X[n]$ is

$$\mu_X[n] = \mathrm{E}[X[n]] \tag{3.35}$$

the autocorrelation of $X[n]$ is[1]

$$\phi_{XX}[n,k] = \tfrac{1}{2}\mathrm{E}\left[X^*[n]X[k]\right] \tag{3.36}$$

and the autocovariance is

$$\lambda_{XX}[n,k] = \phi_{XX}[n,k] - \tfrac{1}{2}\mathrm{E}\left[X^*[n]\right]\mathrm{E}\left[X[k]\right]. \tag{3.37}$$

A WSS discrete-time random process $\{X[n]\}$ satisfies

$$\mu_X[n] = \mu_X \tag{3.38a}$$

$$\phi_{XX}[n,k] = \phi_{XX}[k-n] = \phi_{XX}[\ell]\,,\quad \ell = k-n. \tag{3.38b}$$

Note that the mean is independent of the time index n and the autocorrelation depends only on the discrete-time lag $\ell = k - n$ and not on the time index n. Likewise, the autocovariance of a WSS discrete-time process becomes:

$$\lambda_{XX}[n,k] = \lambda_{XX}[\ell] = \phi_{XX}[\ell] - \tfrac{1}{2}|\mu_X|^2\,,\quad \ell = k-n. \tag{3.39}$$

The power in the random process $\{X[n]\}$ is

$$P = \tfrac{1}{2}\mathrm{E}[|X[n]|^2] = \phi_{XX}[0]. \tag{3.40}$$

3.6.1 Power Density Spectrum

The psd of a WSS discrete-time random process $X[n]$ and its autocorrelation are a discrete-time Fourier transform pair:

[1] The factor of 1/2 is included in the autocorrelation function because $X[n]$ is complex-valued. If $X[n]$ is real-valued, there is no factor of 1/2 and the complex conjugates can be omitted.

$$S_{XX}(f) = \sum_{n=-\infty}^{\infty} \phi_{XX}[n] \mathrm{e}^{-j2\pi f n} \tag{3.41a}$$

$$\phi_{XX}[n] = \int_{-1/2}^{1/2} S_{XX}(f) \mathrm{e}^{j2\pi f n} \mathrm{d}f \tag{3.41b}$$

Note that the psd $S_{XX}(f)$ is periodic in frequency variable f with a period of unity, i.e., $S_{XX}(f) = S_{XX}(f+k)$ for any integer k. This is a characteristic of any discrete-time random process.

3.6.2 Filtering of Discrete-Time Random Processes

Suppose that a WSS complex-valued discrete-time random process $\{X[n]\}$ with mean μ_X and autocorrelation $\phi_{XX}[n]$ is input to a discrete-time linear time-invariant filter having a real-valued impulse response $h[n]$. The transfer function of the filter is given by the discrete-time Fourier transform of its impulse response:

$$H(f) = \sum_{n=-\infty}^{\infty} h[n] \mathrm{e}^{-j2\pi f n}. \tag{3.42}$$

The filter input, $X[n]$, and output, $Y[n]$, are related by the linear convolution sum:

$$Y[n] = \sum_{k=-\infty}^{\infty} h[k] X[n-k]. \tag{3.43}$$

The output mean is

$$\mu_Y = \mathrm{E}[Y[n]] = \sum_{k=-\infty}^{\infty} h[k] \mathrm{E}[X[n-k]] = \mu_X \sum_{k=-\infty}^{\infty} h[k] = \mu_X H(0). \tag{3.44}$$

Once again, the output mean is equal to the input mean μ_X multiplied by the dc gain of the filter $H(0)$.

The output autocorrelation is

$$\begin{aligned} \phi_{YY}[k] &= \tfrac{1}{2} \mathrm{E}[Y^*[n] Y[n+k]] \\ &= \frac{1}{2} \mathrm{E}\left[\sum_{\ell=-\infty}^{\infty} h[\ell] X^*[n-\ell] \sum_{m=-\infty}^{\infty} h[m] X[n+k-m] \right] \end{aligned}$$

$$
\begin{aligned}
&= \sum_{\ell=-\infty}^{\infty} \sum_{m=-\infty}^{\infty} h[\ell]h[m]\tfrac{1}{2}\mathrm{E}[X^*[n-\ell]X[n+k-m]] \\
&= \sum_{\ell=-\infty}^{\infty} \sum_{m=-\infty}^{\infty} h[\ell]h[m]\phi_{XX}[k+\ell-m] \\
&= \sum_{m=-\infty}^{\infty} h[m] \sum_{\ell=-\infty}^{\infty} h[\ell]\phi_{XX}[k+\ell-m] \\
&= h[k] * \left\{ \sum_{\ell=-\infty}^{\infty} h[\ell]\phi_{XX}[k+\ell] \right\} \\
&= h[k] * \phi_{XX}[k] * h[-k].
\end{aligned} \tag{3.45}
$$

where the convolution operation $*$ is understood to be a discrete-time linear convolution.

The output psd may be obtained by taking the discrete-time Fourier transform of both sides of (3.45), resulting in

$$
\begin{aligned}
S_{YY}(f) &= H(f)S_{XX}(f)H^*(f) \\
&= |H(f)|^2 S_{XX}(f).
\end{aligned} \tag{3.46}
$$

Once again, $S_{YY}(f)$ is periodic in f with a period of unity, i.e., $S_{YY}(f) = S_{YY}(f + k)$ for any integer k.

3.7 Cyclostationary Random Processes

Consider the complex-valued random process:

$$
\tilde{s}(t) = \sum_n x_n h_a(t - nT), \tag{3.47}
$$

where $\{x_n\}$ is a WSS discrete-time sequence of complex-valued random variables having mean $\mu_x = \mathrm{E}[x_n]$ and discrete-time autocorrelation $\phi_{xx}(k) = \frac{1}{2}\mathrm{E}[x_n^* x_{n+k}]$, and $h_a(t)$ is an arbitrary real-valued finite energy pulse with Fourier transform $H_a(f)$.

The mean of $\tilde{s}(t)$ is

$$
\mu_{\tilde{s}}(t) = \sum_n \mathrm{E}[x_n]h_a(t - nT) = \mu_x \sum_n h_a(t - nT) \tag{3.48}
$$

which is a periodic function in the time variable t with period T.

The autocorrelation of $\tilde{s}(t)$ is

$$\begin{aligned}
\phi_{\tilde{s}\tilde{s}}(t, t+\tau) &= \tfrac{1}{2}\mathrm{E}[\tilde{s}^*(t)\tilde{s}(t+\tau)] \\
&= \frac{1}{2}\mathrm{E}\left[\sum_{n=-\infty}^{\infty} x_n^* h_a(t-nT) \sum_{m=-\infty}^{\infty} x_m h_a(t+\tau-mT)\right] \\
&= \sum_{n=-\infty}^{\infty}\sum_{m=-\infty}^{\infty} \tfrac{1}{2}\mathrm{E}\left[x_n^* x_m\right] h_a(t-nT) h_a(t+\tau-mT) \\
&= \sum_{n=-\infty}^{\infty}\sum_{m=-\infty}^{\infty} \phi_{xx}(m-n) h_a(t-nT) h_a(t+\tau-mT). \qquad (3.49)
\end{aligned}$$

which is also a periodic function in the time variable t with period T. To observe this periodicity, consider the following:

$$\begin{aligned}
&\phi_{\tilde{s}\tilde{s}}(t+T, t+T+\tau) \\
&= \sum_{n=-\infty}^{\infty}\sum_{m=-\infty}^{\infty} \phi_{xx}(m-n) h_a(t+T-nT) h_a(t+T+\tau-mT) \\
&= \sum_{n=-\infty}^{\infty}\sum_{m=-\infty}^{\infty} \phi_{xx}(m-n) h_a(t+-(n-1)T) h_a(t++\tau-(m-1)T) \\
&= \sum_{n\prime=-\infty}^{\infty}\sum_{m\prime=-\infty}^{\infty} \phi_{xx}(m\prime-n\prime) h_a(t+-n\prime T) h_a(t++\tau-m\prime T) \\
&= \phi_{\tilde{s}\tilde{s}}(t, t+\tau)
\end{aligned}$$

where the second last line follows from a change in the indices of summation with $n\prime = n-1$ and $m\prime = m-1$. Hence, the autocorrelation $\phi_{\tilde{s}\tilde{s}}(t, t+\tau)$ is periodic in the time variable t with period T. Random processes that have a mean $\mu_{\tilde{s}}(t)$ and autocorrelation $\phi_{\tilde{s}\tilde{s}}(t, t+\tau)$ that are periodic functions of time t with period T are said to be *periodic WSS* or *cyclostationary* random processes.

Due to the cyclostationary behavior, the autocorrelation of $\tilde{s}(t)$ is obtained by time averaging $\phi_{\tilde{s}\tilde{s}}(t, t+\tau)$ over one period T as

$$\begin{aligned}
\phi_{\tilde{s}\tilde{s}}(\tau) &= \frac{1}{T}\int_0^T \phi_{\tilde{s}\tilde{s}}(t, t+\tau)\mathrm{d}t \\
&= \frac{1}{T}\int_0^T \sum_{n=-\infty}^{\infty}\sum_{m=-\infty}^{\infty} \phi_{xx}(m-n) h_a(t-nT) h_a(t+\tau-mT)\mathrm{d}t
\end{aligned}$$

$$
\begin{aligned}
&= \frac{1}{T}\sum_{n=-\infty}^{\infty}\sum_{m=-\infty}^{\infty}\phi_{xx}(m-n)\int_{0}^{T} h_a(t-nT)h_a(t+\tau-mT)\mathrm{d}t \\
&= \frac{1}{T}\sum_{n=-\infty}^{\infty}\sum_{m=-\infty}^{\infty}\phi_{xx}(m-n)\int_{-nT}^{-nT+T} h_a(z)h_a(z+\tau-(m-n)T)\mathrm{d}z \\
&= \frac{1}{T}\sum_{n=-\infty}^{\infty}\sum_{k=-\infty}^{\infty}\phi_{xx}(k)\int_{-nT}^{-nT+T} h_a(z)h_a(z+\tau-kT)\mathrm{d}z \\
&= \frac{1}{T}\sum_{k=-\infty}^{\infty}\phi_{xx}(k)\sum_{n=-\infty}^{\infty}\int_{-nT}^{-nT+T} h_a(z)h_a(z+\tau-kT)\mathrm{d}z \\
&= \frac{1}{T}\sum_{k=-\infty}^{\infty}\phi_{xx}(k)\int_{-\infty}^{\infty} h_a(z)h_a(z+\tau-kT)\mathrm{d}z
\end{aligned}
\tag{3.50}
$$

The psd of $\tilde{s}(t)$ is obtained by taking a Fourier transform of $\phi_{\tilde{s}\tilde{s}}(\tau)$ above with respect to the delay variable τ as follows:

$$
\begin{aligned}
S_{\tilde{s}\tilde{s}}(f) &= \frac{1}{T}\sum_{k=-\infty}^{\infty}\phi_{xx}(k)\int_{-\infty}^{\infty}\int_{-\infty}^{\infty} h_a(z)h_a(z+\tau-kT)\mathrm{e}^{-j2\pi f\tau}\mathrm{d}z\mathrm{d}\tau \\
&= \frac{1}{T}\sum_{k=-\infty}^{\infty}\phi_{xx}(k)\int_{-\infty}^{\infty}\int_{-\infty}^{\infty} h_a(z)h_a(s)\mathrm{e}^{-j2\pi f(s-z+kT)}\mathrm{d}z\mathrm{d}s \\
&= \frac{1}{T}\sum_{k=-\infty}^{\infty}\phi_{xx}(k)\mathrm{e}^{-j2\pi kfT}\int_{-\infty}^{\infty} h_a(s)\mathrm{e}^{-j2\pi fs}\mathrm{d}s\int_{-\infty}^{\infty} h_a(z)\mathrm{e}^{j2\pi fz}\mathrm{d}z \\
&= \frac{1}{T}\sum_{k=-\infty}^{\infty}\phi_{xx}(k)\mathrm{e}^{-j2\pi kfT}H_a(f)H_a^*(f) \\
&= \frac{1}{T}S_{xx}(f)\,|H_a(f)|^2
\end{aligned}
\tag{3.51}
$$

where

$$
S_{xx}(f) = \sum_{k=-\infty}^{\infty}\phi_{xx}(k)\mathrm{e}^{-j2\pi kfT} \tag{3.52}
$$

is the discrete-time Fourier transform of the autocorrelation $\phi_{xx}(k)$ except that the frequency variable f is replaced with the normalized frequency variable fT. This means that $S_{xx}(f)$ is periodic in the frequency variable f with period $1/T$. The power spectrum of $\tilde{s}(t)$ in (3.51) is shown to depend on the autocorrelation of the sequence $\{x_n\}$, $\phi_{xx}(k)$, and the choice of shaping pulse $h_a(t)$.

Discrete-Time Cyclostationary Random Processes

Similar to a continuous-time cyclostationary random process, a discrete-time cyclostationary random process is periodic in the time index n, with period N. Analogous to (3.47), consider the complex-valued random process:

$$\tilde{s}[n] = \sum_k x[k]h_a[n - kN], \tag{3.53}$$

where $\{x[k]\}$ is a complex-valued wide-sense stationary discrete-time random process having mean $\mu_x = \mathrm{E}[x[n]]$ and autocorrelation $\phi_{xx}[k] = \frac{1}{2}\mathrm{E}[x^*[n]x[n+k]]$, and where $h_a[n]$ is a real-valued finite-energy discrete-time pulse having discrete-time Fourier transform $H_a(f)$.

The mean of $\tilde{s}[n]$ is

$$\mu_{\tilde{s}}[n] = \mu_x \sum_k h_a[n - kN] \tag{3.54}$$

which is a periodic function in the time index n with period N. The autocorrelation of $\tilde{s}[n]$ is

$$\begin{aligned}
\phi_{\tilde{s}\tilde{s}}[n, n+k] &= \tfrac{1}{2}\mathrm{E}\left[\tilde{s}^*[n]\tilde{s}[n+k]\right] \\
&= \frac{1}{2}\mathrm{E}\left[\sum_i x^*[i]h_a[n - iN]\sum_j x[j]h_a[n+k-jN]\right] \\
&= \sum_i \sum_j \tfrac{1}{2}\mathrm{E}\left[x^*[i]x[j]\right] h_a[n - iN]h_a[n+k-jN] \\
&= \sum_i \sum_j \phi_{xx}[j-i]h_a[n - iN]h_a[n+k-jN].
\end{aligned} \tag{3.55}$$

The autocorrelation is periodic in the time index n with period N. To see this consider the following:

$$\begin{aligned}
&\phi_{\tilde{s}\tilde{s}}[n+N, n+N+k] \\
&\quad = \sum_i \sum_j \phi_{xx}[j-i]h_a[n+N-iN]h_a[n+N+k-jN] \\
&\quad = \sum_i \sum_j \phi_{xx}[j-i]h_a[n-(i-1)N]h_a[n+k-(j-1)N] \\
&\quad = \sum_{i'} \sum_{j'} \phi_{xx}[j'-i']h_a[n-i'N]h_a[n+k-j'N] \\
&\quad = \phi_{\tilde{s}\tilde{s}}[n, n+k]
\end{aligned}$$

Similar to the continuous-time cyclostationary random process, the autocorrelation of $\tilde{s}[n]$ is obtained by averaging $\phi_{\tilde{s}\tilde{s}}[n, n+k]$ over one period:

$$\begin{aligned}
\phi_{\tilde{s}\tilde{s}}[k] &= \frac{1}{N}\sum_{n=0}^{N-1}\phi_{\tilde{s}\tilde{s}}[n, n+k] \\
&= \frac{1}{N}\sum_{n=0}^{N-1}\sum_{i}\sum_{j}\phi_{xx}[j-i]h_a[n-iN]h_a[n+k-jN] \\
&= \frac{1}{N}\sum_{i}\sum_{j}\phi_{xx}[j-i]\sum_{n=0}^{N-1}h_a[n-iN]h_a[n+k-jN] \\
&= \frac{1}{N}\sum_{i}\sum_{j}\phi_{xx}[j-i]\sum_{\ell=iN}^{-iN+N-1}h_a[\ell]h_a[\ell+k-(j-i)N] \\
&= \frac{1}{N}\sum_{i}\sum_{m}\phi_{xx}[m]\sum_{\ell=iN}^{-iN+N-1}h_a[\ell]h_a[\ell+k-mN] \\
&= \frac{1}{N}\sum_{m}\phi_{xx}[m]\sum_{\ell}h_a[\ell]h_a[\ell+k-mN]
\end{aligned} \tag{3.56}$$

The psd of $\tilde{s}[n]$ is obtained by taking the discrete-time Fourier transform of its autocorrelation function in (3.56). This leads to the following:

$$\begin{aligned}
S_{\tilde{s}\tilde{s}}(f) &= \frac{1}{N}\sum_{m}\phi_{xx}[m]\sum_{k}\sum_{\ell}h_a[\ell]h_a[\ell+k-mN]\mathrm{e}^{-j2\pi kf} \\
&= \frac{1}{N}\sum_{m}\phi_{xx}[m]\sum_{n}\sum_{\ell}h_a[\ell]h_a[n]\mathrm{e}^{-j2\pi(n-\ell+mN)f} \\
&= \frac{1}{N}\sum_{m}\phi_{xx}[m]\mathrm{e}^{-j2\pi mfN}\sum_{n}h_a[n]\mathrm{e}^{-j2\pi nf}\sum_{\ell}h_a[\ell]\mathrm{e}^{j2\pi\ell f} \\
&= \frac{1}{N}S_{xx}(f)H_a(f)H_a^*(f) \\
&= \frac{1}{N}S_{xx}(f)\,|H_a(f)|^2
\end{aligned} \tag{3.57}$$

where $H_a(f) = \sum_n h_a[n]\mathrm{e}^{-j2\pi nf}$ is the discrete-time Fourier transform of $h_a[n]$ and

$$S_{xx}(f) = \sum_{m}\phi_{xx}[m]\mathrm{e}^{-j2\pi mfN} \tag{3.58}$$

is the discrete-time Fourier transform of the autocorrelation function $\phi_{xx}[m]$ where the digital frequency variable f is replaced with the normalized digital frequency variable fN. This means that $S_{xx}(f)$ is periodic in the digital frequency variable f with period $1/N$.

Problems

3.1 Which of the following functions are suitable autocorrelations? If a function is not suitable, tell why it is not ($A > 0$):

1. $A\exp(-\alpha|\tau|), \quad \alpha > 0$
2. $A \sin \omega_o \tau$
3. $A\exp(-\beta|\tau|)u(\tau), \quad \beta > 0$
4. $A\text{rect}(\tau/T)$
5. $A\text{sinc}(2W\tau)$
6. $A\exp - (\tau/(2\tau_o))^2)$
7. $A\tau^2\exp(-\alpha|\tau|)$
8. $A\log(|\tau|)$
9. $A \sin(\omega_o|\tau|)$
10. $A\delta(\tau)\exp(\alpha|\tau|), \quad \alpha > 0$

3.2 A random process is defined as

$$X(t) = A\cos(2\pi f_o t)\ , \quad \infty \le t \le \infty$$

where f_o is a constant. The amplitude A is uniformly distributed with the pdf:

$$f_A(a) = \begin{cases} 1, & 0 \le a \le 1 \\ 0, & \text{otherwise} \end{cases}$$

(a) Find the mean and variance of $X(t)$.
(b) Is $X(t)$ WSS?

3.3 A random process is defined as

$$X(t) = A\cos(2\pi f_o t + \Theta)\ , \quad \infty \le t \le \infty$$

where the amplitude A and frequency f_o are constants.
The random phase Θ has the pdf:

$$f_\Theta(\theta) = \begin{cases} \frac{2}{\pi}, & 0 \le \theta \le \frac{\pi}{2} \\ 0, & \text{otherwise} \end{cases}$$

(a) Find the mean and variance of $X(t)$.
(b) Obtain the autocorrelation of $X(t)$.
(c) Is $X(t)$ WSS? Cyclostationary?

3.4 A random process $X(t)$ is defined by

$$X(t) = A\cos(2\pi f_c t)$$

where the amplitude A is a Gaussian random variable $A \sim N(0, \sigma_A^2)$, and the frequency f_c is a constant.

(a) Find the mean of $X(t)$.
(b) Find the autocorrelation of $X(t)$.
(c) Is $X(t)$ WSS? Why?
(d) Is $X(t)$ ergodic? Why?

3.5 Consider the random process $X(t) = A + Bt$ where A and B are independent random variables each uniformly distributed on the interval $[-1, +1]$:

(a) Find the mean of $X(t)$.
(b) Find the autocorrelation of $X(t)$.
(c) Is $X(t)$ WSS? Why?

3.6 A WSS random process $X(t)$ has autocorrelation:

$$\phi_{XX}(\tau) = c_1 \exp\{-c_2|\tau|\}\cos(2\pi f_c \tau), \quad -\infty < \tau < \infty$$

(a) The fact that this is a valid autocorrelation imposes certain restrictions on c_1 and c_2. What are the ranges of possible values for c_1 and c_2?
(b) What is the power in $X(t)$?
(c) Consider the random process $Y(t) = 3X(4t), -\infty < t < \infty$. Is $Y(t)$ WSS? Justify your answer.
(d) Find the autocorrelation of $Y(t)$.

3.7 Let X and Y be statistically independent standard normal random variables. Define the Gaussian random process:

$$Z(t) = X\cos(2\pi f_c t) + Y\sin(2\pi f_c t)$$

(a) Determine the autocorrelation of $Z(t)$.
(b) Determine the joint pdf of the random variables $Z(t_1)$ and $Z(t_2)$ obtained by observing $Z(t)$ at times t_1 and t_2, respectively.
(c) Is $Z(t)$ WSS? Strictly stationary? Why?

3.8 Consider a WSS random process $X(t)$ having mean μ_X and autocorrelation $\phi_{XX}(\tau)$ and power spectrum $S_{XX}(f)$. The random process $X(t)$ is used to construct another random process:

$$Z(t) = X(t) - a_1 X(t-T) - a_2 X(t-2T)$$

where all quantities are real valued.

(a) What is the mean μ_Z?
(b) What is the autocorrelation $\phi_{ZZ}(\tau)$?
(c) Using part b) write down an expression for the psd $S_{ZZ}(f)$. You do not need to simplify.

3.9 The psd of a non-zero mean WSS random process $X(t)$ is given by

$$S_{XX}(f) = 5\delta(f) + 10\mathrm{sinc}^2(5f)$$

(a) Find and sketch the autocorrelation $\phi_{XX}(\tau)$.
(b) What is the power in the dc component of $X(t)$?
(c) What is the total power in $X(t)$?
(d) What is the smallest τ for which the random variables $X_1 = X(t)$ and $X_2 = X(t+\tau)$ are uncorrelated.

3.10 Suppose $X(t)$ and $Y(t)$ are a pair of uncorrelated zero-mean WSS Gaussian random processes both having the identical autocorrelation:

$$\phi_{XX}(\tau) = \phi_{YY}(\tau) = \sigma^2 \mathrm{e}^{-\alpha|\tau|}$$

Consider the random process:

$$Z(t) = X(t)\cos(2\pi f_c t) + Y(t)\sin(2\pi f_c t)$$

(a) Find the autocorrelation of $Z(t)$.
(b) Find the psd of $Z(t)$.

3.11 Suppose $X(t)$ and $Y(t)$ are a pair zero-mean WSS random processes. If $X(t)$ and $Y(t)$ are independent, find the autocorrelation for $Z(t)$ in terms of the autocorrelations of $X(t)$ and $Y(t)$ in each of the cases that follow. In each case, determine whether the random process $Z(t)$ is WSS:

(a) $Z(t) = cX(t)Y(t) + d$, where c and d are deterministic constants.
(b) $Z(t) = X(t)\cos(2\pi f_c t) + Y(t)\sin(2\pi f_c t)$

3.12 Suppose V is a zero-mean Gaussian random variable, and define the pair of random processes $X(t) = Vt$ and $Y(t) = V^2 t$ for $-\infty < t < \infty$:

(a) Find their cross-correlation.
(b) Are these random processes jointly WSS?

3.13 Consider a pair of WSS random processes $X(t)$ and $Y(t)$ with autocorrelations $\phi_{XX}(\tau)$ and $\phi_{YY}(\tau)$. Show that their cross-correlations $\phi_{XY}(\tau)$ and $\phi_{YX}(\tau)$ satisfy the following:

(a) $\phi_{XY}(\tau) = \phi_{YX}(-\tau)$.
(b) $|\phi_{XY}(\tau)| \leq \frac{1}{2}[\phi_{XX}(0) + \phi_{YY}(0)]$.

3.14 The random process

$$X(t) = A\cos(2\pi f_o t + \Phi) + W(t)$$

describes a sinusoid with random phase that is corrupted by an additive white Gaussian noise process $W(t)$ with psd $S_{WW}(f) = N_o/2$. The phase Φ is a uniformly distributed random variable having the pdf:

$$f_\Phi(\phi) = \begin{cases} \frac{1}{2\pi} & , \text{ for } -\pi \le \phi \le \pi \\ 0 & , \text{ otherwise} \end{cases}$$

and is independent of $W(t)$. The amplitude A and frequency f_o are known constants:

(a) Determine the autocorrelation of $X(t)$.
(b) Determine the psd of $X(t)$.

3.15 A pair of noise processes $n_1(t)$ and $n_2(t)$ are related by

$$n_2(t) = n_1(t)\cos(2\pi f_c t + \Theta) - n_1(t)\sin(2\pi f_c t + \Theta)$$

where f_c is a constant and Θ is a random variable with the pdf

$$f_\Theta(\theta) = \begin{cases} \frac{1}{2\pi}, & 0 \le \theta \le 2\pi \\ 0, & \text{elsewhere} \end{cases}$$

The noise process $n_1(t)$ is WSS and its psd is shown in Fig. 3.5.

(a) Derive the autocorrelation of $n_2(t)$, $\phi_{n_2n_2}(\tau)$, in terms of the autocorrelation of $n_1(t)$, $\phi_{n_1n_1}(\tau)$.
(b) Find and plot the corresponding psd of $n_2(t)$.

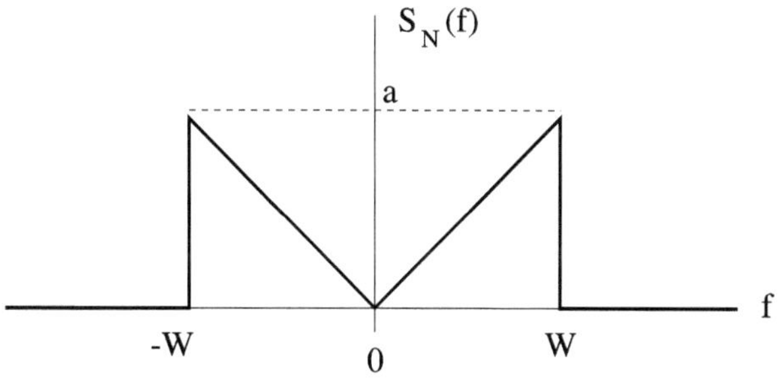

Fig. 3.5 psd for $n_1(t)$ in Problem 3.15

3.16 A random process is defined as $Y(t) = X(t) + X(t - T)$, where $X(t)$ is a WSS random process with autocorrelation $\phi_{XX}(\tau)$ and psd $S_{XX}(f)$:

(a) Find the autocorrelation $\phi_{YY}(\tau)$.
(b) Find the psd $S_{YY}(f)$, and simplify your expression to a single term.
(c) If $X(t)$ has autocorrelation $\phi_{xx}(\tau) = 5\Lambda(\tau)$, where $\Lambda(\tau)$ is the unit-area triangular function shown in Fig. 3.6, and $T = 0.5$ seconds, find and sketch the psd of $Y(t)$ showing all important values.

3.17 Consider the random process:

$$X(t) = W(t) + aW(t - t_o),$$

where $W(t)$ is a white Gaussian noise process with psd $N_o/2$ watts/Hz, and the parameters a and t_o are constants.

(a) Determine the autocorrelation of $X(t)$ and sketch it.
(b) Determine the psd of $X(t)$ and sketch it.
(c) What is the power in $X(t)$?
(d) What is the power in the dc component of $X(t)$?

3.18 The WSS random process $X(t)$ with psd $S_{XX}(f)$ is input to the linear filter shown in Fig. 3.7. Let $Y(t)$ be the random process at the output of the filter.

(a) Is $Y(t)$ WSS? Why?
(b) Find the frequency response, $H(f)$, of the filter.
(c) If $X(t)$ is a white noise process with psd $N_o/2$, find the psd of $Y(t)$.
(d) What frequency components cannot be present in $Y(t)$? Explain.

3.19 The input $X(t)$ to a low-pass filter with impulse response

$$h(t) = \exp(-10t)u(t)$$

is a white Gaussian noise process with $N_o = 1$ watts/Hz and the output is $Y(t)$.

(a) Find the mean of $Y(t)$.
(b) Find the autocovariance of $Y(t)$.

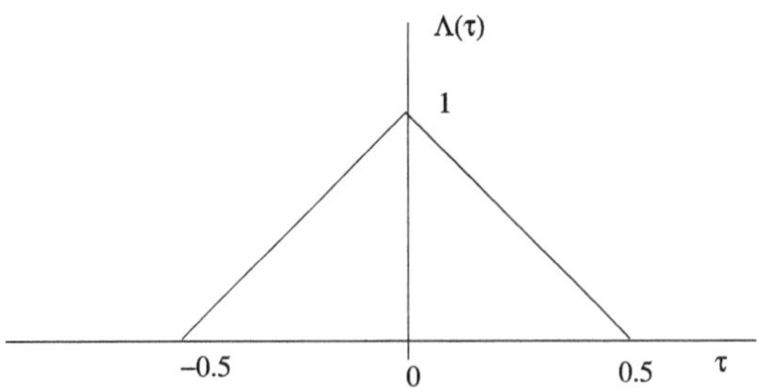

Fig. 3.6 Triangular function for Problem 3.16

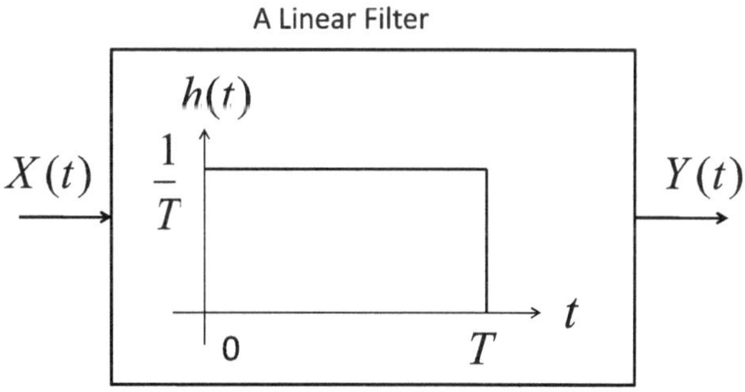

Fig. 3.7 Linear filter for Problem 3.18

(c) Find the pdf of the random variable $Y(t_o)$.
(d) Find the joint pdf of $Y(t_o)$ and $Y(t_o + 0.1)$.

3.20 A WSS Gaussian random process $X(t)$ with autocorrelation function $\phi_{XX}(\tau)$ and psd $S_{XX}(f)$ is input to the linear time-invariant system shown in Fig. 3.8. The filter $h(t)$ has impulse response:

$$h(t) = \mathrm{e}^{-\alpha t} u(t)$$

(a) Find the overall transfer function, $H_{\text{eff}}(f)$, where $Y(f) = H_{\text{eff}}(f)X(f)$.
(b) Find the psd $S_{YY}(f)$ in terms of $S_{XX}(f)$.
(c) Which frequencies cannot be present in $S_{YY}(f)$?
(d) Is $Y(t)$ a Gaussian random process? Is it WSS?

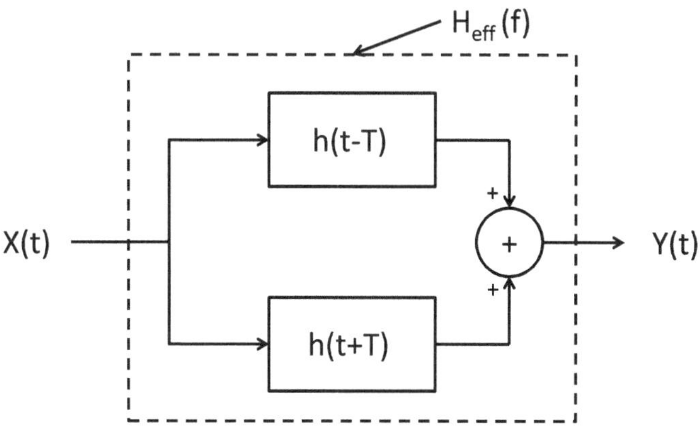

Fig. 3.8 Linear time-invariant system for Problem 3.20

3.21 Suppose a white Gaussian noise process $X(t)$ with psd $N_o/2$ watts/Hz is input to a linear time-invariant filter $h(t)$, having transfer function $H(f)$, and the output is $Y(t)$. The output psd is observed to be

$$S_{YY}(f) = \frac{N_o a}{a^2 + (2\pi f)^2}$$

(a) What is the output autocorrelation function $\phi_{YY}(\tau)$?
(b) Find the filter transfer function $H(f)$ and impulse response $h(t)$.
(c) What is the power in the output process $Y(t)$?
(d) What is the joint pdf of the random variables $Y_1 = Y(t_1)$ and $Y_2 = Y(t_1+1/a)$?

3.22 Consider the system shown in Fig. 3.9. The input $X(t)$ to the filter is a WSS random process with psd $S_{XX}(f)$.

a) Find the psd $S_{YY}(f)$ of the random process $Y(t)$ at the output of the filter $h(t)$ in terms of $S_{XX}(f)$.
b) What frequency components cannot be present in the filter output $Y(t)$?
c) Find the psd $S^d_{YY}(f)$ of the discrete-time random process Y_k at the output of the sampler.

3.23 A running integrator is defined by

$$Y(t) = \int_{t-T}^{t} X(\tau)d\tau$$

where $X(t)$ is the input, $Y(t)$ is the output, and T is the integration period. Both $X(t)$ and $Y(t)$ are sample functions of WSS random processes $X(t)$ and $Y(t)$, respectively. Show that the psd of the integrator output is related to that of the integrator input as

$$S_{YY}(f) = T^2\text{sinc}^2(fT)S_{XX}(f)$$

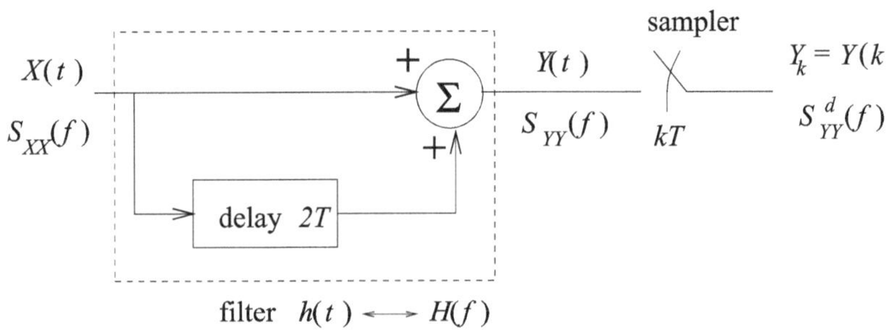

Fig. 3.9 Linear time-invariant system for Problem 3.22

3.24 The psd of a Gaussian random process $X(t)$ is shown in Fig. 3.10, consisting of a delta function at $f = 0$ and a triangular component.

(a) Determine and sketch the autocorrelation of $X(t)$.
(b) What is the DC power contained in $X(t)$?
(c) What is the AC power contained in $X(t)$?
(d) Suppose that the waveform $X(t)$ is sampled to yield the sample sequence $\{X_n\}$, where $X_n = X(nT_s)$ and T_s is the sample spacing. What sampling rates will give uncorrelated samples? Are the samples statistically independent?

3.25 Obtain the autocorrelations and psds of the outputs of the following systems $h(t) \leftrightarrow H(f)$ having the input autocorrelation functions $\phi_{XX}(\tau)$ or psds $S_{XX}(f)$ as the case may be:

(a)

$$H(f) = \text{rect}(f/2B) \qquad \phi_{XX}(\tau) = \frac{N_o}{2}\delta(\tau)$$

where B and N_o are positive constants.

(b)

$$h(t) = Ae^{-\alpha t}u(t) \qquad S_{XX}(f) = \frac{B}{1 + (2\pi\beta f)^2}$$

where A, B, β, and α are positive constants.

3.26 Consider a WSS random process $X(t)$ with autocorrelation $\phi_{XX}(\tau)$. A sample function of $X_i(t)$ is observed on the interval $0 \le t \le T$. An estimate of $\phi_{XX}(\tau)$ is obtained as follows:

$$\hat{\phi}_{XX}(\tau) = \begin{cases} \frac{1}{T}\int_0^{T-|\tau|} X_i(t)X_i(t+|\tau|)dt \ , & |\tau| \le T \\ 0 \ , & \text{elsewhere} \end{cases}$$

(a) Show that the expected value of $\hat{\phi}_{XX}(\tau)$ approaches the autocorrelation $\phi_{XX}(\tau)$ as the observation interval length T approaches infinity.

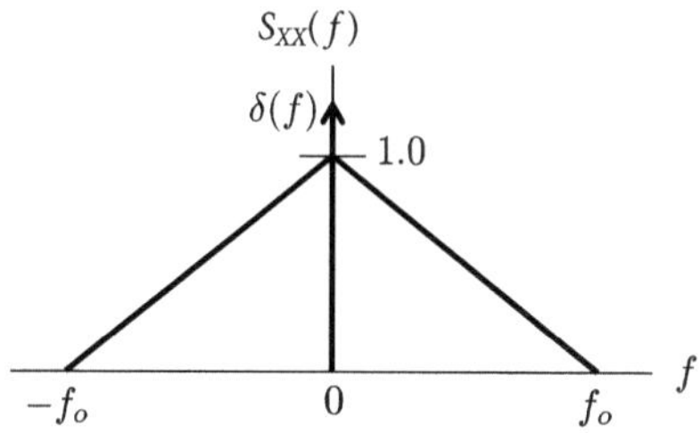

Fig. 3.10 psd of $X(t)$ in Problem 3.24

(b) Show that $\hat{\phi}_{XX}(\tau)$ is equal to the time-averaged autocorrelation of the truncated sample function:

$$X_i^T(t) = \begin{cases} X_i(t) \ , & 0 \le t \le T \\ 0 \ , & \text{elsewhere} \end{cases}$$

3.27 A Gaussian random process $X(t)$ having autocorrelation $\phi_{XX}(\tau) = \sigma^2\delta(\tau)$ is input to a linear time-invariant filter $h(t)$. It is desirable to generate an output process $Y(t)$ having the autocorrelation:

$$\phi_{YY}(\tau) = \sigma^2 e^{-\alpha|\tau|}$$

(a) What is the required transfer function $H(f)$ and impulse response $h(t)$ of the filter?
(b) Two samples are taken at the filter output spaced 1 second apart. What is their joint pdf?
(c) What is the ac and dc power at the filter output?

3.28 Suppose a WSS Gaussian random process $X(t)$ is input to the filter shown in Fig. 3.11. The autocorrelation of $X(t)$ is

$$\Phi_{XX}(\tau) = \exp\{-\alpha\tau^2\}$$

(a) Find the psd of the output, $S_{YY}(f)$.
(b) What frequency components are not present in $S_{YY}(f)$?
(c) What is the output autocorrelation $\phi_{YY}(\tau)$?
(d) What is the power in $Y(t)$?

3.29 Consider a WSS Gaussian random process $X(t)$ having the autocorrelation

$$\phi_{XX}(\tau) = \text{sinc}(4\tau/T)$$

that is input to a linear time-invariant filter having the impulse response

$$h(t) = A\text{sinc}((t - 4t_o)/T)$$

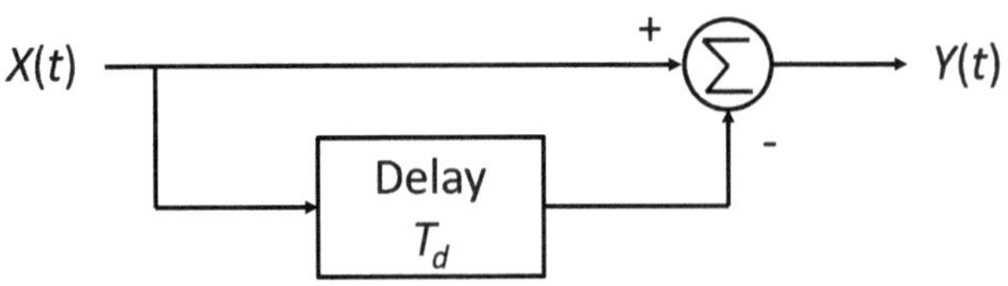

Fig. 3.11 Filter for Problem 3.28

(a) What is the input mean μ_X?
(b) What is the power in $X(t)$?
(c) What is the output autocorrelation $\phi_{YY}(\tau)$?
(d) For what sampling rates, R, are the output samples $Y_k = Y(k/R)$ statistically independent?
(e) What is joint pdf of the random variables $Y_1 = Y(t)$ and $Y_2 = Y(t + T/2)$?

3.30 Let $X(t)$ be a zero-mean WSS Gaussian random process with autocorrelation

$$\phi_{XX}(\tau) = 10^{-5} B_w \mathrm{sinc}^2(B_w \tau), \qquad B_w = 10^6 \text{ Hz}$$

The process $X(t)$ is input to a linear filter having the impulse response:

$$h(t) = B\mathrm{sinc}(2Bt), \qquad B = 10^5 \text{ Hz}.$$

(a) Sketch and fully dimension the psd of the filter output $Y(t)$.
(b) What is the pdf of the random variable $Y(t_1)$ at any time instant t_1?
(c) What is the probability that the output $Y(t_1)$ lies in the range $[-0.1, 0.1]$?

3.31 A WSS random process $X(t)$ having mean μ_X and autocorrelation $\phi_{XX}(\tau)$ is applied to the filters $g(t)$ and $h(t)$ as shown in Fig. 3.12. The outputs are $Y(t)$ and $Z(t)$.

(a) Find the output means μ_Y and μ_Z in terms of μ_X.
(b) Find the cross-correlation function $\phi_{YZ}(\tau)$ in terms of $\phi_{XX}(\tau)$, $g(t)$ and $h(t)$.

3.32 Consider two linear time-invariant filters $g(t)$ and $h(t)$ cascaded as shown in Fig. 3.13. Let the input process $X(t)$ be a WSS random process with autocorrelation $\phi_{XX}(\tau)$ and psd $S_{XX}(f)$. The random process appearing at the output of the first filter is $Y(t)$ and the random process appearing at the output of the second filter is $Z(t)$.

(a) Find the autocorrelation of $Z(t)$.
(b) Find the cross-correlation of the random processes $Y(t)$ and $Z(t)$, $\phi_{YZ}(\tau)$.
(c) What is the psd of $Z(t)$?

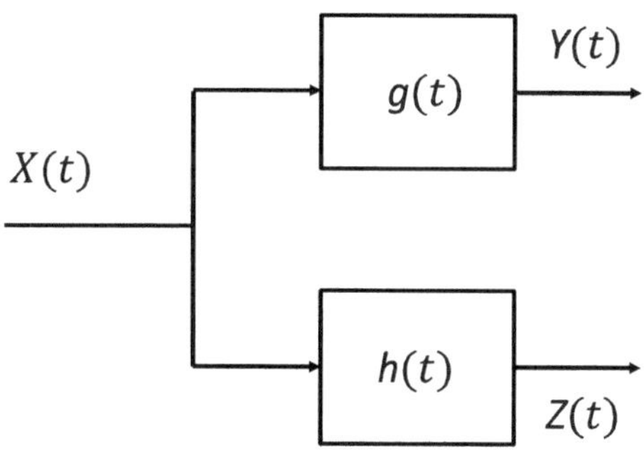

Fig. 3.12 Filter arrangement for Problem 3.31

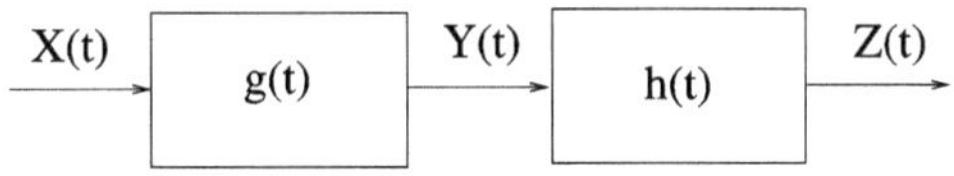

Fig. 3.13 Filter arrangement for Problem 3.32

(d) If the input process $X(t)$ happens to be a zero-mean WSS Gaussian random process, what is the pdf of the random variable $Z = Z(t_1)$?

3.33 A WSS random process $X(t)$ having autocorrelation

$$\phi_{XX}(\tau) = P\,\mathrm{e}^{-\alpha|\tau|}$$

is applied to the filter $h(t)$ shown in Fig. 3.14 to produce the output $Y(t)$.

(a) What is the filter impulse response, $h(t)$?
(b) What is the filter transfer function, $H(f)$?
(c) What is the psd of the output waveform, $S_{YY}(f)$?

3.34 A WSS random process $X(t)$ having autocorrelation

$$\phi_{XX}(\tau) = \sigma^2 \mathrm{sinc}^2(B\tau)$$

is applied to an ideal low-pass filter with cutoff frequency W and dc gain $H(0) = 1$ as shown in Fig. 3.15. Find filter bandwidth W such that the power of the output $Y(t)$ is 3 dB less than the power of the input $X(t)$.

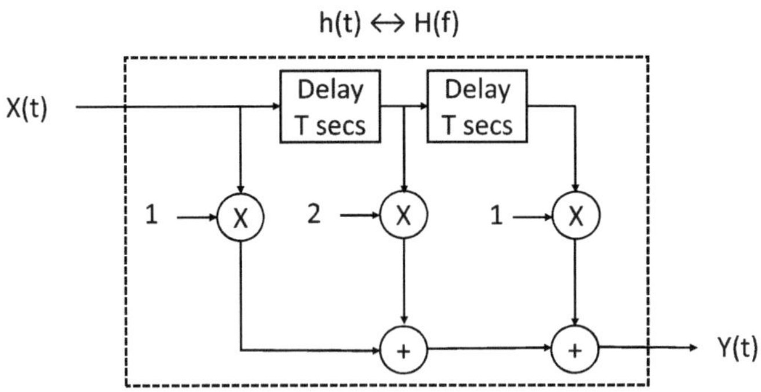

Fig. 3.14 Filter for Problem 3.33

Fig. 3.15 Ideal low-pass filter for Problem 3.34

3.35 Two statistically independent WSS random processes $X(t)$ and $Y(t)$ having autocorrelations

$$\phi_{XX}(\tau) = A\cos(2\pi f_o \tau)$$
$$\phi_{YY}(\tau) = B\text{sinc}(2W\tau)$$

with $f_o > W$ are input to a multiplier as shown in Fig. 3.16.

(a) Find the autocorrelation function of the multiplier output, $\phi_{ZZ}(\tau)$.
(b) Obtain and plot the psd of the multiplier output, $S_{ZZ}(f)$.

3.36 Two statistically independent WSS random processes $X(t)$ and $Y(t)$ are input to the multiplier in Fig. 3.16, where $X(t)$ and $Y(t)$ both have the same psd:

$$S_{XX}(f) = S_{YY}(f) = A\text{rect}\left(\frac{f}{2W}\right)$$

(a) Find an expression for the autocorrelation and psd of the output $Z(t)$ in terms of the autocorrelation and psd of $X(t)$ and $Y(t)$.
(b) Suppose that $X(t)$ and $Y(t)$ both have the same psd:

$$S_{XX}(f) = S_{YY}(f) = A\text{rect}\left(\frac{f}{2W}\right)$$

The output $Z(t)$ is sampled at intervals of T_s seconds to yield the sample sequence $\{Z_k\}$, where $Z_k = Z(kT_s)$. What sample spacing T_s will ensure that the Z_k are mutually independent random variables?

3.37 Let $h_a(t)$ be a pulse on the interval $[0, T]$ having energy E, i.e.,

$$\int_0^T h_a^2(t)dt = E$$

and let $n(t)$ be a zero-mean Gaussian random process having psd $N_o/2$ watts/Hz.

Suppose that $n(t)$ is correlated with the pulse $h_a(t)$ to form the random variable $X = \int_0^T w(t)h_a(t)dt$. Find the mean and variance of the random variable X.

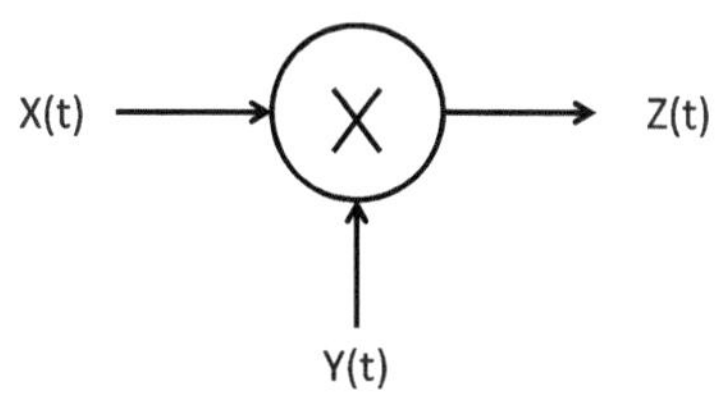

Fig. 3.16 Multiplier arrangement for Problem 3.35

Chapter 4
Communication Signals, Systems, and Noise

Communication signals, systems, and noise are all random processes. For carrier-modulated communication systems, they are all bandpass random processes. Section 4.1 begins the chapter with a mathematical framework for bandpass-modulated signals and presents various mathematical representations for bandpass signals. Section 4.2 considers bandpass systems and their complex low-pass equivalents. Bandpass random signals and their power spectrum are considered in Sect. 4.3. Section 4.4 concludes the chapter with a discussion of thermal noise that appears in all communication receivers.

4.1 Bandpass Signals

Bandpass signals are characterized by having a signal bandwidth W that is much less than the center or carrier frequency f_c. A bandpass waveform $s(t)$ has three representations, namely, the complex envelope representation, the quadrature representation, and the amplitude-phase representation.

The *complex envelope representation* of bandpass waveform $s(t)$ is

$$s(t) = \mathrm{Re}\left[\tilde{s}(t)\mathrm{e}^{j2\pi f_c t}\right], \qquad (4.1)$$

where

$$\tilde{s}(t) = \tilde{s}_I(t) + j\tilde{s}_Q(t) \qquad (4.2)$$

is the *complex envelope* having real and imaginary parts $\tilde{s}_I(t)$ and $\tilde{s}_Q(t)$, respectively, and f_c is the carrier frequency.

G. Stuber, *Introductory Digital Wireless Communications*,
https://doi.org/10.1007/978-3-032-01559-4_4

The *quadrature representation* may be obtained from (4.1) by using the Euler identity:

$$\mathrm{e}^{j\theta} = \cos(\theta) + j\sin(\theta) \tag{4.3}$$

to obtain

$$s(t) = \tilde{s}_I(t)\cos(2\pi f_c t) - \tilde{s}_Q(t)\sin(2\pi f_c t). \tag{4.4}$$

The real and imaginary parts of the complex envelope, $\tilde{s}_I(t)$ and $\tilde{s}_Q(t)$, modulate or multiply the cosine and sine components of the carrier, $\cos(2\pi f_c t)$ and $\sin(2\pi f_c t)$, respectively, which are in quadrature or separated 90° in phase, hence the label quadrature representation. Note the minus sign in (4.4) due to the $j^2 = -1$ term.

The *amplitude-phase representation* is obtained by converting the complex envelope $\tilde{s}(t)$ from Cartesian to polar form, viz.,

$$\tilde{s}(t) = a(t)\mathrm{e}^{j\phi(t)}$$

where

$$a(t) = |\tilde{s}(t)| = \sqrt{\tilde{s}_I^2(t) + \tilde{s}_Q^2(t)} \tag{4.5a}$$

$$\phi(t) = \begin{cases} \mathrm{Tan}^{-1}\left(\frac{\tilde{s}_Q(t)}{\tilde{s}_I(t)}\right) & \text{if } \tilde{s}_I(t) \geq 0 \\ \mathrm{Tan}^{-1}\left(\frac{\tilde{s}_Q(t)}{\tilde{s}_I(t)}\right) \pm \pi & \text{if } \tilde{s}_I(t) < 0 \end{cases}, \tag{4.5b}$$

and where $a(t)$ is the amplitude and $\phi(t)$ is the phase. Then

$$\begin{aligned} s(t) &= \mathrm{Re}\left[a(t)\mathrm{e}^{j\phi(t)}\mathrm{e}^{j2\pi f_c t}\right] \\ &= a(t)\cos\left(2\pi f_c t + \phi(t)\right). \end{aligned} \tag{4.6}$$

4.1.1 Signal Energy

An arbitrary finite-energy bandpass pulse $s(t)$ has an energy equal to

$$\begin{aligned} E &= \int_{-\infty}^{\infty} s^2(t)\mathrm{d}t \\ &= \int_{-\infty}^{\infty} \left(\mathrm{Re}\left[\tilde{s}(t)\mathrm{e}^{j2\pi f_c t}\right]\right)^2 \mathrm{d}t \end{aligned} \tag{4.7}$$

where the second line substitutes the complex envelope representation in (4.1). The bandpass pulse $s(t)$ usually causal and with a finite duration; however, the integral is taken from $-\infty$ to ∞ to avoid any loss of generality. Using the identity

$$\mathrm{Re}[z] = \frac{z + z^*}{2} \tag{4.8}$$

gives the result:

$$\begin{aligned} E &= \int_{-\infty}^{\infty} \left(\frac{\tilde{s}(t)\mathrm{e}^{j2\pi f_c t} + \tilde{s}^*(t)\mathrm{e}^{-j2\pi f_c t}}{2} \right)^2 \mathrm{d}t \\ &= \frac{1}{2}\int_{-\infty}^{\infty} |\tilde{s}(t)|^2 \,\mathrm{d}t + \frac{1}{4}\int_{-\infty}^{\infty} \left(\tilde{s}^2(t)\mathrm{e}^{j4\pi f_c t} + \tilde{s}^{*2}(t)\mathrm{e}^{-j4\pi f_c t} \right) \mathrm{d}t \\ &= \frac{1}{2}\int_{-\infty}^{\infty} |\tilde{s}(t)|^2 \,\mathrm{d}t + \frac{1}{4}\int_{-\infty}^{\infty} \left(|\tilde{s}(t)|^2 \,\mathrm{e}^{j2\phi(t)}\mathrm{e}^{j4\pi f_c t} + |\tilde{s}(t)|^2 \,\mathrm{e}^{-j2\phi(t)}\mathrm{e}^{-j4\pi f_c t} \right) \mathrm{d}t \\ &= \frac{1}{2}\int_{-\infty}^{\infty} |\tilde{s}(t)|^2 \,\mathrm{d}t + \frac{1}{2}\int_{-\infty}^{\infty} |\tilde{s}(t)|^2 \cos(4\pi f_c t + 2\phi(t))\mathrm{d}t \\ &\doteq \frac{1}{2}\int_{-\infty}^{\infty} |\tilde{s}(t)|^2 \,\mathrm{d}t \end{aligned} \tag{4.9}$$

where the last line follows under the bandpass signal assumption that the bandwidth of $\tilde{s}(t)$, W, is much smaller than the carrier frequency f_c such that the integral involving the cosine with frequency $2f_c$ tends to zero. Viewed another way, the envelope $|\tilde{s}(t)|$ varies much more slowly with time compared to the cosine term with frequency $2f_c$. Hence, the positive and negative contributions in the integral cancel and the result approaches zero. This important result shows that the energy in the bandpass pulse $s(t)$ is one-half the energy in its corresponding complex envelope $\tilde{s}(t)$.

Example 5.1 Consider a bandpass signal having the complex envelope:

$$\tilde{s}(t) = \sqrt{\frac{2E}{T}} u_T(t)$$

where

$$u_T(t) \triangleq u(t) - u(t - T)$$

(continued)

Example 5.1 (continued)
is defined as a unit amplitude rectangular pulse on the interval $(0, T]$. In this case, the complex envelope happens to be real-valued and the corresponding bandpass waveform is

$$s(t) = \sqrt{\frac{2E}{T}} \cos(2\pi f_c t) u_T(t).$$

The energy in $s(t)$ is

$$\begin{aligned} E &= \frac{2E}{T} \int_0^T \cos^2(2\pi f_c t) \mathrm{d}t \\ &= \frac{2E}{T} \int_0^T \left(\frac{1}{2} + \frac{1}{2} \cos(4\pi f_c t) \right) \mathrm{d}t \\ &= E + E \frac{\sin(4\pi f_c T)}{4\pi f_c T} \\ &= E \qquad \text{if } f_c T \gg 1 \end{aligned}$$

The last equality uses the fact that the magnitude of the sine function is bounded by unity, and the denominator contains the term $f_c T$ with $f_c T \gg 1$. If T is interpreted as the baud interval, then the condition $f_c T \gg 1$ implies that there are many cycles of the cosine function $\cos(4\pi f_c t)$ having frequency $2 f_c$ in the baud interval T. For the case when $4\pi f_c T = k\pi$ or $T = k/(4 f_c)$ where k is a positive integer, the second term is exactly zero regardless of whether or not the condition $f_c T \gg 1$ holds true.

4.1.2 Signal Spectrum

The spectrum of a bandpass signal $s(t)$ is obtained by taking its Fourier transform. Using (4.8) and (4.1),

$$\begin{aligned} S(f) &= \int_{-\infty}^{\infty} s(t) \mathrm{e}^{-j2\pi f t} \mathrm{d}t \\ &= \frac{1}{2} \int_{-\infty}^{\infty} \left(\tilde{s}(t) \mathrm{e}^{j2\pi f_c t} + \tilde{s}^*(t) \mathrm{e}^{-j2\pi f_c t} \right) \mathrm{e}^{-j2\pi f t} \mathrm{d}t \\ &= \frac{1}{2} \int_{-\infty}^{\infty} \tilde{s}(t) \mathrm{e}^{-j2\pi (f - f_c) t} \mathrm{d}t + \frac{1}{2} \int_{-\infty}^{\infty} \tilde{s}^*(t) \mathrm{e}^{-j2\pi (f + f_c) t} \mathrm{d}t \end{aligned}$$

$$= \frac{1}{2}\int_{-\infty}^{\infty} \tilde{s}(t)\mathrm{e}^{-j2\pi(f-f_c)t}\mathrm{d}t + \frac{1}{2}\left(\int_{-\infty}^{\infty} \tilde{s}(t)\mathrm{e}^{j2\pi(f+f_c)t}\mathrm{d}t\right)^*$$
$$= \tfrac{1}{2}\left(\tilde{S}(f-f_c) + \tilde{S}^*(-f-f_c)\right) \tag{4.10}$$

where $\tilde{S}(f)$ is the Fourier transform of the complex envelope $\tilde{s}(t)$.

Once $\tilde{S}(f)$ is specified, the bandpass signal spectrum $S(f)$ can be obtained in a straightforward fashion as illustrated in Fig. 4.1. Since the bandpass signal $s(t)$ is real-valued, its spectrum exhibits a conjugate symmetry about $f = 0$ as noted by the phase negation at negative frequencies.

4.2 Bandpass Systems and Their Low-Pass Equivalent

Some communication channels, such as land mobile radio channels, are modeled as a linear time-variant bandpass filter having impulse response $h(t, \tau)$, where t is the *time* variable and τ is the *time delay* variable. The bandpass waveforms at the input and output of such a band pass channel, $s(t)$ and $r(t)$, respectively, are related through the convolution integral:

$$r(t) = \int_0^t h(t,\tau)s(t-\tau)\mathrm{d}\tau. \tag{4.11}$$

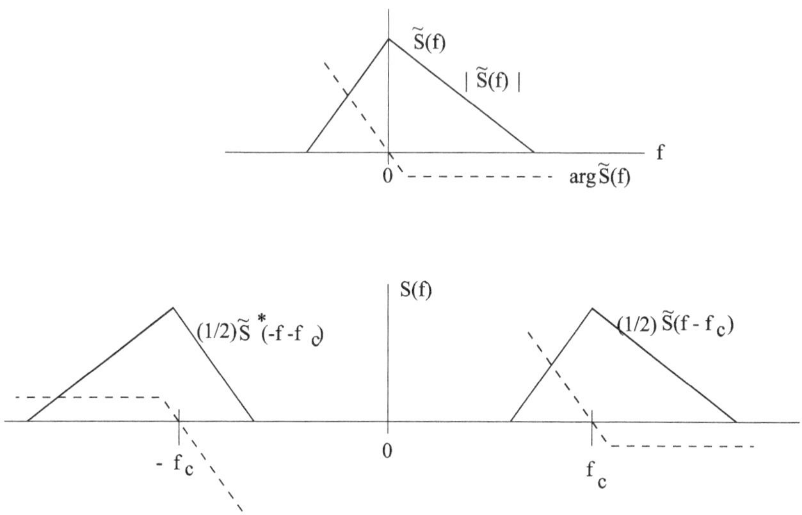

Fig. 4.1 Relationship between the spectrum of the bandpass waveform $S(f)$ and the spectrum of its complex envelope $\tilde{S}(f)$

Here it is assumed that the bandpass channel is causal so that the lower limit of the integral in (4.11) is zero. Moreover, it is assumed that the input signal has $s(t) = 0, t < 0$, so that the upper limit of the integral is t. Note that the integral in (4.11) is taken with respect to the *time delay* variable τ in the time-variant channel impulse response $h(t, \tau)$. The Fourier transform of $h(t, \tau)$ with respect to the delay variable τ yields the time-variant transfer function or time-variant frequency response $H(t, f)$.

Other types of bandpass communication channels are time-invariant and may be modeled as a linear time-invariant filter having impulse response $h(t)$. In this case, the channel input and output are related through the convolution integral:

$$r(t) = \int_0^t h(\tau)s(t - \tau)\mathrm{d}\tau. \tag{4.12}$$

Note that the integral in (4.12) is taken again with respect to the *time delay* variable τ. Although the variable t appears in the time-invariant impulse response $h(t)$ by convention, the channel impulse response is not time-varying. In the convolution integral in (4.12), the time variable t is substituted with the time delay variable τ. The Fourier transform of time-invariant impulse response $h(t)$ with respect to the t variable gives the time-invariant transfer function $H(f)$. From the convolution property of Fourier transforms

$$R(f) = S(f)H(f) \tag{4.13}$$

For the remainder of this section, a linear time-invariant bandpass channel is considered having impulse response $h(t)$. The bandpass input and output waveforms $s(t)$ and $r(t)$, respectively, can be expressed in terms of their complex envelopes, $\tilde{s}(t)$ and $\tilde{r}(t)$, respectively, as

$$s(t) = \mathrm{Re}\left[\tilde{s}(t)\mathrm{e}^{j2\pi f_c t}\right] \tag{4.14a}$$

$$r(t) = \mathrm{Re}\left[\tilde{r}(t)\mathrm{e}^{j2\pi f_c t}\right]. \tag{4.14b}$$

Considering $r(t)$ and applying the identity in (4.8), it follows that

$$r(t) = \tfrac{1}{2}\left(\tilde{r}(t)\mathrm{e}^{j2\pi f_c t} + \tilde{r}^*(t)\mathrm{e}^{-j2\pi f_c t}\right) \tag{4.15}$$

Taking the Fourier transform of both sides gives:

$$R(f) = \tfrac{1}{2}\left(\tilde{R}(f - f_c) + \tilde{R}^*(-f - f_c)\right) \tag{4.16}$$

Likewise,

$$S(f) = \frac{1}{2}\left(\tilde{S}(f - f_c) + \tilde{S}^*(-f - f_c)\right) \tag{4.17}$$

To relate the spectrum of the output complex envelope $\tilde{R}(f)$ to the spectrum of the input complex envelope $\tilde{S}(f)$, consider the following useful property of bandpass systems:

$$\begin{aligned} H(f) &= \int_{-\infty}^{\infty} h(t)e^{-j2\pi ft}dt \\ H^*(f) &= \int_{-\infty}^{\infty} h(t)e^{j2\pi ft}dt \\ H^*(-f) &= \int_{-\infty}^{\infty} h(t)e^{-j2\pi ft}dt = H(f) \end{aligned} \tag{4.18}$$

Now define:

$$\tilde{H}(f - f_c) \triangleq \begin{cases} H(f) \ , & f > 0 \\ 0 \ , & f < 0 \end{cases}$$

such that

$$\tilde{H}^*(-f - f_c) = \begin{cases} 0 \ , & f > 0 \\ H^*(-f) \ , & f < 0 \end{cases}$$

It follows from (4.18) that

$$H(f) = \tilde{H}(f - f_c) + \tilde{H}^*(-f - f_c) \tag{4.19}$$

The relationship between $H(f)$ and $\tilde{H}(f)$ is illustrated in Fig. 4.2. Since the bandpass system is real-valued, its transfer function exhibits a conjugate symmetry similar to the real-valued bandpass signal spectrum. Note, however, the factor of 1/2 that appears in the bandpass signal spectrum $S(f)$ does not appear in the bandpass transfer function $H(f)$.

Taking the inverse Fourier transform of both sides of (4.19) converts to the time domain as follows:

$$\begin{aligned} F^{-1}\{\tilde{H}(f - f_c)\} &= \int_{-\infty}^{\infty} \tilde{H}(f - f_c)e^{j2\pi ft}df \\ &= \int_{-\infty}^{\infty} \tilde{H}(\hat{f})e^{j2\pi \hat{f}t}e^{j2\pi f_c t}d\hat{f} \ , \qquad \hat{f} = f - f_c \\ &= \tilde{h}(t)e^{j2\pi f_c t} \end{aligned}$$

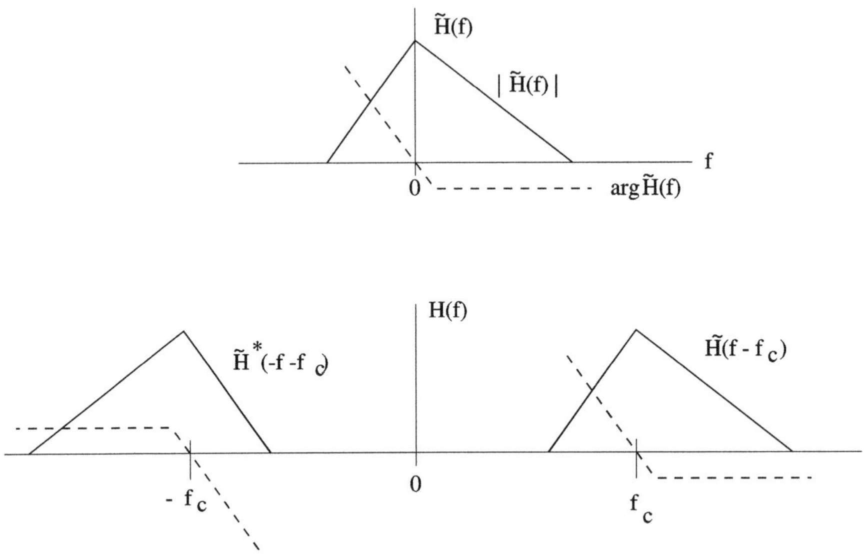

Fig. 4.2 Relationship between the transfer function of the bandpass channel $H(f)$ and its complex low-pass representation $\tilde{H}(f)$

and

$$\begin{aligned} F^{-1}\{\tilde{H}^*(-f-f_c)\} &= \int_{-\infty}^{\infty} \tilde{H}^*(-f-f_c)e^{j2\pi ft}df \\ &= \int_{-\infty}^{\infty} \tilde{H}^*(\hat{f})e^{-j2\pi\hat{f}t}e^{-j2\pi f_c t}d\hat{f}\,, \qquad \hat{f} = -f-f_c \\ &= \tilde{h}^*(t)e^{-j2\pi f_c t} \end{aligned}$$

Therefore,

$$\begin{aligned} h(t) &= \tilde{h}(t)e^{j2\pi f_c t} + \tilde{h}^*(t)e^{-j2\pi f_c t} \\ &= \mathrm{Re}\left[2\tilde{h}(t)e^{j2\pi f_c t}\right] \end{aligned} \tag{4.20}$$

Comparing the result in (4.20) with the complex envelope representation for a bandpass signal in (4.1), there is an extra factor of 2 relating the bandpass impulse response of a channel $h(t)$ to its complex low-pass impulse response $\tilde{h}(t)$.

Using (4.16), (4.17), and (4.19) in (4.13)

$$
\begin{aligned}
&\tfrac{1}{2}\left(\tilde{R}(f - f_c) + \tilde{R}^*(-f - f_c)\right) \\
&= \tfrac{1}{2}\left(\tilde{S}(f - f_c) + \tilde{S}^*(-f - f_c)\right)\left(\tilde{H}(f - f_c) + \tilde{H}^*(-f - f_c)\right) \\
&= \tfrac{1}{2}\Big[\tilde{S}(f - f_c)\tilde{H}(f - f_c) + \tilde{S}^*(-f - f_c)\tilde{H}^*(-f - f_c) \\
&\quad + \tilde{S}(f - f_c)\tilde{H}^*(-f - f_c) + \tilde{S}^*(-f - f_c)\tilde{H}(f - f_c)\Big] \\
&= \tfrac{1}{2}\left(\tilde{S}(f - f_c)\tilde{H}(f - f_c) + \tilde{S}^*(-f - f_c)\tilde{H}^*(-f - f_c)\right) \qquad (4.21)
\end{aligned}
$$

The last equality follows because the spectra of $\tilde{S}(f - f_c)$ and $\tilde{H}^*(-f - f_c)$, and the spectra of $\tilde{S}^*(-f - f_c)$ and $\tilde{H}(f - f_c)$, do not overlap so their products are zero. This can be seen from the example bandpass spectrum shown in Fig. 4.1 and the example channel transfer function shown in Fig. 4.2. Consequently, from (4.21), the relationship between the input and output spectra and the corresponding input and output time-domain waveforms of a bandpass system is

$$\tilde{R}(f) = \tilde{H}(f)\tilde{S}(f) \qquad (4.22)$$

$$\tilde{r}(t) = \int_{-\infty}^{\infty} \tilde{h}(\tau)\tilde{s}(t - \tau)\mathrm{d}\tau \qquad (4.23)$$

The input-output signal representations for a bandpass system and its equivalent complex low-pass representation are summarized in Fig. 4.3.

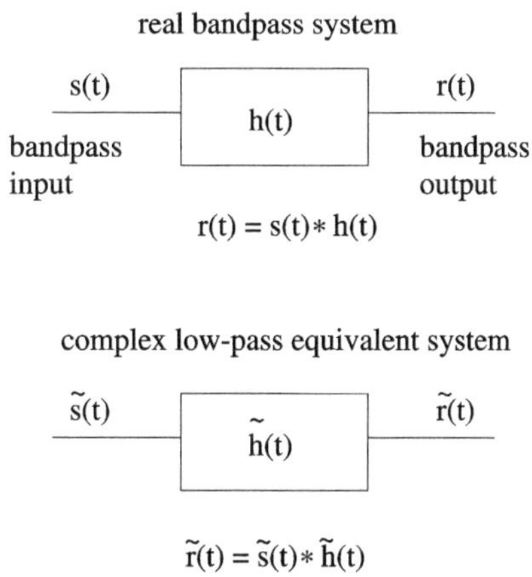

Fig. 4.3 Bandpass linear system and its equivalent complex low-pass representation

4.3 Bandpass Random Signals

Bandpass-modulated waveforms and bandpass noise are examples of bandpass random processes. In this section, the properties of wide-sense stationary bandpass random processes are examined.

Consider the quadrature representation of a bandpass signal $s(t)$ in (4.4), where the complex envelope $\tilde{s}(t) = \tilde{s}_I(t) + j\tilde{s}_Q(t)$ is a wide-sense stationary complex-valued random process. The autocorrelation of $s(t)$ may be derived in terms of the statistical properties of its complex envelope as follows:

$$\begin{aligned}
\phi_{ss}(\tau) &= \mathrm{E}[s(t)s(t+\tau)] \\
&= \mathrm{E}\Big[\big(\tilde{s}_I(t)\cos(2\pi f_c t) - \tilde{s}_Q(t)\sin(2\pi f_c t)\big) \\
&\quad \times \big(\tilde{s}_I(t+\tau)\cos(2\pi f_c(t+\tau)) - \tilde{s}_Q(t+\tau)\sin(2\pi f_c(t+\tau))\big)\Big] \\
&= \mathrm{E}\big[\tilde{s}_I(t)\tilde{s}_I(t+\tau)\cos(2\pi f_c t)\cos(2\pi f_c(t+\tau)) \\
&\quad -\tilde{s}_Q(t)\tilde{s}_I(t+\tau)\sin(2\pi f_c t)\cos(2\pi f_c(t+\tau)) \\
&\quad -\tilde{s}_I(t)\tilde{s}_Q(t+\tau)\cos(2\pi f_c t)\sin(2\pi f_c(t+\tau)) \\
&\quad +\tilde{s}_Q(t)\tilde{s}_Q(t+\tau)\sin(2\pi f_c t)\sin(2\pi f_c(t+\tau))\big] \\
&= \mathrm{E}\big[\tilde{s}_I(t)\tilde{s}_I(t+\tau)\big]\cos(2\pi f_c t)\cos(2\pi f_c(t+\tau)) \\
&\quad -\mathrm{E}\big[\tilde{s}_Q(t)\tilde{s}_I(t+\tau)\big]\sin(2\pi f_c t)\cos(2\pi f_c(t+\tau)) \\
&\quad -\mathrm{E}\big[\tilde{s}_I(t)\tilde{s}_Q(t+\tau)\big]\cos(2\pi f_c t)\sin(2\pi f_c(t+\tau)) \\
&\quad +\mathrm{E}\big[\tilde{s}_Q(t)\tilde{s}_Q(t+\tau)\big]\sin(2\pi f_c t)\sin(2\pi f_c(t+\tau)) \\
&= \phi_{\tilde{s}_I\tilde{s}_I}(\tau)\cos(2\pi f_c t)\cos(2\pi f_c(t+\tau)) \\
&\quad -\phi_{\tilde{s}_Q\tilde{s}_I}(\tau)\sin(2\pi f_c t)\cos(2\pi f_c(t+\tau)) \\
&\quad -\phi_{\tilde{s}_I\tilde{s}_Q}(\tau)\cos(2\pi f_c t)\sin(2\pi f_c(t+\tau)) \\
&\quad +\phi_{\tilde{s}_Q\tilde{s}_Q}(\tau)\sin(2\pi f_c t)\sin(2\pi f_c(t+\tau))
\end{aligned}$$

where E[·] is the ensemble average operator, and

$$\phi_{\tilde{s}_I\tilde{s}_I}(\tau) \triangleq \mathrm{E}[\tilde{s}_I(t)\tilde{s}_I(t+\tau)] \tag{4.24a}$$

$$\phi_{\tilde{s}_I\tilde{s}_Q}(\tau) \triangleq \mathrm{E}[\tilde{s}_I(t)\tilde{s}_Q(t+\tau)]. \tag{4.24b}$$

By using common trig identities, $\phi_{ss}(\tau)$ may be arranged into the form:

$$\begin{aligned}
\phi_{ss}(\tau) = \phi_{\tilde{s}_I\tilde{s}_I}(\tau)\tfrac{1}{2}\Big(\cos(2\pi f_c\tau) + \cos(2\pi f_c(2t+\tau))\Big) \\
-\phi_{\tilde{s}_Q\tilde{s}_I}(\tau)\tfrac{1}{2}\Big(\sin(2\pi f_c(2t+\tau)) + \sin(-2\pi f_c\tau)\Big)
\end{aligned}$$

$$
\begin{aligned}
&-\phi_{\tilde{s}_I\tilde{s}_Q}(\tau)\tfrac{1}{2}\Big(\sin(2\pi f_c(2t+\tau)) - \sin(-2\pi f_c\tau)\Big) \\
&+\phi_{\tilde{s}_Q\tilde{s}_Q}(\tau)\tfrac{1}{2}\Big(\cos(2\pi f_c\tau) - \cos(2\pi f_c(2t+\tau))\Big) \\
=&\ \tfrac{1}{2}\left(\phi_{\tilde{s}_I\tilde{s}_I}(\tau) + \phi_{\tilde{s}_Q\tilde{s}_Q}(\tau)\right)\cos(2\pi f_c\tau) \\
&+\tfrac{1}{2}\left(\phi_{\tilde{s}_Q\tilde{s}_I}(\tau) - \phi_{\tilde{s}_I\tilde{s}_Q}(\tau)\right)\sin(2\pi f_c\tau) \\
&+\tfrac{1}{2}\left(\phi_{\tilde{s}_I\tilde{s}_I}(\tau) - \phi_{\tilde{s}_Q\tilde{s}_Q}(\tau)\right)\cos(2\pi f_c(2t+\tau)) \\
&-\tfrac{1}{2}\left(\phi_{\tilde{s}_Q\tilde{s}_I}(\tau) + \phi_{\tilde{s}_I\tilde{s}_Q}(\tau)\right)\cos(2\pi f_c(2t+\tau))
\end{aligned}
$$

For $s(t)$ to be a WSS random process, the terms involving time variable t in its autocorrelation must vanish, implying that

$$\phi_{\tilde{s}_I\tilde{s}_I}(\tau) = \phi_{\tilde{s}_Q\tilde{s}_Q}(\tau) \tag{4.25a}$$

$$\phi_{\tilde{s}_Q\tilde{s}_I}(\tau) = -\phi_{\tilde{s}_I\tilde{s}_Q}(\tau) \tag{4.25b}$$

Hence, the autocorrelation of $s(t)$ has the quadrature representation:

$$\phi_{ss}(\tau) = \phi_{\tilde{s}_I\tilde{s}_I}(\tau)\cos 2\pi f_c\tau - \phi_{\tilde{s}_I\tilde{s}_Q}(\tau)\sin 2\pi f_c\tau. \tag{4.26}$$

The autocorrelation of the complex envelope $\tilde{s}(t)$ is

$$
\begin{aligned}
\phi_{\tilde{s}\tilde{s}}(\tau) &\triangleq \tfrac{1}{2}\mathrm{E}\left[\tilde{s}^*(t)\tilde{s}(t+\tau)\right] \\
&= \tfrac{1}{2}\mathrm{E}\left[(\tilde{s}_I(t) - j\tilde{s}_Q(t))(\tilde{s}_I(t+\tau) + j\tilde{s}_Q(t+\tau))\right] \\
&= \tfrac{1}{2}\mathrm{E}\left[(\tilde{s}_I(t)\tilde{s}_I(t+\tau)\right] + \tfrac{1}{2}\mathrm{E}\left[(\tilde{s}_Q(t)\tilde{s}_Q(t+\tau)\right] \\
&\quad +\tfrac{j}{2}\mathrm{E}\left[(\tilde{s}_I(t)\tilde{s}_Q(t+\tau)\right] - \tfrac{j}{2}\mathrm{E}\left[(\tilde{s}_Q(t)\tilde{s}_I(t+\tau)\right] \\
&= \phi_{\tilde{s}_I\tilde{s}_I}(\tau) + j\phi_{\tilde{s}_I\tilde{s}_Q}(\tau).
\end{aligned} \tag{4.27}
$$

Note the factor of 1/2 by definition in the autocorrelation of the complex envelope of a random bandpass signal $s(t)$. It follows that

$$\phi_{ss}(\tau) = \mathrm{Re}\left\{\phi_{\tilde{s}\tilde{s}}(\tau)\mathrm{e}^{j2\pi f_c\tau}\right\}. \tag{4.28}$$

The autocorrelation $\phi_{ss}(\tau)$ evaluated at $\tau = 0$ gives the *bandpass power* in the bandpass random signal $s(t)$ as

$$P = \phi_{ss}(0) = \phi_{\tilde{s}_I\tilde{s}_I}(0) = \phi_{\tilde{s}_Q\tilde{s}_Q}(0). \tag{4.29}$$

The *envelope power* of a WSS bandpass random process is defined as its mean-square complex envelope:

$$\Omega_p \triangleq \mathrm{E}[|\tilde{s}(t)|^2] = \mathrm{E}[\tilde{s}_I^2(t)] + \mathrm{E}[\tilde{s}_Q^2(t)] = \phi_{\tilde{s}_I\tilde{s}_I}(0) + \phi_{\tilde{s}_Q\tilde{s}_Q}(0) = 2P. \quad (4.30)$$

Hence, the bandpass signal power is one-half the envelope power, $P = \Omega_p/2$.

4.3.1 Power Spectral Density of Bandpass Random Signals

The psd and autocorrelation of the complex envelope $\tilde{s}(t)$ are Fourier transform pairs:

$$S_{\tilde{s}\tilde{s}}(f) = \int_{-\infty}^{\infty} \phi_{\tilde{s}\tilde{s}}(\tau)\mathrm{e}^{-j2\pi f\tau}\mathrm{d}\tau \quad (4.31a)$$

$$\phi_{\tilde{s}\tilde{s}}(\tau) = \int_{-\infty}^{\infty} S_{\tilde{s}\tilde{s}}(f)\mathrm{e}^{j2\pi f\tau}\mathrm{d}f, \quad (4.31b)$$

From (4.27), the complex envelope of a WSS bandpass random process has the psd

$$S_{\tilde{s}\tilde{s}}(f) = S_{\tilde{s}_I\tilde{s}_I}(f) + jS_{\tilde{s}_I\tilde{s}_Q}(f). \quad (4.32)$$

and the envelope power is

$$\Omega_p = \phi_{\tilde{s}\tilde{s}}(0) = \int_{-\infty}^{\infty} S_{\tilde{s}\tilde{s}}(f)\mathrm{d}f \quad (4.33)$$

From (4.32), it may appear that the psd of the complex envelope is complex valued. However, the autocorrelation $\phi_{\tilde{s}_I\tilde{s}_I}(\tau)$, as with any autocorrelation, is an even valued function such that $\phi_{\tilde{s}_I\tilde{s}_I}(\tau) = \phi_{\tilde{s}_I\tilde{s}_I}(-\tau)$. Hence, its Fourier transform $S_{\tilde{s}_I\tilde{s}_I}(f)$ is real-valued. Moreover, the cross-correlation $\phi_{\tilde{s}_I\tilde{s}_Q}(\tau)$ in (4.27), as with any cross-correlation, satisfies the property:

$$\phi_{\tilde{s}_I\tilde{s}_Q}(\tau) = \phi_{\tilde{s}_Q\tilde{s}_I}(-\tau). \quad (4.34)$$

Combining (4.34) with (4.25b) yields the property:

$$\phi_{\tilde{s}_I\tilde{s}_Q}(\tau) = -\phi_{\tilde{s}_I\tilde{s}_Q}(-\tau). \quad (4.35)$$

This means that the cross-correlation $\phi_{\tilde{s}_I\tilde{s}_Q}(\tau)$ is an odd function of τ and, hence, its Fourier transform $S_{\tilde{s}_I\tilde{s}_Q}(f)$ is purely imaginary valued. Therefore, the psd in (4.32) is always real-valued.

The psd of a bandpass waveform $s(t)$ in (11.32) can be related to the psd of its complex envelope $\tilde{s}(t)$. From (4.28), and using the identity in (4.8) and the property $\phi_{\tilde{s}\tilde{s}}(\tau) = \phi^*_{\tilde{s}\tilde{s}}(-\tau)$, the bandpass psd is

$$S_{ss}(f) = \frac{1}{2}\Big(S_{\tilde{s}\tilde{s}}(f - f_c) + S_{\tilde{s}\tilde{s}}(-f - f_c)\Big). \tag{4.36}$$

The psd of the complex envelope $S_{\tilde{s}\tilde{s}}(f)$ is real-valued but not necessarily even. However, the psd of the corresponding bandpass waveform, $S_{ss}(f)$, is always real-valued and even.

4.4 Thermal Noise

Thermal noise is present in all communication receivers and arises due to the random motion of electrons in a conductor. From fundamental physics which is outside the scope of this text, thermal noise $n(t)$ can be modeled as a Gaussian random process having the psd:

$$S_{nn}(f) = \frac{h|f|}{2\left(e^{h|f|/kT} - 1\right)} \quad \text{watts/Hz} \tag{4.37}$$

where

$$\begin{aligned}
h &= 6.62 \times 10^{-34} \text{ Joules} = \text{Plank's constant} \\
k &= 1.38 \times 10^{-23} \text{ Joules/degree} = \text{Boltzmann's constant} \\
T &= \text{effective temperature in degrees Kelvin (degrees Celsius + 273)}
\end{aligned}$$

Using the Taylor series expansion $e^x = 1 + x + x^2/2! + x^3/3! + \cdots$ and keeping the first two terms gives:

$$\begin{aligned}
S_{nn}(f) &\approx \frac{h|f|}{2(1 + h|f|/kT - 1)} \\
&= kT/2 \text{ watts/Hz}.
\end{aligned} \tag{4.38}$$

Following (4.38), bandpass thermal noise is sometimes described in terms of the psd of its complex envelope, $\tilde{n}(t)$, equal to $S_{\tilde{n}\tilde{n}}(f) = kT$ watts/Hz. Moreover, the psd may be expressed in units of dBm/Hz, such that $S_{\tilde{n}\tilde{n}}(f) = 10\log_{10}(1000kT)$ dBm/Hz. For example, when $T = 293^{\circ}$ K (20° C), $S_{\tilde{n}\tilde{n}}(f) = -174$ dBm/Hz.

4.4.1 White Gaussian Noise

White Gaussian noise $w(t)$ is modeled as having a two-sided psd $N_o/2$ watts/Hz for all frequencies, where $N_o = kT$. Similar to white light, white Gaussian noise has all frequency components present with the same density. The autocorrelation of $w(t)$ is

$$\phi_{ww}(\tau) = \mathcal{F}^{-1}\left\{N_o/2\right\} = \frac{N_o}{2}\delta(\tau). \tag{4.39}$$

Due to the Dirac delta function, $w(t)$ is uncorrelated with $w(t+\tau)$ for any $\tau \neq 0$. Since $w(t)$ is a Gaussian random process, $w(t)$ and $w(t+\tau)$ are also statistically independent for any $\tau \neq 0$.

4.4.2 Low-Pass Filtered White Noise

Suppose that white noise is passed through a linear time-invariant filter having impulse $h(t)$ and transfer function $H(f)$ as shown in Fig. 4.4.

If the input noise $w(t)$ is white noise with $S_{ww}(f) = N_o/2$ watts/Hz, then from (3.34) the output noise $n(t)$ has psd:

$$S_{nn}(f) = \frac{N_o}{2}\left|H(f)\right|^2.$$

Consider an ideal low-pass filter having the transfer function:

$$H(f) = \mathrm{rect}\left(\frac{f}{2W}\right) = \begin{cases} 1, & |f| \leq W \\ 0, & \text{elsewhere} \end{cases},$$

where W is the cutoff frequency of the low-pass filter. It follows that

$$S_{nn}(f) = \frac{N_o}{2}\mathrm{rect}\left(\frac{f}{2W}\right).$$

The autocorrelation of the ideal low-pass filtered white noise is obtained by taking the inverse Fourier transform of $S_{nn}(f)$ as

$$\phi_{nn}(\tau) = N_o W \mathrm{sinc}(2W\tau). \tag{4.40}$$

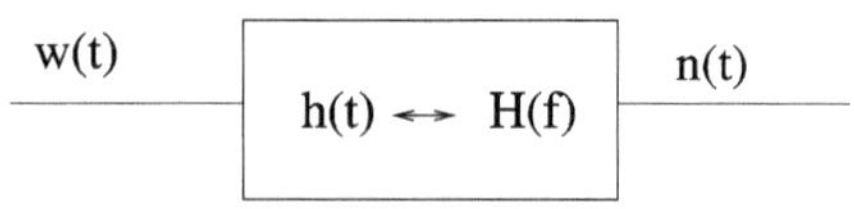

Fig. 4.4 White noise applied to a time-invariant linear filter

The output noise power is equal to

$$P_n = \phi_{nn}(0) = N_0 W \quad \text{watts}$$

The autocorrelation in (4.40) is plotted in Fig. 4.5. Samples of output noise process $n(t)$ spaced $T = 1/(2W)$ seconds apart, or separated by any multiple of $1/(2W)$ seconds, are uncorrelated due to zero crossings in the sinc function. Since the input noise process to the ideal low-pass filter is Gaussian, the output noise process is also Gaussian, and it follows that the output noise samples are also statistically independent. It is notable that a sample spacing of $T = 1/(2W)$ seconds corresponds to the Nyquist-Shannon sampling rate of $2W$ samples per second, for a waveform having bandwidth W.

4.4.3 Bandpass Filtered White Noise

Suppose that white noise $w(t)$ is passed through an ideal bandpass filter having impulse $h(t)$ and transfer function $H(f)$ as shown in Fig. 4.6. The transfer function of the bandpass filter is

$$H(f) = \text{rect}\left(\frac{f - f_c}{2W}\right) + \text{rect}\left(\frac{f + f_c}{2W}\right).$$

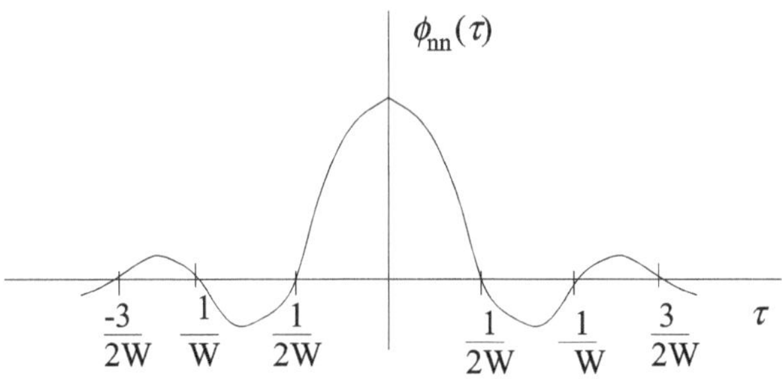

Fig. 4.5 Autocorrelation of white noise passed through an ideal low-pass filter

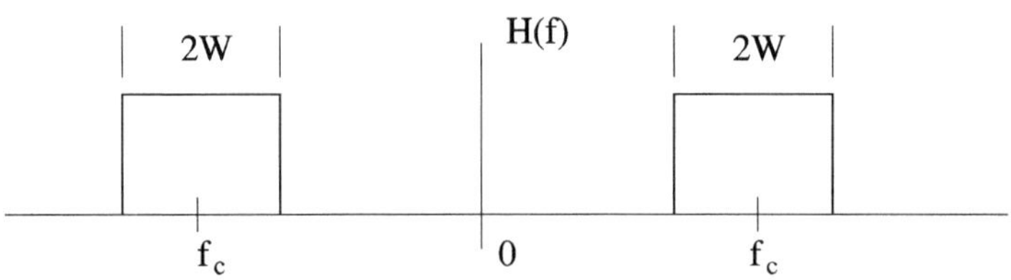

Fig. 4.6 White noise applied to an ideal bandpass filter

Therefore, the output noise process $n(t)$ has psd:

$$S_{nn}(f) = \frac{N_o}{2}\left[\text{rect}\left(\frac{f-f_c}{2W}\right) + \text{rect}\left(\frac{f+f_c}{2W}\right)\right].$$

It follows that the autocorrelation of the output noise, $n(t)$, is

$$\begin{aligned}\phi_{nn}(\tau) &= \frac{N_o}{2} \cdot 2W\text{sinc}(2W\tau) \cdot 2\cos(2\pi f_c \tau) \\ &= 2N_o W\text{sinc}(2W\tau) \cdot \cos(2\pi f_c \tau)\end{aligned} \tag{4.41}$$

The total output noise power is $P_n = 2N_o W$ watts.

Once again, samples of $n(t)$ taken $1/(2W)$ seconds apart are uncorrelated due to the $\text{sinc}(2W\tau)$ term, and, since they are Gaussian, they are also statistically independent. Moreover, due to the $\cos(2\pi f_c \tau)$ term, samples of $n(t)$ taken $\tau = 1/(4f_c)$ seconds apart are also uncorrelated and independent.

4.4.4 Noise Equivalent Bandwidth

Consider an arbitrary linear time-invariant filter having transfer function $H(f)$ as shown in Fig. 4.7. If the input to the filter $H(f)$ is white noise $w(t)$ having a psd $N_o/2$ watts/Hz, then the noise power at the output of the filter is

$$\begin{aligned}N_{\text{out}} &= \frac{N_o}{2}\int_{-\infty}^{\infty} |H(f)|^2 \mathrm{d}f \\ &= N_o \int_0^{\infty} |H(f)|^2 \mathrm{d}f.\end{aligned} \tag{4.42}$$

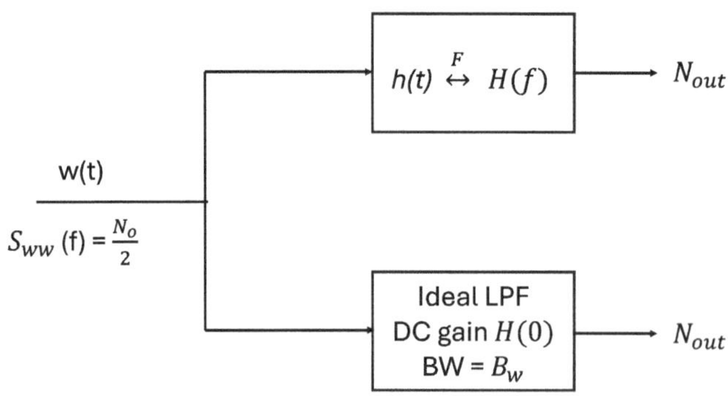

Fig. 4.7 Noise equivalent bandwidth

Now suppose that the same white noise process $w(t)$ is applied to an ideal low-pass filter having bandwidth B_w and dc filter gain $H(0)$ as shown in Fig. 4.7. The noise power at the output of such an ideal low-pass filter is

$$N_{\text{out}} = N_o B_w H^2(0). \tag{4.43}$$

Equating the above two equations and solving for B_w gives the *noise equivalent bandwidth* of the filter $H(f)$, defined as

$$B_w \triangleq \frac{\int_0^\infty |H(f)|^2 \mathrm{d}f}{H^2(0)}. \tag{4.44}$$

Problems

4.1 Suppose that white Gaussian noise $w(t)$ having psd $S_{ww}(f) = N_o/2$ watts/Hz is passed through an anti-aliasing filter that is modeled as an ideal low-pass filter having a cutoff frequency of W Hz. The transfer function of the filter is

$$H(f) = \text{rect}\left(\frac{f}{2W}\right) = \begin{cases} 1, & |f| \leq W \\ 0, & \text{elsewhere} \end{cases}$$

The output of the filter is *oversampled* with a sampling rate of $4W$ samples per second. What is the joint pdf of two successive samples taken at the output of the ideal low-pass filter?

4.2 Suppose that additive white Gaussian noise $w(t)$ having psd $S_{ww}(f) = N_o/2 = 10^{-12}$ watts/Hz is passed through a low-pass filter having the impulse response

$$h(t) = \alpha \mathrm{e}^{-\alpha t} u(t)$$

where $u(t)$ is the unit step function and $\alpha = 2\pi \times 10^6$:

(a) What is the autocorrelation of the noise waveform $n(t)$ at the output of the filter?
(b) What is the noise power at the output of the filter in dBm units?
(c) What is the noise equivalent bandwidth B_w of the filter $h(t)$?

4.3 Suppose that additive white Gaussian noise $w(t)$ having the psd $S_{ww}(f) = N_o/2 = 10^{-15}$ watts/Hz is passed through a low-pass filter having the impulse response

$$h(t) = W \text{sinc}^2(Wt)\,, \quad -\infty < t < \infty$$

where $W = 1 \times 10^6$.

(a) What is the noise equivalent bandwidth B_w of the filter $h(t)$?

(b) What is the power in the noise waveform $n(t)$ at the output of the filter $h(t)$ in dBm units?

(c) If the noise waveform $n(t)$ output of the filter $h(t)$ is sampled at some arbitrary time instant t_1 *and its magnitude is taken* to generate the statistic $y = |n(t_1)|$, what is the pdf of y?

Chapter 5
Communication Waveforms

Modulation is the process whereby message information is embedded into a sinusoidal waveform for transmission over a communication channel. Some types of modulation are implemented at baseband, not having a radio frequency carrier as in impulse radio or ultra wideband (UWB) communication systems. However, most communication systems are carrier modulated, using a microwave or millimeter wave RF carrier, where the message information is embedded in the amplitude, frequency, or phase of the RF carrier, or their combination.

Section 5.1 begins the chapter with a vector space representation of modulated waveforms, including orthogonal expansions, signal space representations, the Gram-Schmidt procedure for obtaining signal vectors, and the properties of signal vectors. Section 5.2 then considers binary modulated waveforms, while Sect. 5.3 considers nonbinary modulated waveforms. Section 5.4 discusses Nyquist pulse shaping for ISI-free transmission. Finally, Sect. 5.5 considers the psd of modulated signals.

5.1 Vector Space Representations

For digital bandpass modulation schemes, the bandpass signal that is transmitted at each baud epoch belongs to a finite-sized set of real-valued finite-energy waveforms. Let $\mathcal{S} = \{s_1(t), s_2(t), \ldots, s_M(t)\}$ be the set of such bandpass waveforms, where M denotes the size of the signal set. Let $\tilde{\mathcal{S}} = \{\tilde{s}_1(t), \tilde{s}_2(t), \ldots, \tilde{s}_M(t)\}$ be the corresponding set of complex envelopes. Usually, an integer number of bits are contained in each message, so $M = 2^k$ for some k, i.e., $M = 2, 4, 8, 16$, etc. Every T seconds, $k = \log_2 M$ bits are mapped to one of the M waveforms for transmission over the waveform channel. The baud rate or modulation symbol rate is $R = 1/T$, and the information bit rate and baud rate are related as $R_b = kR$.

G. Stuber, *Introductory Digital Wireless Communications*,
https://doi.org/10.1007/978-3-032-01559-4_5

A vector space representation may be obtained by using either the set of bandpass waveforms or their corresponding set of complex envelopes. In the former case, the vector space is real-valued, whereas in the latter case, it is generally complex-valued. However, the dimension of the vector space (length of vectors) may be different for some sets of waveforms where information is embedded in the signal amplitude and phase; in these cases, the dimension of the complex-valued vector space is often one-half the dimension of the real-valued vector space. For wireless channels that randomly affect both the amplitude and phase of the transmitted waveform, it is mathematically more convenient to work with complex envelopes and a complex-valued vector space representation. However, both the real- and complex-valued vector space representations will be considered in this textbook.

5.1.1 Orthogonal Expansions

Consider a set of complex-valued functions $\{\varphi_n(t), n = 1, \ldots, N\}$ having the property:

$$\int_{-\infty}^{\infty} \varphi_n(t)\varphi_k^*(t)\mathrm{d}t = \delta_{kn} \tag{5.1}$$

where

$$\delta_{kn} = \begin{cases} 1\ , & k = n \\ 0\ , & k \neq n \end{cases} \tag{5.2}$$

The $\varphi_n(t)$ are all mutually orthogonal and all have unit energy and, hence, the $\varphi_n(t)$ are said to be *orthonormal*. As a result, the $\varphi_n(t)$ may be used as the basis for an N-dimensional complex-valued vector space. The complex-valued vector space is spanned by taking linear combinations of the basis functions:

$$\tilde{s}(t) = \sum_{n=1}^{N} \tilde{s}_n \varphi_n(t) \tag{5.3}$$

where the $\{\tilde{s}_n, n = 1, \ldots, N\}$ are complex-valued coefficients.

A vector space may be similarly defined that is spanned by a set of N real-valued orthonormal basis functions $\{\varphi_n(t), n = 1, \ldots, N\}$ having the property:

$$\int_{-\infty}^{\infty} \varphi_n(t)\varphi_k(t)dt = \delta_{kn} \tag{5.4}$$

In this case

$$s(t) = \sum_{n=1}^{N} s_n \varphi_n(t) \tag{5.5}$$

where the $\{s_n, n = 1, \ldots, N\}$ are real-valued coefficients. Note that (5.1) and (5.4) are the same except for the conjugation of one term in the integrand. Many sets of orthonormal functions may be defined and two examples are given below.

Example 6.1 Consider the set of real-valued basis functions $\{\varphi(t)\}$ defined by

$$\varphi_k(t) = \sqrt{\frac{N}{T}} u_{T/N}(t - kT/N)\ , \quad k = 0, 1, \ldots, N-1$$

where $u_{T/N}(t) = u(t) - u(t - T/N)$. The set $\{\varphi_k(t)\}$ consists of nonoverlapping, square pulses of width T/N, having amplitude $\sqrt{\frac{N}{T}}$. Hence, the $\{\varphi(t)\}$ are orthonormal and will span an N dimensional real-valued vector space.

Example 6.2 Consider the set $\{\varphi_k(t)\}$ of complex-valued exponentials:

$$\varphi_k(t) = \sqrt{\frac{1}{T}} \exp\left\{-j\frac{2\pi kt}{T}\right\} u_T(t)\ , \quad k = 0, 1, \ldots, N-1\ ,$$

having frequencies that are multiples of $1/T$, and where

$$u_T(t) = u(t) - u(t - T)$$

For this set of waveforms

$$\begin{aligned} \int_{-\infty}^{\infty} \varphi_n(t)\varphi_k^*(t)\mathrm{d}t &= \frac{1}{T}\int_0^T \exp\left\{-j\frac{2\pi(n-k)t}{T}\right\} \mathrm{d}t \\ &= \delta_{nk} \end{aligned}$$

The $\{\varphi_n(t), n = 1, \ldots, N\}$ constitute a set of N complex-valued orthonormal basis functions that will span an N dimensional complex-valued vector space.

Consider an arbitrary bandpass waveform $s(t)$ having complex envelope $\tilde{s}(t)$ and energy:

$$E = \int_{-\infty}^{\infty} s^2(t)dt = \frac{1}{2}\int_{-\infty}^{\infty} |\tilde{s}(t)|^2 \mathrm{d}t \tag{5.6}$$

In practice, the signal $\tilde{s}(t)$ is causal such that $\tilde{s}(t) = 0, t < 0$ and $s(t)$ will have a finite duration. However, the integral in (5.6) is extended from $-\infty$ to ∞ for analytical convenience.

Consider a given set of N complex orthonormal basis functions $\{\varphi_n(t), n = 1, \ldots, N\}$ and suppose the complex envelope $\tilde{s}(t)$ is approximated as the weighted linear sum:

$$\hat{s}(t) \approx \sum_{k=1}^{N} \tilde{s}_k \varphi_k(t) \tag{5.7}$$

The problem at hand is to determine the set of coefficients $\tilde{s}_k, k = 1, \ldots, N$ that provides the best possible approximation. One method is to minimize the mean-squared error:

$$\varepsilon = \frac{1}{2}\int_{-\infty}^{\infty} \left|\tilde{s}(t) - \hat{s}(t)\right|^2 \mathrm{d}t = \frac{1}{2}\int_{-\infty}^{\infty} \left|\tilde{s}(t) - \sum_{k=1}^{N} \tilde{s}_k \varphi_k(t)\right|^2 \mathrm{d}t \tag{5.8}$$

Observe that the mean-square error is one half the energy in the difference signal $\tilde{s}(t) - \hat{s}(t)$.

To minimize the mean-square error, partial derivatives can be taken with respect to each of the complex coefficients $\tilde{s}_n$ and the result set equal to zero, i.e., for the nth term, solve:

$$\frac{\partial \varepsilon}{\partial \tilde{s}_n} = 0, \quad n = 1, \ldots, N \ . \tag{5.9}$$

Taking the partial derivative of ε with respect to the complex coefficient $\tilde{s}_n = \tilde{s}_{n,R} + j\tilde{s}_{n,I}$ is accomplished by taking the partial derivatives with respect to the real and imaginary components of $\tilde{s}_n$ as

$$\frac{\partial \varepsilon}{\partial \tilde{s}_n} = \frac{\partial \varepsilon}{\partial \tilde{s}_{n,R}} + j\frac{\partial \varepsilon}{\partial \tilde{s}_{n,I}} \tag{5.10}$$

Using $|z|^2 = z \cdot z^*$ in (5.8)

$$\varepsilon = \frac{1}{2}\int_{-\infty}^{\infty}\left(\tilde{s}(t) - \sum_{k=1}^{N}\tilde{s}_k\varphi_k(t)\right)\left(\tilde{s}^*(t) - \sum_{k=1}^{N}\tilde{s}_k^*\varphi_k^*(t)\right)\mathrm{d}t$$

$$= \frac{1}{2}\int_{-\infty}^{\infty}\left(\tilde{s}(t) - \sum_{k=1}^{N}(\tilde{s}_{k,R} + j\tilde{s}_{k,I})\varphi_k(t)\right)\left(\tilde{s}^*(t) - \sum_{k=1}^{N}(\tilde{s}_{k,R} - j\tilde{s}_{k,I})\varphi_k^*(t)\right)\mathrm{d}t$$

and taking the partial derivative with respect to $\tilde{s}_{n,R}$ gives:

$$\begin{aligned}
\frac{\partial\varepsilon}{\partial\tilde{s}_{n,R}} &= -\frac{1}{2}\int_{-\infty}^{\infty}\left[\left(\tilde{s}(t) - \sum_{k=1}^{N}(\tilde{s}_{k,R} + j\tilde{s}_{k,I})\varphi_k(t)\right)\varphi_n^*(t)\right. \\
&\quad \left. + \left(\tilde{s}^*(t) - \sum_{k=1}^{N}(\tilde{s}_{k,R} - j\tilde{s}_{k,I})\varphi_k^*(t)\right)\varphi_n(t)\right]\mathrm{d}t \\
&= -\frac{1}{2}\left[\int_{-\infty}^{\infty}\tilde{s}(t)\varphi_n^*(t)\mathrm{d}t - \sum_{k=1}^{N}(\tilde{s}_{k,R} + j\tilde{s}_{k,I})\int_{-\infty}^{\infty}\varphi_k(t)\varphi_n^*(t)\mathrm{d}t\right. \\
&\quad \left. + \int_{-\infty}^{\infty}\tilde{s}^*(t)\varphi_n(t)\mathrm{d}t - \sum_{k=1}^{N}(\tilde{s}_{k,R} - j\tilde{s}_{k,I})\int_{-\infty}^{\infty}\varphi_k^*(t)\varphi_n(t)\mathrm{d}t\right] \\
&= -\frac{1}{2}\left[\int_{-\infty}^{\infty}\tilde{s}(t)\varphi_n^*(t)\mathrm{d}t - \tilde{s}_n + \int_{-\infty}^{\infty}\tilde{s}^*(t)\varphi_n(t)\mathrm{d}t - \tilde{s}_n^*\right]
\end{aligned}$$

Likewise, taking the partial derivative with respect to $\tilde{s}_{n,I}$ gives:

$$\begin{aligned}
\frac{\partial\varepsilon}{\partial\tilde{s}_{n,I}} &= \frac{j}{2}\int_{-\infty}^{\infty}\left[\left(\tilde{s}(t) - \sum_{k=1}^{N}(\tilde{s}_{k,R} + j\tilde{s}_{k,I})\varphi_k(t)\right)\varphi_n^*(t)\right. \\
&\quad \left. - \left(\tilde{s}^*(t) - \sum_{k=1}^{N}(\tilde{s}_{k,R} - j\tilde{s}_{k,I})\varphi_k^*(t)\right)\varphi_n(t)\right]\mathrm{d}t \\
&= \frac{j}{2}\left[\int_{-\infty}^{\infty}\tilde{s}(t)\varphi_n^*(t)\mathrm{d}t - \sum_{k=1}^{N}(\tilde{s}_{k,R} + j\tilde{s}_{k,I})\int_{-\infty}^{\infty}\varphi_k(t)\varphi_n^*(t)\mathrm{d}t\right. \\
&\quad \left. - \int_{-\infty}^{\infty}\tilde{s}^*(t)\varphi_n(t)\mathrm{d}t + \sum_{k=1}^{N}(\tilde{s}_{k,R} - j\tilde{s}_{k,I})\int_{-\infty}^{\infty}\varphi_k^*(t)\varphi_n(t)\mathrm{d}t\right] \\
&= \frac{j}{2}\left[\int_{-\infty}^{\infty}\tilde{s}(t)\varphi_n^*(t)\mathrm{d}t - \tilde{s}_n - \int_{-\infty}^{\infty}\tilde{s}^*(t)\varphi_n(t)\mathrm{d}t + \tilde{s}_n^*\right]
\end{aligned}$$

Hence, (5.9) gives the minimum mean-square error solution:

$$\tilde{s}_n = \int_{-\infty}^{\infty} \tilde{s}(t)\varphi_n^*(t)\mathrm{d}t, \quad n = 1, \ldots, N \ . \tag{5.11}$$

Likewise, if the bandpass waveform $s(t)$ is approximated as

$$\hat{s}(t) \approx \sum_{k=1}^{N} s_k \varphi_k(t) \tag{5.12}$$

then the mean-square error is

$$\varepsilon = \int_{-\infty}^{\infty} \left(s(t) - \hat{s}(t)\right)^2 \mathrm{d}t = \int_{-\infty}^{\infty} \left(s(t) - \sum_{k=1}^{N} s_k \varphi_k(t)\right)^2 \mathrm{d}t \tag{5.13}$$

Taking the partial derivative of (5.13) with respect to each of the s_n and setting the result equal to zero gives:

$$\begin{aligned}
\frac{\partial \varepsilon}{\partial s_n} &= -2 \int_{-\infty}^{\infty} \left(s(t) - \sum_{k=1}^{N} s_k \varphi_k(t)\right) \varphi_n(t)\mathrm{d}t \\
&= -2 \int_{-\infty}^{\infty} s(t)\varphi_n(t)\mathrm{d}t + 2 \sum_{k=1}^{N} s_k \int_{-\infty}^{\infty} \varphi_k(t)\varphi_n(t)\mathrm{d}t \\
&= -2 \int_{-\infty}^{\infty} s(t)\varphi_n(t)\mathrm{d}t + 2s_n = 0
\end{aligned}$$

This yields the minimum mean-square error solution:

$$s_n = \int_{-\infty}^{\infty} s(t)\varphi_n(t)\mathrm{d}t, \quad n = 1, \ldots, N \ . \tag{5.14}$$

Substituting (5.11) back into (5.8) gives minimum mean-square error:

$$\begin{aligned}
\varepsilon_{\min} &= \frac{1}{2} \int_{-\infty}^{\infty} \left| \tilde{s}(t) - \sum_{k=1}^{N} \tilde{s}_k \varphi_k(t) \right|^2 \mathrm{d}t \\
&= \frac{1}{2} \int_{-\infty}^{\infty} \left(\tilde{s}(t) - \sum_{k=1}^{N} \tilde{s}_k \varphi_k(t)\right) \left(\tilde{s}^*(t) - \sum_{\ell=1}^{N} \tilde{s}_\ell^* \varphi_\ell^*(t)\right) \mathrm{d}t
\end{aligned}$$

$$
\begin{aligned}
&= \frac{1}{2}\int_{-\infty}^{\infty} |\tilde{s}(t)|^2 \mathrm{d}t - \frac{1}{2}\int_{-\infty}^{\infty} \tilde{s}(t) \sum_{\ell=1}^{N} \tilde{s}_\ell^* \varphi_\ell^*(t) \mathrm{d}t - \frac{1}{2}\int_{-\infty}^{\infty} \tilde{s}^*(t) \sum_{k=1}^{N} \tilde{s}_k \varphi_k(t) \mathrm{d}t \\
&+ \frac{1}{2}\int_{-\infty}^{\infty} \sum_{k=1}^{N} \tilde{s}_k \varphi_k(t) \sum_{\ell=1}^{N} s_\ell^* \varphi_\ell^*(t) \mathrm{d}t \\
&= \frac{1}{2}\int_{-\infty}^{\infty} |\tilde{s}(t)|^2 \mathrm{d}t - \frac{1}{2}\sum_{k=1}^{N} \tilde{s}_k \int_{-\infty}^{\infty} s^*(t)\varphi_k(t) \mathrm{d}t - \frac{1}{2}\sum_{\ell=1}^{N} \tilde{s}_\ell^* \int_{-\infty}^{\infty} \tilde{s}(t)\varphi_\ell^*(t) \mathrm{d}t \\
&+ \frac{1}{2}\sum_{k=1}^{N}\sum_{\ell=1}^{N} \tilde{s}_k \tilde{s}_\ell^* \int_{-\infty}^{\infty} \varphi_k(t)\varphi_\ell^*(t) \mathrm{d}t \\
&= \frac{1}{2}\int_{-\infty}^{\infty} |\tilde{s}(t)|^2 \mathrm{d}t - \frac{1}{2}\sum_{k=1}^{N} |\tilde{s}_k|^2 \\
&= E - \frac{1}{2}\sum_{k=1}^{N} |\tilde{s}_k|^2
\end{aligned}
\tag{5.15}
$$

Likewise, substituting (5.14) back into (5.13) gives minimum mean-square error:

$$
\begin{aligned}
\varepsilon_{\min} &= \int_{-\infty}^{\infty} \left(s(t) - \sum_{k=1}^{N} s_k \varphi_k(t) \right)^2 \mathrm{d}t \\
&= E - \sum_{k=1}^{N} s_k^2
\end{aligned}
\tag{5.16}
$$

If the weighted linear sum in (5.7), with coefficients obtained in (5.11), represents the complex envelope exactly, then $\varepsilon = 0$. Likewise, if the weighted linear sum in (5.12) with coefficients obtained in (5.14) represents the bandpass waveform exactly, then $\varepsilon = 0$ as well. If the complex envelope $\tilde{s}(t)$ is represented by the N-dimensional complex-valued vector

$$
\tilde{s}(t) \leftrightarrow \tilde{\mathbf{s}} = (\tilde{s}_1, \tilde{s}_2, \ldots, \tilde{s}_N)
\tag{5.17}
$$

and $\varepsilon = 0$, then the energy of the corresponding bandpass waveform $s(t)$ is equal to one-half the squared length of the complex-valued vector $\tilde{\mathbf{s}}$ from the origin, i.e.,

$$
E = \tfrac{1}{2}\|\tilde{\mathbf{s}}\|^2 = \frac{1}{2}\sum_{k=1}^{N} |\tilde{s}_k|^2 \ .
\tag{5.18}
$$

Likewise, if the bandpass waveform $s(t)$ is represented by the N-dimensional real-valued vector

$$s(t) \longleftrightarrow \mathbf{s} = (s_1, s_2, \ldots, s_N) \tag{5.19}$$

and $\varepsilon = 0$, then the energy of $s(t)$ is given by the squared length of the real-valued vector $\mathbf{s}$ from the origin, i.e.,

$$E = \|\mathbf{s}\|^2 = \sum_{k=1}^{N} s_k^2 \ . \tag{5.20}$$

5.1.2 Signal Space Representation

With digital modulation, the bandpass waveform transmitted at each channel use belongs to a finite set of finite-energy waveforms. Let $\mathcal{S} = \{s_1(t), s_2(t), \ldots, s_M(t)\}$ be the set of bandpass waveforms, and let set $\tilde{\mathcal{S}} = \{\tilde{s}_1(t), \tilde{s}_2(t), \ldots, \tilde{s}_M(t)\}$ be their complex envelopes, where M denotes the size of the message set.

Consider an N-dimensional complex-valued vector space that is spanned by a set of N orthonormal basis functions $\{\varphi_1(t), \varphi_2(t), \ldots, \varphi_N(t)\}$. Each complex envelope $\tilde{s}_m(t)$ in the set $\tilde{\mathcal{S}}$ can be projected onto the set of basis functions to yield a *complex message vector*:

$$\tilde{\mathbf{s}}_m = (\tilde{s}_{m,1},\ \tilde{s}_{m,2},\ \ldots,\ \tilde{s}_{m,N})\ ,\ \ m = 1, \ldots, M\ , \tag{5.21}$$

where

$$\tilde{s}_{m,i} = \int_{-\infty}^{\infty} \tilde{s}_m(t)\varphi_i^*(t)\mathrm{d}t,\ \ i = 1,\ \ldots,\ N\ . \tag{5.22}$$

The collection of N orthonormal basis functions constitutes a *complete set* if each waveform in the set $\tilde{\mathcal{S}}$ can be represented *exactly* as a linear combination of the basis functions. That is, $\varepsilon = 0$ in (5.8) for each waveform in $\tilde{\mathcal{S}}$ and

$$\tilde{s}_m(t) = \sum_{i=1}^{N} \tilde{s}_{m,i}\varphi_i(t)\ ,\ \ m = 1,\ \ldots,\ M\ . \tag{5.23}$$

Likewise, consider an N-dimensional real-valued vector space that is spanned by a set of N real-valued orthonormal basis functions $\{\varphi_1(t), \varphi_2(t), \ldots, \varphi_N(t)\}$. Each bandpass waveform $s_m(t)$ in the set $\mathcal{S}$ can be projected onto this set of basis functions to yield a *bandpass message vector*:

$$\mathbf{s}_m = (s_{m,1},\ s_{m,2},\ \ldots,\ s_{m,N})\ ,\ \ m = 1, \ldots, M\ , \tag{5.24}$$

where

$$s_{m,i} = \int_{-\infty}^{\infty} s_m(t)\varphi_i(t)\mathrm{d}t, \quad i = 1, \ \ldots, \ N \ . \tag{5.25}$$

Again, the collection of N orthonormal basis functions constitutes a *complete set* if each bandpass waveform in the set $\mathcal{S}$ can be expressed exactly as a linear combination of the orthonormal basis functions. That is, $\varepsilon = 0$ in (5.13) for each waveform in $\mathcal{S}$ and

$$s_m(t) = \sum_{i=1}^{N} s_{m_i}\varphi_i(t), \quad m = 1, \ \ldots, \ M \ . \tag{5.26}$$

Note the complete set of basis functions $\{\varphi_1(t), \varphi_2(t), \ldots, \varphi_N(t)\}$ and corresponding dimension N for the set of bandpass waveforms $\mathcal{S}$ and their corresponding set of complex envelopes $\tilde{\mathcal{S}}$ are different as will be seen from various examples in the sequel. A systematic procedure for constructing a complete set of orthonormal basis functions $\{\varphi_1(t), \varphi_2(t), \ldots, \varphi_N(t)\}$ for the message sets $\mathcal{S}$ and $\tilde{\mathcal{S}}$ is now described.

5.1.3 Gram-Schmidt Orthonormalization Procedure

Define the inner product between two complex-valued functions $u(t)$ and $v(t)$ as

$$(u, v) \triangleq \int_{-\infty}^{\infty} u(t)v^*(t)\mathrm{d}t \tag{5.27}$$

and define the norm of $u(t)$ as

$$\|u\| \triangleq \sqrt{(u, u)} \ . \tag{5.28}$$

Note that the squared norm of $u(t)$

$$\|u\|^2 \triangleq (u, u) = \int_{-\infty}^{\infty} |u(t)|^2\mathrm{d}t \tag{5.29}$$

is the energy contained in $u(t)$. In the case of real-valued functions $u(t)$ and $v(t)$, the complex conjugate in (5.27) is omitted.

Given the set of complex envelopes $\tilde{\mathcal{S}}$, a complete set of orthonormal basis functions $\{\varphi_1(t), \varphi_2(t), \ldots, \varphi_N(t)\}$ may be constructed by using the following systematic procedure, known as the Gram-Schmidt orthonormalization procedure:

1. Using $\tilde{s}_1(t)$, let

$$g_1(t) = \tilde{s}_1(t)$$

and define

$$\varphi_1(t) = \frac{g_1(t)}{\|g_1\|} \, .$$

2. Using $\tilde{s}_2(t)$, let

$$g_2(t) = \tilde{s}_2(t) - \tilde{s}_{21}\varphi_1(t)$$

where

$$\tilde{s}_{21} = (\tilde{s}_2, \varphi_1)$$

and define

$$\varphi_2(t) = \frac{g_2(t)}{\|g_2\|} \, .$$

3. Using $\tilde{s}_i(t)$, let

$$g_i(t) = \tilde{s}_i(t) - \sum_{j=0}^{i-1} \tilde{s}_{ij}\varphi_j(t)$$

where

$$\tilde{s}_{ij} = (\tilde{s}_i, \varphi_j)$$

and define

$$\varphi_i(t) = \frac{g_i(t)}{\|g_i\|} \, .$$

4. Repeat Step 3 in a recursive fashion until all messages in the set $\tilde{S}$ have been used.

If $g_i(t) = 0$ in the ith step, then $\tilde{s}_i(t)$ can be expressed exactly using the set of basis functions $\{\varphi_j(t),\ j = 1, \ldots, i-1\}$ already generated. Consequently, the complex envelope $\tilde{s}_i(t)$ will not yield an additional basis function, and the procedure continues with the next complex envelope in the set $\tilde{S}$, $\tilde{s}_{i+1}(t)$. In the end, a complete set of N, $1 \le N \le M$ orthonormal basis functions $\{\varphi_1(t), \varphi_2(t), \ldots, \varphi_N(t)\}$ corresponding to the non-zero $g_i(t)$ will be obtained. Note that the complete set of orthonomal basis functions produced by the Gram-Schmidt orthonormalization procedure is not unique and will depend on the ordering on the waveforms in the set $\tilde{S}$; however, all complete orthonormal basis will have the same dimension N and any will do.

The dimension of the complex-valued vector space N is equal to M if and only if the set of waveforms $\tilde{S}$ are linearly independent, meaning that no complex envelope in the set is a linear combination of the other complex envelopes in the set. If the set of complex envelopes $\tilde{S}$ are linearly dependent, then there exists a set of complex-valued coefficients $a_1, a_2 \ldots, a_M$, not all zero, such that

$$a_1\tilde{s}_1(t) + a_2\tilde{s}_2(t) + \cdots + a_M\tilde{s}_M(t) = 0 \ .$$

Suppose, without loss of generality, that $a_M \neq 0$. If $a_M = 0$, then the set $\tilde{S}$ can be permuted so that $a_M \neq 0$ and

$$\tilde{s}_M(t) = -\left(\frac{a_1}{a_M}\tilde{s}_1(t) + \frac{a_2}{a_M}\tilde{s}_2(t) + \cdots + \frac{a_{M-1}}{a_M}\tilde{s}_{M-1}(t)\right) \ .$$

Next consider the reduced message set $\{\tilde{s}_i(t)\}_{i=1}^{M-1}$. If this reduced set of complex envelopes is linearly dependent, then there exists another set of complex-valued coefficients $\{b_i\}_{i=1}^{M-1}$, not all zero, such that

$$b_1\tilde{s}_1(t) + b_2 v_2(t) + \cdots + b_{M-1}\tilde{s}_{M-1}(t) = 0 \ .$$

Once again, assuming that $b_{M-1} \neq 0$, this will yield the further reduced message set $\{\tilde{s}_i(t)\}_{i=1}^{M-2}$. The above procedure can be continued until a set $\{\tilde{s}_i(t)\}_{i=1}^{N}$ of N linearly independent waveforms is obtained. If $N < M$, then the set of linearly independent waveforms $\{\tilde{s}_i(t)\}_{i=1}^{N}$ is not unique and depends upon which complex envelopes are removed from the set $\tilde{S}$, and each distinct set will yield a different set of complete basis functions. However, all such complete sets of orthonormal basis functions will have the same dimension N and any complete set will do.

The set $\tilde{S}$ may be checked for linear dependencies and a reduced set of N linearly independent complex envelopes $\{\tilde{s}_i(t)\}_{i=1}^{N}$ be obtained before applying the Gram-Schmidt orthonormalization procedure. Sometimes, this reduction is intuitive for the problem at hand and saves effort. Applying the Gram-Schmidt procedure to the reduced set $\{\tilde{s}_i(t)\}_{i=1}^{N}$ will yield the N required basis functions. Finally, the Gram-Schmidt orthonormalization procedure is a foolproof, but often tedious, method to obtain the required complete set of basis functions. However, a complete set of basis functions can often be obtained for the problem at hand fairly easily by using a procedure that proceeds by intuition.

After applying the Gram-Schmidt procedure, the messages of the message set $\tilde{S}$ may be expressed as weighted linear combinations of complete set of N orthonomal basis functions, viz.,

$$\begin{aligned}
\tilde{s}_1(t) &= \tilde{s}_{1,1}\varphi_1(t) \\
\tilde{s}_2(t) &= \tilde{s}_{2,1}\varphi_1(t) + \tilde{s}_{2,2}\varphi_2(t) \\
\tilde{s}_3(t) &= \tilde{s}_{3,1}\varphi_1(t) + \tilde{s}_{3,2}\varphi_2(t) + \tilde{s}_{3,3}\varphi_3(t) \\
\vdots &= \vdots \\
\tilde{s}_N(t) &= \tilde{s}_{N,1}\varphi_1(t) + \cdots + \tilde{s}_{N,N}\varphi_N(t)
\end{aligned} \tag{5.30}$$

For the remaining linearly dependent complex envelopes $\tilde{s}_i(t)$, $i = N+1, \ldots, M$, in $\tilde{S}$

$$\tilde{s}_i(t) = \sum_{k=1}^{N} \tilde{s}_{i,k}\varphi_k(t) \tag{5.31}$$

where

$$\tilde{s}_{i,k} = \int_0^T \tilde{s}_i(t)\varphi_k(t)dt \tag{5.32}$$

It follows that the message set $\tilde{S}$ can be represented by a set of complex message vectors $\tilde{\mathbf{s}}_i$, $i = 1, \ldots, M$ in an N-dimensional complex-valued vector space, i.e.,

$$\begin{aligned}
\tilde{s}_1(t) \leftrightarrow \tilde{\mathbf{s}}_1 &= (\tilde{s}_{1,1}, \tilde{s}_{1,2}, \ldots, \tilde{s}_{1,N}) \\
\tilde{s}_2(t) \leftrightarrow \tilde{\mathbf{s}}_2 &= (\tilde{s}_{2,1}, \tilde{s}_{2,2}, \ldots, \tilde{s}_{2,N}) \\
\vdots &= \vdots \\
\tilde{s}_M(t) \leftrightarrow \tilde{\mathbf{s}}_M &= (\tilde{s}_{M,1}, \tilde{s}_{M,2}, \ldots, \tilde{s}_{M,N})
\end{aligned} \tag{5.33}$$

Example 6.3 Consider a set S of four complex envelopes $\tilde{S}$ shown in Fig. 5.1. In this example, all the complex envelopes happen to be real-valued, such that the bandpass waveforms are

$$s_i(t) = \mathrm{Re}\left[\tilde{s}_i(t)\mathrm{e}^{j2\pi f_c t}\right] = \tilde{s}_i(t)\cos(2\pi f_c t)\ ,\ \ i = 1, \ldots, 4$$

The energy in each of the bandpass waveforms is the same and equal to

$$E = \int_0^T s_i^2(t)\mathrm{d}t = \frac{1}{2}\int_0^T |\tilde{s}_i(t)|^2\mathrm{d}t = \frac{A^2T}{2}$$

(continued)

Example 6.3 (continued)
Complex Message Vectors
Apply the Gram-Schmidt orthonormalization procedure to the set $\tilde{S}$:

1.

$$g_1(t) = \tilde{s}_1(t)$$

$$\|g_1\| = \sqrt{\int_0^T |\tilde{s}_1(t)|^2 dt} = A\sqrt{T}$$

$$\begin{aligned}\varphi_1(t) &= \frac{g_1(t)}{\|g_1\|} \\ &= \begin{cases} \sqrt{1/T} \ , & 0 \le t \le T \\ 0 \ , & \text{else} \end{cases}\end{aligned}$$

2. Let

$$g_2(t) = \tilde{s}_2(t) - \tilde{s}_{2,1}\varphi_1(t)\mathrm{d}t$$

where

$$\tilde{s}_{2,1} = \int_0^T \tilde{s}_2(t)\varphi_1(t)\mathrm{d}t = -A\sqrt{T} = -\sqrt{2E}$$

However $g_2(t) = 0$, meaning that $\tilde{s}_2(t)$ is linearly dependent with $\tilde{s}_1(t)$ and, therefore, $\tilde{s}_2(t)$ will not generate an additional basis function.

3. Let

$$g_3(t) = \tilde{s}_3(t) - \tilde{s}_{3,1}\varphi_1(t)$$

where

$$s_{3,1} = \int_0^T \tilde{s}_3(t)\varphi_1(t)\mathrm{d}t = 0 \ .$$

Then

$$\|g_3\| = \sqrt{\int_0^T \|\tilde{s}_3(t)\|^2 \, dt} = A\sqrt{T}$$

(continued)

Example 6.3 (continued)

and

$$\varphi_2(t) = \frac{g_3(t)}{\|g_3\|}$$
$$= \begin{cases} \sqrt{1/T} \ , & 0 \le t \le T/2 \\ -\sqrt{1/T} \ , & T/2 \le t \le T \end{cases} .$$

4. Let

$$g_4(t) = \tilde{s}_4(t) - \tilde{s}_{4,1}\varphi_1(t) - \tilde{s}_{4,2}\varphi_2(t) \ ,$$

where

$$s_{4,1} = \int_0^T \tilde{s}_4(t)\varphi_1(t)\mathrm{d}t = 0$$
$$s_{4,2} = \int_0^T \tilde{s}_4(t)\varphi_2(t)\mathrm{d}t = -A\sqrt{T} = -\sqrt{2E}$$

However $g_4(t) = 0$ and, therefore, $\tilde{s}_4(t)$ does not yield an additional basis function.

The complete set of $N = 2$ orthonormal basis functions is shown in Fig. 5.2, and they define a 2-D complex-valued vector space.

Referring to Fig. 5.1, the linear dependencies $\tilde{s}_1(t) = -\tilde{s}_2(t)$ and $\tilde{s}_3(t) = -\tilde{s}_4(t)$ are obvious by inspection. Moreover, it is obvious by inspection that $\tilde{s}_1(t)$ and $\tilde{s}_2(t)$ are orthogonal to $\tilde{s}_3(t)$ and $\tilde{s}_4(t)$. Hence, the required set of orthonormal basis functions can be easily constructed by choosing either $\tilde{s}_1(t)$ or $\tilde{s}_2(t)$ and either $\tilde{s}_3(t)$ or $\tilde{s}_4(t)$ and normalizing each choice of waveform to have unit energy. Choosing $\tilde{s}_1(t)$ and $\tilde{s}_3(t)$ and normalizing each to have unit energy will yield the exact same complete orthonormal basis shown in Fig. 5.2 and avoids the tedious Gram-Schmidt orthonormalization procedure.

Each complex envelope in the signal set can be expressed as a linear combination of the basis functions, according to (5.23), and the corresponding complex-valued signal vectors in (5.21) and (5.22) can be constructed. For this example, the complex message vectors are

$$\tilde{\mathbf{s}}_1 = (\sqrt{2E}, 0)$$
$$\tilde{\mathbf{s}}_2 = (-\sqrt{2E}, 0)$$
$$\tilde{\mathbf{s}}_3 = (0, \sqrt{2E})$$
$$\tilde{\mathbf{s}}_4 = (0, -\sqrt{2E}) \ ,$$

(continued)

Example 6.3 (continued)
where $E = A^2T/2$ is the energy in each bandpass waveform. The complex message vectors happen to be real-valued because the set of complex envelopes $\tilde{\mathcal{S}}$ are real-valued. From the complex message vectors, the energy in each waveform in the set $\mathcal{S}$ is the same and equal to $\frac{1}{2}\|\tilde{\mathbf{s}}_i\|^2 = E = A^2T/2$.

Bandpass Message Vectors
The Gram-Schmidt orthonormalization procedure or other intuitive method may be followed to obtain a complete orthonormal basis and vector representation for the bandpass signals. Normalizing the orthogonal bandpass waveforms $s_1(t)$ and $s_3(t)$ to have unit energy yields the $N = 2$-dimensional orthonormal basis:

$$\varphi_1(t) = \begin{cases} \sqrt{2/T}\cos(2\pi f_c t) \ , & 0 \le t \le T \\ 0 \ , & \text{else} \end{cases}$$

$$\varphi_2(t) = \begin{cases} \sqrt{2/T}\cos(2\pi f_c t) & , \ 0 \le t \le T/2 \\ -\sqrt{2/T}\cos(2\pi f_c t) \ , & T/2 \le t \le T \end{cases} .$$

and the bandpass message vectors are

$$\begin{aligned} \tilde{\mathbf{s}}_1 &= (\sqrt{E}, 0) \\ \tilde{\mathbf{s}}_2 &= (-\sqrt{E}, 0) \\ \tilde{\mathbf{s}}_3 &= (0, \sqrt{E}) \\ \tilde{\mathbf{s}}_4 &= (0, -\sqrt{E}) \ , \end{aligned}$$

where, once again, $E = A^2T/2$ is the energy in each bandpass waveform, such that $\|\mathbf{s}_i\|^2 = E = A^2T/2$.

The bandpass message vectors are plotted in a two-dimensional vector space, defined by the orthonormal basis $\{\varphi_1(t), \varphi_2(t)\}$, as shown in Fig. 5.3. Such a plot is called a *signal constellation* because it resembles a constellation of stars.

5.1.4 Signal Correlation and Euclidean Distance

The inner (dot) product between two length-N complex-valued vectors $\mathbf{u}$ and $\mathbf{v}$ is defined as

$$\mathbf{u} \cdot \mathbf{v}^* \triangleq \sum_{i=1}^{N} u_i v_i^* \tag{5.34}$$

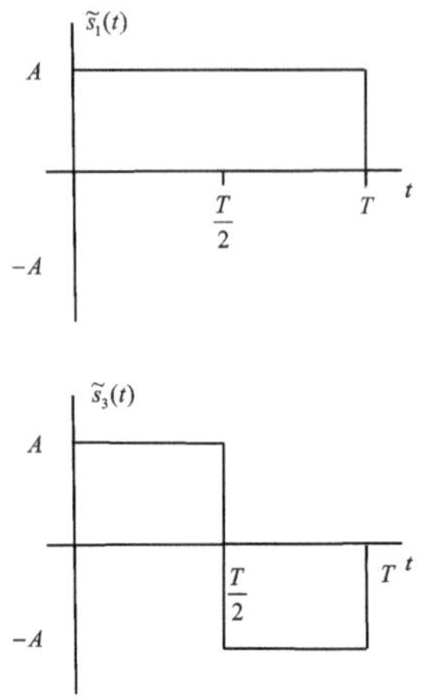

Fig. 5.1 Signal set $\tilde{S} = \{\tilde{s}_1(t), \tilde{s}_2(t), \tilde{s}_3(t), \tilde{s}_4(t)\}$

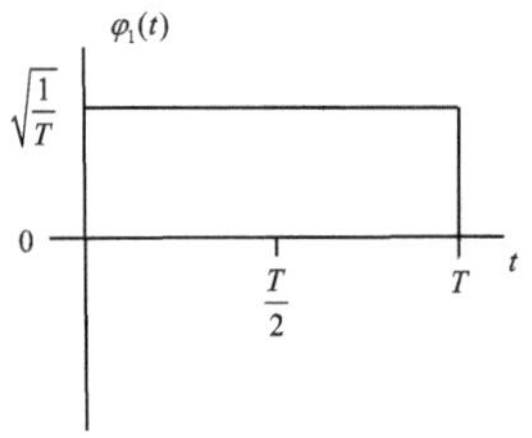

Fig. 5.2 Orthonormal basis for Example 6.3

Fig. 5.3 Signal vectors in the 2-D vector space

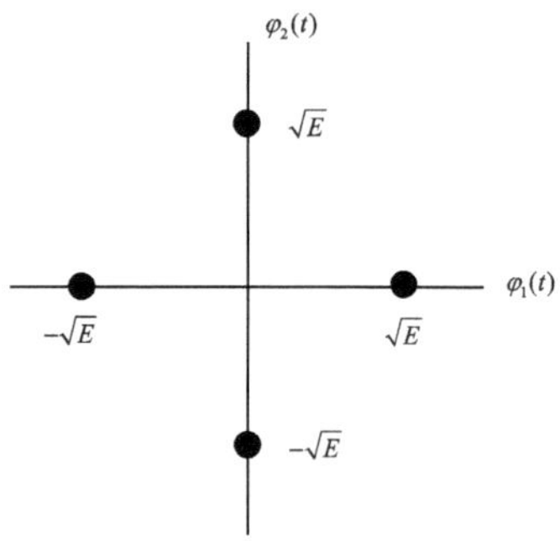

and the norm (length) of the vector $\mathbf{u}$ is defined as

$$\|\mathbf{u}\| \triangleq \sqrt{\mathbf{u} \cdot \mathbf{u}^*} = \sqrt{\sum_{i=1}^{N} |u_i|^2} \ . \tag{5.35}$$

For real-valued vectors, the conjugates in (5.34) and (5.35) may be omitted.

The *signal correlation* between the bandpass waveforms $s_j(t)$ and $s_k(t)$ is defined as

$$\rho_{jk} \stackrel{\Delta}{=} \frac{1}{\sqrt{E_j E_k}} \int_{-\infty}^{\infty} s_j(t) s_k(t) \mathrm{d}t \tag{5.36}$$

$$\begin{aligned}
&= \frac{1}{\sqrt{E_j E_k}} \int_{-\infty}^{\infty} \sum_{n=1}^{N} s_{j,n} \varphi_n(t) \sum_{m=1}^{N} s_{k,m} \varphi_m(t) \\
&= \frac{1}{\sqrt{E_j E_k}} \sum_{n=1}^{N} \sum_{m=1}^{N} s_{j,n} s_{k,m} \int_{-\infty}^{\infty} \varphi_n(t) \varphi_m(t) \\
&= \frac{1}{\sqrt{E_j E_k}} \sum_{n=1}^{N} s_{j,n} s_{k,n} \\
&= \frac{\mathbf{s}_j \cdot \mathbf{s}_k}{\|\mathbf{s}_j\| \|\mathbf{s}_k\|} \ .
\end{aligned} \tag{5.37}$$

The signal correlation between the bandpass waveforms $s_j(t)$ and $s_k(t)$ is equal to the dot product of their corresponding bandpass message vectors $\mathbf{s}_j$ and $\mathbf{s}_k$ divided by their lengths.

The signal correlation between $s_j(t)$ and $s_k(t)$ may also be obtained from their complex envelopes and complex message vectors as follows (details are left to the reader):

$$\begin{aligned}
\rho_{jk} &= \frac{1}{\sqrt{E_j E_k}} \int_{-\infty}^{\infty} s_j(t) s_k(t) \mathrm{d}t \\
&= \mathrm{Re} \left\{ \frac{1}{2\sqrt{E_j E_k}} \int_{-\infty}^{\infty} \tilde{s}_j(t) \tilde{s}_k^*(t) \mathrm{d}t \right\} \\
&= \mathrm{Re} \left\{ \frac{\tilde{\mathbf{s}}_j \cdot \tilde{\mathbf{s}}_k^*}{\|\tilde{\mathbf{s}}_j\| \|\tilde{\mathbf{s}}_k\|} \right\} \ .
\end{aligned} \tag{5.38}$$

The signal correlation gives a measure of the similarity between two message waveforms. If $\rho_{jk} = 0$, then $s_j(t)$ and $s_k(t)$ are orthogonal. If $\rho_{jk} = \pm 1$, then $s_j(t)$ and $s_k(t)$ are linearly dependent, e.g., $s_k(t) = \pm c s_j(t)$ where c is a positive scalar.

The *squared Euclidean distance* between the bandpass waveforms $s_j(t)$ and $s_k(t)$ is defined as

$$d_{jk}^2 \stackrel{\Delta}{=} \int_{-\infty}^{\infty} \left(s_j(t) - s_k(t) \right)^2 \mathrm{d}t \tag{5.39}$$

$$= \int_{-\infty}^{\infty} \left(\sum_{n=1}^{N} (s_{j,n} - s_{k,n}) \varphi_n(t) \right)^2 \mathrm{d}t$$

$$
\begin{aligned}
&= \int_{-\infty}^{\infty} \left(\sum_{n=1}^{N} (s_{j,n} - s_{k,n}) \varphi_n(t) \right) \left(\sum_{m=1}^{N} (s_{j,m} - s_{k,m}) \varphi_m(t) \right) \mathrm{d}t \\
&= \sum_{n=1}^{N} \sum_{m=1}^{N} (s_{j,n} - s_{k,n})(s_{j,m} - s_{k,m}) \int_{-\infty}^{\infty} \varphi_n(t) \varphi_m(t) \mathrm{d}t \\
&= \sum_{n=1}^{N} (s_{j,n} - s_{k,n})^2 \\
&= \|\mathbf{s}_j - \mathbf{s}_k\|^2 \ . \qquad (5.40)
\end{aligned}
$$

The squared Euclidean distance between the waveforms $s_j(t)$ and $s_k(t)$ can be interpreted as the energy in the difference signal $s_j(t) - s_k(t)$, and it is also equal to the squared length of the difference vector $\mathbf{s}_j - \mathbf{s}_k$.

The squared Euclidean distance between the bandpass waveforms $s_j(t)$ and $s_k(t)$ may also be expressed in terms of their complex envelopes and complex message vectors as follows (details are left to the reader):

$$
\begin{aligned}
d_{jk}^2 &= \int_{-\infty}^{\infty} \left(s_j(t) - s_k(t) \right)^2 \mathrm{d}t \\
&= \frac{1}{2} \int_{-\infty}^{\infty} \left| \tilde{s}_j(t) - \tilde{s}_k(t) \right|^2 \mathrm{d}t \\
&= \tfrac{1}{2} \|\tilde{\mathbf{s}}_j - \tilde{\mathbf{s}}_k\|^2 \ . \qquad (5.41)
\end{aligned}
$$

5.2 Binary Modulated Waveforms

This section introduces basic binary digital modulation schemes that transmit information in the RF carrier phase and frequency.

5.2.1 Binary PSK (BPSK)

Binary phase shift keying (BPSK) transmits information in the carrier phase. The bandpass waveform set consists of two sinusoids having a relative phase difference of π radians. That is

$$
\begin{aligned}
s_1(t) &= \sqrt{\frac{2E}{T}} \cos(2\pi f_c t) u_T(t) \\
s_2(t) &= \sqrt{\frac{2E}{T}} \cos(2\pi f_c t + \pi) u_T(t) \qquad (5.42)
\end{aligned}
$$

where $u_T(t) = u(t) - u(t - T)$ and $u(t)$ is the unit-step function. The amplitude $A = \sqrt{2E/T}$ is expressed in terms of the bandpass message energy E and baud interval T. Applying a trigonometric identity

$$\begin{aligned} s_2(t) &= \sqrt{\frac{2E}{T}} \cos(2\pi f_c t + \pi) u_T(t) \\ &= \sqrt{\frac{2E}{T}} \{\cos(2\pi f_c t)\cos(\pi) - \sin(2\pi f_c t)\sin(\pi)\} u_T(t) \\ &= -\sqrt{\frac{2E}{T}} \cos(2\pi f_c t) u_T(t) = -s_1(t) \end{aligned}$$

The BPSK waveforms are antipodal, meaning they are equal but opposite in sign. Under the assumption $f_c T \gg 1$, the waveforms have energy E. The corresponding complex envelopes are

$$\tilde{s}_1(t) = \sqrt{\frac{2E}{T}} u_T(t) = -\tilde{s}_2(t) \ . \tag{5.43}$$

The vector representation of BPSK waveforms requires only a single basis function because of the antipodal property.

Complex Message Vectors
From the set of complex envelopes $\tilde{\mathcal{S}}$, the required basis function can be readily obtained by normalizing $\tilde{s}_1(t)$ to have unit energy. Since the bandpass energy E is one-half the energy in the complex envelope setting, $E = 1/2$ in $\tilde{s}_1(t)$ yields the required basis function:

$$\varphi_1(t) = \sqrt{\frac{1}{T}} u_T(t) \ . \tag{5.44}$$

The complex message vectors are the scalars

$$\tilde{s}_1 = \sqrt{2E} = -\tilde{s}_2 \ . \tag{5.45}$$

Recall that the energy in a bandpass waveform, E, is equal to one-half the squared length of its complex message vector. Figure 5.4 shows the signal constellation for BPSK, plotting its complex message vectors.

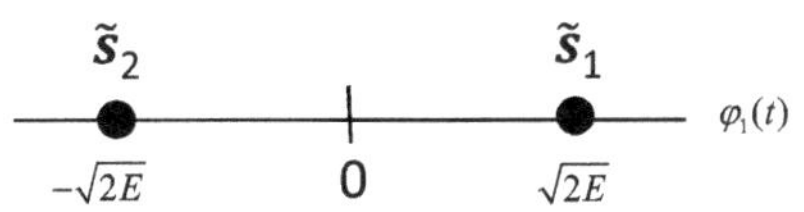

Fig. 5.4 BPSK signal constellation showing complex message vectors

Bandpass Message Vectors

For the bandpass waveforms, the required basis function can be obtained by setting $E = 1$ in the bandpass waveform $s_1(t)$, yielding

$$\varphi_1(t) = \sqrt{\frac{2}{T}} \cos(2\pi f_c t) u_T(t) \ . \tag{5.46}$$

The bandpass message vectors are

$$s_1 = \sqrt{E} = -s_2 \ . \tag{5.47}$$

5.2.2 *Binary FSK (BFSK)*

Binary frequency shift keying (BFSK) transmits information in the frequency of the RF carrier. Two sinusoids are used that have different frequencies, viz.,

$$s_1(t) = \sqrt{\frac{2E}{T}} \cos(2\pi f_c t) u_T(t) \tag{5.48a}$$

$$s_2(t) = \sqrt{\frac{2E}{T}} \cos(2\pi (f_c + \Delta_f) t) u_T(t) \tag{5.48b}$$

where Δ_f is the frequency difference between the two sinusoids. Note that $s_1(t)$ and $s_2(t)$ both have energy E under the assumption $f_c T \gg 1$, and depending on the choice of Δ_f, $s_1(t)$ and $s_2(t)$ may or may not be orthogonal. The corresponding complex envelopes are

$$\tilde{s}_1(t) = \sqrt{\frac{2E}{T}} u_T(t) \tag{5.49a}$$

$$\tilde{s}_2(t) = \sqrt{\frac{2E}{T}} \exp\left\{ j2\pi \Delta_f t \right\} u_T(t) \tag{5.49b}$$

Orthogonal BFSK is an important special case where $s_1(t)$ and $s_2(t)$ are orthogonal. For orthogonal waveforms the signal correlation $\rho = 0$. Using the definition of signal correlation for bandpass waveforms in (5.36) yields the following:

$$\begin{aligned}
\rho &= \frac{1}{\sqrt{E_1 E_2}} \int_0^T s_1(t) s_2(t) \mathrm{d}t \\
&= \frac{2}{T} \int_0^T \cos(2\pi f_c t) \cos(2\pi (f_c + \Delta_f) t) \mathrm{d}t
\end{aligned}$$

$$
\begin{aligned}
&= \frac{1}{T}\int_0^T \cos(2\pi\Delta_f)t\mathrm{d}t + \frac{1}{T}\int_0^T \cos(2\pi(2f_c+\Delta_f)t)\mathrm{d}t \\
&= \frac{1}{T}\int_0^T \cos(2\pi\Delta_f)t\ \mathrm{d}t \\
&= \frac{1}{T}\left.\frac{\sin(2\pi\Delta_f t)}{2\pi\Delta_f}\right|_0^T \\
&= \frac{\sin(2\pi\Delta_f T)}{2\pi\Delta_f T} \\
&= \mathrm{sinc}(2\Delta_f T)
\end{aligned}
\tag{5.50}
$$

Note that the cosine term having frequency $2f_c + \Delta_f$ integrates to zero in the third line under the condition $f_c T \gg 1$. For $s_1(t)$ and $s_2(t)$ to be orthogonal, the sinc(x) function must go to zero. This happens when $\Delta_f = k/(2T)$ where k is a non-zero integer. The minimum frequency separation needed to make $s_1(t)$ and $s_2(t)$ orthogonal over an interval of length T is $\Delta_f = 1/(2T)$.

Complex Message Vectors

For BFSK, two complex basis functions are needed. In the case of orthogonal BFSK with $\Delta_f = 1/(2T)$, the two complex-valued basis functions can be readily obtained by normalizing the pair of complex envelopes to have unit energy. Setting $E = 1/2$ gives:

$$\varphi_1(t) = \sqrt{\frac{1}{T}} u_T(t) \tag{5.51a}$$

$$\varphi_2(t) = \sqrt{\frac{1}{T}} \exp\left\{ j\frac{\pi t}{T} \right\} u_T(t) \tag{5.51b}$$

The complex message vectors are two-dimensional and equal to

$$\tilde{\mathbf{s}}_1 = (\sqrt{2E}, 0), \qquad \tilde{\mathbf{s}}_2 = (0, \sqrt{2E}) \tag{5.52}$$

Note that the dot product of the message vectors is zero meaning that they are orthogonal according to (5.37). Figure 5.5 plots the BFSK signal constellation, showing its complex message vectors.

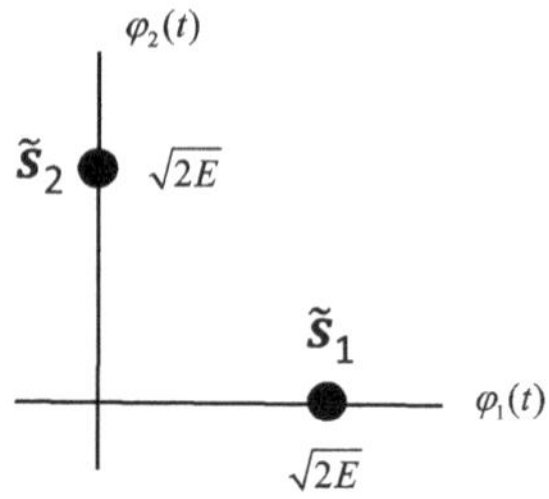

Fig. 5.5 Orthogonal BFSK signal constellation showing complex message vectors

Bandpass Message Vectors

In the case of orthogonal BFSK, the two required basis functions can be readily obtained from the pair of orthogonal bandpass waveforms by setting $E = 1$, giving

$$\varphi_1(t) = \sqrt{\frac{2}{T}} \cos(2\pi f_c t) u_T(t) \tag{5.53a}$$

$$\varphi_2(t) = \sqrt{\frac{2}{T}} \cos\left(2\pi\left(f_c + \frac{1}{2T}\right)t\right) u_T(t) \tag{5.53b}$$

The orthogonal BFSK bandpass message vectors are

$$\mathbf{s}_1 = (\sqrt{E}, 0), \qquad \mathbf{s}_2 = (0, \sqrt{E}) \tag{5.54}$$

5.3 *M*-ary Modulated Waveforms

Nonbinary or M-ary modulation schemes transmit more than one bit per channel use. Referring to the discussion of channel capacity in Chap. 1, some types of M-ary modulation schemes are bandwidth efficient with $R/W > 1$, while others are power efficient with $R/W \leq 1$.

5.3.1 *M-ary Phase Shift Keying (MPSK)*

M-ary phase shift keying (MPSK) transmits information in the RF carrier phase which takes on M possible values. The bandpass signal set S consists of M bandpass waveforms defined by

$$s_m(t) = \sqrt{\frac{2E}{T}} \cos(2\pi f_c t + \theta_m) u_T(t), \quad m = 1, \ldots, M, \tag{5.55}$$

where the random phase θ_m takes on values from the set

$$\theta_m \in \left\{ 2\pi \frac{(m-1)}{M}, \quad m = 1, \ldots, M \right\} \tag{5.56}$$

The corresponding set $\tilde{\mathcal{S}}$ of MPSK complex envelopes is given by

$$\tilde{s}_m(t) = \sqrt{\frac{2E}{T}} \mathrm{e}^{j\theta_m} u_T(t) \ , \quad m = 1, \ldots, M \ . \tag{5.57}$$

Complex Signal Vectors

The MPSK complex envelopes are all linearly dependent, and, hence, a single complex-valued basis function is required that can be obtained by normalizing $\tilde{s}_1(t)$ to have unit energy, yielding

$$\varphi_1(t) = \sqrt{\frac{1}{T}} u_T(t) \tag{5.58}$$

From (5.57), the complex signal vectors are readily obtained as the scalars

$$\tilde{s}_m = \sqrt{2E} \mathrm{e}^{j\theta_m} \ , m = 1, \ldots, M \tag{5.59}$$

The signal constellation for 8-PSK is shown in Fig. 5.6 and consists of 8 message vectors uniformly spaced on a circle of radius $\sqrt{2E}$. Recall the energy in a bandpass waveform is equal to one-half the squared length of its complex signal vector. The PSK bandpass waveforms all have energy equal to E.

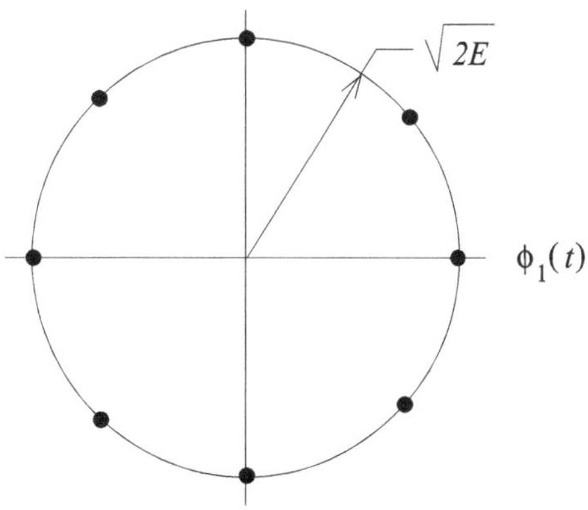

Fig. 5.6 8-PSK signal constellation showing complex message vectors

Bandpass Signal Vectors

The amplitude-phase representation in (5.55) can be expanded into the quadrature form:

$$s_m(t) = \sqrt{\frac{2E}{T}} \Big(\cos(\theta_m)\cos(2\pi f_c t) - \sin(\theta_m)\sin(2\pi f_c t)\Big)\, u_T(t) \ . \tag{5.60}$$

Since $\cos(2\pi f_c t)u_T(t)$ and $\sin(2\pi f_c t)u_T(t)$ are in quadrature or orthogonal, it is apparent that a complete orthonormal basis for MPSK is

$$\varphi_1(t) = \sqrt{\frac{2}{T}}\cos(2\pi f_c t)u_T(t) \tag{5.61a}$$

$$\varphi_2(t) = -\sqrt{\frac{2}{T}}\sin(2\pi f_c t)u_T(t) \tag{5.61b}$$

Note that the minus sign in (5.60) has been absorbed into the definition of $\varphi_2(t)$. The MPSK bandpass signal vectors are

$$\mathbf{s_m} = \sqrt{E}\Big(\cos(\theta_m),\ \sin(\theta_m)\Big) \ , m = 1, \ldots, M \ . \tag{5.62}$$

5.3.2 *M-ary Frequency Shift Keying (MFSK)*

M-ary frequency shift keying (MFSK) transmits information in the frequency of the RF carrier which takes on one of M possible values. The bandpass signal set $\mathcal{S}$ consists of M bandpass waveforms defined by

$$s_m(t) = \sqrt{\frac{2E}{T}}\cos\left(2\pi(f_c + f_m)t\right) u_T(t) \ , \quad m = 1, \ldots, M \tag{5.63}$$

where the frequency f_m takes on values from the set

$$f_m = (m-1)\Delta_f \ , \quad m = 1, \ldots, M \tag{5.64}$$

and Δ_f is the frequency separation. The corresponding MFSK complex envelopes are

$$\tilde{s}_m(t) = \sqrt{\frac{2E}{T}} e^{j2\pi f_m t} u_T(t) \ , \quad m = 1, \ldots, M \ . \tag{5.65}$$

The discussion of BFSK showed that the choice $\Delta_f = 1/(2T)$ yields waveforms that are mutually orthogonal leading to an important class of waveforms called *M-ary orthogonal FSK*.

Complex Message Vectors
The M-ary orthogonal FSK complex envelopes $\tilde{s}_m(t)$ are all mutually orthogonal, and, hence, M complex-valued basis function can be obtained by simply normalizing the $\tilde{s}_m(t)$ to have unit energy, yielding

$$\varphi_m(t) = \sqrt{\frac{1}{T}} \exp\left\{ j\frac{\pi(m-1)t}{T} \right\} u_T(t) \ , m = 1, \ldots, M \ . \tag{5.66}$$

The M-ary orthogonal FSK complex message vectors are

$$\tilde{\mathbf{s}}_m = \sqrt{2E}\, \mathbf{e}_m \ , m = 1, \ldots, M \ . \tag{5.67}$$

where $\mathbf{e}_m = (e_1, e_2, \ \ldots, e_M), \ e_j = \delta_{jm}$, is a length-$M$ vector with a "1" in the mth coordinate and zeros elsewhere.

Bandpass Message Vectors
For M-ary orthogonal FSK, the M basis functions for the bandpass waveforms can be readily obtained by setting $E = 1$ in (5.63), giving

$$\varphi_m(t) = \sqrt{\frac{2}{T}} \cos\left(2\pi \left(f_c + \frac{m-1}{2T}\right) t\right) u_T(t) \ , \quad m = 1, \ldots, M \ . \tag{5.68}$$

The M-ary orthogonal FSK bandpass signal vectors are

$$\tilde{\mathbf{s}}_m = \sqrt{E}\, \mathbf{e}_m \ , \quad m = 1, \ldots, M \ . \tag{5.69}$$

5.3.3 *Pulse Amplitude Modulation (PAM)*

Pulse amplitude modulation (PAM) transmits information in the signed amplitude of the RF carrier that takes on one of M possible values. The set $\mathcal{S}$ of M bandpass waveforms is

$$s_m(t) = \sqrt{\frac{2E_o}{T}} A_m \cos(2\pi f_c t) u_T(t) \ , \quad m = 1, \ldots, M \ , \tag{5.70}$$

where

$$A_m \in \{\pm 1, \pm 3, \pm 5, \ldots, \pm(M-1)\} \tag{5.71}$$

and E_o is the energy of the bandpass signal having the lowest amplitude, i.e., when $A_m = \pm 1$. Notice that A_m is chosen from the set of odd integers. This is necessary so the center of mass of the PAM signal constellation will be at the origin. The PAM complex envelopes are all real-valued

$$\tilde{s}_m(t) = \sqrt{\frac{2E_o}{T}} A_m u_T(t)\ , \quad m = 1, \ldots, M\ . \tag{5.72}$$

Complex Message Vectors
The PAM complex envelopes are all linearly dependent and, hence, a single basis function will suffice, given by

$$\varphi_1(t) = \sqrt{\frac{1}{T}} u_T(t) \tag{5.73}$$

The PAM complex signal vectors are readily obtained as the scalars:

$$\tilde{s}_m = \sqrt{2E_o} A_m\ , m = 1, \ldots, M\ . \tag{5.74}$$

Bandpass Message Vectors
Once again, only a single basis function is required:

$$\varphi_1(t) = \sqrt{\frac{2}{T}} \cos(2\pi f_c t) u_T(t)\ . \tag{5.75}$$

and the PAM bandpass signal vectors are readily obtained as

$$s_m = \sqrt{E_o} A_m\ , m = 1, \ldots, M\ . \tag{5.76}$$

Unlike MPSK and MFSK, the PAM messages do not have equal energy. The average energy E in a message is an important quantity since it appears in the communication link budget later in Sect. 13.4. Recall:

$$E = PT = P/R \tag{5.77}$$

where P is the power. It is useful to relate E_o to E. For this purpose, the set of A_m in (5.71) may be generated as $A_m = (2m - 1 - M)$, $m = 1, \ldots, M$. Hence, the PAM complex message vectors may be expressed as

$$\tilde{s}_m = \sqrt{2E_o}(2m - 1 - M)\ , \quad m = 1, \ldots, M \tag{5.78}$$

and the energy in $\tilde{s}_m$ is

$$E_m = \tfrac{1}{2}\tilde{s}_m^2 = E_o(2m - 1 - M)^2 \ . \tag{5.79}$$

Assuming equally likely messages, the average energy in a message is

$$\begin{aligned} E &= E_o \frac{1}{M} \sum_{m=1}^{M} (2m - 1 - M)^2 \\ &= E_o \frac{1}{M} \left(4 \sum_{m=1}^{M} m^2 - 4(M+1) \sum_{m=1}^{M} m + M(M+1)^2 \right) \ . \end{aligned} \tag{5.80}$$

Using the identities

$$\sum_{k=1}^{n} k = \frac{n(n+1)}{2} \ , \qquad \sum_{k=1}^{n} k^2 = \frac{n(n+1)(2n+1)}{6} \tag{5.81}$$

for the summation terms and simplifying yields the result

$$E = E_o \frac{M^2 - 1}{3} \quad \text{or} \quad E_o = E \frac{3}{M^2 - 1} \ . \tag{5.82}$$

Hence, the PAM complex signal vectors can be written in terms of the average message energy E as

$$\tilde{s}_m = \sqrt{\frac{6E}{M^2 - 1}} A_m \ , m = 1, \ldots, M \tag{5.83}$$

5.3.4 *Quadrature Amplitude Modulation (QAM)*

Quadrature amplitude modulation (QAM) is a bandwidth-efficient modulation scheme that is used in numerous modern communication standards. With QAM, information is transmitted in the amplitude and phase of the RF carrier. When M is a power of 4, QAM can be thought of as two independent $\sqrt{M}$-PAM message streams on the inphase (cosine) and quadrature (sine) components of the RF carrier. The set S consists of M bandpass waveforms defined by

$$s_m(t) = \sqrt{\frac{2E_o}{T}} \left(A_m^c \cos(2\pi f_c t) - A_m^s \sin(2\pi f_c t) \right) u_T(t) \ , \quad m = 1, \ldots, M \tag{5.84}$$

where

$$A_m^{\{c,s\}} \in \{\pm 1, \pm 3, \pm 5, \ldots \pm (\sqrt{M} - 1)\} \tag{5.85}$$

and $2E_o$ is the energy of the waveform having the lowest amplitude, i.e., when $A_m^c, A_m^s = \pm 1$. Note that the data $A_m^{\{c,s\}}$ modulates the amplitude of the quadrature components of the RF carrier $\cos(2\pi f_c t)u_T(t)$ and $\sin(2\pi f_c t)u_T(t)$ that are in quadrature, hence the name quadrature amplitude modulation. By defining the complex-valued data symbol

$$A_m \triangleq A_m^c + jA_m^s \tag{5.86}$$

the QAM complex envelopes are equal to

$$\tilde{s}_m(t) = \sqrt{\frac{2E_o}{T}} A_m u_T(t) \ , \quad m = 1, \ldots, M \ . \tag{5.87}$$

Complex Message Vectors
The set of QAM complex envelopes $\tilde{S}$ are all linearly dependent, and, hence, a single basis function is needed:

$$\varphi_1(t) = \sqrt{\frac{1}{T}} u_T(t) \tag{5.88}$$

The QAM complex message vectors are readily obtained as the complex scalars:

$$\tilde{s}_m = \sqrt{2E_o} A_m \ , \quad m = 1, \ldots, M \tag{5.89}$$

Bandpass Message Vectors
For QAM bandpass messages, two basis functions are required that can be obtained intuitively as

$$\varphi_1(t) = \sqrt{\frac{2}{T}} \cos(2\pi f_c t) u_T(t) \tag{5.90a}$$

$$\varphi_2(t) = -\sqrt{\frac{2}{T}} \sin(2\pi f_c t), u_T(t) \tag{5.90b}$$

and the QAM bandpass signal vectors are

$$\mathbf{s_m} = \sqrt{E_o} \left(A_m^c, A_m^s \right) \ . \tag{5.91}$$

Similar to PAM, QAM signals do not have equal energy. When M is a power of 4, the average energy of the independent $\sqrt{M}$-PAM symbols on the inphase and quadrature branches of the RF carrier are equal and from (5.82), the average energy per QAM symbol is

$$E = \frac{\left(\sqrt{M}\right)^2 - 1}{3} E_o = \frac{M-1}{3} E_o \quad \text{or} \quad E_o = \frac{3E}{M-1} \; . \tag{5.92}$$

Hence, the QAM complex message vectors may be rewritten as

$$\tilde{s}_m = \sqrt{\frac{6E}{M-1}} A_m \; , \quad m = 1, \ldots, M \tag{5.93}$$

QAM Signal Constellations

When M is a power of 4, square-shaped QAM signal constellations are obtained. The signal constellations for 4, 16, and 64 QAM are shown in Fig. 5.7 where complex message vectors are plotted. Notice that the minimum Euclidean distance between any two complex message vectors is $2\sqrt{2E_o}$. When M is an odd power of 2, cross-shaped QAM signal constellations are obtained. The signal constellations for 32- and 128-QAM are shown in Fig. 5.8 where, again, complex message vectors are plotted. The signal constellations for the QAM bandpass waveforms appear similar except that the minimum Euclidean distance between any two message vectors is $2\sqrt{E_o}$.

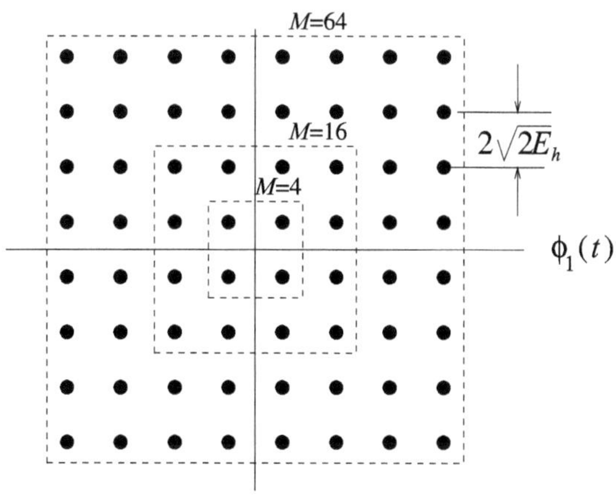

Fig. 5.7 Square-shaped signal constellations for 4-, 16-, and 64-QAM showing complex message vectors

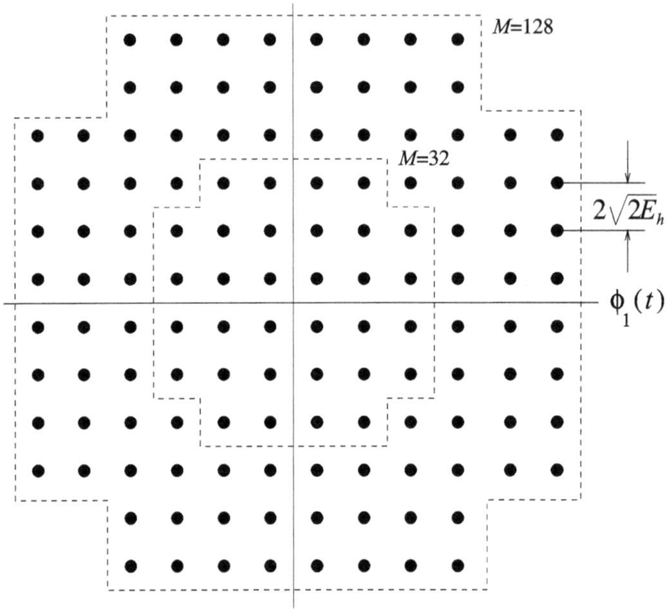

Fig. 5.8 Cross-shaped signal constellations for 32- and 128-QAM showing complex message vectors

5.4 Nyquist Pulse Shaping

Up to this point, all the modulated waveforms have used the unit-energy rectangular amplitude shaping pulse $h_a(t) = \sqrt{1/T}u_T(t)$, where $u_T(t) = u(t)-u(t-T)$. Such a pulse ensures that the modulation symbols are transmitted as waveforms having a duration equal to the baud period T. Hence, the waveforms representing successive modulation symbols will not overlap in the time domain. However, the spectrum of the rectangular shaping pulse given by its Fourier transform

$$\sqrt{1/T}u_T(t) \leftrightarrow \sqrt{T}\mathrm{sinc}(fT)\exp\{-j\pi fT\}$$

which is not bandlimited and has low-bandwidth efficiency due to the high sidelobes of the sinc(fT) function. For high-bandwidth efficiency, shaping pulses are desired that are strictly bandlimited. However, a strictly bandlimited pulse will not be time-limited and can potentially lead to intersymbol interference (ISI) because the waveforms of successive data symbols will overlap. It is desirable to construct a family of shaping pulses, $h_a(t)$, which are both bandlimited and free of ISI. A strictly bandlimited pulse is strictly speaking noncausal in the time domain, and, therefore, it must be approximated in practice by finite-length causal pulse having sufficient length.

As detailed in Chap. 6, when the transmitter uses a shaping pulse $h_a(t)$ of duration $T_o = LT$, the receiver will employ a *matched filter* having the form $h_r(t) = h_a(T_o - t)$, followed by a sampler and further digital signal processing. The overall arrangement is shown in Fig. 5.9. The matched filter $h_a(T_o - t)$ minimizes the effects of receiver thermal noise by maximizing the peak pulse signal-to-noise ratio at the output of the receiver filter $h_r(t)$ at the sampling instant T_o. The parameter $T_o = LT$ is the length of the shaping pulse $h_a(t)$ and also represents the delay needed to make the matched filter a causal filter.

Assuming a linear time-invariant channel, the combination of the transmitter filter $h_a(t)$, channel $\tilde{h}(t)$, and receiver matched filter $h_r(t) = h_a(T_o - t)$ arranged in cascade results in the overall pulse

$$p(t) = h_a(t) * \tilde{h}(t) * h_r(t) \tag{5.94}$$

Consider the illustrative case of QAM and let:

$$x_n \triangleq \sqrt{\frac{2E_o}{T}} A_n \ , \tag{5.95}$$

where n is the discrete-time index and x_n represents the nth complex-valued QAM symbol. Ignoring noise, the continuous-time complex envelope at the output of the receiver filter is

$$\tilde{y}(t) = \sum_n x_n p(t - nT) \ . \tag{5.96}$$

The complex envelope $\tilde{y}(t)$ is sampled every T seconds to produce decision variables that are used to decide on the sequence of transmitted messages. With perfect sampler timing phase (explained in more detail later), samples of $\tilde{y}(t)$ are taken at time instants $t = kT$. This yields the complex-valued sample sequence $\{\tilde{y}_k\}$, where

$$\tilde{y}_k = \tilde{y}(kT) = \sum_n x_n p(kT - nT) = \sum_n x_n p((k - n)T) \ . \tag{5.97}$$

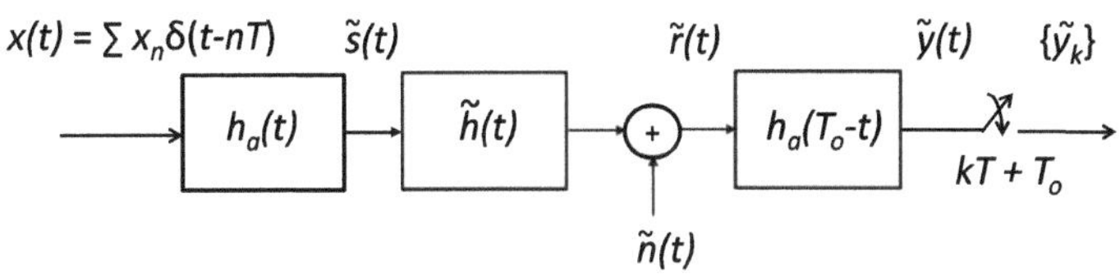

Fig. 5.9 Receiver processing with a filter matched to the transmitted pulse

Defining $p_k \stackrel{\Delta}{=} p(kT)$ as the sampled continuous-time pulse $p(t)$, the above can be written as

$$\begin{aligned}\tilde{y}_k &= \sum_n x_n p_{k-n} \\ &= x_k p_0 + \sum_{n \neq k} x_n p_{k-n} \ ,\end{aligned} \tag{5.98}$$

The first term in (5.98) is the kth message x_k scaled by p_0. The energy in the QAM symbols can be maintained as defined in (5.95) by normalizing $p_0 = 1$. The second term in (5.98) is the contribution of all data symbols other than the kth data symbol to the sample $\tilde{y}_k$. This term is called *intersymbol interference* (ISI). To avoid ISI, the sampled overall pulse response $\{p_k\}$ must satisfy the condition:

$$p_k = \begin{cases} 1 & k = 0 \\ 0 & k \neq 0 \end{cases} = \delta_{k0} \ , \tag{5.99}$$

where δ_{jk} is the Dirac delta function. This requirement is known as the *(first) Nyquist criterion*. If the first Nyquist criterion is satisfied and $p_0 = 1$, then

$$\tilde{y}_k = x_k \ . \tag{5.100}$$

It follows that any continuous-time pulse $p(t)$ satisfying (5.99) results in zero ISI. Conversely, if $p(t)$ does not satisfy (5.99), there will be ISI. An equivalent requirement in the frequency domain is given by the following important theorem due to Nyquist.

Theorem 5.1 *A continuous-time pulse $p(t)$ satisfies the condition $p_k = \delta_{k0}$ if and only if*

$$P_\Sigma(f) \stackrel{\Delta}{=} \frac{1}{T} \sum_{n=-\infty}^{\infty} P\left(f + \frac{n}{T}\right) = 1 \ . \tag{5.101}$$

The function $P_\Sigma(f)$ is called the folded spectrum, *and the ISI is zero if and only if the folded spectrum is* flat, *i.e., is a constant value in the frequency variable f.*

Proof To prove the above theorem, the inverse Fourier transform leads to

$$\begin{aligned}p_k = p(kT) &= \int_{-\infty}^{\infty} P(f) \mathrm{e}^{j2\pi f kT} \mathrm{d}f \\ &= \sum_{n=-\infty}^{\infty} \int_{(2n-1)/2T}^{(2n+1)/2T} P(f) \mathrm{e}^{j2\pi kfT} \mathrm{d}f\end{aligned}$$

$$
\begin{aligned}
&= \sum_{n=-\infty}^{\infty} \int_{-1/2T}^{1/2T} P\left(f' + \frac{n}{T}\right) e^{j2\pi k(f'+\frac{n}{T})T} \mathrm{d}f' \\
&= \int_{-1/2T}^{1/2T} \left[\sum_{n=-\infty}^{\infty} P\left(f + \frac{n}{T}\right) \right] e^{j2\pi f kT} \mathrm{d}f \\
&= T \int_{-1/2T}^{1/2T} P_{\Sigma}(f) e^{j2\pi f kT} \mathrm{d}f \ .
\end{aligned} \tag{5.102}
$$

Since $P_{\Sigma}(f)$ is periodic in f with period $1/T$, it follows that the last line in (5.102) represents a Fourier series analysis equation except for the sign of the exponential term. Note that the Fourier series in this case is being used to represent a periodic function in the frequency domain variable f instead of the time-domain variable t.

It follows that $\{p_{-k}\}$ and $P_{\Sigma}(f)$ are a Fourier series pair, and $P_{\Sigma}(f)$ can be constructed from $\{p_{-k}\}$ by using the Fourier series synthesis equation, viz.,

$$
P_{\Sigma}(f) = \sum_{k=-\infty}^{\infty} p_{-k} e^{j2\pi kfT} = \sum_{k=-\infty}^{\infty} p_k e^{-j2\pi kfT} \ . \tag{5.103}
$$

To prove that (5.101) is a sufficient condition for ISI-free transmission, suppose that (5.101) holds true. Then $P_{\Sigma}(f) = 1$ and from the last line of (5.102)

$$
p_k = T \int_{-1/2T}^{1/2T} e^{j2\pi f kT} \mathrm{d}f = \frac{\sin \pi k}{\pi k} = \delta_{k0} \ . \tag{5.104}
$$

To prove that (5.101) is a necessary condition for ISI-free transmission, suppose that $p_k = \delta_{k0}$ holds true. Then from (5.103) $P_{\Sigma}(f) = 1$, and the folded spectrum $P_{\Sigma}(f)$ is flat. □

The necessity of a flat folded spectrum in (5.101) can be exploited to design pulses in the frequency domain that will yield ISI-free transmission. First consider the pulse $p_N(t) \leftrightarrow P_N(f)$ where

$$
P_N(f) = T\mathrm{rect}(fT) \ , \tag{5.105}
$$

and

$$
\mathrm{rect}(fT) = \begin{cases} 1 \ , & |f| \leq \frac{1}{2T} \\ 0 \ , & \text{elsewhere} \end{cases} \ . \tag{5.106}
$$

This pulse has a flat folded spectrum as shown in Fig. 5.10. The corresponding time domain pulse is

$$
p_N(t) = \mathrm{sinc}(t/T) \tag{5.107}
$$

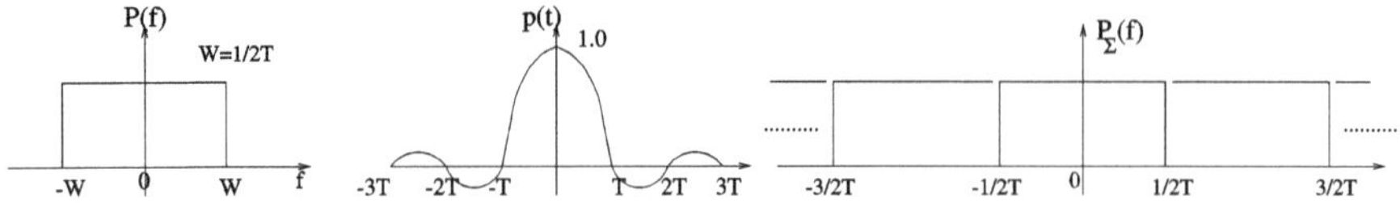

Fig. 5.10 The ideal Nyquist pulse

which yields zero ISI because it has equally spaced zero crossings at T second intervals. The pulse $p_N(t)$ happens to achieve zero ISI while occupying the smallest possible bandwidth $W = 1/2T$. Hence, $p_N(t)$ is called an *ideal Nyquist pulse*. The edge frequency $W = 1/2T = R/2$ is called the *Nyquist frequency*. Recall that the Nyquist or Shannon sampling rate for a signal having highest frequency W is $R = 2W$.

Although samples at the output of the receiver matched filter are taken every T seconds, they must also be taken at the correct time instants to avoid the introduction ISI. This is done in practice through a process known as symbol or baud timing synchronization. Baud timing synchronization will never be perfect, and there will exist a sampler timing phase error t_o, $-T/2 < t_o \leq T/2$, the smaller the better. The effect of this timing error may be illustrated when the overall pulse $p(t)$ is the ideal Nyquist pulse $p_N(t)$ in (5.107). Figure 5.11 shows a sample function of $\tilde{y}(t)$ for the case of BPSK or binary PAM with the data sequence $\{+1, -1, +1\}$. There is no ISI at the correct sampling instants $t = kT$ because of the T-spaced zero crossings of the sinc function. However, with a sampler timing offset t_o, samples are taken instead at time intervals $t = kT + t_o$ yielding

$$\begin{aligned}\tilde{y}_k &= \sum_n x_n \text{sinc}((kT + t_o - nT)/T) \\ &= x_k \text{sinc}(t_o/T) + \sum_{n \neq k} x_n \text{sinc}((kT + t_o - nT)/T)\end{aligned}$$

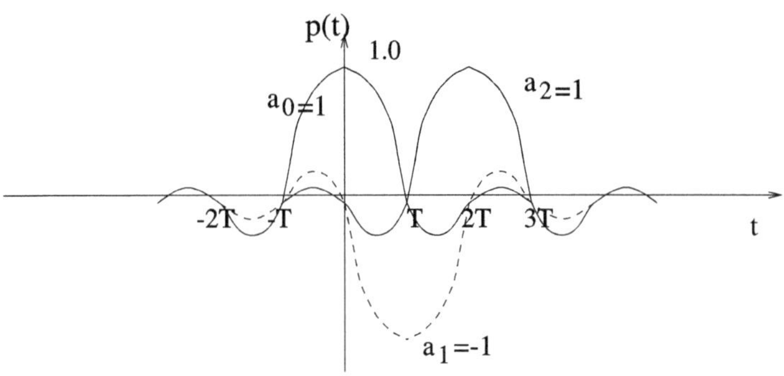

Fig. 5.11 Overlapping ideal Nyquist pulses

$$= x_k \mathrm{sinc}(t_o/T) + \sum_{n \neq k} x_n \frac{\sin(\pi(t_o + (k-n)T)/T)}{\pi(t_o + (k-n)T)/T}$$

$$= x_k \mathrm{sinc}(t_o/T) + \sum_{n \neq k} x_n \frac{\sin(\pi t_o/T)\cos(\pi(k-n)) - \cos(\pi t_o/T)\sin(\pi(k-n))}{\pi(t_o + (k-n)T)/T}$$

$$= x_k \mathrm{sinc}(t_o/T) + \sum_{n \neq k} x_n \frac{(-1)^{k-n}\sin(\pi t_o/T)}{\pi t_o/T + \pi(k-n)}$$

$$= x_k \mathrm{sinc}(t_o/T) + \frac{\sin(\pi t_o/T)}{\pi} \sum_{n \neq k} x_n \frac{(-1)^{k-n}}{t_o/T + (k-n)} \tag{5.108}$$

Observe that the sampler timing phase error introduces two harmful effects. First, the symbol to be detected, x_k, is multiplied by the $\mathrm{sinc}(t_o/T)$ term which for non-zero t_o is less than unity. This reduces the effective received symbol energy by the factor $\mathrm{sinc}^2(t_o/T)$. Second, when t_o is non-zero, ISI is introduced because the summation term is non-zero. With an ideal Nyquist pulse, the summation is not always absolutely summable because the tails of the ideal Nyquist pulse $p_N(t)$ decay as $1/t$. Recall that the infinite series $\sum_{k=1}^{\infty} 1/k$ does not converge. To reduce the effect of ISI due to sampler timing phase errors, it is desirable that $p(t)$ satisfy the first Nyquist criterion while having tails that decay faster than $1/t$.

Nyquist Pulse Design

The construction of Nyquist pulses begins with the ideal Nyquist pulse, $P_N(f)$, as shown by example in Fig. 5.12. To the pulse $P_N(f)$, a "transmittance" function

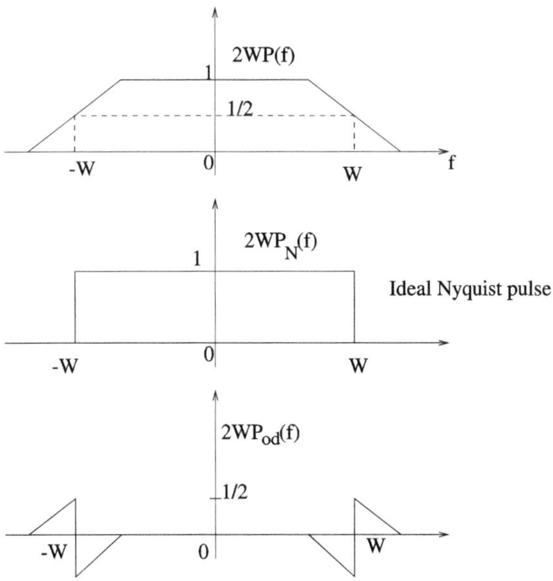

Fig. 5.12 Construction of pulses satisfying the first Nyquist criterion

$P_{od}(f)$ is added as shown. The transmittance function has *skew* symmetry about the Nyquist frequency W, and any function with skew symmetry will do. For the particular transmittance function in Fig. 5.12, the resulting Nyquist pulse $P(f)$ has a trapezoidal shape as shown. The folded spectrum $P_\Sigma(f)$ corresponding to $P(f)$ in Fig. 5.12 is shown in Fig. 5.13 and is observed to be flat. Hence, the corresponding continuous-time pulse $p(t)$ has T-spaced zero crossings, a fact that can be established without having to take the inverse Fourier transform of $P(f)$.

The folded spectrum $P_\Sigma(f)$ is flat if the transmittance function is skew symmetric about the Nyquist frequency W. The corresponding time-domain pulse $p(t)$ can be obtained by taking the inverse Fourier transform of $P(f)$. Notice that the pulse $P(f)$ consumes additional bandwidth beyond the Nyquist frequency, so there is a cost in spectral efficiency. However, the additional bandwidth will yield a time-domain pulse $p(t)$ having tails that decay faster with time than $1/t$ with the ideal Nyquist pulse and, hence, be less susceptible to sampler timing phase errors.

5.4.1 Raised Cosine Pulse

The raised cosine pulse has the transmittance function:

$$2W P_{\text{od}}(f) = \begin{cases} -\frac{1}{2} - \frac{1}{2}\sin\left(\frac{\pi(|f|-W)}{2f_x}\right) & W - f_x \leq |f| \leq W \\ \frac{1}{2} - \frac{1}{2}\sin\left(\frac{\pi(|f|-W)}{2f_x}\right) & W \leq |f| \leq W + f_x \end{cases} \tag{5.109}$$

where f_x Hz is the bandwidth expansion and $f_x/W \times 100\ \%$ is the excess bandwidth. The raised cosine pulse is

$$2WP(f) = \begin{cases} 1 & 0 \leq |f| \leq W - f_x \\ \frac{1}{2} - \frac{1}{2}\sin\left(\frac{\pi(|f|-W)}{2f_x}\right) & W - f_x \leq |f| \leq W + f_x \\ 0 & |f| \geq W + f_x \end{cases} \tag{5.110}$$

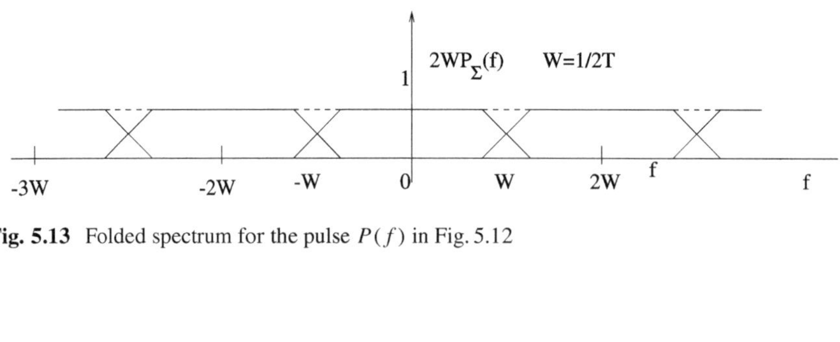

Fig. 5.13 Folded spectrum for the pulse $P(f)$ in Fig. 5.12

The raise cosine pulse is often defined in terms of the *roll off factor* $\beta \triangleq f_x/W,\ 0 \leq \beta \leq 1$, as

$$2WP(f) = \begin{cases} 1 & , \ 0 \leq |f| \leq (1-\beta)W \\ \frac{1}{2}\left[1 - \sin\left(\frac{\pi|f|}{2W\beta} - \frac{\pi}{2\beta}\right)\right] & , \ (1-\beta)W \leq |f| \leq (1+\beta)W \\ 0 & , \ |f| \geq (1+\beta)W \end{cases} . \tag{5.111}$$

The bandwidth of the raised cosine pulse is $(1+\beta)W$.

As illustrated in Fig. 5.14, the term "raised cosine" comes from the fact that pulse spectrum $P(f)$ with $\beta = 1$ reduces to

$$2WP(f) = \frac{1}{2}\left[1 + \cos\left(\frac{\pi f}{2W}\right)\right], \ \ 0 \leq |f| \leq 2W \ . \tag{5.112}$$

which has the shape of a "raised cosine" function defined on the interval $0 \leq |f| \leq 2W$.

The inverse Fourier transform of $P(f)$ in (5.111) can be computed by starting with the inverse Fourier transform integral and simplifying to give the time-domain pulse

$$p(t) = \frac{\sin(\pi t/T)}{\pi t/T} \frac{\cos(\beta\pi t/T)}{1-(2\beta t/T)^2} \tag{5.113}$$

which is plotted in Fig. 5.15. For $\beta = 0$, $p(t)$ reduces to the ideal Nyquist pulse $p_N(t) = \text{sinc}(t/T)$ as expected. Observe that the tails of the raised cosine pulse with $\beta = 1$ decay as $1/t^3$ which is much faster than the $1/t$ decay of the ideal Nyquist pulse. This means that sampler timing phase errors will introduce much less ISI.

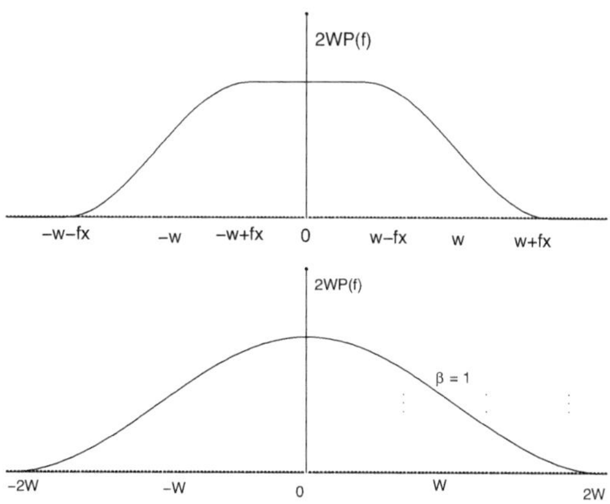

Fig. 5.14 Raised cosine pulse $P(f)$ for $0 < \beta < 1$ and $\beta = 1$

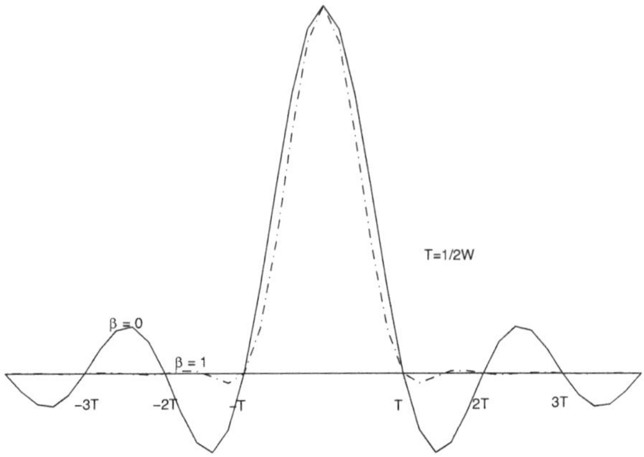

Fig. 5.15 Raised cosine pulse $p(t)$ for $\beta = 0$ and $\beta = 1$

5.4.2 Root-Raised Cosine Pulse

Consider again the arrangement in Fig. 5.9 with an ideal channel. For the time being, set the receiver filter time delay $T_o = 0$. In this case, the overall pulse produced by the cascade of the transmitter and receiver matched filters is $p(t) = h_a(t) * h_a(-t)$. It follows that the Fourier transform of $p(t)$ is $P(f) = H_a(f)H_a^*(f) = |H_a(f)|^2$. Hence, the transmitter pulse shaping filter and receiver matched filter both have the same magnitude response $|H_a(f)| = |P(f)|^{1/2}$. If $P(f)$ is a Nyquist pulse, $|H_a(f)| = |P(f)|^{1/2}$ is said to be a root Nyquist pulse.

For the raised cosine pulse with $P(f)$ defined in (5.111), the pulse $H_a(f)$ is said to be a root-raised cosine pulse. Taking the inverse Fourier transform of the frequency domain pulse $|H_a(f)| = |P(f)|^{1/2}$ with $P(f)$ defined in (5.111) yields the root-raised cosine pulse time-domain pulse:

$$h_a(t) = \begin{cases} 1 - \beta + 4\beta/\pi & , \ t = 0 \\ (\beta/\sqrt{2})\left((1 + 2/\pi)\sin(\pi/4\beta) + (1 - 2/\pi)\cos(\pi/4\beta)\right) & , \ t = \pm T/4\beta \\ \frac{4\beta(t/T)\cos((1+\beta)\pi t/T)+\sin((1-\beta)\pi t/T)}{\pi(t/T)(1-(4\beta t/T)^2)} & , \ \text{elsewhere} \end{cases} \tag{5.114}$$

The values at $t = 0$ and $t = \pm T/4\beta$ may be obtained by applying l'Hôpital's rule. For $\beta = 0$, the root-raised cosine pulse reduces to the ideal root Nyquist pulse:

$$h_a(t) = (1/\sqrt{T})\text{sinc}(t/T) \ . \tag{5.115}$$

While the raised cosine pulse is a Nyquist pulse, the root-raised cosine pulse by itself is not a Nyquist pulse and does not have equally spaced zero crossings except when $\beta = 0$.

The root-raised cosine pulse as defined in (5.114) is symmetric about $t = 0$, and is noncausal and not physically realizable. In practice, a truncated and time-shifted approximation of the pulse may be used instead, where the time domain pulse $h_a(t)$ in (5.114) is truncated, for example, with a length-LT rectangular window, and then right-shifted by $LT/2$ seconds to create the causal pulse:

$$\hat{h}_a(t) = h_a(t - LT/2)\text{rect}\left(\frac{t - LT/2}{LT}\right) . \tag{5.116}$$

Figure 5.16 shows the result for the case when $L = 6$. The convolution of two such length-LT root-raised cosine pulses will approximate a length-$2LT$ raised cosine pulse, $\hat{p}(t)$, on the interval $[0, 2LT)$, having a peak at time LT. The sampled impulse response $\hat{p}_k = \hat{p}(kT)$ has a peak at $k = L$.

Taking the inverse Fourier transform of the pulse in (5.116) results in

$$\hat{H}_a(f) = H_a(f) * LT\text{sinc}(LTf)\text{e}^{j\pi LTf} . \tag{5.117}$$

Due to the frequency domain convolution of the bandlimited pulse $H_a(f)$ with the $LT\text{sinc}(LTf)$ function, the pulse $\hat{H}_a(f)$ is no longer strictly bandlimited. However, as the truncation length LT tends to infinity, $LT\text{sinc}(LTf)$ approaches a Dirac delta function $\delta(f)$ so that $\hat{H}_a(f) = H_a(f)$ as expected.

Finally, if a complex-vauled white Gaussian noise process $\tilde{w}(t)$ having the one-sided noise spectral density $S_{ww}(f) = N_o$ watts/Hz, is input to any root Nyquist

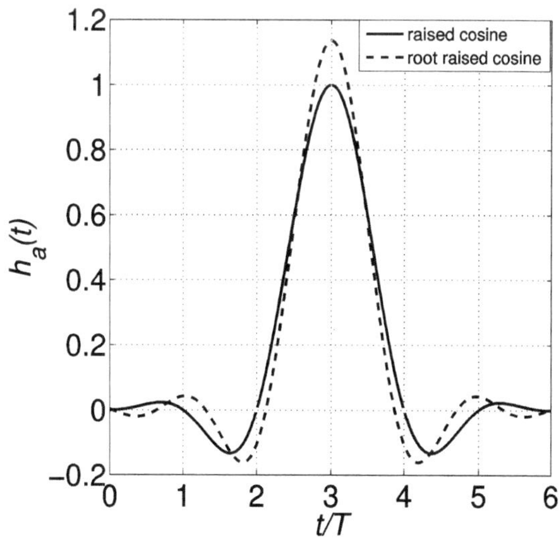

Fig. 5.16 Raised cosine $p(t)$ and root-raised cosine $h_a(t)$ pulses with roll-off factor $\beta = 0.5$. The pulses are truncated to length $6T$ and right time-shifted by $3T$ seconds to yield causal pulses

filter $h_a(t)$, the complex-valued Gaussian noise at the filter output, $\tilde{n}(t)$, has power spectral density:

$$S_{\tilde{n}\tilde{n}}(f) = S_{ww}(f)\,|H_a(f)|^2 = N_o P(f)\,. \tag{5.118}$$

The autocorrelation of the output noise $\tilde{n}(t)$ is

$$\phi_{\tilde{n}\tilde{n}}(\tau) = N_o p(\tau)\,. \tag{5.119}$$

Since $p(\tau)$ is a Nyquist pulse, T-spaced samples of the output noise $\tilde{n}_k = \tilde{n}(kT)$ are uncorrelated and, because they are complex Gaussian, are statistically independent.

5.4.3 Eye Diagram

One practical and effective method to observe and assess the effect of ISI is the *eye diagram*. Consider the binary antipodal waveform:

$$\tilde{s}(t) = \sum_n x_n h_a(t - nT)$$

where $x_n \in \{-1, +1\}$ and $h_a(t)$ is a root-raised cosine pulse having roll-off factor β. Assuming an ideal and noiseless channel, the waveform at the output of the receiver matched filter $h(t) = h_a(T_o - t)$ is

$$\tilde{y}(t) = \sum_n x_n p(t - nT)$$

where $p(t) = h_a(t) * h_a(T_o - t)$ is the corresponding raised cosine pulse.

Suppose the waveform $\tilde{y}(t)$ is probed using a virtual oscilloscope where the horizontal sweep rate is set at a multiple of the bit rate $R = 1/T$. What is displayed is called an eye diagram. For illustrative purposes, the eye diagram is shown in Figs. 5.17, 5.18, 5.19, and 5.20 for the roll-off factors $\beta = 0, 0.35, 0.5, 1$ and a sweep rate of $1/2T$. The plots are comprised of many disjoint $2T$-length segments of the waveform $\tilde{y}(t)$ that are overlapped or superimposed. The portion in the middle of each figure resembles and eye, and the eye diagram is periodic with period T.

The eye diagram shows that samples taken at the correct sampling instants $t = kT$ have values ± 1 for any roll-off factor β since $p(t)$ is a Nyquist pulse. However, the eye becomes more open with increasing β. This will reduce the ISI caused by sampler timing phase errors and make the overall system more robust.

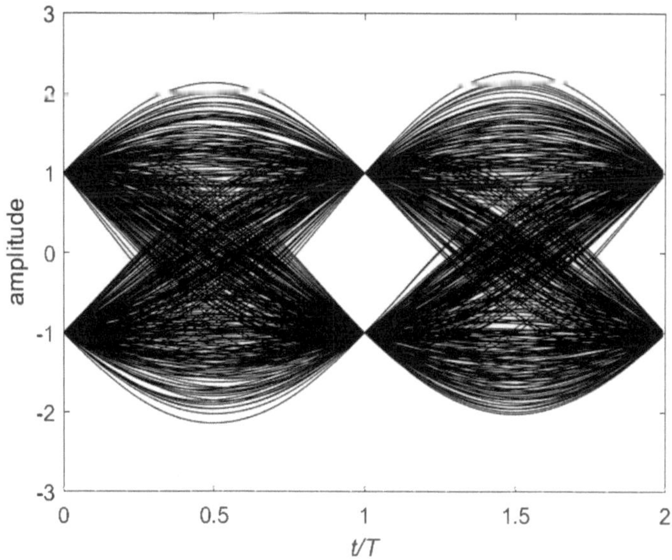

Fig. 5.17 Eye diagram for an ideal Nyquist pulse

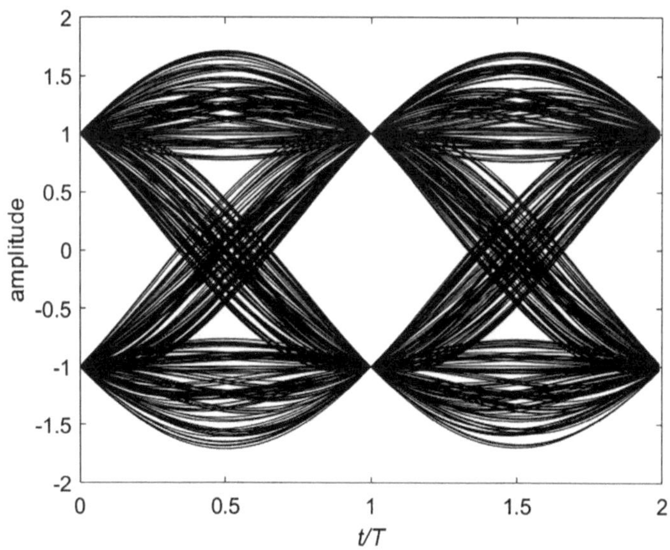

Fig. 5.18 Eye diagram for raised cosine pulse shaping with $\beta = 0.35$

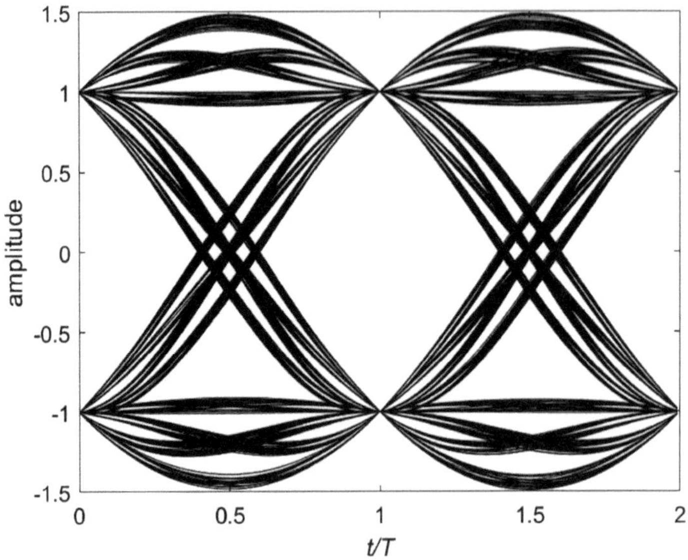

Fig. 5.19 Eye diagram for raised cosine pulse shaping with $\beta = 0.5$

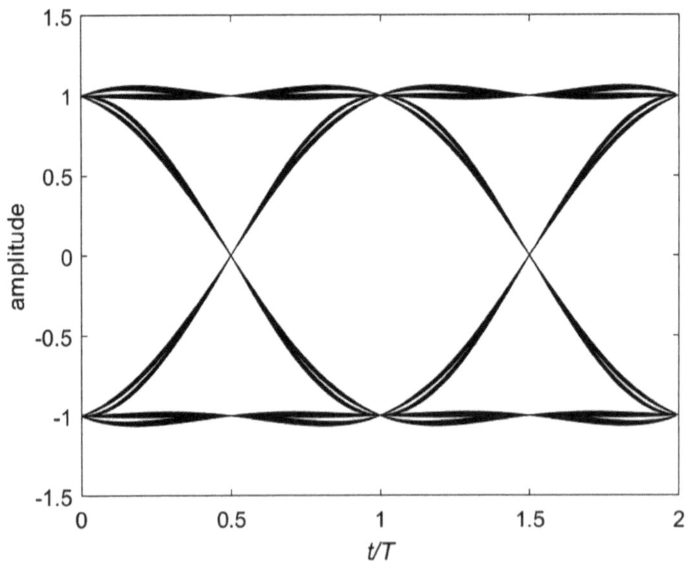

Fig. 5.20 Eye diagram for raised cosine pulse shaping with $\beta = 1.0$

5.5 Power Spectral Density

The power spectral density (psd) of modulated waveforms is important in the context of bandwidth efficiency. Consider the complex envelope:

$$\tilde{s}(t) = \sum_n x_n h_a(t - nT) \ , \tag{5.120}$$

where n is the discrete-time index, $h_a(t)$ is the amplitude shaping pulse and x_n depends on the particular modulation scheme being used. The shaping pulse is assumed to have unit energy:

$$\int_{-\infty}^{\infty} h_a^2(t)\mathrm{d}t = \int_{-\infty}^{\infty} |H_a(f)|^2 \, \mathrm{d}t = 1$$

For PSK, PAM, and QAM, x_n takes on the values:

$$x_n = \begin{cases} \sqrt{2E}\mathrm{e}^{j\theta_n} & \text{BPSK, QPSK, MPSK} \\ \sqrt{\frac{6E}{M^2-1}}A_n & \text{PAM} \\ \sqrt{\frac{6E}{M-1}}A_n & \text{QAM} \\ \sqrt{2E}\mathrm{e}^{j2\pi f_n t} & \text{MFSK} \end{cases} \tag{5.121}$$

The complex envelope is a cyclostationary random process, and its autocorrelation and psd have been derived in Sect. 3.7.

Uncorrelated Data

For the case of an uncorrelated data sequence $\{x_n\}$

$$\phi_{xx}[k] = \begin{cases} \sigma_x^2 + \frac{1}{2}|\mu_x|^2 & k = 0 \\ \frac{1}{2}|\mu_x|^2 & k \neq 0 \end{cases} \tag{5.122}$$

where $\sigma_x^2 = \frac{1}{2}\mathrm{E}[|x_k|^2]$ is the modulation symbol variance. In this case, the psd in (3.51) is separable into two terms, viz.,

$$S_{\tilde{s}\tilde{s}}(f) = \frac{\sigma_x^2}{T}P(f) + \frac{|\mu_x|^2}{2T}P(f)\sum_{k=-\infty}^{\infty} \mathrm{e}^{-j2\pi kfT} \ , \tag{5.123}$$

where $P(f) = |H_a(f)|^2$. The following identity is now applied:

$$T\sum_{k=-\infty}^{\infty} \mathrm{e}^{-j2\pi fkT} = \sum_{n=-\infty}^{\infty} \delta\left(f - \frac{n}{T}\right) \tag{5.124}$$

to write

$$S_{\tilde{s}\tilde{s}}(f) = \frac{\sigma_x^2}{T} P(f) + \frac{|\mu_x|^2}{2T^2} \sum_{n=-\infty}^{\infty} P(f)\delta\left(f - \frac{n}{T}\right) \tag{5.125}$$

The psd in (5.125) has both continuous and discrete (line spectrum) components. The discrete line spectrum has tones that are spaced at $1/T$ Hz apart.

Generally, the discrete line spectrum is undesirable since it does not result in good power or bandwidth efficiency. However, the discrete component vanishes when the mean of the signal constellation is zero, i.e., $\mu_x = 0$. If $\mu_x = 0$, then $\sigma_x^2 = E$ where $E = PT$ with P being power, and the psd reduces to the simple form:

$$S_{\tilde{s}\tilde{s}}(f) = P \cdot P(f) \ . \tag{5.126}$$

For example, if $h_a(t) = \sqrt{1/T} u_T(t)$, then

$$S_{\tilde{s}\tilde{s}}(f) = P\left(\frac{\sin(\pi f T)}{\pi f T}\right)^2 \ . \tag{5.127}$$

For root-raised cosine pulse shaping with $h_a(t)$ in (5.15), $P(f)$ is the raised cosine defined in (5.111). For MPSK, PAM, and QAM, the complex envelop has bandwidth $B = (1+\beta)/T$, where $0 \le \beta \le 1$ is the roll-off factor. The smallest possible RF bandwidth is $W = 1/T = R$ Hz, where R is the baud rate. Meanwhile, the bit rate R_b and baud rate are related as $R_b = (\log_2 M)R$ bits/s. Dividing the two gives the bandwidth efficiency as $R/W = \log_2 M$ bits/s/Hz which grows with the logarithm of M. Hence, MPSK, PAM, and QAM are said to be bandwidth efficient.

As mentioned earlier, the root-raised cosine pulse is non-causal. When the practical form in (5.117) is used instead, appropriately normalized to have unit energy, then

$$S_{\tilde{s}\tilde{s}}(f) = P\,|H_a(f) * LT\mathrm{sinc}(LTf)|^2 \ . \tag{5.128}$$

Figure 5.21 plots the psd for the waveform in (5.120) with the pulse $h_a(t)$ in (5.117), showing that the pulse truncation leads to side lobe regeneration. Note that the psd in this case is symmetric about $f = 0$, but only the positive frequencies are shown.

5.5.1 Orthogonal MFSK

Orthogonal MFSK waveforms $s_i(t), \ , i = 1, \ldots, M$ have frequencies separated by $\Delta_f = 1/2T$ Hz, the minimum frequency separation necessary to ensure orthogonal waveforms. To center the psd of the MFSK complex envelope at the origin, the set

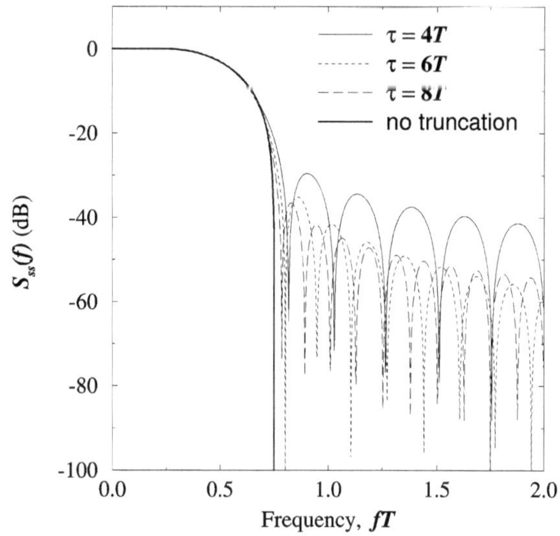

Fig. 5.21 Psd of linear modulation with uncorrelated zero-mean data and a truncated square root-raised cosine amplitude shaping pulse with various truncation lengths; $\beta = 0.5$. Truncation of the amplitude shaping pulse leads to side lobe regeneration

of frequencies $k/2T, k = 0, 1, \ldots, M-1$ can be left-shifted by $(M-1)/4T$ as

$$f_k = \frac{k}{2T} - \frac{(M-1)}{4T}, \quad k = 0, 1, \ldots, M-1$$

MFSK typically uses rectangular pulse shaping $h_a(t) = \sqrt{1/T} u_T(t)$ such that $|H_a(f)|^2 = T \text{sinc}^2(fT)$ which is not bandlimited and has nulls in its spectrum at $k/T,\ k \neq 0$. Hence, the psd for the individual MFSK messages will overlap and, for equally likely messages, the psd of the MFSK complex envelope is

$$\begin{aligned} \tilde{S}(f) &= P\frac{1}{M}\sum_{k=0}^{M-1} T\text{sinc}^2((f-f_k)T) \\ &= P\frac{1}{M}\sum_{k=0}^{M-1} T\text{sinc}^2\left(fT - \frac{k}{2} + \frac{(M-1)}{4}\right) \end{aligned} \quad (5.129)$$

The MFSK psd is plotted in Fig. 5.22 for the case $M = 16$.

For MFSK, the required bandwidth grows linearly with M as $W \propto M/2T = MR/2$ Hz, where R is the baud rate. Meanwhile, $R_b = (\log_2 M)R$ bits/s. Their ratio gives the MFSK bandwidth efficiency:

$$R/W = \frac{2(\log_2 M)}{M} \text{ bits/s/Hz} \quad (5.130)$$

which is seen to decrease with M. The next chapter will show that orthogonal MFSK is power efficient.

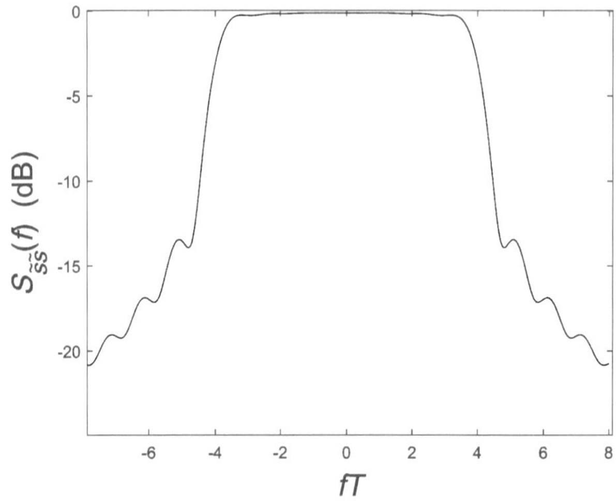

Fig. 5.22 Psd of MFSK with $M = 16$

Problems

5.1 A given set of three functions is

$$\phi_0(t) = 1; \qquad \phi_1(t) = t; \qquad \phi_2(t) = \frac{3}{2}t^2 - \frac{1}{2}$$

(a) Show that these functions are orthogonal over the interval $(-1, 1)$.
(b) Represent the signal:

$$s(t) = \begin{cases} t & t > 0 \\ 0 & t < 0 \end{cases}$$

over the interval $(-1, 1)$ using this set of functions, i.e., obtain $\hat{s}(t)$, a minimum mean-square error approximation to $s(t)$.
(c) Roughly sketch $s(t)$ and $\hat{s}(t)$ obtained in part (b) over the interval $(-1, 1)$ on the same graph.

5.2 Find a complete set of orthonormal basis functions for the following three complex envelopes and write down the three complex signal vectors:

$$\tilde{s}_1(t) = u(t) - u(t-1)$$
$$\tilde{s}_2(t) = u(t-2) - u(t-3)$$
$$\tilde{s}_3(t) = u(t) - u(t-3)$$

5.3 Using the Gram-Schmidt orthonormalization procedure, obtain an orthonormal basis for the following set of three real-valued waveforms and write down the three signal vectors:

$$s_1(t) = e^{-t}u(t)$$
$$s_2(t) = e^{-2t}u(t)$$
$$s_3(t) = e^{-3t}u(t)$$

5.4 Consider the following set of three real-valued waveforms:

$$s_1(t) = u_T(t)$$
$$s_2(t) = \sin(2\pi t/T)u_T(t)$$
$$s_3(t) = \cos(4\pi t/T)u_T(t)$$

where $u_T(t) = u(t) - u(t - T)$ and $u(t)$ is the unit step function:

(a) Find a complete set of orthonormal basis functions.
(b) Find the signal vectors $\mathbf{s}_1$, $\mathbf{s}_2$ and $\mathbf{s}_3$ and plot them.
(c) What is the Euclidean distance between $\mathbf{s}_2$ and $\mathbf{s}_3$?
(d) What is the signal correlation between $\mathbf{s}_1$ and $\mathbf{s}_3$?

5.5 Consider the following two bandpass waveforms:

$$s_1(t) = V\cos(2\pi f_c t)u_T(t)$$
$$s_2(t) = V\cos(2\pi f_c t + \theta)u_T(t)$$

where $f_c = k/2T$ and k is an integer.

(a) Find the energies of both signals. Determine the value of V for which both waveforms have unit energy.
(b) Determine the correlation coefficient:

$$\rho = \frac{1}{\sqrt{E_1}\sqrt{E_2}} \int_0^T s_1(t)s_2(t)dt$$

(c) Plot ρ as a function of θ over the range $0 \leq \theta \leq 2\pi$. What is the value of θ that makes the two waveforms orthogonal?
(d) Verify that the Euclidean distance between the two waveforms is $d = \sqrt{2E(1-\rho)}$. What value of θ maximizes the Euclidean distance?

5.6 Consider a set of M orthogonal bandpass signal waveforms $s_m(t)$, $1 \le m \le M$, $0 \le t \le T$, all of which have the same energy E. Define a new set of M *simplex* waveforms as

$$s'_m(t) = s_m(t) - \frac{1}{M}\sum_{k=1}^{M} s_k(t), \quad 1 \le m \le M, 0 \le t \le T$$

Show that the M simplex waveforms $s'_m(t)$ have equal energy, given by

$$E' = (M-1)E/M$$

and are equally correlated with correlation coefficient

$$\rho_{mn} = \frac{1}{E'}\int_0^T s'_m(t)s'_n(t)dt = -\frac{1}{M-1}$$

5.7 Consider the set of bandpass waveforms having complex envelopes $\{\tilde{s}_i(t)\}_{i=1}^{M}$. The complex envelopes have the complex signal vector representations:

$$\tilde{\mathbf{s}}_i = [\tilde{s}_{i1},\ \tilde{s}_{i2},\ \ldots,\ \tilde{s}_{iN}],\quad i = 1,\ 2,\ \ldots, M$$

where in each coordinate

$$\tilde{s}_{ij} = \begin{cases} +1 \\ \text{or} & \text{for all } i,\ j \\ -1 \end{cases}$$

The complex signal vectors form the vertices of a hypercube:

(a) How many waveforms M are in there in N dimensions?
(b) Sketch the signal vectors in two ($N = 2$) and three ($N = 3$) dimensions.
(c) For the case $N = 3$, specify a set of complex envelopes $\{\tilde{s}_i(t)\}_{i=1}^{M}$ that will have these signal vectors. The solution is not unique.

5.8 Consider a binary FSK system where

$$s_1(t) = \sqrt{\frac{2E}{T}}\cos(2\pi f_c t + \Phi_1)u_T(t)$$

$$s_2(t) = \sqrt{\frac{2E}{T}}\cos(2\pi (f_c + \Delta_f)t + \Phi_2)u_T(t)$$

and where Φ_1 and Φ_2 are independent random phases that can take on any values on the interval $[-\pi, \pi)$. Assuming that $f_c T \gg 1$, find all values of Δ_f so that $s_1(t)$ and $s_2(t)$ are orthogonal.

5.9 Quadrature multiplexing is used to transmit two binary data streams $d_I(t)$ and $d_Q(t)$ as

$$s(t) = d_I(t)\cos(2\pi f_c t) + d_Q(t)\sin(2\pi f_c t)$$

where

$$d_I(t) = A_1 \sum_n d_{I,n} u_{T_{bI}}(t - nT_{bI})$$

$$d_Q(t) = A_2 \sum_n d_{Q,n} u_{T_{bQ}}(t - nT_{bQ})$$

where $d_{I,n}, d_{Q,n} \in \{-1, +1\}$ are the data bits and $R_{bI} = 1/T_{bI}$ and $R_{bQ} = 1/T_{bQ}$ are the corresponding bit rates. If the bit rates R_{bI} and R_{bQ} differ by a factor of 10, what must be the ratio of the amplitudes be if the energy per data bit E_b is the same for the two data streams?

5.10 A computer outputs binary data at a rate of 56 Mb/s. The data bits are transmitted using QPSK modulation with a raised cosine pulse. Determine the transmission bandwidth required for each of the following roll-off factors: $\beta = 0.25, 0.5, 0.75, 1$.

5.11 Consider the frequency domain triangular pulse $P(f)$ shown in Fig. 5.23.

(a) Can this pulse be used to communicate with zero-ISI and, if so, at what baud rate R?

(b) Suppose this pulse is used to construct a transmit filter $h_a(t)$ and corresponding receiver matched filter $h_r(t)$. Find an equation for $|H_a(f)|$ and sketch.

(c) What is the energy in the pulse $h_a(t)$ obtained in part (b)?

5.12 An available wireless bandpass channel passes the frequencies in the band from 3450 MHz to 3475 MHz. It is desired to design a wireless modem that transmits information at a symbol rate of 20 Mbaud (20×10^6 symbols/second) and a bit rate of 80 Mbps (80×10^6 bits/second).

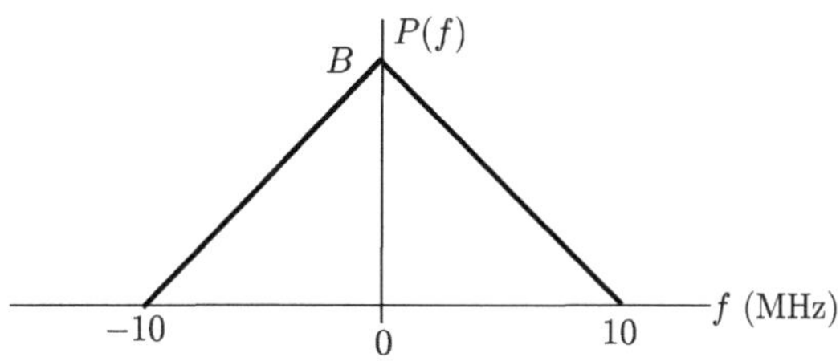

Fig. 5.23 Triangular pulse for Problem 5.11

(a) Select an appropriate QAM signal constellation size M, carrier frequency f_c, and roll-off factor β of a spectral raised cosine pulse that uses the entire frequency band.

(b) Based on your results in part (a), what are the *complex-valued* frequency responses of the transmit filter $h_a(t)$ and receiver matched filter $h_r(t)$ that you would implement?

5.13 A baseband pulse $p(t)$ has a transfer function $P(f)$ that is a trapezoid as shown in Fig. 5.24.

(a) Show that the pulse $p(t)$ must have zeroes at all non-zero multiples of $1/(B_0 + B_1)$.

(b) If the pulse $p(t)$ is passed through the matched filter $p^*(-t)$, show that the filter output also must have zeroes at all non-zero multiples of $1/(B_0 + B_1)$.

5.14 Suppose a digital communication system operates over an ideal channel and employs an overall pulse $p(t)$ that has the Gaussian-shaped form:

$$p(t) = e^{-\pi a^2 t^2}$$

(a) Explain why $p(t)$ does not admit ISI-free transmission.

(b) To manage the level of ISI, suppose the condition is imposed that $p(T) = 0.01$, where T is the modulation symbol period. The bandwidth of $p(t)$ is defined as that value of W for which $P(W)/P(0) = 0.01$, where $P(f)$ is the Fourier transform of $p(t)$. Determine the value W and compare this value to that of a raised cosine spectrum with 100 percent roll-off.

5.15 Consider the overall pulse responses $P(f)$ shown in Fig. 5.25.

(a) For each $P(f)$, determine if one can transmit data with zero ISI and, if so, find the maximum baud rate R in terms of the system bandwidth B.

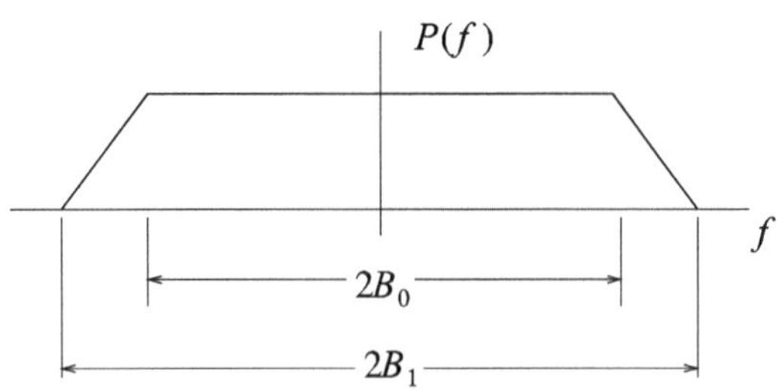

Fig. 5.24 $P(f)$ for Problem 5.13

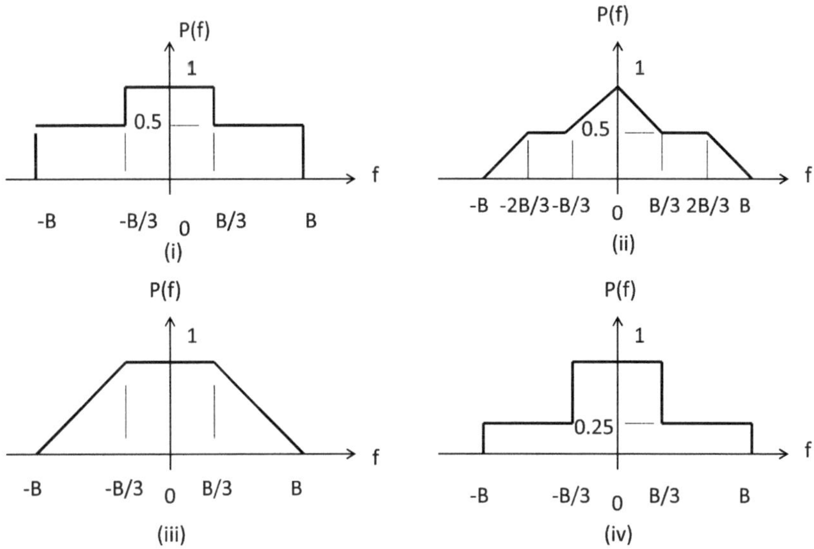

Fig. 5.25 Pulses $P(f)$ for Problem 5.15

Chapter 6
Communication Signal Detection

This chapter considers the detection of digitally modulated signals on ideal channels that are corrupted with additive white Gaussian noise (AWGN). In Sect. 6.1, the received bandpass signal $r(t)$ is processed with a quadrature demodulator to extract its complex envelope $\tilde{r}(t)$, which is then projected onto the complete set of N orthonomal basis functions to generate the length-N received vector $\tilde{\mathbf{r}}$ for each message sent. Section 6.2 considers coherent detection where the complex channel gain $\tilde{h}$ is assumed known to the receiver. Afterwards, the error rate performance of various modulation schemes with coherent detection is considered. Section 6.4 considers noncoherent detection where no attempt is made to estimate the received carrier phase and applied to orthogonal MFSK signals. Finally, Sect. 6.5 considers various bounds on error probability, bit versus symbol error probabilities and the effects that rotations and translations of a signal constellation have on the error probability.

6.1 Vector Space Representation of Received Signals

The received bandpass waveform in quadrature form is

$$r(t) = \tilde{r}_I(t)\cos(2\pi f_c t) - \tilde{r}_Q(t)\sin(2\pi f_c t), \tag{6.1}$$

where $\tilde{r}(t) = tr_I(t) + j\tilde{r}_Q(t)$ is the received complex envelope. The receiver first performs quadrature RF demodulation, functionally as shown in Fig. 6.1, to extract the real and imaginary components of the received complex envelope $\tilde{r}(t)$. The low-pass filter in each RF demodulator branch rejects the double frequency image

G. Stuber, *Introductory Digital Wireless Communications*,
https://doi.org/10.1007/978-3-032-01559-4_6

Fig. 6.1 Quadrature demodulator

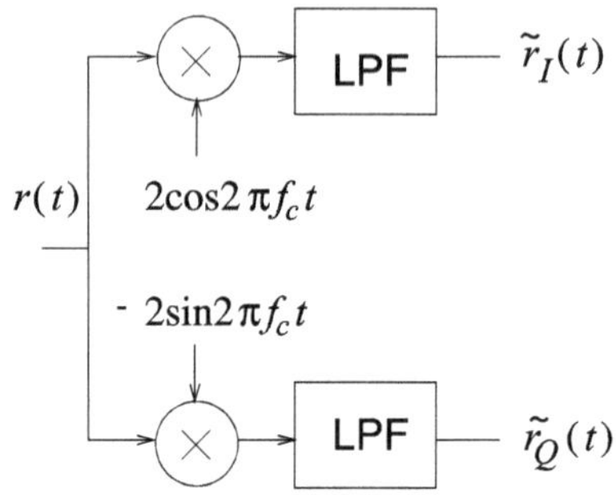

centered at $2f_c$, while having a cutoff frequency that is high enough to admit all frequencies in the complex envelope of the signal waveform. It follows that

$$[r(t) \cdot 2\cos(2\pi f_c t)]_{\text{LP}} = \tilde{r}_I(t) \tag{6.2a}$$

$$[-r(t) \cdot 2\sin(2\pi f_c t)]_{\text{LP}} = \tilde{r}_Q(t), \tag{6.2b}$$

where $[\,\cdot\,]_{\text{LP}}$ indicates low-pass filtering.

In a digital baseband implementation, $\tilde{r}_I(t)$ and $\tilde{r}_Q(t)$ are each passed through an anti-aliasing filter which approximates an ideal low-pass filter and applied to an analog-to-digital converter (ADC) that is typically clocked at the baud rate or a multiple of the baud rate. However, in this textbook continuous-time processing is considered, with the understanding that the anti-aliasing filters and ADCs may be included so as to perform equivalent digital baseband processing.

Consider a signaling scheme where one of a set of M bandpass waveforms, $S = \{s_i(t)\}_{i=1}^{M}$, having the corresponding set of complex envelopes $\tilde{S} = \{\tilde{s}_i(t)\}_{i=1}^{M}$ is transmitted over an ideal channel that attenuates and delays the transmitted waveform. The received complex envelope is

$$\tilde{r}(t) = \tilde{h}\tilde{s}_n(t) + \tilde{w}(t) \tag{6.3}$$

where $\tilde{h} = \alpha e^{j\phi}$ is the attenuation α and phase shift ϕ introduced by the channel, and $\tilde{w}(t)$ is the zero-mean complex-valued additive white Gaussian noise (AWGN) having a power spectral density of N_o watts/Hz. By observing the received waveform $\tilde{r}(t)$, the receiver must decide which message waveform from the set $\tilde{S}$ (or S) was sent at each baud epoch.

Various types of detectors may be employed depending on how the message information is embedded into the RF carrier. If information is only transmitted in RF carrier phase such as PSK, then a coherent detector may be used that has knowledge of the phase ϕ as obtained from a phase-locked loop. If information is embedded in both the RF amplitude and phase such as QAM, then the coherent detector requires knowledge of the entire complex channel gain $\tilde{h}$. If no information is embedded in the RF carrier phase, then a noncoherent detector may be used that does not require any knowledge of the phase ϕ. Moreover, if no information is embedded in either the amplitude or phase of the RF carrier, such as MFSK, then no knowledge of α or ϕ is required. First, consider coherent detection.

Chapter 5 showed that the set of waveforms $\{\tilde{s}_i(t)\}_{i=1}^M$ can be represented as a set vectors $\{\tilde{\mathbf{s}}_i\}_{i=1}^M$ in an N-dimensional vector space now referred to as the signal space. The signal vectors and the associated complete set of orthonormal basis functions $\{\varphi_i(t)\}_{i=1}^N$ may be obtained by using the Gram-Schmidt orthonormalization procedure or intuitive means. The received complex envelope $\tilde{r}(t)$ in (6.3) may be projected onto the same set of basis functions $\{\varphi_i(t)\}_{i=1}^N$ to give the representation

$$\tilde{r}(t) = \sum_{i=1}^{N} \tilde{r}_i \varphi_i(t) + \tilde{z}(t), \tag{6.4}$$

where

$$\tilde{r}_i = \int_{-\infty}^{\infty} \tilde{r}(t)\varphi_i^*(t)\mathrm{d}t \tag{6.5a}$$

$$= \tilde{h}\int_{-\infty}^{\infty} \tilde{s}_n(t)\varphi_i^*(t)\mathrm{d}t + \int_{-\infty}^{\infty} \tilde{w}(t)\varphi_i^*(t)\mathrm{d}t$$

$$= \tilde{h}\tilde{s}_{n,i} + \tilde{w}_i \ , \quad i = 1, \ldots, N \tag{6.5b}$$

and

$$\tilde{z}(t) = \tilde{w}(t) - \sum_{i=1}^{N} \tilde{w}_i \varphi_i(t) \tag{6.6}$$

is a noise "remainder" process, which is the portion of the noise process $\tilde{w}(t)$ that lies outside of the signal space spanned by the complete set of orthonormal basis functions. The above process yields the complex-valued received vector:

$$\tilde{\mathbf{r}} = \tilde{h}\tilde{\mathbf{s}}_n + \tilde{\mathbf{w}}, \tag{6.7}$$

where

$$\tilde{\mathbf{r}} = (\tilde{r}_1, \tilde{r}_2, \ldots, \tilde{r}_N) \tag{6.8a}$$

$$\tilde{\mathbf{s}}_n = (\tilde{s}_{n_1}, \tilde{s}_{n_2}, \ldots, \tilde{s}_{n_N}) \tag{6.8b}$$

$$\tilde{\mathbf{w}} = (\tilde{w}_1, \tilde{w}_2, \ldots, \tilde{w}_N). \tag{6.8c}$$

For an AWGN channel, the $\tilde{w}_k, k = 1, \ldots, N$ are complex Gaussian random variables having a joint pdf that is completely determined by their means and covariances. The means are

$$\mathrm{E}[\tilde{w}_k] = \int_{-\infty}^{\infty} \mathrm{E}[\tilde{w}(t)]\varphi_k(t)\mathrm{d}t = 0 \tag{6.9}$$

and the covariances are[1]

$$\begin{aligned}
\phi_{\tilde{w}_j \tilde{w}_k} = \tfrac{1}{2}\mathrm{E}[\tilde{w}_j^* \tilde{w}_k] &= \int_{-\infty}^{\infty}\int_{-\infty}^{\infty} \tfrac{1}{2}\mathrm{E}[\tilde{w}^*(t)\tilde{w}(s)]\varphi_j(t)\varphi_k^*(s)\mathrm{d}t\mathrm{d}s \\
&= N_o \int_{-\infty}^{\infty}\int_{-\infty}^{\infty} \delta(t-s)\varphi_j(t)\varphi_k^*(s)\mathrm{d}t\mathrm{d}s \\
&= N_o \int_{-\infty}^{\infty} \varphi_j(t)\varphi_k^*(t)\mathrm{d}t \\
&= N_o \delta_{jk}.
\end{aligned} \tag{6.10}$$

It follows that the $\tilde{w}_k$ are i.i.d. zero-mean complex Gaussian random variables with variance N_o. From Sect. 2.7.3, the vector $\tilde{\mathbf{w}}$ has the multivariate complex Gaussian pdf:

$$\begin{aligned}
f(\tilde{\mathbf{w}}) &= \prod_{i=1}^{N} f(n_i) \\
&= \prod_{i=1}^{N} \frac{1}{2\pi N_o} \exp\left\{-\frac{1}{2N_o}|\tilde{w}_i|^2\right\} \\
&= \frac{1}{(2\pi N_o)^N} \exp\left\{-\frac{1}{2N_o}\sum_{i=1}^{N}|\tilde{w}_i|^2\right\} \\
&= \frac{1}{(2\pi N_o)^N} \exp\left\{-\frac{1}{2N_o}\|\tilde{\mathbf{w}}\|^2\right\}.
\end{aligned} \tag{6.11}$$

The joint pdf $f(\tilde{\mathbf{w}})$ is circularly symmetric, appearing as a hyperspherical cloud centered at the origin in the N-dimensional complex-valued vector space.

Later, the joint conditional pdf $f(\tilde{\mathbf{r}}|\tilde{h}, \tilde{\mathbf{s}}_m)$ will be needed. Since $\tilde{\mathbf{r}} = \tilde{h}\tilde{\mathbf{s}}_m + \tilde{\mathbf{w}}$ and $\tilde{\mathbf{w}}$ has the joint pdf in (6.11), it follows that the vector $\tilde{h}\tilde{\mathbf{s}}_m$ just changes the mean of the multivariate complex Gaussian pdf in (6.11), such that

$$f(\tilde{\mathbf{r}}|\tilde{h}, \tilde{\mathbf{s}}_m) = \frac{1}{(2\pi N_o)^N} \exp\left\{-\frac{1}{2N_o}\|\tilde{\mathbf{r}} - \tilde{h}\tilde{\mathbf{s}}_m\|^2\right\}. \tag{6.12}$$

[1] Since the $\tilde{w}_k$ have zero mean, their covariances $\lambda_{\tilde{w}_j \tilde{w}_k}$ are equal to their autocorrelations $\phi_{\tilde{w}_j \tilde{w}_k}$.

6.1.1 Wozencraft's Irrelevance Theorem

The remainder process $\tilde{z}(t)$ in (6.8) lies outside the signal space that is spanned by the set of basis functions $\{\varphi_n(t)\}_{n=1}^{N}$. It is shown below that $\tilde{z}(t)$ is uncorrelated with received vector $\tilde{\mathbf{r}}$, viz.,

$$
\begin{aligned}
\tfrac{1}{2}\mathrm{E}[\tilde{z}(t)r_j^*] &= \tfrac{1}{2}\mathrm{E}[\tilde{z}(t)]\tilde{h}\tilde{s}_{m_j}^* + \mathrm{E}[\tilde{z}(t)\tilde{w}_j^*] \\
&= \tfrac{1}{2}\mathrm{E}[\tilde{z}(t)\tilde{w}_j^*] \\
&= \frac{1}{2}\mathrm{E}\left[\left(\tilde{w}(t) - \sum_{n=1}^{N}\tilde{w}_n\varphi_n(t)\right)\tilde{w}_j^*\right] \\
&= \int_{-\infty}^{\infty}\tfrac{1}{2}\mathrm{E}[\tilde{w}(t)\tilde{w}^*(\tau)]\varphi_j(\tau)\mathrm{d}\tau - \sum_{n=1}^{N}\tfrac{1}{2}\mathrm{E}[\tilde{w}_n\tilde{w}_j^*]\varphi_n(t) \\
&= N_o\varphi_j(t) - N_o\varphi_j(t) = 0\,,\;\; j = 1, \ldots, N.
\end{aligned}
$$

Hence, $\tilde{z}(t)$ is uncorrelated with the received vector $\tilde{\mathbf{r}}$. This property implies that the remainder process $\tilde{z}(t)$ is irrelevant when making the decision as to which signal waveform was transmitted, a result known as Wozencraft's irrelevance theorem [35]. In other words, the received vector $\tilde{\mathbf{r}}$ provides *sufficient statistics* for determining which message was transmitted.

6.2 Detection of Known Signals in Noise

Based on the observation of the noisy received vector $\tilde{\mathbf{r}}$ in (6.7), the receiver decides which message was transmitted at each baud interval such that the probability of decision error is minimized. Assume a receiver with perfect knowledge of the complex channel gain $\tilde{h}$. Consider the set of a posteriori probabilities:

$$
\mathrm{P}[\mathbf{s}_i \text{ sent } |\tilde{h}, \tilde{\mathbf{r}}], \quad i = 1, \ldots, M, \tag{6.13}
$$

abbreviated as $\mathrm{P}[\mathbf{s}_i|\tilde{h}, \tilde{\mathbf{r}}]$. The maximum *a posteriori* probability (MAP) receiver chooses the message $\tilde{\mathbf{s}}_m$ having the largest *a posteriori* probability $\mathrm{P}[\tilde{\mathbf{s}}_m|\tilde{h}, \tilde{\mathbf{r}}]$. That is, the MAP decision rule is

$$
\text{choose } \tilde{\mathbf{s}}_m \text{ if } \mathrm{P}[\tilde{\mathbf{s}}_m|\tilde{h}, \tilde{\mathbf{r}}] \geq \mathrm{P}[\tilde{\mathbf{s}}_i|\tilde{h}, \tilde{\mathbf{r}}] \;\; \forall\, i \neq m. \tag{6.14}
$$

The probability of decision error is

$$
\begin{aligned}
\mathrm{P}_e &= \mathrm{P}[\mathbf{s}_m \text{ not sent } |\tilde{h}, \tilde{\mathbf{r}}] \\
&= 1 - \mathrm{P}[\mathbf{s}_m|\tilde{h}, \tilde{\mathbf{r}}].
\end{aligned} \tag{6.15}
$$

Since the MAP receiver always decides in favor of the message $\mathbf{s}_m$ having the largest *a posteriori* probability $\mathrm{P}[\mathbf{s}_m|\tilde{h}, \tilde{\mathbf{r}}]$ for any received vector $\tilde{\mathbf{r}}$ and channel gain $\tilde{h}$, the probability of error is minimized.

By using Bayes' theorem, the *a posteriori* probability $\mathrm{P}[\tilde{\mathbf{s}}_m|\tilde{h}, \tilde{\mathbf{r}}]$ can be expressed in the form

$$\mathrm{P}[\tilde{\mathbf{s}}_m|\tilde{h}, \tilde{\mathbf{r}}] = \frac{f(\tilde{\mathbf{r}}|\tilde{h}, \tilde{\mathbf{s}}_m)P_m}{f(\tilde{\mathbf{r}})}, \quad m = 1, \cdots, M, \tag{6.16}$$

where the joint conditional pdf $f(\tilde{\mathbf{r}}|\tilde{h}, \tilde{\mathbf{s}}_m)$ is the likelihood of the received vector $\tilde{\mathbf{r}}$ given the transmitted message $\tilde{\mathbf{s}}_m$ and complex channel gain $\tilde{h}$, and P_m is the prior probability of sending message $\tilde{\mathbf{s}}_m$. Since the denominator $f(\tilde{\mathbf{r}})$ in (6.16) does not depend on the message index m, the MAP receiver chooses the vector $\tilde{\mathbf{s}}_m$ that maximizes $f(\tilde{\mathbf{r}}|\tilde{h}, \tilde{\mathbf{s}}_m)\mathrm{P}_m$. In other words, the MAP decision rule is equivalent to

$$\text{choose } \tilde{\mathbf{s}}_m \text{ if } \quad f(\tilde{\mathbf{r}}|\tilde{h}, \tilde{\mathbf{s}}_m)P_m \geq f(\tilde{\mathbf{r}}|\tilde{h}, \tilde{\mathbf{s}}_i)P_i \quad \forall\, i \neq m. \tag{6.17}$$

Substituting the joint conditional pdf in (6.12) into the above equation and eliminating common terms on both sides of the inequality, the MAP decision rule for an AWGN channel simplifies to

$$\text{choose } \mathbf{s}_m \text{ if } \quad \|\mathbf{r} - \tilde{h}\mathbf{s}_i\|^2 - \|\mathbf{r} - \tilde{h}\mathbf{s}_m\|^2 \geq N_o \ln\left(\frac{P_i}{P_m}\right), \quad \forall\, i \neq m. \tag{6.18}$$

A receiver that chooses the message $\tilde{\mathbf{s}}_m$ to maximize $f(\tilde{\mathbf{r}}|\tilde{h}, \tilde{\mathbf{s}}_m)$ regardless of the prior message probabilities is called a maximum likelihood (ML) receiver. The ML decision rule is

$$\text{choose } \tilde{\mathbf{s}}_m \text{ if } \quad f(\tilde{\mathbf{r}}|\tilde{h}, \tilde{\mathbf{s}}_m) \geq f(\tilde{\mathbf{r}}|\tilde{h}, \tilde{\mathbf{s}}_i) \quad \forall\, i \neq m. \tag{6.19}$$

If the prior message probabilities are all equal, i.e., $P_m = 1/M, \ m = 1, \cdots, M$, then the message $\tilde{\mathbf{s}}_m$ that maximizes the likelihood $f(\tilde{\mathbf{r}}|\tilde{h}, \tilde{\mathbf{s}}_m)$ also maximizes $\mathrm{P}(\tilde{\mathbf{s}}_m|\tilde{h}, \tilde{\mathbf{r}})$. Under this condition the ML receiver minimizes the error probability. In Sect. 7.4.2, it will be shown that a uniform input distribution achieves the channel capacity for a symmetric coding channel, such as a binary symmetric channel (BSC). Often a uniform input distribution is used even if the coding channel is not symmetric.

From (6.12), it is apparent that the message $\tilde{\mathbf{s}}_m$ that maximizes the likelihood $f(\tilde{\mathbf{r}}|\tilde{h}, \tilde{\mathbf{s}}_m)$ also minimizes the exponent in (6.12). Hence, the ML receiver may simply choose the message $\tilde{\mathbf{s}}_m$ that minimizes the metric

$$\mu_1(\tilde{\mathbf{s}}_m) = \|\tilde{\mathbf{r}} - \tilde{h}\tilde{\mathbf{s}}_m\|^2, \quad m = 1, \ldots, M. \tag{6.20}$$

From (6.20), the ML receiver chooses the scaled message $\tilde{h}\tilde{\mathbf{s}}_m$ that is closest in squared Euclidean distance to the received vector $\tilde{\mathbf{r}}$. Such a receiver is said to make *minimum distance decisions*.

An alternative form of the ML or minimum distance receiver can be derived by expanding (6.20) as

$$\begin{aligned}
\mu_1(\tilde{\mathbf{s}}_m) &= \|\tilde{\mathbf{r}} - \tilde{h}\tilde{\mathbf{s}}_m\|^2 \\
&= \sum_{n=1}^{N} |\tilde{r}_n - \tilde{h}\tilde{s}_{m,n}|^2 \\
&= \sum_{n=1}^{N} |\tilde{r}_n|^2 - \sum_{n=1}^{N} \left(\tilde{r}_n \tilde{h}^* \tilde{s}_{m,n}^* + \tilde{r}_n^* \tilde{h}\tilde{s}_{m,n}\right) + |\tilde{h}|^2 \sum_{n=1}^{N} |\tilde{s}_{m,n}|^2 \\
&= \sum_{n=1}^{N} |\tilde{r}_n|^2 - 2\mathrm{Re}\left\{\sum_{n=1}^{N} \tilde{r}_n \tilde{h}^* \tilde{s}_{m,n}^*\right\} + |\tilde{h}|^2 \sum_{n=1}^{N} |\tilde{s}_{m,n}|^2 \\
&= \|\tilde{\mathbf{r}}\|^2 - 2\mathrm{Re}\left\{\tilde{\mathbf{r}} \cdot \tilde{h}^* \tilde{\mathbf{s}}_m^*\right\} + |\tilde{h}|^2 \|\tilde{\mathbf{s}}_m\|^2.
\end{aligned} \tag{6.21}$$

Observe that $\|\tilde{\mathbf{r}}\|^2$ is irrelevant and $\|\tilde{\mathbf{s}}_m\|^2 = 2E_m$, where E_m is the energy in the bandpass modulated waveform corresponding to the message $\tilde{\mathbf{s}}_m$. Hence, ignoring $\|\tilde{\mathbf{r}}\|^2$ and dividing the R.H.S. of (6.21) through by -2, the ML receiver may equivalently choose the message $\tilde{\mathbf{s}}_m$ to *maximize* the metric

$$\mu_2(\tilde{\mathbf{s}}_m) = \mathrm{Re}\left\{\tilde{\mathbf{r}} \cdot \tilde{h}^* \tilde{\mathbf{s}}_m^*\right\} - |\tilde{h}|^2 E_m \quad , m = 1, \ldots, M. \tag{6.22}$$

The above decision metric leads to the integral form

$$\mu_3(\tilde{\mathbf{s}}_m) = \mathrm{Re}\left\{\int_{-\infty}^{\infty} \tilde{r}(t)\tilde{h}^* \tilde{s}_m^*(t)\mathrm{d}t\right\} - |\tilde{h}|^2 E_m \quad , m = 1, \ldots, M, \tag{6.23}$$

where the received complex envelope $\tilde{r}(t)$ is correlated directly with the set of scaled and conjugated complex signal envelopes $\{\tilde{h}^* \tilde{s}_m^*(t)\}$. Note that (6.22) and (6.23) are not exactly the same, as $\tilde{r}(t)$ in (6.23) includes the noise remainder process $\tilde{z}(t)$ in (6.6) that was shown to be irrelevant by Wozencraft's irrelevance theorem.

Some simplifications can be made for certain types of signal sets. If the messages are equally likely and have equal energy, E, such as PSK signals, then $E_m = E$ and the bias term $|\tilde{h}|^2 E_m$ in (6.22) and (6.23) may be neglected. In this case, the ML receiver may choose the message $\tilde{\mathbf{s}}_m$ or $\tilde{s}_m(t)$ to maximize the metrics

$$\mu_2(\tilde{\mathbf{s}}_m) = \mathrm{Re}\left\{\tilde{\mathbf{r}} \cdot \tilde{h}^* \tilde{\mathbf{s}}_m^*\right\} \tag{6.24a}$$

$$\mu_3(\tilde{\mathbf{s}}_m) = \mathrm{Re}\left\{\int_{-\infty}^{\infty} \tilde{r}(t)\tilde{h}^* \tilde{s}_m^*(t)\mathrm{d}t\right\}, \quad m = 1, \ldots, M. \tag{6.24b}$$

After quadrature demodulation, there are various receiver structures that are functionally equivalent, but differ in their method of implementation and complexity. As shown in Fig. 6.2, one possibility is to generate the observation vector $\tilde{\mathbf{r}}$ by correlating the received complex envelope with each of the N basis functions used to define the signal space. The integrators are implemented as sliding window correlators, where the lower integral limit is at the beginning of a received modulation symbol and the upper integral limit is T_o seconds later, where T_o is the length of the shaping pulse $h_a(t)$. This receiver structure is called a correlation detector.

After obtaining the received vector $\tilde{\mathbf{r}}$ corresponding to each modulation symbol, the metric computer in Fig. 6.3 takes a dot product with received signal vectors $\tilde{h}\tilde{\mathbf{s}}_m$ and subtracts the bias terms to produce M decision metrics $\mu_2(\tilde{\mathbf{s}}_m)$, $m = 1, \ldots, M$ in (6.22). Note that an estimate of the channel gain $\tilde{h}$ is needed, assumed here to be perfect. The decision is made in favor of the message $\tilde{\mathbf{s}}_m$ having the largest decision metric. Alternatively, from (6.23), the receiver may be implemented as a bank of M complex correlators, one for each scaled message waveform $\tilde{h}\tilde{s}_m(t)$. If the dimension of the signal space N is much less than M, e.g., $N = 1$, then this latter approach will add complexity due to the larger number of complex correlators, but it does not require the additional metric computer in Fig. 6.3 that takes dot products and subtracts bias terms.

An alternative to the correlation detector in Fig. 6.2 is the *matched filter detector* shown in Fig. 6.4. The received complex envelope is filtered with a bank of N filters having impulse responses $\varphi_i^*(T_o - t)$ that are matched to the orthonormal basis functions $\varphi_i(t)$, hence the name matched filter detector. If the received modulation symbol starts at time t_a as determined by a separate symbol timing estimator, then the matched filter outputs are sampled at time $t_a + T_o$, where T_o is the duration of the shaping pulse $h_a(t)$. Thereafter, samples are taken every T seconds, i.e., at time instants $t_a + T_o + kT$.

Fig. 6.2 Correlator detector

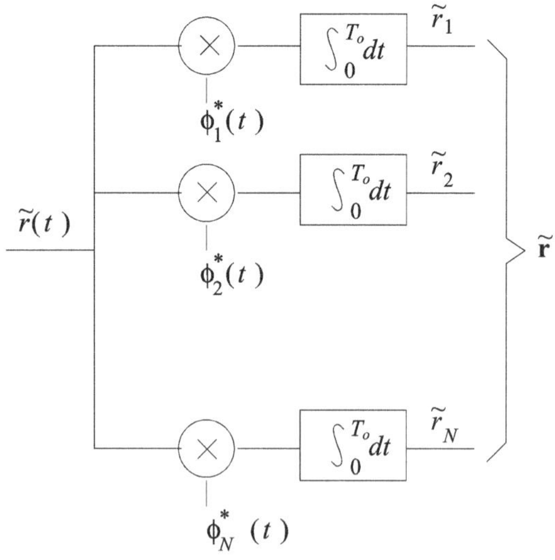

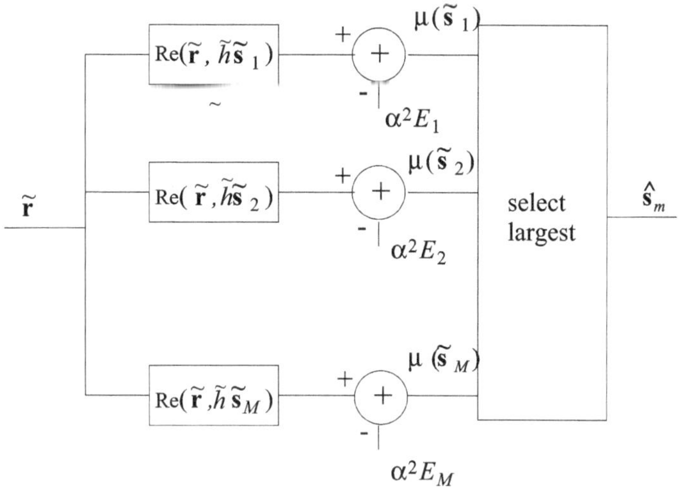

Fig. 6.3 Metric computer

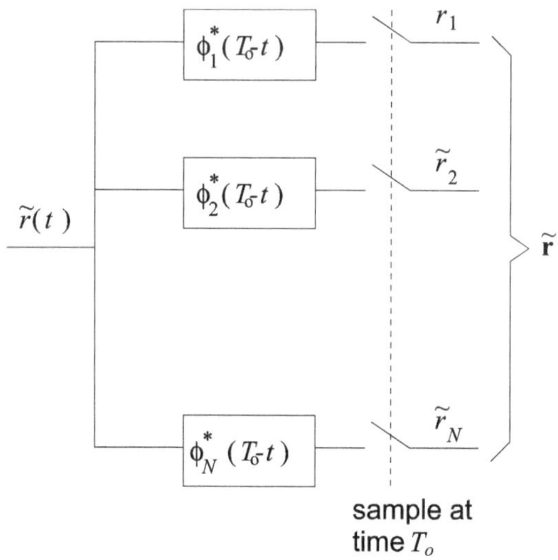

Fig. 6.4 Matched filter detector

To show equivalence of the correlation and matched filter detectors in Figs. 6.2 and 6.4, respectively, let $h_i(t) = \varphi_i^*(T_o - t)$ denote the filter that is matched to $\varphi_i(t)$ and assume $t_a = 0$. Then the output of the matched filter is the convolution

$$\begin{aligned}\tilde{y}(t) &= \int_0^t \tilde{r}(\tau)h_i(t-\tau)\mathrm{d}\tau \\ &= \int_0^t \tilde{r}(\tau)\varphi_i^*(T_o - t + \tau)\mathrm{d}\tau. \qquad (6.25)\end{aligned}$$

Sampling the matched filter output at time T_o gives

$$\tilde{y}(T_o) = \int_0^{T_o} \tilde{r}(\tau)\varphi_i^*(\tau)\mathrm{d}\tau \tag{6.26}$$

which is exactly the same as the output of the correlation detector in (6.5a). However, the outputs of the correlation and matched filter detectors will differ at time instants other than the correct sampling instants.

6.2.1 Matched Filter Alternate Derivation

Consider the complex-valued binary baseband communication system shown in Fig. 6.5 that transmits a sequence of binary messages $\{x_n\}$, $x_n \in \{-\sqrt{2E}, +\sqrt{2E}\}$ over an ideal channel by using a real-valued unit-energy amplitude shaping pulse $h_a(t)$ that is causal and of finite duration T_o. Each message waveform $\tilde{x}(t) = x_n h_a(t)$ is corrupted by complex additive white Gaussian noise $\tilde{w}(t)$ having power spectrum N_o watts/Hz. To recover the data sequence, $\tilde{r}(t)$ is passed through a filter having impulse response $h_r(t)$ to maximize the output signal-to-noise ratio at the sampling instants.

Suppose the message $x_0 = +\sqrt{2E}$ is sent. The received complex envelope is

$$\tilde{r}(t) = \sqrt{2E}h_a(t) + \tilde{w}(t) \tag{6.27}$$

and the output of the filter $h_r(t)$ is

$$\tilde{y}(t) = \sqrt{2E}p(t) + \tilde{n}(t) \tag{6.28}$$

where

$$\begin{aligned} p(t) &= h_a(t) * h_r(t) \\ \tilde{n}(t) &= \tilde{w}(t) * h_r(t) \end{aligned} \tag{6.29a}$$

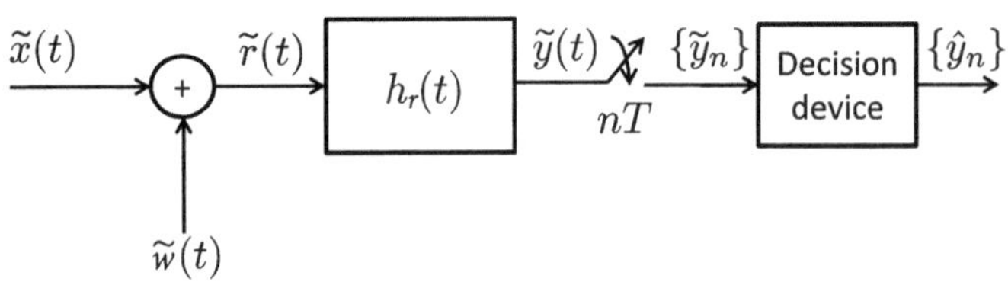

Fig. 6.5 Basic baseband binary communication system

The filter output $\tilde{y}(t)$ is sampled at time T_o to yield the sample

$$\tilde{y}_0 = \sqrt{2E}\,p(T_o) + \tilde{n}(T_o). \tag{6.30}$$

The filter $h_r(t)$ is chosen to maximize the *peak pulse signal-to-noise ratio* at the sampling instant, T_o, defined by

$$\eta \triangleq \frac{\frac{1}{2}|\sqrt{2E}\,p(T_o)|^2}{\frac{1}{2}\mathrm{E}[|\tilde{n}(T_o)|^2]} = \frac{\text{peak signal power}}{\text{average noise power}}. \tag{6.31}$$

From the basic properties of filtered random processes

$$S_{\tilde{n}\tilde{n}}(f) = |H_r(f)|^2 S_{\tilde{w}\tilde{w}}(f) = N_o|H_r(f)|^2.$$

It follows that $\tilde{n}(t)$ is a zero-mean complex Gaussian random process. Sampling $\tilde{n}(t)$ at any time instant yields a zero-mean complex Gaussian random variable having variance

$$\begin{aligned}\tfrac{1}{2}\mathrm{E}[|\tilde{n}(T_o)|^2] = \phi_{\tilde{n}\tilde{n}}(0) &= \int_{-\infty}^{\infty} S_{\tilde{n}\tilde{n}}(f)\mathrm{d}f \\ &= N_o \int_{-\infty}^{\infty} |H_r(f)|^2 \mathrm{d}f.\end{aligned} \tag{6.32}$$

The sampled overall pulse, $p(T_o)$, can be obtained by taking the inverse Fourier transform of $P(f) = H_a(f)H_r(f)$ and sampling at $t = T_o$. This gives

$$p(T_o) = \int_{-\infty}^{\infty} H_a(f)H_r(f)e^{j2\pi f T_o}\mathrm{d}f. \tag{6.33}$$

Using (6.32) and (6.33) in (6.31), it follows that η can be rewritten as

$$\eta = \frac{E|\int_{-\infty}^{\infty} H_a(f)H_r(f)e^{j2\pi f T_o}\mathrm{d}f|^2}{N_o \int_{-\infty}^{\infty} |H_r(f)|^2 \mathrm{d}f}. \tag{6.34}$$

The problem is to find $H_r(f)$ to maximize η. The solution is obtained by applying a variant of the *Schwartz inequality* to the numerator in (6.34), given here without proof.

Lemma 6.1 (Schwartz Inequality) *Consider two complex-valued functions $A(f)$ and $B(f)$. The Schwartz inequality states that*

$$\left|\int_{-\infty}^{\infty} A(f)B(f)\mathrm{d}f\right|^2 \le \int_{-\infty}^{\infty} |A(f)|^2\mathrm{d}f \int_{-\infty}^{\infty} |B(f)|^2\mathrm{d}f \tag{6.35}$$

with equality if and only if $A(f) = kB^*(f)$, where k is an arbitrary scalar constant.

Returning to the problem at hand, let $A(f) = H_a(f)e^{j2\pi f T_o}$ and $B(f) = H_r(f)$ and apply the Schwartz inequality to give

$$\left|\int_{-\infty}^{\infty} H_a(f)H_r(f)e^{j2\pi f T_o}\mathrm{d}f\right|^2 \leq \int_{-\infty}^{\infty} |H_a(f)|^2\mathrm{d}f \int_{-\infty}^{\infty} |H_r(f)|^2\mathrm{d}f \tag{6.36}$$

with equality if and only if

$$H_r(f) = H_{\mathrm{opt}}(f) = kH_a^*(f)\mathrm{e}^{-j2\pi f T_o}, \tag{6.37}$$

where k is an arbitrary scalar constant. Hence,

$$\eta \leq \frac{E}{N_o}\int_{-\infty}^{\infty} |H_a(f)|^2\mathrm{d}f. \tag{6.38}$$

Since k is arbitrary, set $k = 1$ and take the inverse Fourier transform of $H_{\mathrm{opt}}(f)$:

$$\begin{aligned} h_{\mathrm{opt}}(t) &= \int_{-\infty}^{\infty} H_a^*(f)\mathrm{e}^{-j2\pi f T_o}\mathrm{e}^{j2\pi f t}\mathrm{d}f \\ &= \int_{-\infty}^{\infty} H_a^*(f)\mathrm{e}^{-j2\pi f (T_o - t)}\mathrm{d}f. \end{aligned}$$

Conjugating both sides gives

$$\begin{aligned} h_{\mathrm{opt}}^*(t) &= \int_{-\infty}^{\infty} H_a(f)\mathrm{e}^{j2\pi f (T_o - t)}\mathrm{d}f \\ &= h_a(T_o - t). \end{aligned}$$

Since $h_a(t)$ is a real-valued amplitude shaping pulse, it follows that

$$h_{\mathrm{opt}}(t) = h_a^*(T_o - t) = h_a(T_o - t) \tag{6.39}$$

which is the time-reversed and delayed version of $h_a(t)$ otherwise known as the *matched filter* to $h_a(t)$. Using the matched filter and sampling its output at the correct sampling instant attains the upper bound for η in (6.38).

Rayleigh's energy theorem implies that the energy E in the shaping pulse $h_a(t)$ is related to its Fourier transform $H_a(f)$ as

$$\int_{-\infty}^{\infty} h_a^2(t)\mathrm{d}t = \int_{-\infty}^{\infty} |H_a(f)|^2\mathrm{d}f = 1. \tag{6.40}$$

Hence,

$$\eta_{\max} = \frac{E}{N_o}\int_{-\infty}^{\infty} |H_a(f)|^2\mathrm{d}f = \frac{E}{N_o}. \tag{6.41}$$

Example 7.1 Suppose that $h_a(t) = (1/\sqrt{T})u_T(t) = (1/\sqrt{T})(u(t) - u(t-T))$ as shown in Fig. 6.6. The corresponding matched filter $h_{\text{opt}}(t) = (1/\sqrt{T})u_T(T_o - t)$ is shown in Fig. 6.7. In the absence of noise, the matched filter output $y(t)$ is shown in Fig. 6.8. At the correct sampling instant $y(T) = \sqrt{2E}$.

6.3 Probability of Message Error

Consider a signal constellation $\tilde{S} = \{\tilde{\mathbf{s}}_m\}_{m=1}^{M}$ having M complex signal vectors. Assume the messages are equally likely so that $P_m = 1/M$ and the ML receiver achieves the minimum error probability. For an ideal channel with AWGN, the ML receiver uses the metric in (6.20) or the equivalent metric in (6.22) to make minimum distance decisions.

To compute the message or *symbol error probability*, M convex regions R_m, $m = 1, \ldots, M$, called *Voronoi* regions, are first defined around each of the received message vectors $\tilde{h}\tilde{\mathbf{s}}_m$ in the N-D complex-valued signal space, as illustrated in Fig. 6.9. Formally, the decision regions are defined by

$$R_m = \left\{ \tilde{\mathbf{r}} : \|\tilde{\mathbf{r}} - \tilde{h}\tilde{\mathbf{s}}_m\|^2 \leq \|\tilde{\mathbf{r}} - \tilde{h}\tilde{\mathbf{s}}_j\|^2 , \ \forall \, j \neq m \right\} . \tag{6.42}$$

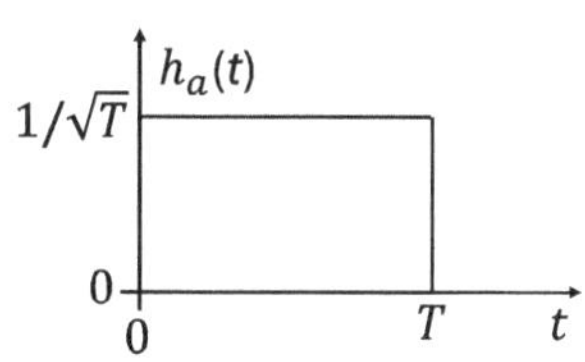

Fig. 6.6 Rectangular pulse $h_a(t) = (1/\sqrt{T})u_T(t)$

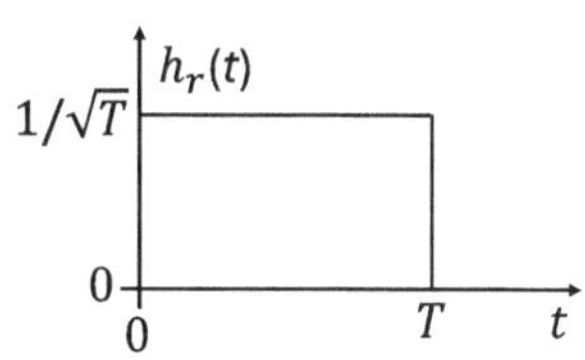

Fig. 6.7 Matched filter $h_{\text{opt}}(t)$ to the pulse $h_a(t) = (1/\sqrt{T})u_T(t)$

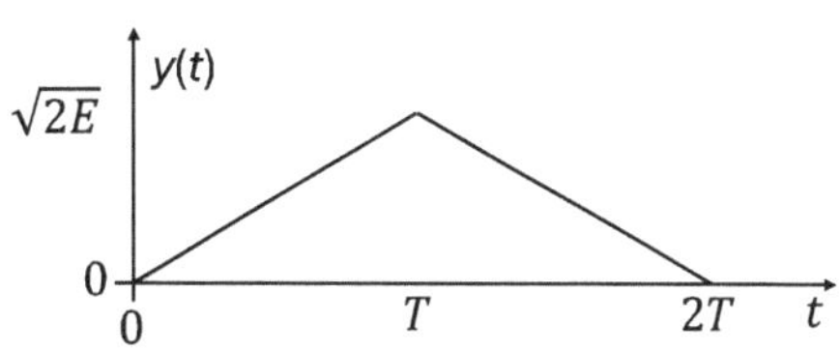

Fig. 6.8 Output of the matched filter $y(t)$

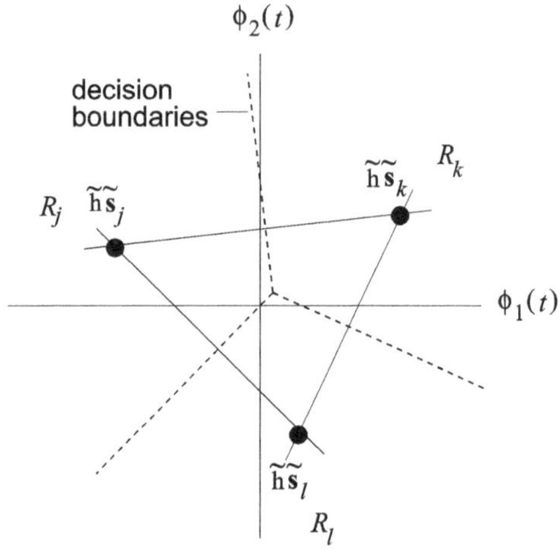

Fig. 6.9 Decision regions in a 2-D complex-valued signal space

Observe that every $\tilde{\mathbf{r}} \in R_m$ is closer to the received message vector $\tilde{h}\tilde{\mathbf{s}}_m$ than to any other received message vector $\tilde{h}\tilde{\mathbf{s}}_i$, $i \neq m$. Hence, the ML decision rule becomes

$$\text{choose } \tilde{\mathbf{s}}_m \text{ if } \tilde{\mathbf{r}} \in R_m \ . \tag{6.43}$$

The decision boundaries are hyperplanes in the N-D signal space that are defined by the locus of points that are equidistant from nearest neighboring received message vectors. In case of complex message vectors laying a complex plane, or bandpass message vectors laying in an $x - y$ plane, the decision boundaries can often be established by drawing perpendicular bisectors between neighboring message vectors as shown in Fig. 6.9.

Given that message $\tilde{\mathbf{s}}_m$ is sent, the probability of message error is

$$\begin{aligned} \mathrm{P}_{e|m} &= \mathrm{P}[\tilde{\mathbf{r}} \notin R_m] \\ &= 1 - \mathrm{P}[\tilde{\mathbf{r}} \in R_m] \\ &= 1 - \mathrm{P}_{c|m}, \end{aligned} \tag{6.44}$$

where $\mathrm{P}_{c|m}$ is the probability of correct message reception given that message $\tilde{\mathbf{s}}_m$ is sent and

$$\mathrm{P}_{c|m} = \int_{R_m} p(\tilde{\mathbf{r}}|\tilde{h}, \tilde{\mathbf{s}}_m)\mathrm{d}\mathbf{r}. \tag{6.45}$$

Averaging (6.44) over the M equally likely messages gives the probability of message error:

$$\mathrm{P}_e = \frac{1}{M}\sum_{i=1}^{M} \mathrm{P}_{e|m}. \tag{6.46}$$

Sometimes it is difficult to obtain the conditional probabilities in (6.45) due to the difficulty in specifying the limits in the N-fold integration that define the Voronoi regions R_m. In this case, various bounds and approximations on the probability of message error are useful that will be discussed later in Sect.6.5.

6.3.1 Pairwise Error Probability

Suppose a pair of complex message vectors $\tilde{\mathbf{s}}_j$ and $\tilde{\mathbf{s}}_k$ is drawn from a signal set $\tilde{S}$ of size M and used for binary communication, ignoring the other $M-2$ complex message vectors. The probability of binary message error in this case with minimum distance decisions is called the *pairwise error probability* because it can be defined for each distinct pair of message vectors drawn from the signal set.

To compute the pairwise error probability, a perpendicular bisector is first placed at the midpoint of a line between the two received message vectors $\tilde{h}\tilde{\mathbf{s}}_j$ and $\tilde{h}\tilde{\mathbf{s}}_k$ to define the two Voronoi regions R_j and R_k as shown in Fig. 6.10. Now assume the message $\tilde{\mathbf{s}}_k$ is sent and let $\mathrm{P}_{e|k,j}$ denote the probability that the receiver chooses the message $\tilde{\mathbf{s}}_j$ instead and thus makes an error. Such an error event happens when the complex AWGN $\tilde{\mathbf{w}}$ causes the received message vector $\tilde{\mathbf{r}} = \tilde{h}\tilde{\mathbf{s}}_k + \tilde{\mathbf{w}}$ to cross the perpendicular decision boundary. Finally, if message $\tilde{\mathbf{s}}_j$ was sent in the first place, then $\mathrm{P}_{e|j,k} = \mathrm{P}_{e|k,j}$.

Due to the circular symmetry of the zero-mean complex AWGN, the pdf of $\tilde{\mathbf{w}}$ is invariant to a rotation around the origin. By suitable rotation, a coordinate axis may be aligned along a line that passes through the two received message vectors $\tilde{h}\tilde{\mathbf{s}}_j$ and $\tilde{h}\tilde{\mathbf{s}}_k$, such that the AWGN along the line appears as a zero-mean Gaussian random variable having variance N_o. All other dimensions in the N-D rotated signal space are irrelevant to the error probability.

The two received message vectors $\tilde{h}\tilde{\mathbf{s}}_j$ and $\tilde{h}\tilde{\mathbf{s}}_k$ are separated by Euclidean distance:

$$\|\tilde{h}\tilde{\mathbf{s}}_j - \tilde{h}\tilde{\mathbf{s}}_k\| = \alpha\|\tilde{\mathbf{s}}_j - \tilde{\mathbf{s}}_k\| = \alpha\tilde{d}_{j,k}$$

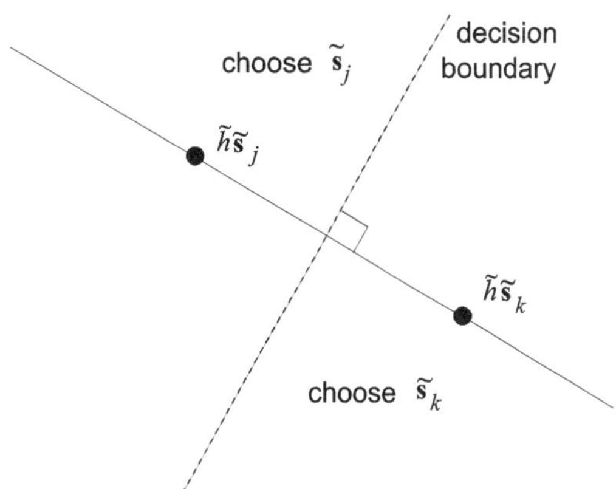

Fig. 6.10 Two received signal points in an N-D signal space

where

$$\tilde{d}_{j,k} = \|\tilde{\mathbf{s}}_j - \tilde{\mathbf{s}}_k\|, \tag{6.47}$$

such that the Euclidean distance from the received message vector $\tilde{h}\tilde{\mathbf{s}}_k$ to the decision boundary is $\alpha\tilde{d}_{j,k}/2$. Using (2.65), the pairwise error probability between the messages $\tilde{\mathbf{s}}_j$ and $\tilde{\mathbf{s}}_k$ is

$$\mathrm{P}_{e|j,k} = \mathrm{P}_{e|k,j} = Q\left(\frac{\alpha\tilde{d}_{j,k}/2}{\sqrt{N_o}}\right) = Q\left(\sqrt{\frac{\alpha^2\tilde{d}_{j,k}^2}{4N_o}}\right). \tag{6.48}$$

6.3.2 *Error Probability with Binary Modulation*

This section considers the error probability of binary modulated waveforms that transmit information in the phase, amplitude, or frequency of the RF carrier.

6.3.2.1 Error Probability of BPSK

Referring to Sect. 5.2.1, the BPSK complex message vectors are

$$\tilde{s}_1 = -\tilde{s}_2 = \sqrt{2E}, \tag{6.49}$$

and the signal constellation for BPSK is shown in Fig. 5.4.

Assume the BPSK messages are used with equal probability and with minimum distance decisions. The bit error probability is given by the pairwise error probability in (6.48), where $\tilde{d}_{12} = 2\sqrt{2E}$. Moreover, the equiprobable BPSK messages have equal energy E and, therefore, the energy per bit is $E_b = \frac{1}{2}\cdot E + \frac{1}{2}\cdot E = E$. The bit error probability becomes

$$P_b(\gamma_b) = Q\left(\sqrt{2\gamma_b}\right) \tag{6.50}$$

where γ_b is the received *bit energy-to-noise ratio*:

$$\gamma_b \triangleq \frac{\alpha^2 E_b}{N_o}. \tag{6.51}$$

The effect of the complex channel gain $\tilde{h} = \alpha e^{j\phi}$ in (6.51) is to scale the bit energy E_b by α^2. This scaling will be relevant for the multipath-fading channels considered in Chap. 11 where α is a random variable. For an AWGN channel, one may generally set $\alpha = 1$ and use the signal constellation, as in Fig. 5.4 for BPSK, to obtain the bit error probability in terms of E_b/N_o. Then replace E_b/N_o with $\gamma_b = \alpha^2 E_b/N_o$ should the complex channel gain $\tilde{h}$ be a random variable. This approach is taken in the sequel

6.3.2.2 Error Probability with On-Off Keying (OOK)

OOK transmits information in the unipolar amplitude of a sinusoidal RF carrier in an on-off fashion. The complex message vectors are

$$\tilde{s}_1(t) = \sqrt{2E} \tag{6.52a}$$

$$\tilde{s}_2(t) = 0. \tag{6.52b}$$

Using (6.48) with the Euclidean distance $d_{12} = \sqrt{2E}$ gives

$$P_b = Q\left(\sqrt{\frac{\alpha^2 E}{2N_o}}\right).$$

With equally likely messages, $E_b = \frac{1}{2} \cdot 0 + \frac{1}{2} \cdot E = E/2$. Therefore,

$$P_b = Q\left(\sqrt{\gamma_b}\right). \tag{6.53}$$

Observe that OOK is 3 dB less power efficient than BPSK. This is shown in Fig. 6.11 that plots the bit error probability in each case.

6.3.2.3 Binary Frequency Shift Keying (BFSK)

Referring to Sect. 5.2.2, consider the case of binary orthogonal FSK where the complex message vectors are

$$\tilde{\mathbf{s}}_1 = (\sqrt{2E}, 0) \tag{6.54a}$$

$$\tilde{\mathbf{s}}_2 = (0, \sqrt{2E}) \tag{6.54b}$$

with the signal constellation shown in Fig. 5.5.

It follows that $\tilde{d}_{12} = 2\sqrt{E}$ and $E_b = E$ since each message has energy E. From (6.48),

$$P_b = Q\left(\sqrt{\gamma_b}\right). \tag{6.55}$$

Orthogonal BFSK has the same performance as OOK and is 3 dB less power efficient that BPSK.

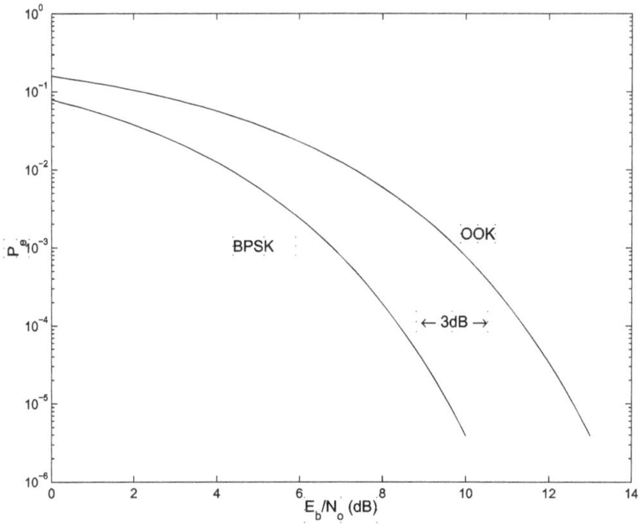

Fig. 6.11 Comparison of probability of bit error for binary antipodal signaling and on-off keying

6.3.3 *Error Probability with Nonbinary Modulation*

This section considers the error probability of nonbinary messages, where information is transmitted in the phase, amplitude, or frequency of an RF carrier. It begins with quadrature phase shift keying (QPSK).

6.3.3.1 Error Probability of QPSK

Consider the set of four QPSK complex signal vectors (scalars in this case):

$$\tilde{s}_1 = -\tilde{s}_3 = (1 + j)\sqrt{E} = \sqrt{2E}\mathrm{e}^{j\pi/4} \tag{6.56a}$$

$$\tilde{s}_2 = -\tilde{s}_4 = (-1 + j)\sqrt{E} = \sqrt{2E}\mathrm{e}^{j3\pi/4}, \tag{6.56b}$$

shown in Fig. 6.12. Note that the signal constellation is rotated 45° such that the coordinate axes constitute the decision boundaries that define the Voronoi regions. Section 6.5.3 will show that the error probability is invariant to a rotation of the signal constellation about the origin.

The noise vector is $\tilde{w} = \tilde{w}_I + j\tilde{w}_Q$, where $\tilde{w}_I$ and $\tilde{w}_Q$ are independent zero-mean Gaussian random variables with variance N_o. With equally likely messages and minimum distance decisions, all messages have the same error probability due

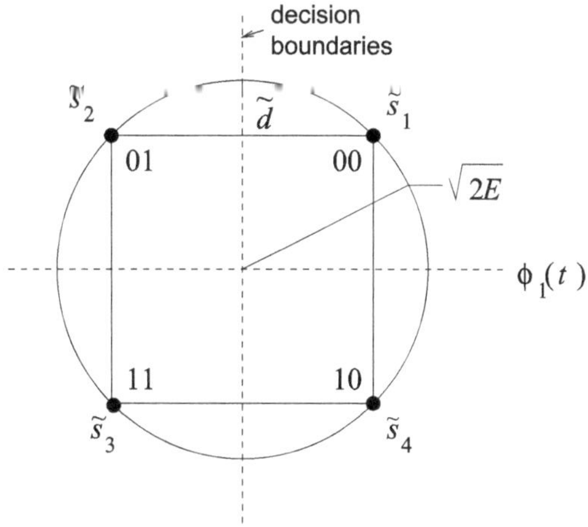

Fig. 6.12 Complex signal-space diagram for QPSK

to symmetry. Assuming $\tilde{s}_1$ is the transmitted message, the probability of message error is

$$\begin{aligned} P_e &= \mathrm{P}_{e|1} \\ &= 1 - \mathrm{P}_{c|1} \\ &= 1 - \mathrm{P}\left[\tilde{w}_I > -\tilde{d}/2, \tilde{w}_Q > -\tilde{d}/2\right] \\ &= 1 - \mathrm{P}\left[\tilde{w}_I > -\tilde{d}/2\right]\mathrm{P}\left[\tilde{w}_Q > -\tilde{d}/2\right] \\ &= 1 - \left(1 - Q\left(\sqrt{\frac{\tilde{d}^2}{4N_o}}\right)\right)^2 . \end{aligned}$$

From Fig. 6.12, $\tilde{d} = 2\sqrt{E}$ and, hence,

$$P_e = 1 - \left(1 - Q\left(\sqrt{\gamma_s}\right)\right)^2 \tag{6.57}$$

where γ_s is the received *symbol energy-to-noise ratio*:

$$\gamma_s \triangleq \frac{\alpha^2 E}{N_o}. \tag{6.58}$$

With QPSK there are two bits per data symbol and the data bits are mapped onto the messages as shown in Fig. 6.12.

Letting P_b denote the probability of bit error, it follows that the probability of correct symbol decision is

$$\mathrm{P}_c = (1 - P_b)^2 \tag{6.59}$$

and the probability of symbol error is

$$P_e = 1 - (1 - P_b)^2. \tag{6.60}$$

Comparing (6.60) with (6.57),

$$P_b = Q\left(\sqrt{\gamma_s}\right). \tag{6.61}$$

Finally, QPSK transmits 2 bits/symbol so $E = 2E_b$, where E_b is the bit energy. Hence, $\gamma_s = 2\gamma_b$ and the probability of bit error is

$$P_b = Q\left(\sqrt{2\gamma_b}\right). \tag{6.62}$$

QPSK and BPSK have identical bit error rate performance as shown in Fig. 6.11. However, QPSK transmits two bits per message, whereas BPSK transmits only one bit per message.

6.3.3.2 Error Probability of MPSK

Consider, for example, the 8-PSK signal constellation and associated Voronoi regions shown in Fig. 6.13. The data bits are mapped to the constellation points with a Gray mapping that be explained further in Sect. 6.5.2.1. Due to symmetry of the signal constellation the error probability of all 8 messages is the same.

Suppose without loss of generality that the message $\tilde{s}_1 = \sqrt{2E}$ is transmitted. The received complex scalar

$$\tilde{r} = \tilde{s}_1 + \tilde{w} \tag{6.63}$$

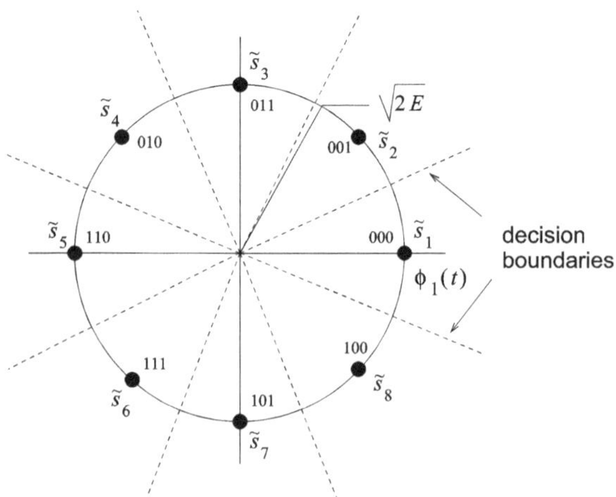

Fig. 6.13 Complex signal-space diagram for 8-PSK along and the associated Voronoi regions. Bits are mapped onto the messages using a Gray mapping

is a complex Gaussian random variable having pdf

$$p_{\tilde{r}}(\tilde{r}) = \frac{1}{2\pi N_o} \exp\left\{-\frac{1}{2N_o}\left|r - \sqrt{2E}\right|^2\right\}. \tag{6.64}$$

The conditional probability of correct symbol reception, $P_{c|1}$, is the probability that $\tilde{r}$ falls in the "pie-shaped" region containing $\tilde{s}_1$. For 8-PSK, this is the probability that the angle $\Theta = \text{Tan}^{-1}[\tilde{r}_Q/\tilde{r}_I]$ falls in the interval $[-\pi/8, \pi/8]$, and $P_c = P_{c|1}$ due to symmetry.

To find the pdf of Θ, first, define the auxiliary random variable:

$$R = \sqrt{\tilde{r}_I^2 + \tilde{r}_Q^2} \tag{6.65}$$

such that

$$\tilde{r}_I = R\cos\Theta\,, \qquad \tilde{r}_Q = R\sin\Theta. \tag{6.66}$$

Section 2.7.6 showed R to have a Rice distribution, as derived from the joint pdf of R and Θ:

$$p_{R,\Theta}(r,\theta) = \frac{r}{2\pi N_o} \mathrm{e}^{-\frac{1}{2N_o}\left(r^2 - 2\sqrt{2E}r\cos\theta + 2E\right)}, \quad r \geq 0, -\pi \leq \theta \leq \pi. \tag{6.67}$$

However, R is irrelevant to the message decision and only the angle Θ matters. The marginal pdf of Θ is

$$\begin{aligned} p_\Theta(\theta) &= \int_0^\infty p_{R,\Theta}(r,\theta)\mathrm{d}r \\ &= \frac{1}{2\pi}\mathrm{e}^{-\gamma_s\sin^2\theta}\int_0^\infty x\mathrm{e}^{-\left(x-\sqrt{2\gamma_s}\cos\theta\right)^2/2}\mathrm{d}x, \end{aligned} \tag{6.68}$$

where $\gamma_s = \alpha^2 E/N_o$ is the symbol energy-to-noise ratio and $\gamma_s = (\log_2 M)\gamma_b$, where γ_b is the bit energy-to-noise ratio. Finally, the MPSK symbol error probability is

$$P_e = 1 - \int_{-\pi/M}^{\pi/M} p(\theta)\mathrm{d}\theta, \tag{6.69}$$

which does not simplify further, except for $M = 2$ BSPK and $M = 4$ QPSK. Figure 6.14 plots the message error probability for PSK with $M = 4, 8$ and 16. PSK is a bandwidth efficient modulation scheme such that the required bit energy-to-noise ratio γ_b for a given message error probability increases with modulation order M.

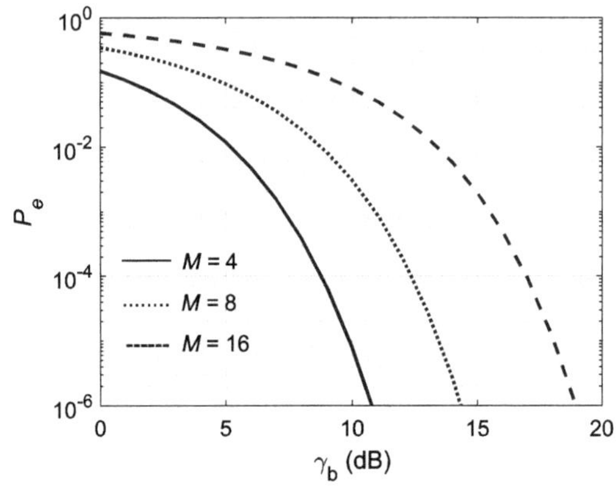

Fig. 6.14 Message error probability P_e versus γ_b for 4-, 8-, and 16-PSK

6.3.3.3 Error Probability of PAM

Consider the PAM signal constellation shown in Fig. 6.15 for $M = 8$.

For the $M - 2$ inner points of the signal constellation, the probability of message error is

$$P_i = 2Q\left(\sqrt{\frac{2\alpha^2 E_o}{N_o}}\right). \tag{6.70}$$

Likewise, for the 2 outer points of the signal constellation, the probability of message error is

$$P_o = Q\left(\sqrt{\frac{2\alpha^2 E_o}{N_o}}\right). \tag{6.71}$$

By averaging over equally likely messages the overall probability of message error is

$$P_e = \frac{M-2}{M}P_i + \frac{2}{M}P_o$$

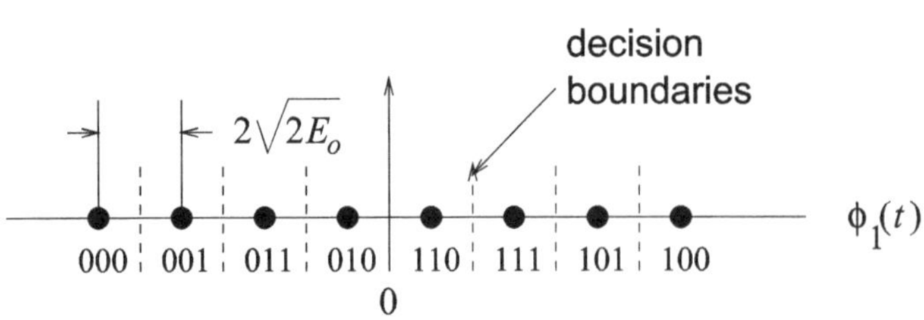

Fig. 6.15 Complex signal-space diagram for 8-PAM. Bits are mapped onto the messages using a Gray code mapping

$$= 2\left(1 - \frac{1}{M}\right) Q\left(\sqrt{\frac{2\alpha^2 E_o}{N_o}}\right). \tag{6.72}$$

From (5.82) E_o is related to the average energy in a message symbol E:

$$P_e = 2\left(1 - \frac{1}{M}\right) Q\left(\sqrt{\frac{6}{M^2 - 1}\gamma_s}\right), \tag{6.73}$$

where

$$\gamma_s = \frac{\alpha^2 E}{N_o} \tag{6.74}$$

is the average symbol energy-to-noise ratio. Substituting $\gamma_s = (\log_2 M)\gamma_b$ gives

$$P_e = 2\left(1 - \frac{1}{M}\right) Q\left(\sqrt{\frac{6(\log_2 M)}{M^2 - 1}\gamma_b}\right). \tag{6.75}$$

6.3.3.4 Error Probability of QAM

Consider a square QAM constellation of size $M = 4^m$ for some integer m. Such a QAM system can be viewed as two $\sqrt{M}$-PAM systems in quadrature, each allocated one-half the power of the QAM system. For example, the 16-QAM system in Fig. 6.16 can be treated as two independent 4-PAM systems in quadrature, each operating with half the power of the 16-QAM system.

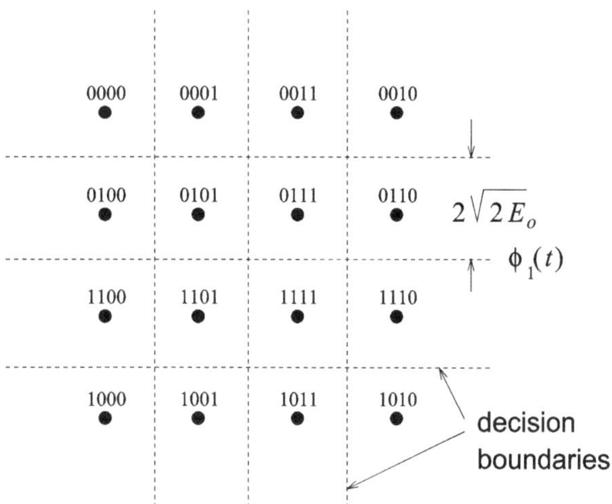

Fig. 6.16 Complex signal-space diagram for 16-QAM constellation. Bits are mapped onto the messages using a Gray code mapping

From (6.73), the message error probability for each $\sqrt{M}$-PAM system is

$$P_{\sqrt{M}} = 2\left(1 - \frac{1}{\sqrt{M}}\right) Q\left(\sqrt{\frac{6(\log_2 M)}{M-1}\frac{\gamma_b}{2}}\right). \tag{6.76}$$

The probability of correct message decision in the QAM system is

$$\mathrm{P}_c = (1 - P_{\sqrt{M}})^2 \tag{6.77}$$

and the probability of message error is

$$P_e = 1 - (1 - P_{\sqrt{M}})^2. \tag{6.78}$$

The QAM message error probability P_e is plotted against γ_b in Fig. 6.17 for 4-, 16-, and 64-QAM. Like PSK, QAM is a bandwidth efficient modulation scheme such that the required bit energy-to-noise ratio γ_b increases with modulation order M for a given message error probability. However, QAM is more power efficient than PSK for the same modulation order M, when $M \geq 8$. For $M = 2, 4$, PSK and QAM have the same message vectors and have the same performance. For QAM cross constellations, such as those in Fig. 5.8, the message error probability can be obtained by defining convex decision regions and using the approach suggested in Sect. 6.3.

6.3.3.5 Error Probability with Orthogonal Messages

Consider an M-ary orthogonal signal set $\tilde{S}$ having equal energy messages:

$$\tilde{\mathbf{s}}_i = \sqrt{2E}\mathbf{e}_m, \quad m = 1, \ldots, M, \tag{6.79}$$

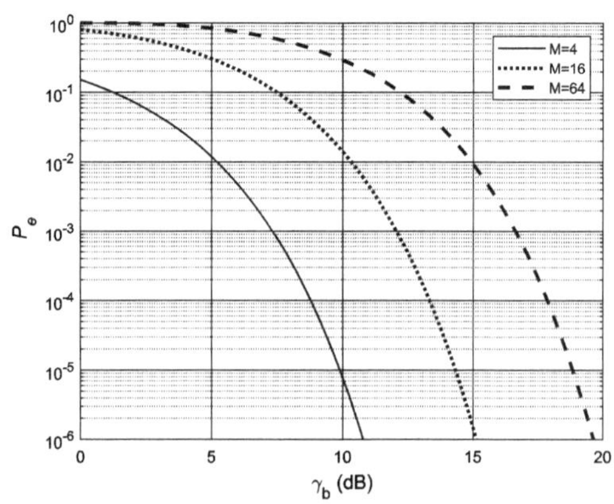

Fig. 6.17 Message error probability P_e versus γ_b for 4-, 16-, and 64-QAM

where $\mathbf{e}_m$ is a length-M unit basis vector with a "1" in the mth coordinate. One such example is the orthogonal MFSK waveforms in Sect. 5.3.2, but the following will apply generally to any set of equal energy orthogonal waveforms.

Since the M-ary orthogonal messages have equal energy, each complex message $\tilde{\mathbf{s}}_i$ lays at the same Euclidean distance $\tilde{d}_{jk} = 2\sqrt{E}$ from every other complex message $\tilde{\mathbf{s}}_j,\ j \neq i$. Hence, all messages will experience the same error probability. Assuming message $\tilde{\mathbf{s}}_1$ is sent, the received complex vector is

$$\tilde{\mathbf{r}} = (\sqrt{2E} + \tilde{w}_1, \tilde{w}_2, \ldots, \tilde{w}_M), \tag{6.80}$$

where the $\tilde{w}_i$ are independent zero-mean complex Gaussian random variables with variance N_o. The ML receiver computes the M decision variables

$$\mu(\tilde{\mathbf{s}}_m) = \text{Re}\left\{\tilde{\mathbf{r}} \cdot \tilde{\mathbf{s}}_m^*\right\}, \quad m = 1, \ldots, M, \tag{6.81}$$

and chooses the message that maximizes $\mu(\tilde{\mathbf{s}}_m)$. It follows that

$$\begin{aligned} \mu(\tilde{\mathbf{s}}_1) &= 2E + \sqrt{2E}\tilde{w}_{I,1} \\ \mu(\tilde{\mathbf{s}}_m) &= \sqrt{2E}\tilde{w}_{I,m}, \quad m = 2, \ldots, M. \end{aligned} \tag{6.82}$$

The $\mu(\tilde{\mathbf{s}}_i), i = 1, \ldots, M$, are independent Gaussian random variables each with variance $2EN_o$; the mean of $\mu(\tilde{\mathbf{s}}_1)$ is $2E$ while the $\mu(\tilde{\mathbf{s}}_m), m \neq 1$, have zero mean.

The probability of correct message may be computed by using the theorem of total probability. The conditional probability of correct message given that $\mu(\tilde{\mathbf{s}}_1) = x$ is the probability that $\mu(\tilde{\mathbf{s}}_m) < x, \forall\, m \neq 1$. This is equal to

$$\text{P}_{c|\mu(\tilde{\mathbf{s}}_1)=x} = \left(\Phi\left(\frac{x}{\sqrt{2EN_o}}\right)\right)^{M-1}. \tag{6.83}$$

Hence,

$$\begin{aligned} \text{P}_c &= \int_{-\infty}^{\infty} \text{P}_{c|\mu(\tilde{\mathbf{s}}_1)=x} f_{\mu(\tilde{\mathbf{s}}_1)}(x)\text{d}x \\ &= \int_{-\infty}^{\infty} \left(\Phi\left(\frac{x}{\sqrt{2EN_o}}\right)\right)^{M-1} \frac{1}{\sqrt{4\pi EN_o}} \exp\left\{-\frac{(x-2E)^2}{4EN_o}\right\} \text{d}x. \end{aligned} \tag{6.84}$$

Now let $y = (x - 2E)/\sqrt{2EN_o}$. Then

$$\text{P}_c = \int_{-\infty}^{\infty} \left(\Phi\left(y + \sqrt{2\gamma_s}\right)\right)^{M-1} \frac{1}{\sqrt{2\pi}} \text{e}^{-y^2/2}\text{d}x, \tag{6.85}$$

where

$$\gamma_s = \frac{\alpha^2 E}{N_o} \tag{6.86}$$

and $\gamma_s = (\log_2 M)\gamma_b$. Finally, the probability of message error is

$$\mathrm{P}_e = 1 - \mathrm{P}_c. \tag{6.87}$$

An alternate expression for the message error probability can be derived by first conditioning on the event that one of the $M-1$ decision variables $\mu(\tilde{\mathbf{s}}_m), m \neq 1$ corresponding to an incorrect message is the largest while all the others are smaller. This gives

$$\begin{aligned}\mathrm{P}_e = (M-1)\int_{-\infty}^{\infty} \Phi\left(\frac{x-2E}{\sqrt{2EN_o}}\right)\left(\Phi\left(\frac{x}{\sqrt{2EN_o}}\right)\right)^{M-2} \\ \times \frac{1}{\sqrt{4\pi EN_o}}\exp\left\{-\frac{x^2}{4EN_o}\right\}\mathrm{d}x,\end{aligned} \tag{6.88}$$

where the factor $(M-1)$ appears in front of the integral since there are $M-1$ choices for the incorrect message that are selected with equal probability. Changing the variable of integration $y = x/\sqrt{2EN_o}$ gives

$$\mathrm{P}_e = (M-1)\int_{-\infty}^{\infty} \Phi\left(y-\sqrt{2\gamma_s}\right)(\Phi(y))^{M-2}\frac{1}{\sqrt{2\pi}}\mathrm{e}^{-y^2/2}\mathrm{d}x, \tag{6.89}$$

where $\gamma_s = (\log_2 M)\gamma_b$. Figure 6.18 plots the probability of message error, P_e, against the bit energy-to-noise ratio, γ_b. Unlike the bandwidth efficient modulation schemes like MPSK and QAM, the power efficiency of coherently detected orthogonal modulation improves as the signal constellation size M increases.

Probability of Bit Error

Orthogonal equal-energy signal constellations have signal vectors all the same Euclidean distance from each other. When a message error occurs, each of the $M-1$

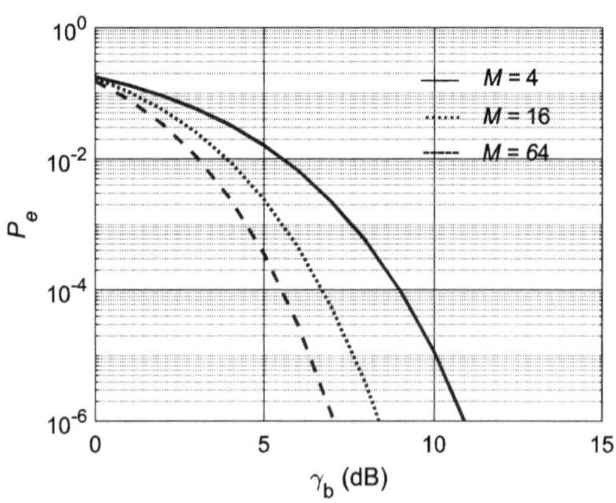

Fig. 6.18 Message error probability P_e versus γ_b for coherently detected 4-, 16-, and 64-orthogonal modulation

incorrect messages is equally likely to be chosen. The set of $M = 2^k$ messages has a one-to-one mapping onto the set of 2^k binary k-tuples as shown in Fig. 6.19. Suppose the all zeroes k-tuple corresponds to the correct message, but the receiver makes an error by choosing the message corresponding to the ith row, $i \neq 1$. There are 2^{k-1} zeros and 2^{k-1} ones in each column, where a "0" corresponds to a correct bit since the all-zeroes k-tuple was assumed sent and a "1" corresponds to a bit error. Hence, the probability of a particular bit position being in error is

$$\mathrm{P}_b = \frac{2^{k-1}}{2^k - 1}\mathrm{P}_e = \frac{M}{2(M-1)}\mathrm{P}_e, \tag{6.90}$$

where P_e is the symbol error probability obtained from either (6.85) and (6.87) or (6.89).

Power Efficiency

As the MPSK, PAM, and QAM constellation sizes M increase, the Euclidean distances between points in their respective signal constellations will decrease for a fixed message energy E. Conversely, to maintain the Euclidean distance between signal points in the constellation, the message energy E must increase as M gets bigger. Hence, the bit energy-to-noise ratio γ_b required to achieve a specified message error probability P_e will increase with the constellation size M.

Orthogonal MFSK signal constellations have messages that are all at Euclidean distance $\tilde{d}_{jk} = 2\sqrt{E}$ from each other no matter the size of the constellation M. With orthogonal MFSK, the bit energy-to-noise ratio γ_b required to achieve a specified message error probability P_e decreases with the constellation size M. This means that orthogonal MFSK is power efficient.

6.4 Noncoherent Detection

If information is transmitted in the amplitude and/or frequency of an RF carrier but not in its phase, then a noncoherent detector may be used since the phase is irrelevant. Noncoherent detectors require no knowledge of the phase ϕ of the complex channel gain $\tilde{h} = \alpha e^{j\phi}$ and are, therefore, easier to implement than

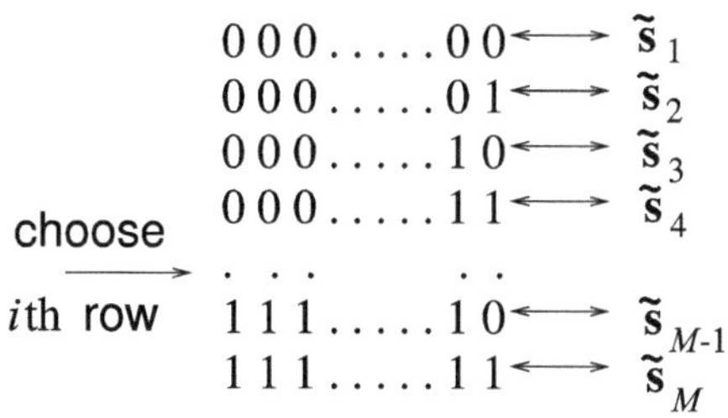

Fig. 6.19 Mapping of binary k-tuples onto M-ary symbols

coherent detectors. If information is transmitted in amplitude the of the RF carrier such as OOK, then knowledge of the magnitude α is still required at the receiver. Noncoherent detectors trade implementation complexity and robustness to channel impairments, e.g., carrier phase estimation error, for power efficiency.

Suppose that one of the M messages in $\tilde{S}$, say $\tilde{s}_i(t)$, is sent and the received complex envelope is

$$\tilde{r}(t) = \tilde{h}\tilde{s}_i(t) + \tilde{w}(t). \tag{6.91}$$

At the receiver $\tilde{r}(t)$ is projected onto the set of basis functions $\{\varphi_n(t)\}_{n=1}^{N}$ to yield the received vector

$$\tilde{\mathbf{r}} = \tilde{h}\tilde{\mathbf{s}}_i + \tilde{\mathbf{w}}, \tag{6.92}$$

where the joint pdf of $\tilde{\mathbf{w}}$ given in (6.11).

The maximum likelihood (ML) noncoherent detector chooses the message vector $\tilde{\mathbf{s}}_m$ that maximizes the joint conditional pdf $f(\tilde{\mathbf{r}}|\alpha, \tilde{\mathbf{s}}_m)$:

$$\text{choose } \tilde{\mathbf{s}}_m \text{ if } \quad f(\tilde{\mathbf{r}}|\alpha, \tilde{\mathbf{s}}_m) \geq f(\tilde{\mathbf{r}}|\alpha, \tilde{\mathbf{s}}_{\hat{m}}) \;\; \forall\, \hat{m} \neq m. \tag{6.93}$$

Having no knowledge of the phase ϕ

$$f(\tilde{\mathbf{r}}|\alpha, \tilde{\mathbf{s}}_m) = \mathrm{E}_{\phi}[f(\tilde{\mathbf{r}}|\tilde{h}, \tilde{\mathbf{s}}_m)] = \int_0^{2\pi} f(\tilde{\mathbf{r}}|\tilde{h}, \tilde{\mathbf{s}}_m) f(\phi) d\phi, \tag{6.94}$$

where $f(\phi)$ is the pdf of ϕ. Using the joint conditional pdf of $f(\tilde{\mathbf{r}}|\tilde{h}, \tilde{\mathbf{s}}_m)$ in (6.12),

$$\begin{aligned} f(\tilde{\mathbf{r}}|\tilde{h}, \tilde{\mathbf{s}}_m) &= \frac{1}{(2\pi N_o)^N} \exp\left\{-\frac{1}{2N_o}\|\tilde{\mathbf{r}} - \tilde{h}\tilde{\mathbf{s}}_m\|^2\right\} \\ &= \frac{1}{(2\pi N_o)^N} \exp\left\{-\frac{\|\tilde{\mathbf{r}}\|^2 + 2\alpha^2 E_m}{2N_o}\right\} \exp\left\{\frac{1}{N_o}\mathrm{Re}\{\tilde{\mathbf{r}} \cdot \tilde{h}^* \tilde{\mathbf{s}}_m^*\}\right\}, \end{aligned} \tag{6.95}$$

where E_m is the energy in the message $\tilde{\mathbf{s}}_m$.

The dot product in (6.95) yields a complex-valued scalar that may be expressed in polar coordinates as

$$\tilde{\mathbf{r}} \cdot \tilde{\mathbf{s}}_m^* = X_m \mathrm{e}^{j\theta_m} \tag{6.96}$$

such that

$$\tilde{\mathbf{r}} \cdot \tilde{h}^* \tilde{\mathbf{s}}_m^* = \tilde{h}^* \tilde{\mathbf{r}} \cdot \tilde{\mathbf{s}}_m^* = \tilde{h}^* X_m^{j\theta_m} = \alpha X_m \mathrm{e}^{j(\theta_m - \phi)}. \tag{6.97}$$

Hence,

$$f(\mathbf{r}|\tilde{h}, \tilde{\mathbf{s}}_m) = \frac{1}{(2\pi N_o)^N} \exp\left\{-\frac{\|\tilde{\mathbf{r}} + 2\alpha^2 E_m\|^2}{2N_o}\right\} \exp\left\{\frac{\alpha X_m}{N_o}\cos(\theta_m - \phi)\right\}. \tag{6.98}$$

In the absence of any prior phase information the random phase ϕ has maximum uncertainty, so its pdf, $f_\phi(\phi)$, is taken to be uniformly distributed on $[-\pi, \pi)$ with density $1/(2\pi)$. Averaging (6.98) over the pdf of the random phase ϕ gives

$$\begin{aligned} f(\tilde{\mathbf{r}}|\alpha, \tilde{\mathbf{s}}_m) &= \frac{1}{(2\pi N_o)^N} \exp\left\{-\frac{2\alpha^2 E_m + \|\tilde{\mathbf{r}}\|^2}{2N_o}\right\} \frac{1}{2\pi} \int_0^{2\pi} \exp\left\{\frac{\alpha X_m}{N_o}\cos(\theta_m - \phi)\right\} \mathrm{d}\phi \\ &= \frac{1}{(2\pi N_o)^N} \exp\left\{-\frac{2\alpha^2 E_m + \|\tilde{\mathbf{r}}\|^2}{2N_o}\right\} I_0\left(\frac{\alpha X_m}{N_o}\right), \end{aligned} \tag{6.99}$$

where $I_0(x)$ is the zero-order modified Bessel function of the first kind defined in (2.101). Since the terms $\|\tilde{\mathbf{r}}\|^2$ and $(2\pi N_o)^N$ are irrelevant in the decision, the message that maximizes the likelihood $p(\tilde{\mathbf{r}}|\alpha, \tilde{\mathbf{s}}_m)$ also maximizes the decision metric:

$$\mu_1(\tilde{\mathbf{s}}_m) = \exp\left\{-\frac{\alpha^2 E_m}{N_o}\right\} I_0\left(\frac{\alpha X_m}{N_o}\right). \tag{6.100}$$

An important case is equal energy messages $E_m = E$, such that the ML receiver can choose $\tilde{\mathbf{s}}_m$ to maximize

$$\mu_2(\tilde{\mathbf{s}}_m) = I_0\left(\frac{\alpha X_m}{N_o}\right). \tag{6.101}$$

However, $I_0(x)$ increases monotonically with x. Therefore, the ML receiver can simply choose $\tilde{\mathbf{s}}_m$ to maximize

$$\mu_3(\tilde{\mathbf{s}}_m) = X_m \;\text{ or }\; \mu_3(\tilde{\mathbf{s}}_m) = X_m^2, \tag{6.102}$$

where

$$X_m^2 = \mathrm{Re}^2(\tilde{\mathbf{r}} \cdot \tilde{\mathbf{s}}_m^*) + \mathrm{Im}^2(\tilde{\mathbf{r}} \cdot \tilde{\mathbf{s}}_m^*). \tag{6.103}$$

In summary, the noncoherent receiver first uses the quadrature demodulator in Fig. 6.1 to extract the real and imaginary components of the complex envelope $\tilde{r}_I(t)$ and $\tilde{r}_Q(t)$. Using the correlator detector in Fig. 6.2 or matched filter detector in Fig. 6.4, it then produces the received vector $\mathbf{r}$. For equal energy messages, the decision variables X_m^2, $m = 1, \ldots, M$ are computed, and the message $\tilde{\mathbf{s}}_m$ is selected having the largest X_m^2. This leads to the detector structure shown in Fig. 6.20,

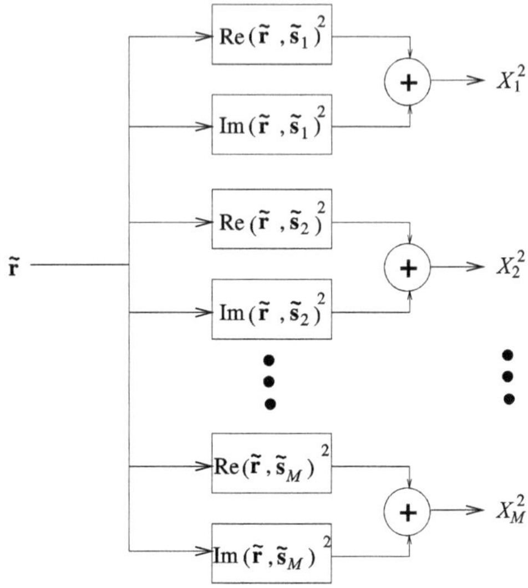

Fig. 6.20 Noncoherent square-law detector for orthogonal equal energy messages

commonly known as a *square-law detector* due to the squaring operations in the branches.

If the messages $\tilde{\mathbf{s}}_m$ are of unequal energy, as, for example, with OOK, then the metric in (6.100) must be used instead. This adds to receiver complexity, because the channel gain α must be estimated and the Bessel function $I_0(x)$ calculated.

6.4.1 Error Probability with Orthogonal Messages

Consider, again, an M-ary orthogonal signal set $\tilde{S}$ having equal energy messages:

$$\tilde{\mathbf{s}}_i = \sqrt{2E}\mathbf{e}_m, \quad m = 1, \ldots, M, \tag{6.104}$$

where $\mathbf{e}_m$ is a length-M unit basis vector with a "1" in the mth coordinate. Once again, all messages will have the same error probability. One important example is the equal energy MFSK waveforms in (5.63). With coherent detection the orthogonal MFSK waveforms must be spaced $\Delta_f = 1/2T$ Hz apart. However, with noncoherent detection the MFSK waveforms will have unknown random phases. In this case, Problem 5.8 shows that the MFSK waveforms must be spaced a multiple of $\Delta_f = 1/T$ Hz to make them mutually orthogonal.

To derive the error probability assume without loss of generality that $\tilde{\mathbf{s}}_1$ is sent such that the received vector $\tilde{\mathbf{r}} = (\tilde{r}_1, \tilde{r}_2, \ldots, \tilde{r}_N)$ has components

$$\tilde{r}_1 = \tilde{h}\sqrt{2E} + \tilde{w}_1 \tag{6.105a}$$

$$\tilde{r}_i = \tilde{w}_i \ , \ \ i = 2, \ldots, M, \tag{6.105b}$$

where, once again, $\tilde{h} = \alpha e^{j\phi}$. For equal-energy M-ary orthogonal signals, the metric in (6.102) may be assumed that produces the magnitudes

$$\begin{aligned} X_1 &= |\tilde{\mathbf{r}} \cdot \tilde{\mathbf{s}}_1^*| \\ &= |2E\tilde{h} + \sqrt{2E}\tilde{w}_1| \\ &= \left|2E\alpha\cos\phi + \sqrt{2E}\tilde{w}_{I,1} + j\left(2E\alpha\sin\phi + \sqrt{2E}\tilde{w}_{Q,1}\right)\right| \end{aligned} \tag{6.106}$$

and

$$\begin{aligned} X_m &= |\tilde{\mathbf{r}} \cdot \tilde{\mathbf{s}}_m^*| \\ &= |\sqrt{2E}\tilde{w}_m| \\ &= \left|\sqrt{2E}\tilde{w}_{I,m} + j\sqrt{2E}\tilde{w}_{Q,m}\right| \ , \quad m = 2, \ldots, M. \end{aligned} \tag{6.107}$$

The receiver will make a correct decision if

$$X_1 > X_i \ , \quad \forall\, i \neq 1. \tag{6.108}$$

From Sect. 2.7.6, X_1 can be shown to have the Rice pdf:

$$p_{X_1}(x) = \frac{x}{2EN_o}\exp\left\{-\frac{x^2 + 4\alpha^2E^2}{4EN_o}\right\} I_0\left(\frac{\alpha x}{N_o}\right). \tag{6.109}$$

Likewise, from Sect. 2.7.4, the X_i, $i \neq 0$, are i.i.d Rayleigh random variables having pdf

$$p_{X_i}(x) = \frac{x}{2EN_o}\exp\left\{-\frac{x^2}{4EN_o}\right\} \ , \quad i = 2, \ldots, M. \tag{6.110}$$

The probability of correct message reception is

$$\begin{aligned} \mathrm{P}_c &= P\left(X_2 < X_1, X_3 < X_1, \ldots, X_M < X_1\right) \\ &= \int_0^\infty \left(\int_0^y \frac{x}{2EN_o}\exp\left\{-\frac{x^2}{4EN_o}\right\} \mathrm{d}x\right)^{M-1} p_{X_1}(y)\mathrm{d}y \\ &= \int_0^\infty \left(1 - \exp\left\{-\frac{y^2}{4EN_o}\right\}\right)^{M-1} p_{X_1}(y)\mathrm{d}y. \end{aligned} \tag{6.111}$$

Next, use the binomial expansion

$$(a+b)^n = \sum_{k=0}^{n} \binom{n}{k} a^k b^{n-k} \tag{6.112}$$

to write

$$(1-x)^n = \sum_{k=0}^{n} \binom{n}{k} (-1)^k x^k$$

where

$$x = \exp\left\{-\frac{y^2}{4EN_o}\right\}.$$

Using the above, (6.111) becomes

$$\mathrm{P}_c = \sum_{k=0}^{M-1} (-1)^k \binom{M-1}{k} \int_0^\infty \exp\left\{-\frac{ky^2}{4EN_o}\right\} p_{X_1}(y)\mathrm{d}y. \tag{6.113}$$

The integral in the above expression is

$$\begin{aligned} \mathcal{I} &= \int_0^\infty \exp\left\{-\frac{ky^2}{4EN_o}\right\} p_{X_1}(y)\mathrm{d}y \\ &= \int_0^\infty \exp\left\{-\frac{ky^2}{4EN_o}\right\} \frac{y}{2EN_o} \exp\left\{-\frac{y^2+4\alpha^2E^2}{4EN_o}\right\} I_0\left(\frac{\alpha y}{N_o}\right)\mathrm{d}y \\ &= \int_0^\infty \frac{y}{2EN_o} \exp\left\{-\frac{(k+1)y^2+4\alpha^2E^2}{4EN_o}\right\} I_0\left(\frac{\alpha y}{N_o}\right)\mathrm{d}y. \end{aligned} \tag{6.114}$$

The trick is to pull a term out of the integral $\mathcal{I}$ such that the remaining integrand is a Ricean pdf that integrates to unity. This may accomplished by making the following substitutions and simplifying

$$N_o' = \frac{2EN_o}{k+1} \qquad E' = \frac{E}{(k+1)} \tag{6.115}$$

and rearranging the integrand. This gives

$$\mathcal{I} = \frac{1}{k+1}\exp\left\{-\frac{2k\alpha^2E'^2}{N_o'}\right\} = \frac{1}{k+1}\exp\left\{-\frac{k\gamma_s}{(k+1)}\right\}, \tag{6.116}$$

where $\gamma_s = \alpha^2 E/N_o$ is the symbol energy-to-noise ratio. Hence, the probability of correct symbol reception is

$$\mathrm{P}_c = \sum_{k=0}^{M-1} \frac{(-1)^k\binom{M-1}{k}}{k+1} \exp\left\{-\frac{k\gamma_s}{(k+1)}\right\} \tag{6.117}$$

and the probability of message error is

$$\mathrm{P}_e = 1 - \mathrm{P}_c = \sum_{k=1}^{M-1} \frac{(-1)^{k+1}\binom{M-1}{k}}{k+1} \exp\left\{-\frac{k\gamma_s}{(k+1)}\right\}. \tag{6.118}$$

Like coherently detected orthogonal MFSK, when a decision error occurs each of the $M-1$ incorrect messages are equally likely to be chosen and the bit error probability is given by (6.90). Further, using $\gamma_s = \gamma_b \log_2 M$ gives the probability of bit error as a function of γ_b:

$$P_b = \frac{M}{2(M-1)} \sum_{k=1}^{M-1} \frac{(-1)^{k+1}\binom{M-1}{k}}{k+1} \exp\left\{-\frac{k(\log_2 M)\gamma_b}{(k+1)}\right\}. \tag{6.119}$$

Figure 6.21 plots the probability of message error, P_e, against the bit energy-to-noise ratio, γ_b. Once again, the power efficiency improves as the signal constellation size M increases with noncoherently detected orthogonal modulation.

6.5 Bounds on Error Probability

Consider again an arbitrary pair of complex message vectors $\tilde{\mathbf{s}}_j$ and $\tilde{\mathbf{s}}_k$ that are drawn from a signal set of size M and ignore the remaining message vectors. Suppose that $\mathbf{s}_k$ is sent and let $\mathcal{E}_j$ denote the event that the receiver chooses $\mathbf{s}_j$ instead. The probability of the event $\mathcal{E}_j$ is the pairwise error probability $\mathrm{P}_{e|j,k}$. The probability of message error given that $\mathbf{s}_k$ is sent is the probability of the union of all events $\mathcal{E}_j,\ j \neq k$:

$$\mathrm{P}_{e|k} = \mathrm{P}\left[\bigcup_{j \neq k} \mathcal{E}_j\right]. \tag{6.120}$$

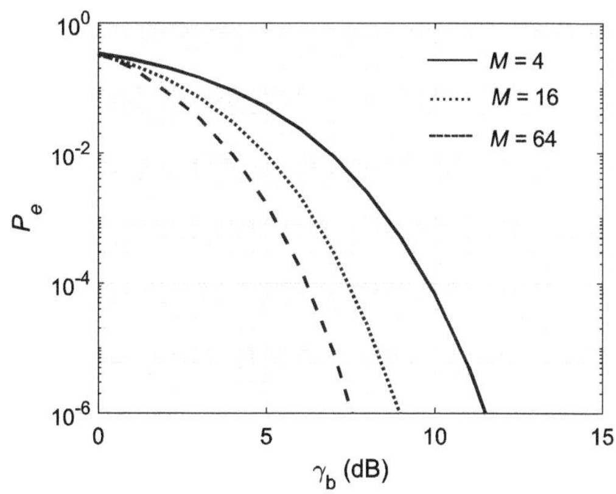

Fig. 6.21 Message error probability P_e versus γ_b for noncoherently detected 4-, 16-, and 64-orthogonal modulation

The events $\mathcal{E}_j$ will often overlap and this complicates the calculation of the conditional error probability $\mathrm{P}_{e|k}$ except in special cases.

An upper bound on the error probability can be obtained by employing the *union bound*:

$$\mathrm{P}\left[\bigcup_{j\neq k}\mathcal{E}_j\right] \leq \sum_{j\neq k}\mathrm{P}[\mathcal{E}_j]. \tag{6.121}$$

This gives the upper bound

$$\mathrm{P}_{e|k} \leq \sum_{j\neq k}\mathrm{P}_{e|j,k}. \tag{6.122}$$

Combining the above result with (6.48) for an AWGN channel gives

$$\mathrm{P}_{e|k} \leq \sum_{j\neq k} Q\left(\sqrt{\frac{\alpha^2\tilde{d}_{jk}^2}{4N_o}}\right). \tag{6.123}$$

Averaging the conditional error probability over all M equally likely messages gives

$$\mathrm{P}_e \leq \frac{1}{M}\sum_{k=1}^{M}\sum_{j\neq k} Q\left(\sqrt{\frac{\alpha^2\tilde{d}_{jk}^2}{4N_o}}\right) \tag{6.124}$$

which only depends on the set of squared Euclidean distances in the complex signal constellation $\{\tilde{d}_{jk}^2\}$ that is readily available.

The union bound can be further simplified by first finding the minimum squared Euclidean distance:

$$\tilde{d}_{\min}^2 = \min_{j,k}\|\tilde{\mathbf{s}}_j - \tilde{\mathbf{s}}_k\|^2. \tag{6.125}$$

Then the pairwise error probability in each term of the union bound is bounded as

$$\mathrm{P}_{e|j,k} = Q\left(\sqrt{\frac{\alpha^2\tilde{d}_{jk}^2}{4N_o}}\right) \leq Q\left(\sqrt{\frac{\alpha^2\tilde{d}_{\min}^2}{4N_o}}\right). \tag{6.126}$$

Hence, (6.124) may be further upper bounded as

$$\mathrm{P}_e \leq (M-1)Q\left(\sqrt{\frac{\alpha^2\tilde{d}_{\min}^2}{4N_o}}\right). \tag{6.127}$$

This upper bound is very simple for any signal constellation but can be quite loose.

Other upper bounds may be obtained by upper bounding the Gaussian Q-function. One such upper bound on $Q(x)$ is

$$Q(x) \leq \tfrac{1}{2}\mathrm{e}^{-x^2/2} \qquad x \geq 0. \tag{6.128}$$

Combining this with the union bound in (6.124) gives

$$\mathrm{P}_e \leq \frac{1}{2M}\sum_{k=1}^{M}\sum_{j\neq k}\exp\left\{-\frac{\alpha^2\tilde{d}_{jk}^2}{8N_o}\right\}. \tag{6.129}$$

Combining (6.128) with the upper bound in (6.127) gives the simplest but loosest upper bound of all

$$\mathrm{P}_e \leq \frac{M-1}{2}\exp\left\{-\frac{\alpha^2\tilde{d}_{\min}^2}{8N_o}\right\}. \tag{6.130}$$

6.5.1 Lower Bound on Error Probability

A useful lower bound on the probability of decision error can be obtained by bounding the error probability:

$$\mathrm{P}_{e|k} \geq \begin{cases} Q\left(\sqrt{\frac{\alpha^2\tilde{d}_{\min}^2}{4N_o}}\right) & \text{, if } \tilde{\mathbf{s}}_k \text{ has at least one neighbor at distance } d_{\min} \\ 0 & \text{, otherwise.} \end{cases} \tag{6.131}$$

Then

$$\mathrm{P}_e = \frac{1}{M}\sum_{k=1}^{M}\mathrm{P}_{e|k} \tag{6.132}$$

$$\geq \frac{w_{\min}}{M}Q\left(\sqrt{\frac{\alpha^2\tilde{d}_{\min}^2}{4N_o}}\right), \tag{6.133}$$

where $w_{\min}$ is the number of message vectors having at least one minimum distance neighbor. Certainly $w_{\min} \geq 2$, so that

$$\mathrm{P}_e \geq \frac{2}{M}Q\left(\sqrt{\frac{\alpha^2\tilde{d}_{\min}^2}{4N_o}}\right). \tag{6.134}$$

6.5.2 *Bit Versus Symbol Error Probabilities*

The discussion so far has mostly considered the probability of decision error otherwise known as the *symbol error probability*, P_e. Also of interest is the *bit error probability*, P_b. For orthogonal MFSK waveforms, the *bit error probability* was neatly related to the symbol error probability in (6.90). In general, the exact bit error probability will depend on the particular mapping between the data bits and modulation symbols and can be complicated to derive even for simple cases. Since each M-ary modulation symbol is mapped onto $\log_2 M$ data bits, the bit error probability can be bounded as

$$\frac{\mathrm{P}_e}{\log_2 M} \le \mathrm{P}_b \le \mathrm{P}_e. \tag{6.135}$$

The lower bound results because each symbol error causes at least one bit error, while the upper bound results because a symbol error causes at most $\log_2 M$ bit errors.

6.5.2.1 Gray Mapping

For signal constellations such as PSK and QAM, the binary data bits may be mapped onto the messages using a Gray mapping, such that the nearest neighboring messages (in Euclidean distance) differ in only one bit position. When γ_b is high, message errors tend to be made onto the nearest neighboring messages that will produce single bit errors. Thus,

$$\mathrm{P}_b \approx \frac{\mathrm{P}_e}{\log_2 M}, \tag{6.136}$$

which is the lower bound.

One method of producing Gray mappings uses Karnough maps that order the cells in a Gray code. For example, consider the 2×4 and 4×4 Karnough maps in Fig. 6.22. For the 8-PSK constellation in Fig. 6.13 the mapping is accomplished by starting with the mapping 000 on $\theta_0 = 0$ and following the line through the cells in the Karnough map as the angle θ_k proceeds counterclockwise. Likewise, for the 8-PAM constellation in Fig. 6.15, start with the mapping 000 assigned to the leftmost message, follow the line through the cells in the Karnough map, and end with the rightmost message. For the 16-QAM constellation in Fig. 6.16, the Gray mapping is obtained by simply superimposing the cells in the 4×4 Karnough map onto the signal constellation. Note that the Gray code mapping is not unique, but any Gray code mapping will do.

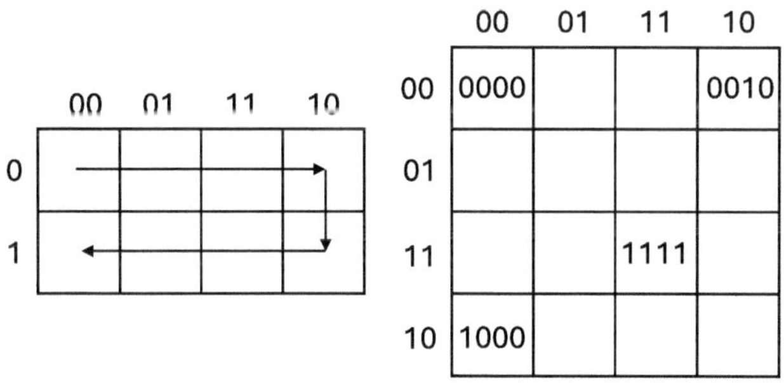

Fig. 6.22 2 × 4 and 4 × 4 Karnough maps for Gray code mapping

Finally, the Gray mapping is optimum for uncoded messages. However, if error control coding is used, other mapping strategies are better. Mapping for coded modulation is discussed further in Chap. 8.

6.5.3 Rotations and Translations

The probability of symbol error in (6.46) is invariant to a rotation of the signal constellation $\{\tilde{\mathbf{s}}_i,\ i = 1, \ldots, M\}$ about the origin. This is a consequence of two properties. First, the probability of message error depends solely on the set of Euclidean distances $\{\tilde{d}_{jk}\}$, $j \neq k$ in the signal constellation. Second, the complex Gaussian components of the noise vector $\tilde{\mathbf{w}}$ are circularly symmetric.

A signal constellation can be rotated about the origin by multiplying each N-dimensional vector $\tilde{\mathbf{s}}_i$ by an $N \times N$ unitary matrix $\mathbf{Q}$. A unitary matrix has the property $\mathbf{Q}\mathbf{Q}^H = \mathbf{Q}^H\mathbf{Q} = \mathbf{I}_{N\times N}$, where $\mathbf{Q}^H$ is the conjugate transpose of $\mathbf{Q}$, and $\mathbf{I}_{N\times N}$ is the $N \times N$ identity matrix. The rotated message vectors are

$$\hat{\mathbf{s}}_i = \tilde{\mathbf{s}}_i\mathbf{Q}, \quad i = 1, \ldots, M. \tag{6.137}$$

Correspondingly, the noise vector $\tilde{\mathbf{w}}$ is replaced with its rotated version:

$$\hat{\mathbf{w}} = \tilde{\mathbf{w}}\mathbf{Q}. \tag{6.138}$$

The rotated complex Gaussian noise vector $\hat{\mathbf{w}}$ can be completely described by its mean and covariance matrix. The mean is

$$\mathrm{E}[\hat{\mathbf{w}}] = \mathrm{E}[\tilde{\mathbf{w}}]\mathbf{Q} = \mathbf{0}. \tag{6.139}$$

The covariance matrix is[2]

$$\begin{aligned}
\boldsymbol{\Phi}_{\hat{\mathbf{w}}\hat{\mathbf{w}}} &= \tfrac{1}{2}\mathrm{E}[\hat{\mathbf{w}}^H\hat{\mathbf{w}}] \\
&= \tfrac{1}{2}\mathrm{E}[(\tilde{\mathbf{w}}\mathbf{Q})^H\tilde{\mathbf{w}}\mathbf{Q}] \\
&= \tfrac{1}{2}\mathrm{E}[\mathbf{Q}^H\tilde{\mathbf{w}}^H\tilde{\mathbf{w}}\mathbf{Q}] \\
&= \mathbf{Q}^H\tfrac{1}{2}\mathrm{E}[\tilde{\mathbf{w}}^H\tilde{\mathbf{w}}]\mathbf{Q} \\
&= N_o\mathbf{Q}^H\mathbf{Q} = N_o\mathbf{I}_{N\times N},
\end{aligned} \tag{6.140}$$

where $\mathbf{x}^H$ is the complex conjugate transpose of the vector $\mathbf{x}$. Since the statistical properties of the noise vector are invariant to rotation, the probability of message error is invariant to rotation of the signal constellation about the origin of the signal space.

Next, consider a translation of the signal set such that

$$\hat{\mathbf{s}}_i = \tilde{\mathbf{s}}_i - \mathbf{a}, \quad i = 1, \ldots, M, \tag{6.141}$$

where $\mathbf{a}$ is a constant vector. In this case, the error probability remains the same since $\hat{d}_{jk} = \tilde{d}_{jk}, j \neq k$. However, the average symbol energy in the signal constellation is altered by the translation and becomes

$$\begin{aligned}
\hat{E} &= \frac{1}{2}\sum_{i=1}^{M}\|\hat{\mathbf{s}}_i\|^2 P_i \\
&= \frac{1}{2}\sum_{i=1}^{M}\|\tilde{\mathbf{s}}_i - \mathbf{a}\|^2 P_i \\
&= \frac{1}{2}\sum_{i=1}^{M}\left\{\|\tilde{\mathbf{s}}_i\|^2 - 2\mathrm{Re}\left\{\tilde{\mathbf{s}}_i \cdot \mathbf{a}^*\right\} + \|\mathbf{a}\|^2\right\} P_i \\
&= \frac{1}{2}\sum_{i=1}^{M}\|\tilde{\mathbf{s}}_i\|^2 P_i - \mathrm{Re}\left\{\sum_{i=1}^{M}\tilde{\mathbf{s}}_i P_i \cdot \mathbf{a}^*\right\} + \frac{1}{2}\|\mathbf{a}\|^2\sum_{i=1}^{M}P_i \\
&= E - \mathrm{Re}\left\{\mathrm{E}[\,\tilde{\mathbf{s}}\,] \cdot \mathbf{a}^*\right\} + \tfrac{1}{2}\|\mathbf{a}\|^2
\end{aligned} \tag{6.142}$$

where E is the average symbol energy and $\mathrm{E}[\,\tilde{\mathbf{s}}\,] = \sum_{i=0}^{M-1}\tilde{\mathbf{s}}_i P_i$ is the centroid or center of mass of the original signal constellation.

[2] Since the vector $\hat{\mathbf{w}}$ has zero mean, its covariance matrix $\boldsymbol{\Lambda}_{\hat{\mathbf{w}}\hat{\mathbf{w}}}$ is equal to its autocorrelation matrix $\boldsymbol{\Phi}_{\hat{\mathbf{w}}\hat{\mathbf{w}}}$.

Taking partial derivatives of (6.142) with respect to elements a_k of the complex vector **a** and setting the result equal to zero will yield the optimum translation that minimizes the average energy in the translated signal constellation

$$\mathbf{a}_{\text{opt}} = \mathrm{E}[\,\tilde{\mathbf{s}}\,]. \tag{6.143}$$

With the optimal translation, the center of mass of the translated signal constellation is moved to the origin and the energy in the translated signal constellation is

$$\hat{E}_{\min} = E - \frac{1}{2}\|\mathbf{a}_{\text{opt}}\|^2. \tag{6.144}$$

Problems

6.1 A bandpass pulse

$$s(t) = \begin{cases} \frac{A}{T}t\cos(2\pi f_c t), & 0 \le t \le T \\ 0, & \text{elsewhere} \end{cases}$$

is to be detected in the presence of additive white Gaussian noise (AWGN).

(a) Design a matched filter for $s(t)$. Sketch the waveform at the output of the matched filter.
(b) Now suppose that a correlator detector is used instead. Sketch the waveform at the output of the correlator detector.
(c) Show the matched filter and correlator outputs are the same at the sampling instant $t = T$.

6.2 Consider the detection of a bandpass signal having the complex envelope

$$\tilde{s}(t) = \sin\left(\frac{8\pi t}{T}\right) u_T(t).$$

The noiseless received complex envelope is $\tilde{r}(t) = \tilde{s}(t)$. Suppose $\tilde{r}(t)$ is passed through either a correlator detector or matched filter detector.

(a) Determine the correlator output over the interval $0 \le t \le T$.
(b) Determine the corresponding matched filter output over the interval $0 \le t \le 2T$, assuming that the matched filter includes a delay T to make it causal.
(c) Hence, show that the two outputs are the same only at the time instant $t = T$.

6.3 Consider the matched filter $h_r(t)$ shown in Fig. 6.23 below

(a) Determine the pulse $a(t)$ to which this filter is matched and sketch it as a function of time.

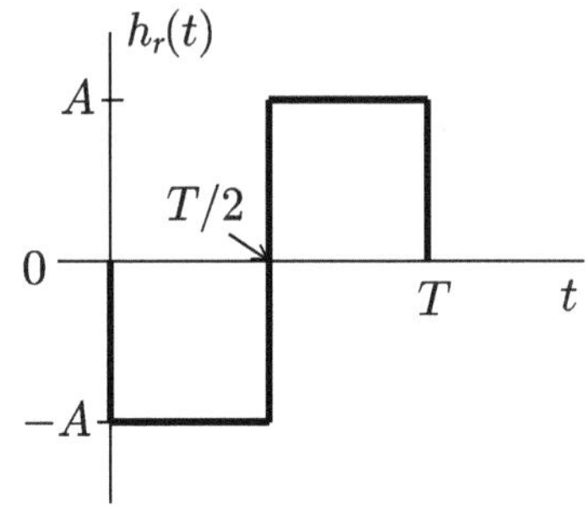

Fig. 6.23 Filter $h_r(t)$ for Problem 6.3

(b) If the pulse $a(t)$ obtained in part (a) is input to the filter $h_r(t)$, plot the output as a function of time.
(c) What is the peak value of the matched filter output in part (b)?

6.4 A matched filter has the frequency response

$$H_r(f) = \frac{1 - e^{-j2\pi f T}}{j2\pi f}.$$

(a) Find the impulse response of the matched filter, $h_r(t)$.
(b) Find the pulse to which this filter is matched, $a(t)$.
(c) Sketch the convolved pulse $a(t) * h_r(t)$ and show all important points.

6.5 Consider a rectangular pulse defined by

$$a(t) = \begin{cases} A \ , & 0 \le t \le T \\ 0 \ , & \text{otherwise.} \end{cases}$$

It is proposed to *approximate* the matched filter for $a(t)$ by an ideal low-pass filter of bandwidth B. Maximization of the peak pulse signal-to-noise ratio η is the primary objective.

(a) Determine the optimal value of the low-pass filter bandwidth B that provides the best approximation to the matched filter, i.e., the value of B that maximizes the peak pulse signal-to-noise ratio η.
(b) By how many decibels is the optimized ideal low-pass filter worse than the true matched filter?

6.6 Find the matched filter impulse response $h_r(t)$ and the corresponding peak pulse signal-to-noise ratio η for the following signal pulses $a(t)$. Assume $N_o/2 = 1$, (A, B, C, and D are constants). Also obtain the transfer function of the matched filter in each case.

(a) $a(t) = A \wedge (t - 1)$

(b) $a(t) = B\text{rect}\left(\frac{t-2}{4}\right)$
(c) $a(t) = C \wedge (t)u(t)$
(d) $a(t) = D[u(t) - 2u(t-1) + 2u(t-2) - u(t-3)]$

Note that

$$\wedge(t) = \begin{cases} 1 - |t|, & |t| \le 1 \\ 0, & \text{otherwise} \end{cases}$$

$$\text{rect}(t) = \begin{cases} 1, & |t| \le 1/2 \\ 0, & \text{otherwise.} \end{cases}$$

6.7 Consider a binary antipodal communication system that transmits information using the half-sinusoidal pulse:

$$a(t) = A\sin\left(\frac{\pi t}{T}\right)u_T(t)$$

according to the mapping rule

$$\text{“0”} \longrightarrow -a(t)$$
$$\text{“1”} \longrightarrow +a(t)$$

The “0”s and “1”s are transmitted with equal probability, and the channel is an AWGN channel, with a two-sided noise power spectral density of $N_o/2$ watts/Hz.

(a) Determine and sketch the filter $h_r(t)$ that is matched to $a(t)$.
(b) Determine and sketch the overall pulse $p(t) = a(t) * h_r(t)$ for the filter you found in part a), labeling all important points.
(c) Determine the pulse energy, E, and the energy per bit, E_b.
(d) What is the probability of bit error in terms of E_b/N_o?

6.8 The two pulses $a_1(t)$ and $a_2(t)$ shown in Fig. 6.24 are used to transmit binary data, such that “0” → $a_1(t)$ and “1” → $a_2(t)$.

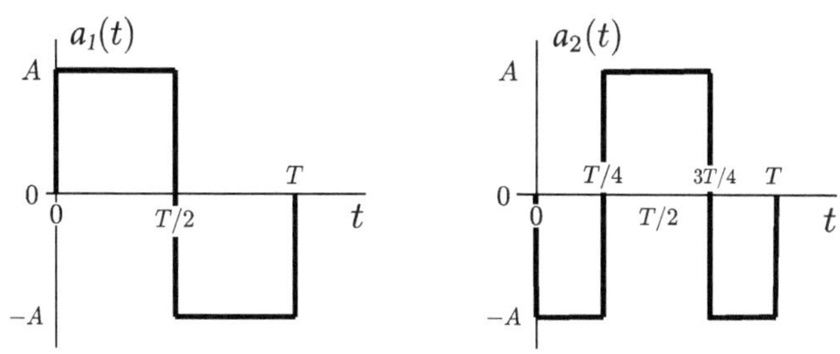

Fig. 6.24 Binary pulses for Problem 6.8

(a) Determine and sketch the corresponding matched filters for $a_1(t)$ and $a_2(t)$.
(b) Sketch the outputs of the two matched filters when the pulse $a_1(t)$ is applied at their input.
(c) The sampled outputs of the matched filters $y_1 \equiv y_1(T)$ and $y_2 \equiv y_2(T)$ are used to make decisions as to whether a "0" or a "1" was sent. Assuming that "0s" and "1s" are sent with equal probability, what simple decision rule based on the observations y_1 and y_2 will minimize the probability of bit error?
(d) What is the probability of bit error in terms of E_b/N_o?

6.9 Suppose the following pulse is used on an additive white Gaussian noise channel with two-sided noise power spectral density $N_o/2$ watts/Hz:

$$a(t) = A\cos\left(\frac{\pi t}{2T}\right)u_T(t)$$

where $u_T(t) = u(t) - u(t-T)$ and $u(t)$ is the unit step function.

(a) What is the bandpass energy contained in the modulated pulse $a(t)\cos(2\pi f_c t)$?
(b) Determine and sketch the impulse response $h_r(t)$ of the matched filter for $h_a(t)$.
(c) What is the peak pulse signal-to-noise ratio η of the sampled output of the matched filter at time T, in terms of A, T, and N_o?
(d) What is the transfer function $H_r(f)$ of the matched filter?
(e) If the pulse $h_a(t)$ is used for bandpass on-off keying with equally likely messages, what is the probability of bit error in terms of the bit-energy-to-noise ratio, E_b/N_o?

6.10 Consider a communication system that uses the following pair of real-valued waveforms with equal probability on an AWGN channel:

$$s_1(t) = u_{T/2}(t) - u_{T/2}(t - T/2)$$
$$s_2(t) = \sin(\pi t/T)u_T(t)$$

where $u_T(t) = u(t) - u(t-T)$ and $u(t)$ is the unit step function.

(a) Find a complete set of orthonormal basis functions to represent these two signals.
(b) Find the signal vectors $\mathbf{s}_1$ and $\mathbf{s}_2$ and plot them in the signal space.
(c) Suppose that the two waveforms are used with equal probability on an AWGN channel with two-sided noise power spectral density $N_o/2$ watts/Hz. What is the probability of bit error in terms of the *bit energy-to-noise ratio*, E_b/N_o?
(d) What E_b/N_o in dB units is required to give a probability of bit error equal to 1×10^{-3}?

6.11 Consider a communication system that uses the following pair of real-valued waveforms with equal probability on an AWGN channel:

$$s_1(t) = Au_T(t)$$

$$s_2(t) = -A \sin\left(\frac{\pi t}{T}\right) u_T(t)$$

where $u_T(t) - u(t) - u(t - T)$ and $u(t)$ is the unit step function.

(a) Obtain a complete set of orthonormal basis functions and sketch them, labeling all important points.
(b) Obtain the complex signal vectors and plot the signal constellation.
(c) If minimum distance decisions are used, obtain an expression for the probability of bit error as a function of the bit-energy-to-noise ratio, E_b/N_o.
(d) What value of E_b/N_o in dB units is required to give a probability of bit error equal to $P_b = 1 \times 10^{-3}$?

6.12 Consider the follow two bandpass waveforms:

$$s_1(t) = A\cos(2\pi f_c t) u_T(t)$$
$$s_2(t) = A\cos(2\pi f_c t + \theta) u_T(t)$$

where $u_T(t) = u(t) - u(t - T)$.

(a) Find a complete set of orthonormal basis functions for this set of signals and write down the corresponding bandpass signal vectors.
(b) Suppose the waveforms are used with equal probability for binary signaling on an AWGN channel with two-sided noise spectral density $N_o/2$ watts/Hz with coherent detection. What is the probability of bit error in terms of the angle θ and E_b/N_o?
(c) What choice of θ will minimize the probability of bit error?

6.13 Consider the binary FSK bandpass signal set:

$$s_1(t) = A\cos(2\pi(f_c - \Delta_f/2)t) u_T(t)$$
$$s_2(t) = A\cos(2\pi(f_c + \Delta_f/2)t) u_T(t)$$

where $f_c = k/2T$ and k is an integer.

(a) Determine and plot the correlation coefficient ρ as a function of Δ_f.
(b) What is the value of Δ_f that will maximize the Euclidean distance?
(c) What is the probability of error for the value of Δ_f obtained in part (b)?
(d) For the value of Δ_f obtained in part (b), determine the increase in E_b/N_o required so that this FSK scheme has the same probability of error as BPSK.

6.14 An 8-PAM signal constellation is shown in Fig. 6.25 where all nearest neighbors are at distance "a." Assume the messages are used with equal probability and the receiver makes minimum distance decisions.

(a) Determine the Voronoi regions and sketch the decision boundaries.
(b) Show a Gray code mapping.

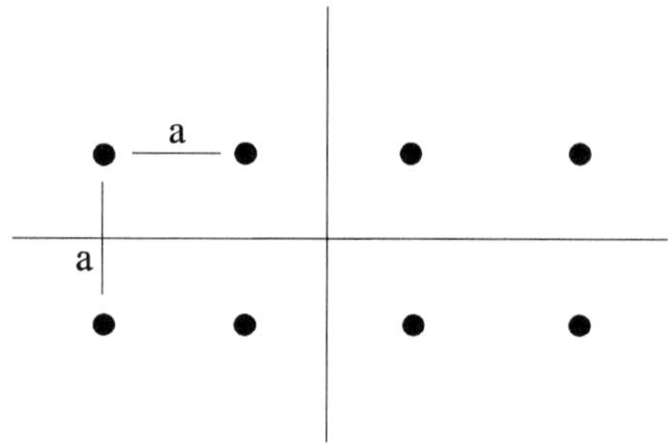

Fig. 6.25 8-PAM constellation for Problem 6.14

(c) Find the *exact* probability of message error in terms of the *bit energy-to-noise ratio*, E_b/N_o.

6.15 Consider the pair of real-valued waveforms $s_0(t)$ and $s_1(t)$ shown in Fig. 6.26.

(a) Using the Gram-Schmidt orthonormalization procedure, find a complete set of orthonormal basis functions.
(b) Find the signal vectors $\mathbf{s}_0$ and $\mathbf{s}_1$ and plot them.
(c) Find the Euclidean distance $d_{12} = \|\mathbf{s}_0 - \mathbf{s}_1\|$.
(d) If the pair of waveforms is used with equal probability on an AWGN channel, write down an expression for the probability of bit error in terms of the bit energy-to-noise ratio E_b/N_o.

6.16 Consider a binary communication system with the signal-space diagram shown in Fig. 6.27.

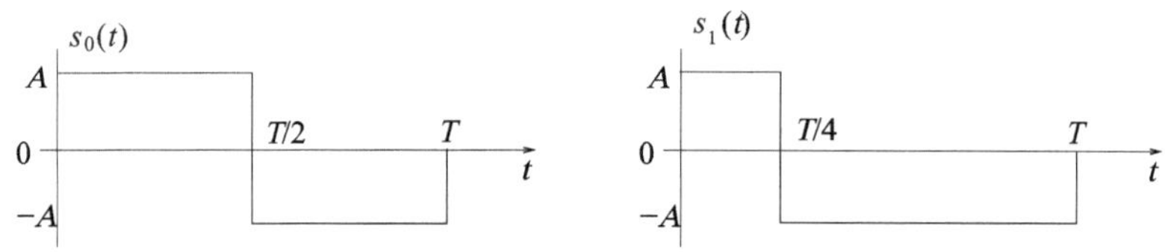

Fig. 6.26 Waveforms for Problem 6.15

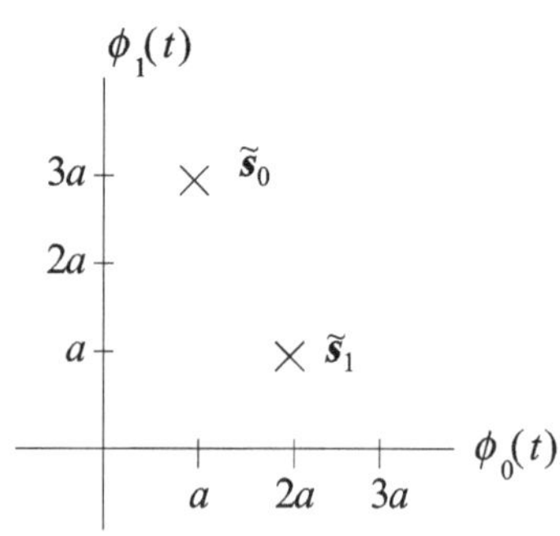

Fig. 6.27 Complex signal-space diagram for Problem 6.16

(a) Calculate the correlation coefficient ρ_{01} between $s_0(t)$ and $s_1(t)$.
(b) What is the average energy per information bit, E_b?
(c) Assuming an AWGN channel with a two-sided noise power spectral density of $N_o/2$ watts/Hz and minimum distance decisions, what is the probability of bit error in terms of E_b/N_o?

6.17 Consider a QAM system on an AWGN channel. Let the maximum symbol rate be $1/T = 2000$ baud.

(a) Determine the required modulation alphabet size M, and the E_b/N_o (in dB) to guarantee a message error probability of $P_e = 10^{-5}$ with a bit rate of $R_b = 4$ kb/s.
(b) Redetermine the above when $R_b = 8$ kb/s and $R_b = 16$ kb/s and draw some conclusions.

6.18 Suppose that the following set of four equally likely bandpass waveforms are used to send messages over an AWGN channel with two-sided noise power spectral density $N_o/2$ watts/Hz with coherent detection at the receiver.

$$s_k(t) = A(k-1)h_a(t)\cos(2\pi f_c t)\,, \quad k = 1, 2, 3, 4$$

where $f_c T \gg 1$ and $\int_{-\infty}^{\infty} h_a^2(t)\mathrm{d}t = 1$.

(a) Represent these signals in a signal space diagram, and determine the optimum decision regions.
(b) Calculate the average message energy E and energy per information bit E_b.
(c) Assuming use of a Gray code, what is the *exact* probability of *bit error* in terms of E_b/N_o? *Do not bound or approximate the bit error probability.*
(d) If the bit rate is $R_b = 1 \times 10^6$ bits/s, and $h_a(t)$ is a root-raised cosine pulse with roll-off factor $\alpha = 0.25$, what is the RF bandwidth used to transmit the signals?

6.19 Consider binary signaling on an AWGN channel. The conditional probability density functions of the matched filter or correlator outputs are

$$f_{y|1}(y|1) = \frac{1}{\sqrt{2\pi}\sigma_w}\exp\left\{\frac{-(y-E)^2}{2\sigma_w^2}\right\},$$

$$f_{y|0}(y|0) = \frac{1}{\sqrt{2\pi}\sigma_w}\exp\left\{\frac{-(y+E)^2}{2\sigma_w^2}\right\},$$

where

$$\sigma_w^2 = \frac{N_o E}{2}.$$

Decisions are made to choose "1" if $y > \lambda$ and to choose "0" if $y < \lambda$.

(a) What is the optimum decision threshold λ if $P_1 = P_0 = 1/2$?

(b) What is the optimum decision threshold λ if $P_1 \neq P_0$?

6.20 Prove the following upper and lower bound on the probability of message error for M-PSK signals, with $M \geq 4$.

$$p \leq P_e \leq 2p$$

where

$$p = Q\left(\sqrt{\frac{2E}{N_o}\sin^2(\pi/M)}\right)$$

and E/N_o is the symbol energy-to-noise ratio.

6.21 A set of M biorthogonal waveforms can be constructed from a set of $M/2$ orthogonal waveforms. The M-ary biorthogonal waveforms have signal vectors

$$\tilde{\mathbf{s}}_i = \begin{cases} \sqrt{2E}\mathbf{e}_i & , \quad i = 1, \ldots, M/2 \\ -\tilde{\mathbf{s}}_{i-M/2} \,, & i = M/2+1, \ldots, M \end{cases},$$

where the unit basis vectors $\mathbf{e}_i$ have length $N = M/2$, and $\mathbf{e}_i = (e_{i1}, e_{i2}, \ldots, e_{iN})$ where $e_{ij} = \delta_{ij}$, is a length-N unit basis vector with a "1" in the ith coordinate and zeroes elsewhere.

Derive a union bound on the probability of message error as a function of the bit energy-to-noise ratio E_b/N_o.

6.22 Consider the set of complex message vectors:

$$\tilde{\mathbf{s}}_i = [\tilde{s}_{i1},\ \tilde{s}_{i2},\ \ldots,\ \tilde{s}_{iN}],\ \ i = 1,\ 2,\ \ldots, M$$

such that

$$\tilde{s}_{ij} = \begin{cases} +d \\ \text{or} \\ -d \end{cases} \quad \text{for all } i,\ j$$

(a) Sketch the signal constellation in two ($N = 2$) and three ($N = 3$) dimensions.
(b) How many messages M are there in N dimensions?
(c) For the case $N = 3$ and an AWGN channel, derive a union bound on the probability of message error as a function of the bit energy-to-noise ratio E_b/N_o.

6.23 Consider the complex signal constellation shown in Fig. 6.28.

(a) Assuming minimum distance decisions, sketch the decision or Voroni regions, R_m, around each signal vector $\tilde{\mathbf{s}}_m$.

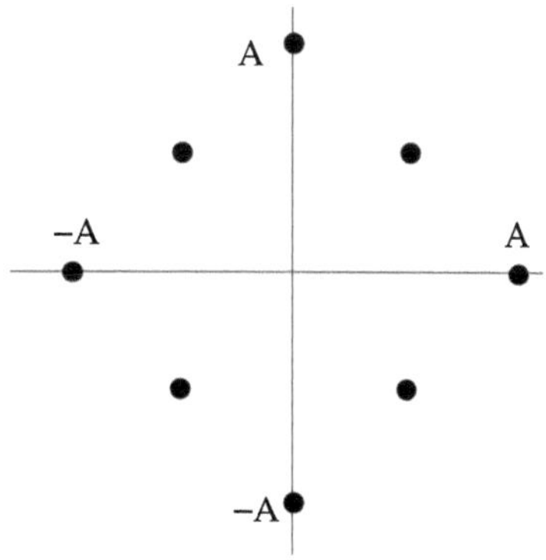

Fig. 6.28 Complex signal constellation for Problem 6.23

(b) Find the *union bound* on the probability of message error. Express your answer in terms of the Euclidean distances between the points in the signal constellation. *Note that the Euclidean distances are a function of the parameter A*.

(c) Express your answer in part (b) in terms of the energy per information bit, E_b.

(d) What is the minimum Euclidean distance, $d_{\min}$, of the signal constellation in terms of E_b?

(e) Using the value of $d_{\min}$ obtained in part (d) write down a simple upper bound on the probability of message error in terms of the bit energy-to-noise ratio E_b/N_o.

6.24 You have been asked to design a modulation scheme that requires an 8-QAM signal constellation. A conventional 8-QAM constellation based on an odd-integer lattice does not use the available power very efficiently. Inspired by a cellular land mobile radio reuse deployment you suggest the signal constellation shown in Fig. 6.29 where complex message vectors are plotted. Assume regular hexagons having sides of length Δ.

(a) Find the *minimum* distance between the message vectors $\tilde{d}_{\min}$ in terms of Δ.

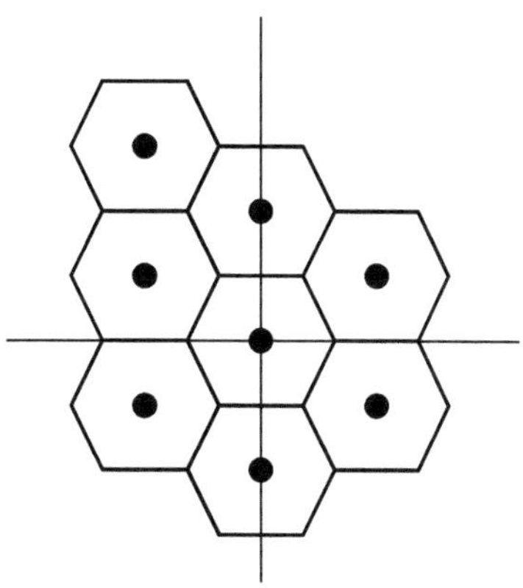

Fig. 6.29 Complex-valued message vectors for Problem 6.24

(b) Find the energy per information bit E_b in terms of Δ.
(c) Assuming equally likely messages, draw the Voronoi regions.
(d) Derive a simple upper bound on the probability of message error based on the minimum distance between signals and the bit energy-to-noise ratio, E_b/N_o.

6.25 Consider the two complex-valued octal signal constellations shown in Fig. 6.30 used to send equally likely messages on an AWGN channel.

(a) Determine and sketch the Voronoi regions in each case.
(b) Find the *minimum Euclidean distance*, $\tilde{d}_{\min}$ of each signal constellation in terms of the average energy per information *bit*, E_b.
(c) Using the minimum distance in part b), write down a simple upper bound on the probability of message error in terms of the bit energy-to-noise ratio E_b/N_o in each case.
(d) Why is the upper bound the same in part (c)?

6.26 Consider the two complex-valued octal signal constellations shown in Fig. 6.31 used to send equally likely messages on an AWGN channel.

(a) Each of the four inner constellation points in the 8-QAM signal constellation have 4 nearest neighbors at minimum distance A. Determine the radii a and b of the inner and outer circles.
(b) The nearest neighbor signal points in the 8-PSK signal constellation are separated by distance A. Determine the radius r of the circle.
(c) Determine the average message or symbol energy for the two signal constellations in terms of A. What is the relative average power advantage of 8-QAM versus 8-PSK (in decibel units) if they both have the same minimum distance A?
(d) For each constellation, derive a simple upper bound on the probability of symbol error on an AWGN channel based on minimum distance. Express your answer in terms of E_b/N_o in each case.

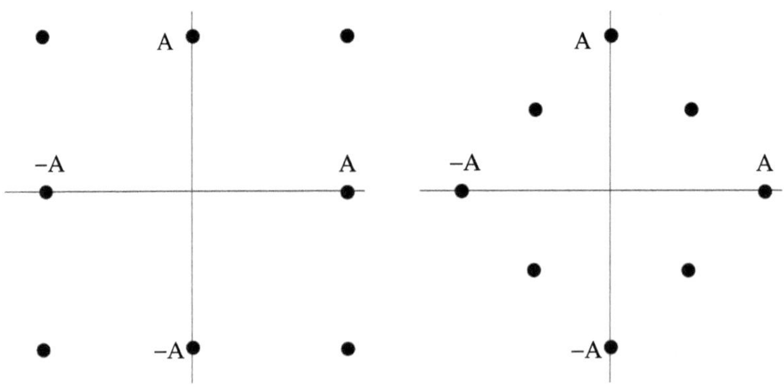

Fig. 6.30 Two complex-valued octal signal constellations for Problem 6.25

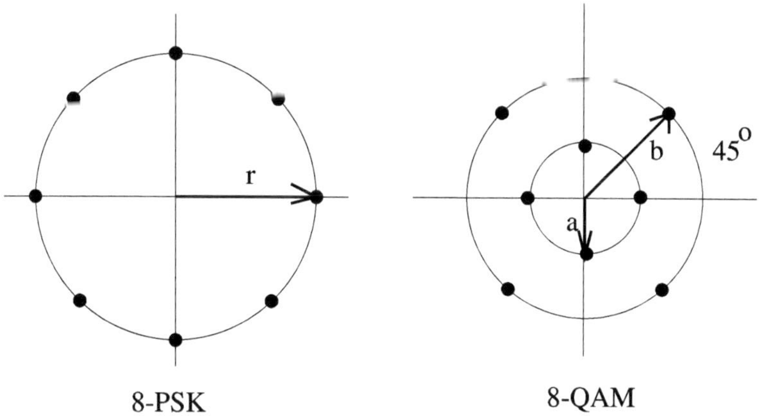

Fig. 6.31 Two complex-valued octal signal constellations for Problem 6.26

6.27 Consider the two complex-valued octal signal constellations in Fig. 6.32 used to send equally likely messages on an AWGN channel.

(a) Find the minimum distance $\tilde{d}_{\min}$ of each signal constellation in terms of the average message energy E.
(b) Using $\tilde{d}_{\min}$, derive a simple upper bound on the probability of message error for each constellation in terms of the bit energy-to-noise ration E_b/N_o.
(c) At high E_b/N_o, which constellation will be better in terms of error probability performance? How much better will it be in terms of the E_b/N_o (in dB units) required to give the same error probability as the other constellation?

6.28 Consider the two complex-valued octal signals constellations in Fig. 6.33 used to send equally likely messages on an AWGN channel. The distance between nearest-neighbor signal points in each constellation is equal to A.

(a) For the constellation on the left, determine the cartesian coordinates of the constellation points.
(b) For the constellation on the right, determine the radii a and b of the inner and outer circles.

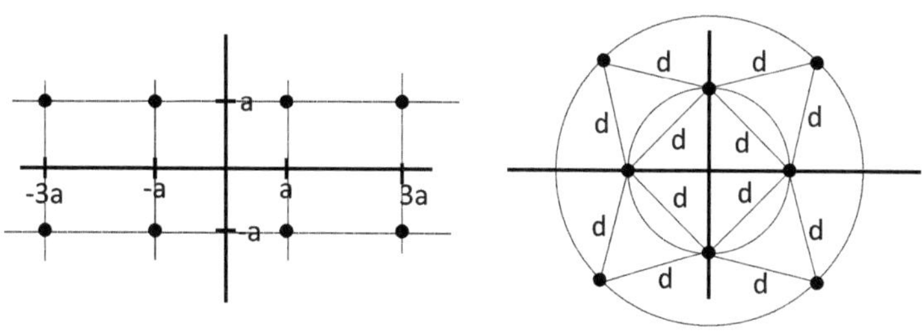

Fig. 6.32 Two complex-valued octal signal constellations for Problem 6.27

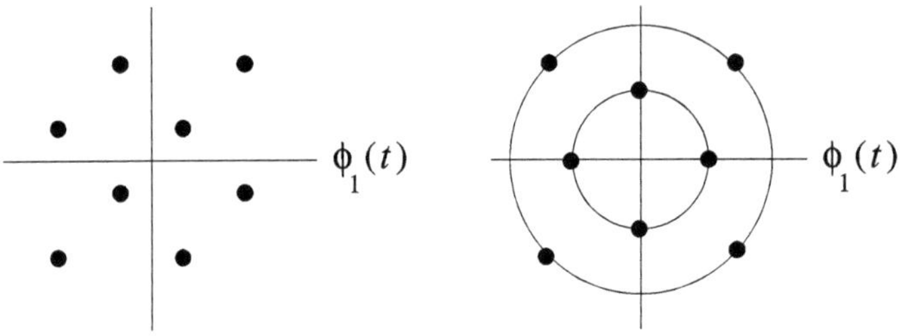

Fig. 6.33 Two complex-valued octal signal constellations for Problem 6.28

(c) Find the energy per message (or modulation symbol) for each of the two signal constellations in terms of A assuming that the messages are used with equal probability.

(d) Which constellation has the lowest energy per message for a given value of A?

6.29 Consider the following set of four signal vectors representing real-valued waveforms.

$$\mathbf{s}_1 = \sqrt{E/5}\,(+1, -1, -1, +1, -1)$$
$$\mathbf{s}_2 = \sqrt{E/5}\,(-1, +1, -1, +1, +1)$$
$$\mathbf{s}_3 = \sqrt{E/5}\,(+1, +1, -1, +1, -1)$$
$$\mathbf{s}_4 = \sqrt{E/5}\,(-1, -1, +1, -1, +1)$$

The messages are used with equal probability on an additive white Gaussian noise channel with two-sided noise power spectral density $N_o/2$ watts/Hz.

(a) Consider minimum distance decisions. Derive a union bound on the probability of message error in terms of E_b/N_o.

(b) Determine the translation that will minimize the average energy in the signal constellation. What is the energy saving in dB units by using the translated signal constellation as compared to the original signal constellation?

(c) Draw a set of four waveforms that would be represented by the original four vectors (not the translated vectors in part [b]).

6.30 Consider the following set of four signal vectors representing real-valued waveforms.

$$\mathbf{s}_1 = \sqrt{E/6}\,(-1, -1, -1, +1, +1, +1)$$
$$\mathbf{s}_2 = \sqrt{E/6}\,(-1, +1, -1, -1, +1, +1)$$
$$\mathbf{s}_3 = \sqrt{E/6}\,(+1, -1, -1, +1, -1, -1)$$
$$\mathbf{s}_4 = \sqrt{E/6}\,(+1, -1, +1, -1, -1, -1)$$

The messages are transmitted with equal probability on an additive white Gaussian noise channel with two-sided noise power spectral density $N_o/2$ watts/Hz.

(a) Consider minimum distance decisions. Derive a union bound on the probability of message error in terms of E_b/N_o.
(b) Derive a simple upper bound on the probability of message error in terms of the minimum distance between signal vectors. Express your answer in terms of E_b/N_o.
(c) Specify a set of time domain waveforms $s_1(t)$, $s_2(t)$, $s_3(t)$, and $s_4(t)$, defined on the interval $0 \le t \le T$ that will have the above signal vectors. *Note: The solution is not unique.*
(d) Determine a translation that will minimize the average energy in the signal constellation. What is the energy saving in dB units by using the translated signal constellation as compared to the original signal constellation?

6.31 Suppose that the following complex signal vectors are used on an AWGN channel with two-sided noise spectral density $N_o/2$ watts/Hz.

$$\tilde{\mathbf{s}}_1 = (a, 2a, -a)$$
$$\tilde{\mathbf{s}}_2 = (a, 0, -a)$$
$$\tilde{\mathbf{s}}_3 = (-a, 2a, -a)$$
$$\tilde{\mathbf{s}}_4 = (-a, 0, -a)$$

(a) Find the *exact* probability of message error in terms of the bit energy-to-noise ratio E_b/N_o.
(b) Obtain the union bound on probability of message error in terms of the bit energy-to-noise ratio E_b/N_o.

6.32 Consider the following 2-D bandpass signal constellations with equally probable messages. Find translated signal constellations in each case so that the message energy is minimized. Plot the signal constellations before and after energy minimization.

(a) $\mathbf{s}_1 = (3, 1),\ \mathbf{s}_2 = (2, 2),\ \mathbf{s}_3 = (-2, -4)$
(b) $\mathbf{s}_1 = (0.2),\ \mathbf{s}_2 = (-1, 3),\ \mathbf{s}_3 = (-4, -4),\ \mathbf{s}_4 = (-2, -4)$
(c) $\mathbf{s}_1 = (3, 3),\ \mathbf{s}_2 = (1, -2),\ \mathbf{s}_3 = (-1, 1),\ \mathbf{s}_4 = (-3, 2)$

6.33 Suppose that the following complex signal constellation is used on an AWGN channel with two-sided noise spectral density $N_o/2$ watts/Hz.

$$\tilde{\mathbf{s}}_1 = (-a, 0, 2a)$$
$$\tilde{\mathbf{s}}_2 = (-a, 2a, 2a)$$
$$\tilde{\mathbf{s}}_3 = (-a, 0, 0)$$
$$\tilde{\mathbf{s}}_4 = (-a, 2a, 0)$$

(a) Find the *exact* probability of message error in terms of the *bit* energy-to-noise ratio, E_b/N_o.
(b) Find the translation $\mathbf{a} = (a_1, a_2, a_3)$ that minimizes the average energy in the signal constellation.
(c) Construct a set of 4 time domain complex envelopes that correspond to the original signal vectors. *The solution is not unique.*

6.34 Suppose that the following complex signal constellation is used on an AWGN channel with two-sided noise spectral density $N_o/2$ watts/Hz.

$$\tilde{\mathbf{s}}_1 = (a, 2a, a)$$
$$\tilde{\mathbf{s}}_2 = (a, 0, a)$$
$$\tilde{\mathbf{s}}_3 = (-a, 2a, a)$$
$$\tilde{\mathbf{s}}_4 = (-a, 0, a)$$

(a) Find the *exact* probability of message error in terms of the *symbol* energy-to-noise ratio. *Hint: Use rotations and translations as appropriate.*
(b) A new set of signal vectors is constructed by translating the vectors as

$$\hat{\mathbf{s}}_i = \tilde{\mathbf{s}}_i - \mathbf{b}\ , \quad i = 1, \ldots 4.$$

What vector $\mathbf{b}$ will minimize the energy in the translated signal constellation $\{\hat{\mathbf{s}}_i\}_{i=1}^4$?
(c) What increase in symbol-energy-to-noise ratio E/N_o n dB units would be required in part (a) to obtain the same error probability as in part (b).

6.35 The two baseband waveforms in Fig. 6.34 are used with equal probability on an AWGN channel with two-sided noise spectral density $N_o/2$ watts/Hz. The parameter δ has the range $0 \leq \delta \leq 1/2$.

(a) Obtain an expression for the probability of bit error as a function of δ and the bit energy-to-noise ratio, E_b/N_o.
(b) What is the error probability when $\delta = 0$?

6.36 A quaternary bandpass signaling scheme has set of four complex envelopes:

$$\tilde{s}_1(t) = 4\varphi_1(t)$$
$$\tilde{s}_2(t) = 2\varphi_1(t) + 2\varphi_2(t)$$
$$\tilde{s}_3(t) = -2\varphi_1(t) - 2\varphi_2(t)$$
$$\tilde{s}_4(t) = -4\varphi_2(t)$$

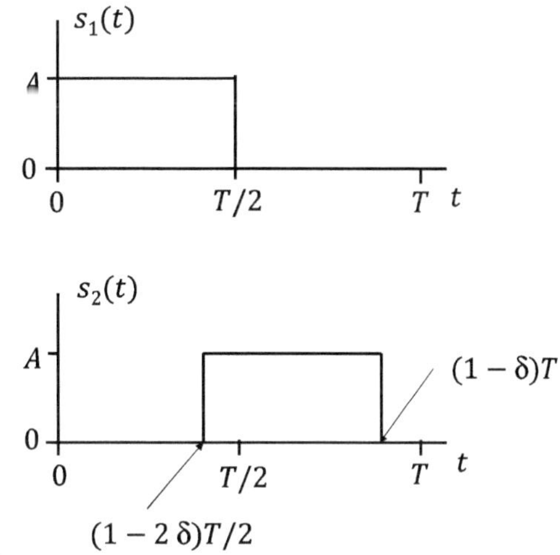

Fig. 6.34 Two baseband waveforms for Problem 6.35

where $\varphi_{(}t)$ and $\varphi_2(t)$ are two orthonormal basis functions. All messages are equiprobable and sent over an AWGN channel with two-sided noise spectral density $N_o/2$ watts/Hz.

(a) Represent these signals in a signal space diagram, and show the Voronoi regions.
(b) What is the *exact* probability of bit error, P_b, for the maximum likelihood receiver in terms of the bit energy-to-noise ratio E_b/N_o? Assume a Gray coding mapping.
(c) Find an equivalent minimum energy signal set.
(d) Determine the reduction in energy in dB units if a translated minimum energy signal set is used instead of the original signal set.

6.37 Consider BSPK modulation with equally likely messages on an AWGN channel with two-sided noise power spectral density $N_o/2$ watts/Hz. The two possible bandpass messages in each bit interval are

$$s_1(t) = -s_2(t) = \sqrt{\frac{2E}{T}}\cos(2\pi f_c t)u_T(t)$$

where E is the bit energy and T is the bit period. The frequency of the receiver local RF oscillator is $f_c + \Delta f$ Hz, where Δ_f is the receiver frequency offset.

(a) Derive the output of the correlator detector as a function of the frequency offset Δf Hz. Assume $f_c T \gg 1$.
(b) What is the probability of bit error in terms of Δf and E_b/N_o?

6.38 Consider BSPK modulation with equally likely messages on an AWGN channel with two-sided noise power spectral density $N_o/2$ watts/Hz. The two possible bandpass messages in each bit interval are

$$s_1(t) = -s_2(t) = \sqrt{\frac{2E}{T}} \cos(2\pi f_c t) u_T(t)$$

where E is the bit energy and T is the bit period. The local RF oscillator has a phase offset φ radians from the correct received carrier phase. Determine the effect of the phase offset φ on the bit error probability.

6.39 Consider the binary OOK system in Sect. 6.3.2.2 operating on an AWGN channel with equally likely messages. Determine the bit error probability under the following scenarios:

(a) Coherent detection
(b) Noncoherent detection. You may leave the error probability in integral form, but simplify as much as possible.

6.40 A multi-hop communication system consists of the serial concatenation of n binary symmetric channels each having crossover probability p as shown in Fig. 6.35, where the number of hops, n, is odd.

(a) What is the end-to-end probability of bit error in terms of n and p?
(b) Evaluate the expression obtained in part a) with $n = 5$ and $p = 0.01$.
(c) Suppose that each hop uses binary OOK in Sect. 6.3.2.2. What is the E_b/N_o in decibel (dB) units needed for $p = 0.01$?

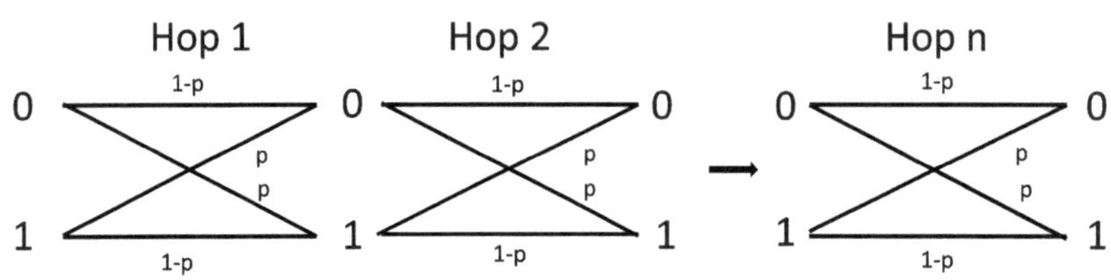

Fig. 6.35 Multi-hop communication system for Problem 6.40

Chapter 7
Information Theory

Information theory was introduced by Claude Shannon in 1948 in his paper "A Mathematical Theory of Communication" [27] which laid the foundation for all modern digital communications. Information theory makes fundamental contributions in many diverse areas including communication theory, probability theory, statistics, physics, and many others.

Consider the basic communication system shown in Fig. 7.1. The encoder is any processing of the source output prior to transmission. The decoder is any processing of the channel output. The objective is to produce an acceptable replica of the source output at the destination or sink. Usually, the encoder/decoder is further partitioned in a communication system as shown in Fig. 7.2. The *data processing theorem* will show that any such partitioning of the encoder and decoder will not impose any fundamental limitations on the end-to-end performance. There are two types of encoding found in communication systems, namely, source coding and channel coding.

Source Coding

There are two types of source encoders/decoders:

- *Non-entropy preserving:* reduces the information content of the source by introducing distortion. For example, low bit rate speech coders
- *Entropy preserving:* represents the source output more efficiently and without any loss of information

A *discrete memoryless source* (DMS) is a source that outputs elements chosen from a finite ensemble:

$$\mathcal{U} = \{(a_1, q_1), (a_2, q_2), \ldots, (a_K, q_K)\}$$

G. Stuber, *Introductory Digital Wireless Communications*,
https://doi.org/10.1007/978-3-032-01559-4_7

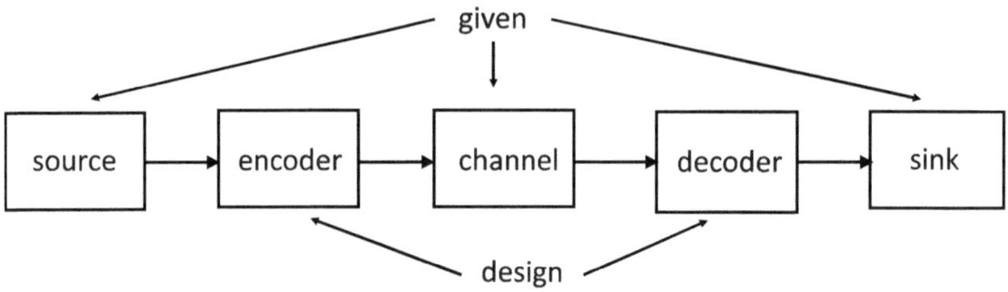

Fig. 7.1 Basic communication system

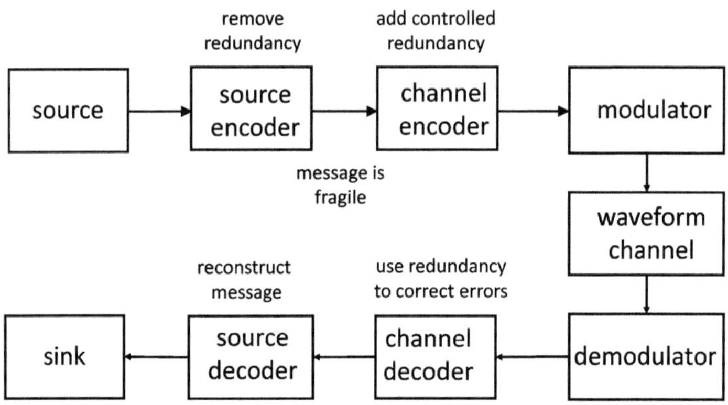

Fig. 7.2 Partitioned communication system

say one every T seconds, where the source element a_i appears with probability q_i and $\sum_{i=1}^{K} q_i = 1$. The outputs are chosen independently from output to output. Thus, the probability that the source outputs the sequence

$$\mathbf{u} = (u_1, u_2, \ldots, u_L)\,, \quad u_i \in \mathcal{U}$$

is

$$\mathrm{P}(\mathbf{u}) = \prod_{i=1}^{L} \mathrm{P}(u_i).$$

The entropy of the DMS is

$$H(U) \triangleq -\sum_{i=1}^{K} q_i \log q_i\,. \tag{7.1}$$

The units of entropy depend on the base used in the log function. If a natural logarithm is used, the units of information are *nats*. If a base-2 logarithm is used instead, the units of information are *bits*. To convert between the two logarithm bases or units of information,

$$\log_2(x) = \frac{\log_e(x)}{\log_e 2}. \tag{7.2}$$

The entropy $H(U)$ of a DMS is the ultimate limit of data compression.

Example 7.1 Suppose that a DMS has a uniform distribution over 16 outputs, i.e.,

$$q_1 = q_2 = \cdots = q_{16} = 1/16.$$

Then

$$H(U) = -\sum_{i=1}^{16} \frac{1}{16}\log_2\frac{1}{16} = 4 \text{ bits}.$$

This is the minimum number of bits needed to specify the output.

Example 7.2 Suppose that a DMS generates 4 possible outputs with

$$q_1 = 1/2,\ q_2 = 1/4,\ q_3 = 1/8,\ q_4 = 1/8.$$

One possible description is

$$u_1 = 00,\ u_2 = 10,\ u_3 = 01,\ u_4 = 11.$$

The average description length is 2 bits.

A more efficient description is

$$u_1 = 1,\ u_2 = 01,\ u_3 = 001,\ u_4 = 000.$$

In this case, the average description length is

$$\bar{\ell} = 1 \times 1/2 + 2 \times 1/4 + 3 \times 1/8 + 3 \times 1/8 = 7/4 \text{ bits}.$$

(continued)

Example 7.2 (continued)
Note that the entropy of the DMS is

$$H(U) = -\tfrac{1}{2}\log_2\tfrac{1}{2} - \tfrac{1}{4}\log_2\tfrac{1}{4} - 2 \times \tfrac{1}{8}\log_2\tfrac{1}{8}$$
$$= 1/2 + 1/2 + 3/4 = 7/4 \text{ bits.}$$

Hence, the second description is as efficient as possible.

Channel Coding

The channel encoder/decoder also has two functions:

- *Error control coding* which allows information to the transmitted reliably through a noisy channel
- *Data translation coding* to control the power spectrum of the modulated signals and to facilitate synchronization and other functions at the receiver

A *discrete memoryless channel* (DMC) accepts as an input a letter from the alphabet:

$$\mathcal{X} = \{(a_1, q_1), (a_2, q_2), \ldots, (a_K, q_K)\}$$

and outputs a letter from the alphabet:

$$\mathcal{Y} = \{(b_1, p_1), (b_2, p_2), \ldots, (b_J, p_J)\}$$

where $K \neq J$ in general, according to the conditional distributions

$$p_{Y|X}(b_j|a_k)\ , \quad j = 1, \ldots, J, k = 1, \ldots, K$$

that are sometimes called the channel transition probabilities.

Each output symbol of a DMC depends only on the current input symbol. For an input sequence

$$\mathbf{x} = (x_1, x_2, \ldots, x_N)\ , \quad x_i \in \mathcal{X}$$

and output sequence

$$\mathbf{y} = (y_1, y_2, \ldots, y_N)\ , \quad y_i \in \mathcal{Y}$$

the memoryless property of the DMC implies

$$p(\mathbf{y}|\mathbf{x}) = \prod_{i=1}^{N} p(y_i|x_i).$$

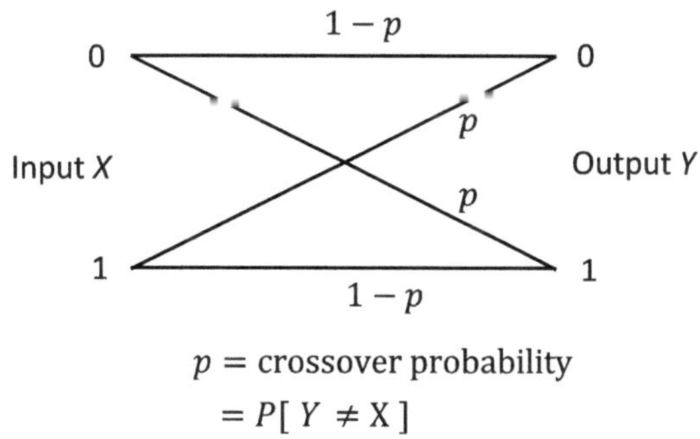

Fig. 7.3 Binary symmetric channel

The capacity of a DMC is

$$C \triangleq \max_{q(x)} I(X;Y) \tag{7.3}$$

where

$$I(X;Y) = \sum_{x\in\mathcal{X}} q(x) \sum_{y\in\mathcal{Y}} p(y|x)\log_2 \frac{p(y|x)}{p(y)} \tag{7.4}$$

and

$$p(y) = \sum_{x\in\mathcal{X}} p(y|x)q(x) \tag{7.5}$$

and $q(x)$ is the input distribution. The channel capacity is the limiting rate of reliable transmission of information over any given channel. However, it is not obvious how to communicate with a rate equal to channel capacity.

Example 7.3 Consider the binary symmetric channel (BSC) shown in Fig. 7.3. For the BSC,

$$\begin{aligned} I(X;Y) = q(0) &\sum_{j=0}^{1} p(j|0)\log_2 \frac{p(j|0)}{p(j)} \\ + q(1) &\sum_{j=0}^{1} p(j|1)\log_2 \frac{p(j|1)}{p(j)} \end{aligned}$$

(continued)

Example 7.3 (continued)

$$= q(0)\left\{(1-p)\log_2\frac{1-p}{p(0)} + p\log_2\frac{p}{p(1)}\right\}$$

$$+ q(1)\left\{(1-p)\log_2\frac{1-p}{p(1)} + p\log_2\frac{p}{p(0)}\right\}.$$

For convenience, let $q(0) \equiv q$ and $q(1) \equiv 1-q$. Then

$$I(X;Y) = q\left\{(1-p)\log_2\frac{1-p}{p(1-q)+(1-p)q} + p\log_2\frac{p}{pq+(1-p)(1-q)}\right\}$$

$$+ (1-q)\left\{(1-p)\log_2\frac{1-p}{pq+(1-p)(1-q)}\right.$$

$$\left. + p\log_2\frac{p}{p(1-q)+(1-p)q}\right\}.$$

For a BSC, the uniform input distribution, i.e., $q = 1/2$, maximizes the mutual information $I(X;Y)$ and achieves the channel capacity

$$C = 1 + p\log_2 p + (1-p)\log_2(1-p).$$

The remainder of this chapter begins with a discussion of entropy in Sect. 7.1 and mutual information in Sect. 7.2. Section 7.3 considers discrete random vectors that are typically found in digital communication systems. Section 7.4 treats the important notion of channel capacity. Finally, Sect. 7.5 extends the above concepts to continuous random variables.

7.1 Entropy

Consider the random variable X with ensemble $\mathcal{X}$ and probability distribution $p(x)$. The entropy of X is

$$H(X) = -\sum_{x\in\mathcal{X}} p(x)\log p(x). \tag{7.6}$$

To derive the properties of $H(X)$, the following fundamental inequality, as illustrated Fig. 7.4, is useful:

$$\log(x) \le x-1 \quad \text{with equality iff } x = 1. \tag{7.7}$$

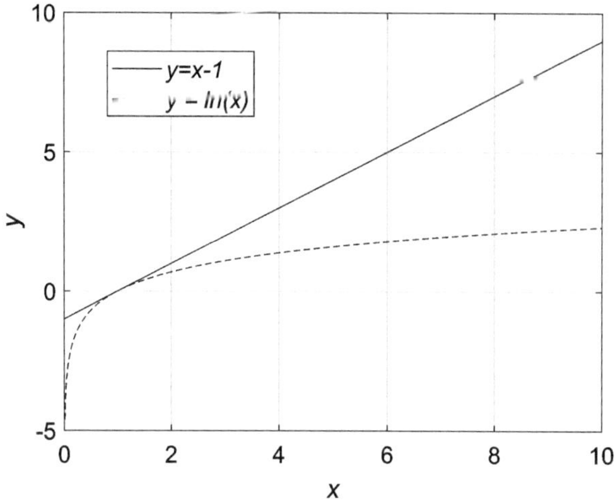

Fig. 7.4 Fundamental bound on the logarithm function

Lemma 7.1 *Let $p(x)$ and $q(x)$ be two distributions on the same ensemble $\mathcal{X}$. Then*

$$-\sum_{x\in\mathcal{X}} p(x)\log p(x) \leq -\sum_{x\in\mathcal{X}} p(x)\log q(x) \tag{7.8}$$

with equality iff $p(x) = q(x)\ \forall\, x \in \mathcal{X}$.

Proof Required to show

$$0 \leq -\sum_{x\in\mathcal{X}} p(x)\log\frac{q(x)}{p(x)} \stackrel{\triangle}{=} D(p||q)$$

where $D(p||q)$ is the relative entropy. But

$$\sum_{x\in\mathcal{X}} p(x)\log\frac{q(x)}{p(x)} \leq \sum_{x\in\mathcal{X}} p(x)\left(\frac{q(x)}{p(x)} - 1\right) = 0.$$

Therefore,

$$-\sum_{x\in\mathcal{X}} p(x)\log\frac{q(x)}{p(x)} \geq 0$$

with equality iff $p(x) = q(x)\ \forall\, x \in \mathcal{X}$ □

Theorem 7.1 *For any random variable X with ensemble $\mathcal{X}$ such that $|\mathcal{X}| = M$*

$$H(X) \leq \log M. \tag{7.9}$$

Proof Using Lemma 7.1 with $q(x) = 1/M$ gives

$$\begin{aligned} H(X) &= -\sum_{x\in\mathcal{X}} p(x)\log p(x) \\ &\leq -\sum_{x\in\mathcal{X}} p(x)\log\frac{1}{M} \\ &= \log M \end{aligned}$$

with equality iff $p(x) = q(x) = 1/M, \ \forall\ x \in \mathcal{X}$ □

Consider two ensembles $\mathcal{X}$ and $\mathcal{Y}$ with distributions $q(x)$ and $p(y)$, respectively, and consider the joint ensemble

$$\mathcal{X} \times \mathcal{Y} = \{(x, y), p(x, y)\}$$

$$q(x) = \sum_{y\in\mathcal{Y}} p(x, y)$$

$$p(y) = \sum_{x\in\mathcal{X}} p(x, y).$$

Theorem 7.2 *For two random variables X and Y, the joint uncertainty of X and Y satisfies*

$$H(X, Y) \leq H(X) + H(Y) \tag{7.10}$$

with equality if and only if X and Y are independent.

Proof

$$\begin{aligned} H(X, Y) &= -\sum_{x\in\mathcal{X}}\sum_{y\in\mathcal{Y}} p(x, y)\log p(x, y) \\ &\leq -\sum_{x\in\mathcal{X}}\sum_{y\in\mathcal{Y}} p(x, y)\log q(x)p(y) \\ &= -\sum_{x\in\mathcal{X}}\sum_{y\in\mathcal{Y}} p(x, y)\log q(x) - \sum_{y\in\mathcal{Y}}\sum_{x\in\mathcal{X}} p(x, y)\log p(y) \\ &= -\sum_{x\in\mathcal{X}} q(x)\log q(x) - \sum_{y\in\mathcal{Y}} p(y)\log p(y) \\ &= H(X) + H(Y) \end{aligned}$$

with equality iff $p(x, y) = q(x)p(y)\ ,\ \ \forall\ x \in \mathcal{X}, y \in \mathcal{Y}$ □

Similarly,

$$H(X_1, X_2, \ldots, X_n) \leq \sum_{i=1}^{n} H(X_i). \tag{7.11}$$

For the joint ensemble, two other functions can be specified:

1. Conditional uncertainty of Y given $X = x$

$$H(Y|X = x) \triangleq -\sum_{y \in \mathcal{Y}} p(y|x) \log p(y|x) \tag{7.12}$$

2. Conditional entropy

$$H(Y|X) \triangleq -\sum_{x \in \mathcal{X}} q(x) H(Y|X = x) = -\sum_{x \in \mathcal{X}} \sum_{y \in \mathcal{Y}} p(x, y) \log p(y|x) \tag{7.13}$$

Theorem 7.3 *For two random variables X and Y,*

$$H(X, Y) = H(X) + H(Y|X) = H(Y) + H(X|Y). \tag{7.14}$$

Proof

$$\begin{aligned}
H(X, Y) &= -\sum_{x \in \mathcal{X}} \sum_{y \in \mathcal{Y}} p(x, y) \log p(x, y) \\
&= -\sum_{x \in \mathcal{X}} \sum_{y \in \mathcal{Y}} p(x, y) \log p(y|x) q(x) \\
&= -\sum_{x \in \mathcal{X}} \sum_{y \in \mathcal{Y}} p(x, y) \log p(y|x) - \sum_{x \in \mathcal{Y}} \sum_{y \in \mathcal{X}} p(x, y) \log q(x) \\
&= H(Y|X) + H(X)
\end{aligned}$$

□

Similarly, for three random variables X, Y and Z,

$$\begin{aligned}
H(X, Y, Z) &= H(X) + H(Y|X) + H(Z|X, Y) \\
&= H(X, Y) + H(Z|X, Y) \\
&= H(X) + H(Y, Z|X).
\end{aligned}$$

Theorem 7.4 *For two random variables X and Y,*

$$H(Y|X) \le H(Y) \tag{7.15}$$

with equality iff X and Y are independent, that is, X can only reduce the uncertainty of Y.

Proof By Theorems 7.2 and 7.3

$$\begin{aligned} H(X) + H(Y) &\ge H(X, Y) = H(X) + H(Y|X) \\ H(Y) &\ge H(Y|X) \end{aligned}$$

□

Similarly,

$$H(Y_1, Y_2, \ldots, Y_m|X_1, X_2, \ldots, X_n) \le H(Y_1, Y_2, \ldots, Y_m). \tag{7.16}$$

7.2 Mutual Information

Information is a reduction in uncertainty. The average mutual information between the random variables X and Y is defined as

$$I(X; Y) \triangleq \sum_{x \in \mathcal{X}} \sum_{y \in \mathcal{Y}} p(x, y) \log \frac{p(x, y)}{q(x)p(y)} \tag{7.17}$$

The mutual information is defined as

$$\mathcal{I}(x, y) \triangleq \log \frac{p(x, y)}{q(x)p(y)} = \log \frac{p(y|x)}{p(y)} = \log \frac{p(x|y)}{q(x)}. \tag{7.18}$$

Mutual information $\mathcal{I}(x, y)$ can be positive or negative. The average mutual information satisfies

$$I(X; Y) = I(Y; X). \tag{7.19}$$

Theorem 7.5 *The average mutual information between two random variables X and Y is non-negative.*

$$I(X; Y) \ge 0 \tag{7.20}$$

with equality iff X and Y are independent.

Proof

$$
\begin{aligned}
I(X,Y) &= \sum_{x\in\mathcal{X}}\sum_{y\in\mathcal{Y}} p(x,y)\log\frac{p(x|y)}{q(x)}\\
&= \sum_{x\in\mathcal{X}}\sum_{y\in\mathcal{Y}} p(x,y)\log p(x|y) - \sum_{x\in\mathcal{X}}\sum_{y\in\mathcal{Y}} p(x,y)\log q(x)\\
&= \sum_{x\in\mathcal{X}}\sum_{y\in\mathcal{Y}} p(x,y)\log p(x|y) - \sum_{x\in\mathcal{Y}} q(x)\log q(x)\\
&= H(X) - H(X|Y)\\
&= H(Y) - H(Y|X).
\end{aligned}
$$

By Theorem 7.4, $H(X|Y) \leq H(X)$ and $H(Y|X) \leq H(Y)$. Therefore $I(X;Y) \geq 0$ □

Consider three random variables X, Y, and Z. The quantity $I(X,Y;Z)$ is equal to the amount of information that Z provides about X *and* Y *or* the amount of information X *and* Y provides about Z. Using the latter viewpoint

$$
I(X,Y;Z) = \sum_{x\in\mathcal{X}}\sum_{y\in\mathcal{Y}}\sum_{z\in\mathcal{Z}} p(x,y,z)\log\frac{p(z|x,y)}{p(z)}. \tag{7.21}
$$

Theorem 7.6 *The following is an important intermediate theorem. For three random variables X, Y, and Z*

$$
I(X,Y;Z) \geq I(Y;Z) \tag{7.22}
$$

with equality if and only if $p(z|x,y) = p(z|y)\ \forall\ (x,y,z)$ *with* $p(x,y,z) > 0$.

Proof

$$
\begin{aligned}
\mathcal{I}(y;z) + \mathcal{I}(z;x|y) &= \log\frac{p(z|y)}{p(z)} + \log\frac{p(z|y,x)}{p(z|y)}\\
&= \log\frac{p(z|y,x)}{p(z)}\\
&= \mathcal{I}(x,y;z).
\end{aligned}
$$

Taking averages

$$
I(X,Y;Z) = I(Y;Z) + I(Z;X|Y).
$$

However, from Theorem 7.5, $I(Z; X|Y) \geq 0$. Therefore,

$$I(X, Y; Z) \geq I(Y; Z)$$

$\square$

Similarly,

$$I(X, Y; Z) = I(X; Z) + I(Z; Y|X) \tag{7.23}$$

and, therefore,

$$I(X, Y; Z) \geq I(X; Z). \tag{7.24}$$

The condition for equality in Theorem 7.6, $p(z|x, y) = p(z|y)\ \forall\ (x, y, z)$ with $p(x, y, z) > 0$, implies that the sequence (X, Y, Z) is a Markov chain, meaning that (X, Y, Z) can be viewed as in the example communication system in Fig. 7.5.

Then by Theorem 7.6

$$I(X; Z) \leq I(X, Y; Z).$$

Since (X, Y, Z) is a Markov chain,

$$I(X, Y; Z) \leq I(Y; Z).$$

Hence,

$$I(X; Z) \leq I(Y; Z).$$

Similarly,

$$I(X; Z) \leq I(X; Y).$$

This leads to the following important theorem.

Theorem 7.7 (Data Processing Theorem) *For a Markov chain* (X, Y, Z)

$$I(X; Z) \leq \begin{cases} I(X; Y) \\ I(Y; Z). \end{cases} \tag{7.25}$$

Paradoxically, this theorem says that data processing can only destroy information.

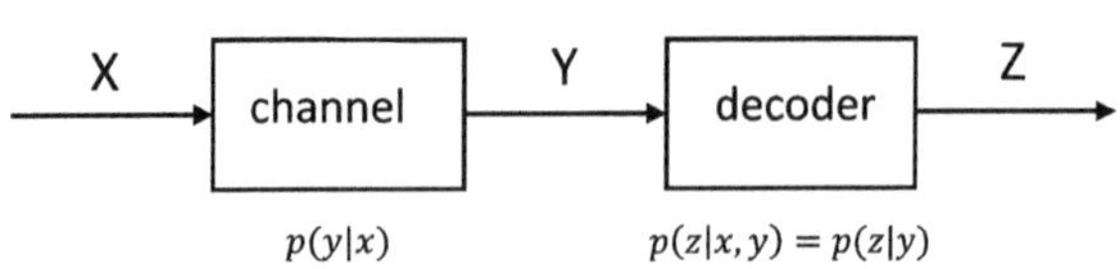

Fig. 7.5 Communication system viewed as a Markov chain

Example 7.4 Consider the concatenation of two identical BSCs as shown in Fig. 7.6. It follows that

$$I(X;Y) = I(Y;Z) = 1 - H(p)$$
$$I(X;Z) = 1 - H(2p(1-p))$$

where

$$H(p) = -p\log p - (1-p)\log(1-p).$$

$I(X;Y)$ and $I(X;Z)$ are plotted in Fig. 7.7 from which it is seen that $I(X;Z) \le I(X;Y)$.

To further develop properties of entropy $H(X)$ and mutual information $I(X;Y)$, some basic knowledge of convex regions and convex functions are required.

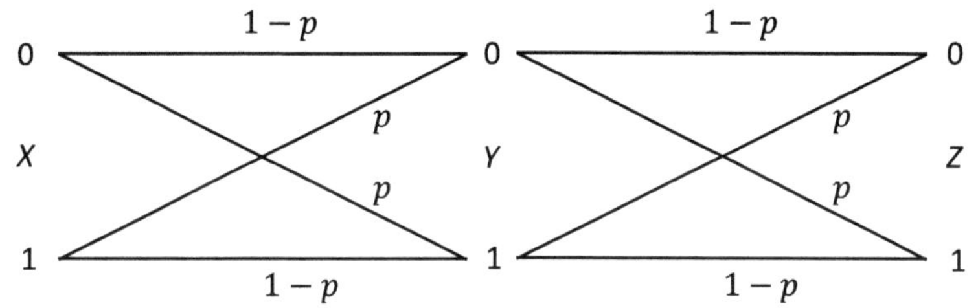

Fig. 7.6 Serial concatenation of two identical BSCs

Fig. 7.7 Mutual information for the arrangement in Fig. 7.6

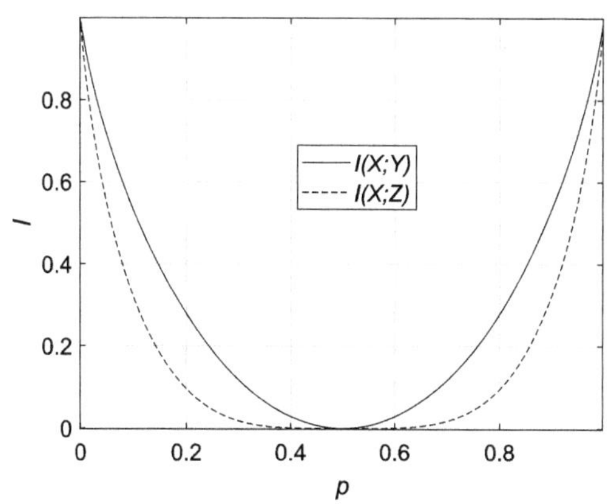

7.2.1 Convex Functions

Let $\mathcal{R}_N$ be the N-dimensional real vector space. A region $\mathcal{V}_N \subset \mathcal{R}_N$ is a *convex region* if for each $\mathbf{x}_1, \mathbf{x}_2 \in \mathcal{V}_N$

$$\theta \mathbf{x}_1 + (1-\theta)\mathbf{x}_2 \in \mathcal{V}_N, \quad \forall\ \theta \in [0, 1]. \tag{7.26}$$

That is, $\mathcal{V}_N$ is convex iff for all $\mathbf{x}_1, \mathbf{x}_2 \in \mathcal{V}_N$, all points on the straight line between $\mathbf{x}_1$ and $\mathbf{x}_2$ are also in $\mathcal{V}_N$. This is illustrated in Fig. 7.8.

Let $\mathcal{P}_N$ be the set of all probability vectors of length N.

$$\mathcal{P}_N = \left\{ \mathbf{x};\, x_n \geq 0, n = 1, \ldots, N; \sum_{n=1}^{M} x_n = 1 \right\}. \tag{7.27}$$

Lemma 7.2 *$\mathcal{P}_N$ is a convex region of $\mathcal{R}_N$, i.e., if $\mathbf{x}_1, \mathbf{x}_2 \in \mathcal{P}_N$, then*

$$\theta \mathbf{x}_1 + (1-\theta)\mathbf{x}_2 \in \mathcal{P}_N, \quad \forall\ \theta \in [0, 1].$$

Proof

$$\begin{aligned} \theta \mathbf{x}_1 &= (\theta x_{11}, \theta x_{12}, \ldots, \theta x_{1N}) \\ (1-\theta)\mathbf{x}_2 &= ((1-\theta)x_{21}, (1-\theta)x_{22}, \ldots, (1-\theta)x_{2N}) \\ \theta \mathbf{x}_1 + (1-\theta)\mathbf{x}_2 &= (\ldots, \theta x_{1i} + (1-\theta)x_{2i}, \ldots). \end{aligned}$$

However,

$$\sum_{i=1}^{N} (\theta x_{1i} + (1-\theta)x_{2i}) = \theta + (1-\theta) = 1$$

□

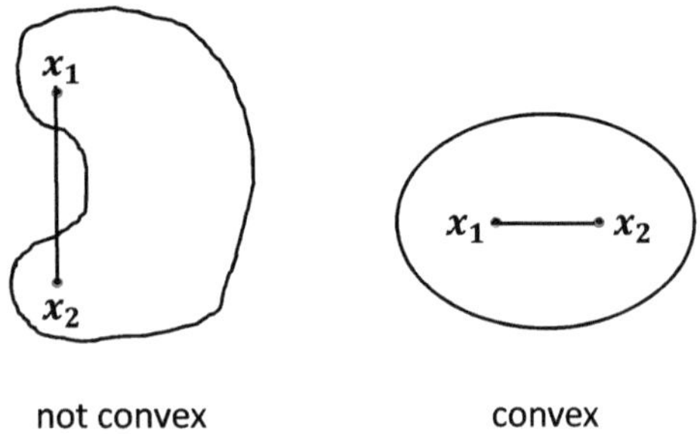

Fig. 7.8 Convex vs. nonconvex regions

A real-valued function $f(\cdot)$ of vectors of dimension N is convex ∩ (cap) over a convex region $\mathcal{V}_N \subset \mathcal{R}_N$ iff

$$\theta f(\mathbf{x}_1) + (1-\theta) f(\mathbf{x}_2) \le f(\theta \mathbf{x}_1 + (1-\theta)\mathbf{x}_2) \ , \quad \forall \mathbf{x}_1, \mathbf{x}_2 \in \mathcal{V}_N \text{ and } \theta \in [0,1]. \tag{7.28}$$

A function $f(\cdot)$ is convex ∪ (cup) if the inequality is reversed. If the inequality is strict, then the function is strictly convex ∩ or ∪. Figure 7.9 illustrates a convex ∩ function.

The following are three important lemmas:

Lemma 7.3 *If $f_1(\mathbf{x}), f_2(\mathbf{x}), \ldots, f_L(\mathbf{x})$ are convex ∩ functions and $c_1, c_2, \ldots, c_L$ are positive numbers, then*

$$f(\mathbf{x}) = \sum_{i=1}^{L} c_i f_i(\mathbf{x})$$

is a convex ∩ with strict convexity if any of the $f_i(\mathbf{x})$ are strictly convex.
Proof

$$\begin{aligned}
\theta f(\mathbf{x}_1) + (1-\theta) f(\mathbf{x}_2) &= \theta \sum_{i=1}^{L} c_i f_i(\mathbf{x}_1) + (1-\theta) \sum_{i=1}^{L} c_i f_i(\mathbf{x}_2) \\
&= \sum_{i=1}^{L} c_i \left(\theta f_i(\mathbf{x}_1) + (1-\theta) f_i(\mathbf{x}_2)\right) \\
&\le \sum_{i=1}^{L} c_i f_i\left(\theta \mathbf{x}_1 + (1-\theta)\mathbf{x}_2\right) \\
&= f(\theta \mathbf{x}_1 + (1-\theta)\mathbf{x}_2)
\end{aligned}$$

□

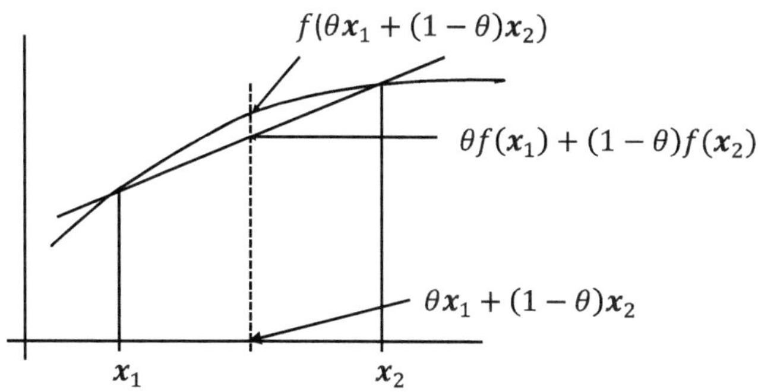

Fig. 7.9 Convex ∩ function

Lemma 7.4 *For a 1-D function $f(x)$, if $f''(x) < 0$ over $\mathcal{V} \subset \mathcal{R}$, then $f(x)$ is convex $\cap$ over $\mathcal{V}$.*

Example 7.5

$$f(x) = \ln x\ , \quad x > 0$$

$$f''(x) = -\frac{1}{x^2} \le 0\ , \quad \forall x > 0.$$

Therefore, $f(x)$ is convex $\cap$ for $x \in (0, \infty)$.

Lemma 7.5 *Let $\mathbf{x}$ be an N-dimensional random vector that can take on values $\{\mathbf{x}_1, \mathbf{x}_2, \ldots, \mathbf{x}_K\}$ and let $f(\mathbf{x})$ be any convex $\cap$ function on $\mathcal{V}_N \subset \mathcal{R}_N$. Let $\boldsymbol{\theta} = (\theta_1, \theta_2, \ldots, \theta_K)$ be a probability vector, i.e.,*

$$\mathrm{P}(\mathbf{x} = \mathbf{x}_i) = \theta_i\,.$$

Then

$$\sum_{i=1}^{K} \theta_i f(\mathbf{x}_i) \le f\left(\sum_{i=1}^{K} \theta_i \mathbf{x}_i\right)$$

or in general

$$\mathrm{E}[f(\mathbf{x})] \le f(\mathrm{E}[\mathbf{x}]) \tag{7.29}$$

a result known as Jensen's inequality.

Proof Proof is by induction. It is true by definition for $K = 2$ that $f(\mathbf{x})$ is convex. Assume it is true for $K - 1$ and prove it is true for K.

By assumption

$$\sum_{i=1}^{K-1} \theta_i f(\mathbf{x}_i) \le f\left(\sum_{i=1}^{K-1} \theta_i \mathbf{x}_i\right).$$

Then

$$\begin{aligned}
\sum_{i=1}^{K} \theta_i f(\mathbf{x}_i) &= \sum_{i=1}^{K-1} \theta_i f(\mathbf{x}_i) + \theta_K f(\mathbf{x}_K) \\
&= (1 - \theta_K) \sum_{i=1}^{K-1} \frac{\theta_i}{1 - \theta_K} f(\mathbf{x}_i) + \theta_K f(\mathbf{x}_K)
\end{aligned}$$

$$\leq (1-\theta_K) f\left(\sum_{i=1}^{K-1} \frac{\theta_i}{1-\theta_K}\mathbf{x}_i\right) + \theta_K f(\mathbf{x}_K) \text{ by assumption}$$

$$\leq f\left((1-\theta_K)\sum_{i=1}^{K-1} \frac{\theta_i}{1-\theta_K}\mathbf{x}_i + \theta_K \mathbf{x}_K\right) \quad f(\,\cdot\,) \text{ is convex } \cap$$

$$= f\left(\sum_{i=1}^{K} \theta_i \mathbf{x}_i\right)$$

□

7.2.2 Further Properties of Entropy and Mutual Information

Theorem 7.8 *The entropy of the random variable X*

$$H(X) = -\sum_{i=1}^{N} p_i \log p_i$$

is a convex ∩ function of the probability vector $\mathbf{p} = (p_1, p_2, \ldots, p_N)$.

Proof Let $f(p_n) = p_n \log p_n,\ n = 1, \ldots, N$. The function $f(p_n)$ is convex ∩ on the positive real line, i.e.,

$$\frac{d^2 f(p_n)}{d{p_n}^2} = -\frac{1}{p_n} \leq 0 \quad \text{for } p_n > 0.$$

By Lemma 7.3 of convex functions $H(X)$ is convex ∩ □

Theorem 7.9 *Consider a DMC where the channel transition probabilities* $p(y|x)$ *are known or fixed functions. Then* $I(X;Y)$ *is a convex ∩ function of the input probabilities* $q(x)$.

Proof Consider two input ensembles $\mathcal{X}_1$ and $\mathcal{X}_2$ with probability distributions $q_1(x)$ and $q_2(x)$. If a new input ensemble $\mathcal{X}$ has a probability distribution $q(x)$ that is a convex combination

$$q(x) = \theta q_1(x) + (1-\theta) q_2(x),$$

it must be shown that

$$\theta I(X_1;Y_1) + (1-\theta) I(X_2;Y_2) \leq I(X;Y).$$

Consider the quantity

$$\begin{aligned}\Delta &= \theta I(X_1; Y_1) + (1-\theta) I(X_2; Y_2) - I(X; Y)\\ &= \sum_{x,y} \theta p_1(x,y) \log \frac{p(y|x)}{p_1(y)} \qquad\qquad p_1(y) = \sum_{\hat{x}} p(y|\hat{x}) q_1(\hat{x})\\ &+ \sum_{x,y} (1-\theta) p_2(x,y) \log \frac{p(y|x)}{p_2(y)} \qquad\qquad p_2(y) = \sum_{\hat{x}} p(y|\hat{x}) q_2(\hat{x})\\ &- \sum_{x,y} (\theta p_1(x,y) + (1-\theta) p_2(x,y)) \log \frac{p(y|x)}{p(y)}\end{aligned}$$

where

$$\begin{aligned}p(y) &= \theta p_1(y) + (1-\theta) p_2(y)\\ &= \sum_{\hat{x}} p(y|\hat{x}) q(\hat{x}).\end{aligned}$$

The top terms in the logs cancel, so

$$\begin{aligned}\Delta &= -\sum_y \theta p_1(y) \log p_1(y) + \sum_y \theta p_1(y) \log p(y)\\ &= -\sum_y (1-\theta) p_2(y) \log p_2(y) + \sum_y (1-\theta) p_2(y) \log p(y).\end{aligned}$$

From Lemma 7.1

$$\sum_y \theta p_1(y) \log p_1(y) \leq \sum_y \theta p_1(y) \log p(y)$$

it follows that $\Delta < 0$ □

Theorem 7.10 *For fixed $q(x)$, $I(X; Y)$ is a convex $\cup$ function in the transition probabilities $p(y|x)$.*

Proof Similar to Theorem 7.9. □

7.3 Discrete Random Vectors

Let **X** and **Y** be arbitrary random vectors and define the functions

$$H(\mathbf{X}), \quad H(\mathbf{Y}|\mathbf{X}), \quad I(\mathbf{X}; \mathbf{Y})$$

where

$$\mathbf{X} = (X_1, X_2, \ldots, X_n)$$
$$\mathbf{Y} = (Y_1, Y_2, \ldots, Y_n).$$

The distribution of **X** is

$$p(\mathbf{x}) = \mathrm{P}(X_1 = x_1, X_2 = x_2, \ldots, X_n = x_n) \tag{7.30}$$

where each x_i is in the range of X_i, i.e, in the ensemble $\mathcal{X}_i$. For example,

$$H(\mathbf{X}) = -\sum_{\mathbf{x}} p(\mathbf{x}) \log p(\mathbf{x}). \tag{7.31}$$

Likewise, the distribution of **Y** is

$$p(\mathbf{y}) = \mathrm{P}(Y_1 = y_1, Y_2 = y_2, \ldots, Y_n = y_n) \tag{7.32}$$

where each y_j is in the range of Y_j, i.e, in the ensemble $\mathcal{Y}_j$.

The generalization of Theorem 7.7 has an important application as illustrated in Fig. 7.10, where

$$\mathbf{U} \to \mathbf{X} \to \mathbf{Y} \to \mathbf{V}$$

forms a Markov chain. From Theorem 7.7, $I(\mathbf{U}; \mathbf{V}) \leq I(\mathbf{X}; \mathbf{Y})$. This implies that encoding and decoding can only destroy information.

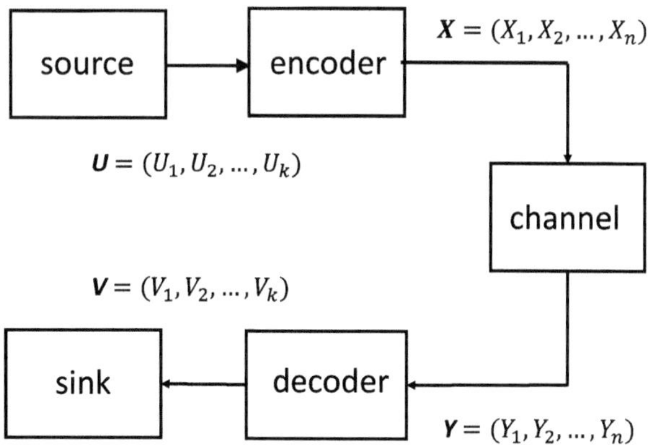

Fig. 7.10 Coded communication system

Theorem 7.11 *If the components of* $\mathbf{X} = (X_1, X_2, \ldots, X_n)$ *are independent (memoryless source), then*

$$I(\mathbf{X}; \mathbf{Y}) \geq \sum_{i=1}^{n} I(X_i; Y_i).$$

This implies that destroying channel memory is bad. Paradoxically, most error control codes are designed to operate on memoryless channels and interleavers are used to destroy the channel memory.

Proof

$$\begin{aligned} I(\mathbf{X}; \mathbf{Y}) &= \sum_{\mathbf{x},\mathbf{y}} p(\mathbf{x}, \mathbf{y}) \log \frac{p(\mathbf{x}|\mathbf{y})}{q(\mathbf{x})} \\ &= \sum_{\mathbf{x},\mathbf{y}} p(\mathbf{x}, \mathbf{y}) \log \frac{p(\mathbf{x}|\mathbf{y})}{q(x_1) \cdots q(x_n)} \quad \text{independent.} \end{aligned}$$

Also,

$$\begin{aligned} \sum_{i=1}^{n} I(X_i; Y_i) &= \sum_{i=1}^{n} \sum_{x_i, y_i} p(x_i, y_i) \log \frac{p(x_i|y_i)}{q(x_i)} \\ &= \sum_{\mathbf{x},\mathbf{y}} p(\mathbf{x}, \mathbf{y}) \log \frac{p(x_1|y_1) \cdots p(x_n|y_n)}{q(x_1) \cdots q(x_n)}. \end{aligned}$$

Then

$$\begin{aligned} I(\mathbf{X}; \mathbf{Y}) - \sum_{i=1}^{n} I(X_i; Y_i) &= \sum_{\mathbf{x},\mathbf{y}} p(\mathbf{x}, \mathbf{y}) \log p(\mathbf{x}|\mathbf{y}) \\ &\quad - \sum_{\mathbf{x},\mathbf{y}} p(\mathbf{x}, \mathbf{y}) \log \left(p(x_1|y_1) \cdots p(x_n|y_n)\right) \\ &= \sum_{\mathbf{y}} p(\mathbf{y}) \sum_{\mathbf{x}} p(\mathbf{x}|\mathbf{y}) \log p(\mathbf{x}|\mathbf{y}) \\ &\quad - \sum_{\mathbf{y}} p(\mathbf{y}) \sum_{\mathbf{x}} p(\mathbf{x}|\mathbf{y}) \log \left(p(x_1|y_1) \cdots p(x_n|y_n)\right) \\ &\geq 0 \end{aligned}$$

where the inequality results from Lemma 7.1 and equality holds if the channel is memoryless and the source is memoryless. □

Theorem 7.12 *If* $\mathbf{X} = (X_1, X_2, \ldots, X_n)$ *and* $\mathbf{Y} = (Y_1, Y_2, \ldots, Y_n)$ *are random vectors and*

$$p(\mathbf{y}|\mathbf{x}) = \prod_{i=1}^{n} p(y_i|x_i) \quad \textit{memoryless channel},$$

then

$$I(\mathbf{X}; \mathbf{Y}) \leq \sum_{i=1}^{n} I(X_i; Y_i)$$

which implies that source correlation is bad for a memoryless channel.

Proof

$$\begin{aligned} I(\mathbf{X}; \mathbf{Y}) &= \sum_{\mathbf{x},\mathbf{y}} p(\mathbf{x}, \mathbf{y}) \log \frac{p(\mathbf{y}|\mathbf{x})}{p(\mathbf{y})} \\ &= \sum_{\mathbf{x},\mathbf{y}} p(\mathbf{x}, \mathbf{y}) \log \frac{p(y_1|x_1) \cdots p(y_n|x_n)}{p(\mathbf{y})}. \end{aligned}$$

Also,

$$\begin{aligned} \sum_{i=1}^{n} I(X_i; Y_i) &= \sum_{i=1}^{n} \sum_{x_i, y_i} p(x_i, y_i) \log \frac{p(y_i|x_i)}{p(y_i)} \\ &= \sum_{\mathbf{x},\mathbf{y}} p(\mathbf{x}, \mathbf{y}) \log \frac{p(y_1|x_1) \cdots p(y_n|x_n)}{p(y_1) \cdots p(y_n)}. \end{aligned}$$

Then

$$\begin{aligned} I(\mathbf{X}; \mathbf{Y}) - \sum_{i=1}^{n} I(X_i; Y_i) &= -\sum_{\mathbf{x},\mathbf{y}} p(\mathbf{x}, \mathbf{y}) \log p(\mathbf{y}) \\ &\quad + \sum_{\mathbf{x},\mathbf{y}} p(\mathbf{x}, \mathbf{y}) \log \left(p(y_1) \cdots p(y_n)\right) \\ &= -\sum_{\mathbf{y}} p(\mathbf{y}) \log p(\mathbf{y}) \\ &\quad + \sum_{\mathbf{y}} p(\mathbf{y}) \log \left(p(y_1) \cdots p(y_n)\right) \\ &\leq 0 \end{aligned}$$

equality holds if the channel is memoryless and the source is memoryless. □

Theorem 7.13 (Fano's Inequality) *Consider the serial concatenation of two DMCs as shown in Fig 7.11. For this system $|\mathcal{X}| = |\mathcal{Z}| = K$ and there is a one-to-one correspondence between X and Z. The decision device decides which x_k has been transmitted based on the observation of y_j.*

The probability of decision error is

$$P_E = \sum_{\ell=1}^{K} \sum_{\substack{m=1 \\ m\neq\ell}}^{K} p(z_m|x_\ell)p(x_\ell) = \sum_{\ell=1}^{K} \sum_{\substack{m=1 \\ m\neq\ell}}^{K} p(x_\ell, z_m).$$

Then

$$H(X|Y) \leq H(P_E) + P_E \log(K-1)$$

where

$$H(P_E) = -P_E \log P_E - (1-P_E)\log(1-P_E).$$

Proof Since $H(X|Z) \geq H(X|Y)$ due to the data processing theorem, it suffices to prove Fano's inequality for $H(X|Z)$. That is, if it works for $H(X|Z)$ it will work for $H(X|Y)$.

Consider the quantity

$$\begin{aligned}
\Delta &= H(X|Y) - H(P_E) - P_E \log(K-1) \\
&= \sum_{x,y} p(x,z)\log p(x|z) + \sum_{\substack{x,z \\ z\neq x}} p(x,z)\log P_E \\
&+ \sum_{\substack{x,z \\ z=x}} p(x,z)\log(1-P_E) - \sum_{\substack{x,z \\ z\neq x}} p(x,z)\log(K-1) \\
&= \sum_{\substack{x,z \\ z\neq x}} p(x,z)\log\frac{P_E}{(K-1)p(x|z)} + \sum_{\substack{x,z \\ z=x}} p(x,z)\log\frac{(1-P_E)}{p(x|z)}
\end{aligned}$$

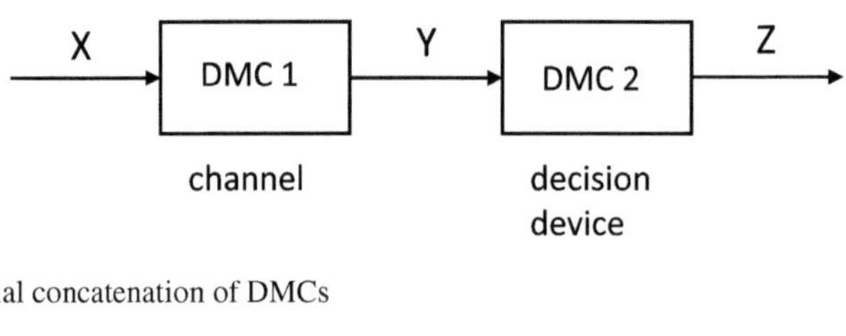

Fig. 7.11 Serial concatenation of DMCs

$$
\begin{aligned}
&\leq \sum_{\substack{x,z\\ z\neq x}} p(x,z)\left(\frac{P_E}{(K-1)p(x|z)}-1\right)+\sum_{\substack{x,z\\ z=x}} p(x,z)\left(\frac{(1-P_E)}{p(x|z)}-1\right)\\
&= \frac{P_E}{K-1}\sum_{\substack{x,z\\ z\neq x}} p(z)+(1-P_E)\sum_{z} p(z)-\sum_{x,z} p(x,z)\\
&= \frac{P_E}{K-1}\sum_{z} p(z)+(1-P_E)-1\\
&= P_E+1-P_E-1=0
\end{aligned}
$$

□

7.4 Channel Capacity

The average mutual information $I(X;Y)$ depends on the channel transition probabilities $p(y|x)$ and the input distribution $q(x)$. The *channel capacity* C is defined as the largest mutual information that can be transmitted over a given channel in one use, maximized over all input probability distributions. Note that the channel capacity only depends on the channel as described by its transition probabilities $p(y|x)$. Theorem 7.9 showed that for any given channel with known transition probabilities $p(y|x)$, the mutual information $I(X;Y)$ is a convex $\cap$ function of the input distribution $q(x)$.

7.4.1 Maximization of Convex Functions

Theorem 7.14 *Let $f(\mathbf{a})$ be a convex $\cap$ function of $\mathbf{a}$ over the convex region $\mathcal{V}_K \in \mathcal{R}_K$, where $\mathbf{a}$ is a probability vector. Assuming that the partial derivatives $\frac{\partial f(\mathbf{a})}{\partial a_k}$ are defined and continuous over $\mathcal{V}_K$, $\mathbf{a}$ maximizes $f(\mathbf{a})$ if and only if*

$$
\begin{aligned}
\frac{\partial f(\mathbf{a})}{\partial a_k} &= \lambda\,, \quad \forall\, k \text{ such that } a_k > 0\\
\frac{\partial f(\mathbf{a})}{\partial a_k} &\leq \lambda\,, \quad \forall\, k \text{ such that } a_k = 0
\end{aligned}
$$

where λ is an appropriate constant.

Proof *Sufficiency*: Assume for some $\mathbf{a}$ and λ, the preceding conditions are satisfied, and show that $\forall\ \mathbf{b} \in V_N$, $f(\mathbf{b}) - f(\mathbf{x}) \leq 0$ and, therefore, $\mathbf{a}$ maximizes $f(\mathbf{a})$. From the convexity of $f(\mathbf{a})$

$$\theta f(\mathbf{b}) + (1-\theta) f(\mathbf{a}) \leq f(\theta\mathbf{b} + (1-\theta)\mathbf{a})$$

or

$$f(\mathbf{b}) - f(\mathbf{a}) \leq \frac{f(\mathbf{a} + \theta(\mathbf{b} - \mathbf{a})) - f(\mathbf{a})}{\theta}$$

for all $\theta \in [0, 1]$.

Since $f(\mathbf{a})$ is differentiable for all $\mathbf{a} \in V_K$, it follows

$$f(\mathbf{a} + h\mathbf{u}) - f(\mathbf{a}) = \nabla f(\mathbf{a}) \cdot h\mathbf{u} + g(h\mathbf{u})$$

where

$$\nabla f(\mathbf{a}) = \left(\frac{\partial f(\mathbf{a})}{\partial a_1}, \frac{\partial f(\mathbf{a})}{\partial a_2}, \ldots, \frac{\partial f(\mathbf{a})}{\partial a_K} \right),$$

$\mathbf{u}$ is a unit vector, and

$$\begin{aligned} \lim_{|h\mathbf{u}| \to 0} \frac{g(h\mathbf{u})}{|h\mathbf{u}|} &= \lim_{|h\mathbf{u}| \to 0} \frac{g(h\mathbf{u})}{|h||\mathbf{u}|} \\ &= \lim_{|h| \to 0} \frac{g(h\mathbf{u})}{|h|} = 0. \end{aligned}$$

Therefore,

$$\begin{aligned} \lim_{h \to 0} \frac{f(\mathbf{a} + h\mathbf{u}) - f(\mathbf{a})}{h} &= \lim_{h \to 0} \frac{\nabla f(\mathbf{a}) \cdot h\mathbf{u}}{h} + \lim_{h \to 0} \frac{g(h\mathbf{u})}{h} \\ &= \nabla f(\mathbf{a}) \cdot \mathbf{u}. \end{aligned}$$

Using the above property, in the limit as $\theta \to 0$

$$\begin{aligned} f(\mathbf{b}) - f(\mathbf{a}) &= \nabla f(\mathbf{a}) \cdot (\mathbf{b} - \mathbf{a}) \\ &= \sum_{k=1}^{K} \frac{\partial f(\mathbf{a})}{\partial a_k} (b_k - a_k). \end{aligned}$$

If $a_k > 0$, then $\frac{\partial f(\mathbf{a})}{\partial a_k} = \lambda$ (by assumption). If $a_k = 0$, then $\frac{\partial f(\mathbf{a})}{\partial a_k} \leq \lambda$ and $b_k = a_k \geq 0$. Therefore,

$$\frac{\partial f(\mathbf{a})}{\partial a_k} (b_k - a_k) \leq \lambda (b_k - a_k).$$

Hence,

$$f(\mathbf{b}) - f(\mathbf{a}) \leq \sum_{k=1}^{K} \lambda(b_k - a_k) = \lambda \sum_{k=1}^{K} b_k - \lambda \sum_{k=1}^{K} a_k = 0$$

establishing that $f(\mathbf{b}) - f(\mathbf{a}) \leq 0$.

Necessity: Let $\mathbf{a}$ maximize $f(\mathbf{a})$. Then $\forall\ \mathbf{b} \in V_N$ and $\theta \in [0, 1]$, $f(\mathbf{a} + \theta(\mathbf{b} - \mathbf{a})) - f(\mathbf{a}) \leq 0$. Dividing both sides by θ and taking the limit when $\theta \to 0$, yields the inequality

$$\nabla f(\mathbf{a}) \cdot (\mathbf{b} - \mathbf{a}) = \sum_{k=1}^{K} \frac{\partial f(\mathbf{a})}{\partial a_k}(b_k - a_k) \leq 0.$$

Since $\mathbf{a}$ is a probability vector, one component is strictly positive, say $a_1 > 0$. Let

$$\mathbf{b} = (a_1 - \varepsilon, a_2, \ldots, a_i + \varepsilon, \ldots, a_K)$$

where $-a_i \leq \varepsilon \leq a_1$. Now

$$\begin{aligned}\sum_{k=1}^{K} \frac{\partial f(\mathbf{a})}{\partial a_k}(b_k - a_k) &= -\varepsilon \frac{\partial f(\mathbf{a})}{\partial a_1} + \varepsilon \frac{\partial f(\mathbf{a})}{\partial a_i} \\ &= \varepsilon \left(\frac{\partial f(\mathbf{a})}{\partial a_i} - \frac{\partial f(\mathbf{a})}{\partial a_1} \right) \leq 0.\end{aligned}$$

If $a_i > 0$, then ε can be positive or negative implying that

$$\frac{\partial f(\mathbf{a})}{\partial a_i} = \frac{\partial f(\mathbf{a})}{\partial a_1} = \lambda.$$

If $a_i = 0$, then $\varepsilon \geq 0$, implying that

$$\frac{\partial f(\mathbf{a})}{\partial a_i} \leq \frac{\partial f(\mathbf{a})}{\partial a_1} = \lambda.$$

□

7.4.2 *Capacity of a DMC*

Usually the channel capacity is extremely difficult to obtain in closed form. However, the following concepts are sometimes useful.

Theorem 7.15 *A necessary and sufficient condition for an input distribution* $\{q(1), q(2), \ldots, q(K)\}$ *to achieve capacity on a DMC is that for some number* C*,*

$$I(X = k; Y) = C \;\forall\; k \text{ with } q(k) > 0$$
$$I(X = k; Y) \leq C \;\forall\; k \text{ with } q(k) = 0.$$

Furthermore, C *is the channel capacity.*

Proof $I(X; Y)$ is a convex $\cap$ function of the input distribution $q(x)$. Hence,

$$\begin{aligned}
\frac{\partial I(X;Y)}{\partial q(k)} &= \frac{\partial}{\partial q(k)} \sum_{i=1}^{K} \sum_{j=1}^{J} p(x_i, y_j) \log \frac{p(y_j|x_i)}{p(y_j)} \\
&= \frac{\partial}{\partial q(k)} \sum_{i=1}^{K} q(i) \sum_{j=1}^{J} p(y_j|x_i) \log \frac{p(y_j|x_i)}{\sum_{m=1}^{K} p(y_j|x_m) q(m)} \\
&= \sum_{j=1}^{J} p(y_j|x_k) \log \frac{p(y_j|x_i)}{p(y_j)} \\
&\quad + \sum_{i=1}^{K} q(i) \sum_{j=1}^{J} p(y_j|x_i) \frac{\partial}{\partial q(k)} \log \frac{p(y_j|x_i)}{\sum_{m=1}^{K} p(y_j|x_m) q(m)} \\
&= \sum_{j=1}^{J} p(y_j|x_k) \log \frac{p(y_j|x_i)}{p(y_j)} \\
&\quad - \sum_{i=1}^{K} q(i) \sum_{j=1}^{J} p(y_j|x_i) \frac{p(y_j|x_i)}{\sum_{m=1}^{K} p(y_j|x_m) q(m)} \cdot \frac{p(y_j|x_k)}{p(y_j|x_i)} \log e \\
&= \sum_{j=1}^{J} p(y_j|x_k) \log \frac{p(y_j|x_i)}{p(y_j)} - \sum_{j=1}^{J} p(y_j) \frac{p(y_j|x_i)}{p(y_j)} \log e \\
&= I(X = k; Y) - \log e.
\end{aligned}$$

Using the Theorem 7.14,

$$\frac{\partial I(X;Y)}{\partial q(k)} = \lambda = I(X = k; Y) - \log e.$$

Hence,

$$I(X = k; Y) = \hat{\lambda} \quad \text{a constant}$$

where

$$\hat{\lambda} = \lambda + \log e.$$

Therefore, the maximizing distribution must have the property

$$I(X = k; Y) = \hat{\lambda} \;\forall\, k \text{ with } q(k) > 0$$
$$I(X = k; Y) \le \hat{\lambda} \;\forall\, k \text{ with } q(k) = 0.$$

Moreover, for the maximizing distribution

$$I(X; Y) = \sum_{k=1}^{K} q(k) I(X = k; Y) = \hat{\lambda} = C.$$

Finding the capacity of an arbitrary DMC is difficult. However, there are two classes of channels for which it can be evaluated easily. The first is a symmetric DMC. □

Symmetric DMCs
A DMC is symmetric if the columns of its transition matrix (rows:inputs, columns:outputs) can be partitioned into subsets such that within each subset each row is a permutation of the first row, and each column (if more than one) is a permutation of the first column. Examples of symmetric DMCs are shown in Fig. 7.12.

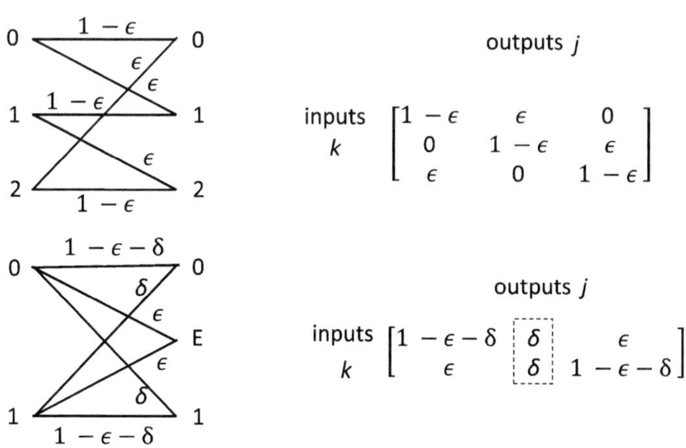

Fig. 7.12 Examples of symmetric DMCs

Lemma 7.6 *For a symmetric DMC, a uniform input distribution $q(x)$ maximizes $I(X;Y)$ and achieves channel capacity, C.*

Proof For equiprobable inputs

$$I(X=k;Y)=\sum_j p(j|k)\log\frac{p(j|l)}{\frac{1}{K}\sum_m p(j|m)}\ ,\quad q(i)=\frac{1}{K}.$$

As shown in Fig. 7.13, within each partition, each column is a permutation of every other column, so that output probability

$$\frac{1}{K}\sum_m p(j|m)$$

is constant. Hence, the terms

$$p(j|k)\log\frac{p(j|l)}{\frac{1}{K}\sum_m p(j|m)}$$

have the same symmetry properties as the $p(j|k)'s$ within each partition. The sum of these terms is the same for each k, i.e., $I(X=k:Y)=$ constant $=C$. Therefore, Theorem 7.15 is satisfied and capacity is achieved □

DMCs with Invertible Channel Transition Matrices

The second category of DMCs for which capacity is easily calculated are those that have invertible channel transition matrices [11]. To be invertible, the channel transition matrix must be square. Let $p(j|k),\ 1\le j,k\le M$ be elements of the channel transition matrix and $\hat{p}(j|k),\ 1\le j,k\le M$ be elements of its inverse. Suppose that

$$d_j=\sum_{k=1}^{M}\hat{p}(j|k)2^{-\sum_{\ell=1}^{M}\hat{p}(\ell|k)H(Y;X=\ell)}\ge 0\ ,\quad j=1,2,\ldots,M. \tag{7.33}$$

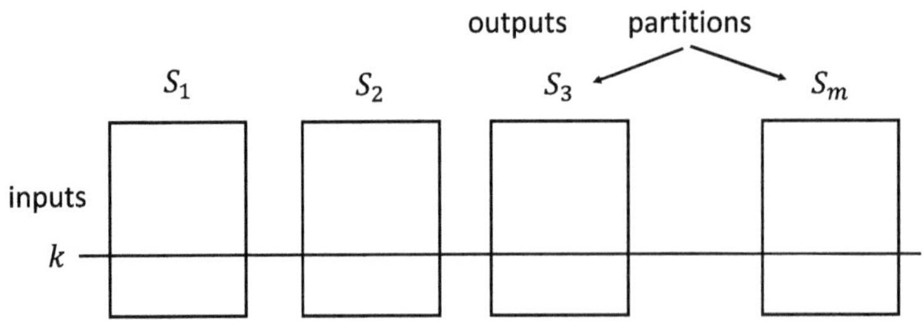

Fig. 7.13 Symmetric DMC partitioned into subsets

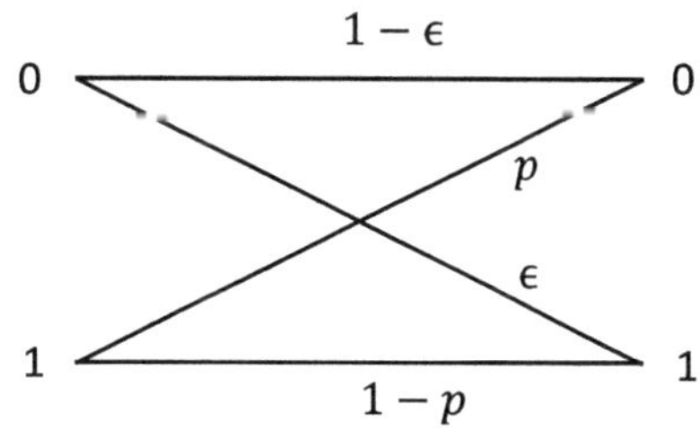

Fig. 7.14 Binary asymmetric channel

Then the channel capacity is achieved with

$$q(k) = 2^{-C} d_k\,, \quad k = 1, \ldots, M \tag{7.34}$$

where

$$C = \log_2 \left\{ \sum_{k=1}^{M} 2^{-\sum_{\ell=1}^{M} \hat{p}(\ell|k) H(Y;X=\ell)} \right\}. \tag{7.35}$$

The proof is omitted for this introductory textbook, but the procedure is illustrated by example.

Example 7.6 Consider the binary asymmetric channel shown in Fig. 7.14. The binary asymmetric channel has channel transition matrix

$$[p(j|k)] = \begin{bmatrix} 1-p & p \\ \epsilon & 1-\epsilon. \end{bmatrix}$$

The inverse of $[p(j|k)]$ may be obtained by direct matrix inversion as

$$[\hat{p}(j|k)] = \frac{1}{\Delta} \begin{bmatrix} 1-\epsilon & -p \\ -\epsilon & 1-p \end{bmatrix}, \quad \Delta = (1-p)(1-\epsilon) - p\epsilon = 1-p-\epsilon, \quad p+\epsilon \neq 1.$$

It follows that

$$\begin{aligned} d_1 &= \hat{p}(1|1) 2^{\{-\hat{p}(1|1)H(Y;X=1) - \hat{p}(2|1)H(Y;X=2)\}} \\ &+ \hat{p}(1|2) 2^{\{-\hat{p}(1|2)H(Y;X=1) - \hat{p}(2|2)H(Y;X=2)\}}. \end{aligned}$$

(continued)

Example 7.6 (continued)
Using

$$\begin{aligned} H(Y; X=1) &= -\sum_y p(y|x=1)\log_2 p(y|x=1) \\ &= -(1-p)\log_2(1-p) - p\log_2 p \\ &= H(p) \\ H(Y; X=2) &= H(\epsilon) \end{aligned}$$

gives

$$d_1 = \frac{1-\epsilon}{\Delta}2^{\left\{-\frac{1-\epsilon}{\Delta}H(p)+\frac{p}{\Delta}H(\epsilon)\right\}} - \frac{\epsilon}{\Delta}2^{\left\{\frac{\epsilon}{\Delta}H(p)-\frac{1-p}{\Delta}H(\epsilon)\right\}}.$$

Likewise,

$$\begin{aligned} d_2 &= \hat{p}(2|1)2^{\{-\hat{p}(1|1)H(Y;X=1)-\hat{p}(2|1)H(Y;X=2)\}} \\ &\quad +\hat{p}(2|2)2^{\{-\hat{p}(1|2)H(Y;X=1)-\hat{p}(2|2)H(Y;X=2)\}} \\ &= -\frac{p}{\Delta}2^{\left\{-\frac{1-\epsilon}{\Delta}H(p)+\frac{p}{\Delta}H(\epsilon)\right\}} + \frac{1-p}{\Delta}2^{\left\{\frac{\epsilon}{\Delta}H(p)-\frac{1-p}{\Delta}H(\epsilon)\right\}}. \end{aligned}$$

Finally, the capacity is

$$C = \log_2\left\{2^{\left\{-\frac{1-\epsilon}{\Delta}H(p)+\frac{p}{\Delta}H(\epsilon)\right\}} + 2^{\left\{\frac{\epsilon}{\Delta}H(p)-\frac{1-p}{\Delta}H(\epsilon)\right\}}\right\}$$

and the optimum input distribution is

$$q(k) = 2^{-C}d_k\ , \quad k=1,2.$$

Consider the special case when $\epsilon = p$, i.e., the BSC. Then

$$\begin{aligned} C &= \log_2\left\{2^{\left\{-\frac{1-p}{\Delta}H(p)+\frac{p}{\Delta}H(p)\right\}} + 2^{\left\{\frac{p}{\Delta}H(p)-\frac{1-p}{\Delta}H(p)\right\}}\right\} \\ &= \log_2\left\{2^{\left\{1-\frac{1-2p}{\Delta}H(p)\right\}}\right\} \ ; \quad \Delta = 1-\epsilon-p = 1-2p \\ &= \log_2\left\{2^{\{1-H(p)\}}\right\} \\ &= 1-H(p). \end{aligned}$$

(continued)

Example 7.6 (continued)
Also,

$$d_1 = 2^{-H(p)} = d_2$$

so that

$$q(1) = q(2) = 2^{-1+H(p)} \cdot 2^{-H(p)} = 2^{-1} = 1/2$$

and the uniform input distribution achieves capacity.

7.4.3 Further Properties of Capacity

This section concludes with two more lemmas.

Lemma 7.7 *For any input probability assignment that achieves capacity, the output probabilities are all strictly positive. Here it is assumed that each output can be reached from at least one input.*

Proof From Theorem 7.15, it is required

$$I(X = k; Y) \le C\ , \quad \forall\, k \quad \text{such that } q(k) = 0.$$

That is,

$$I(X = k; Y) = \sum_y p(y|k) \log \frac{p(y|k)}{p(y)} \le C.$$

If for some $p(y) = 0$, any input that reaches output y with nonzero transition must have $q(k) = 0$. For such a k, the L.H.S. of the above equation is infinite, establishing a contradiction □

Lemma 7.8 *The output vector that achieves capacity is unique. All input probability vectors with the proper zero components that give rise to that output vector also reach capacity.*

Proof The conditions in Theorem 7.15 depend only on $q(k)$ though the output probabilities

$$p(y) = \sum_k p(y|X = k)q(k).$$

Hence, any input probability vector giving rise to this output probability vector $p(y)$ satisfies the conditions in Theorem 7.15 and, thus, reaches capacity □

7.5 Continuous Random Variables

The concepts of entropy, mutual information, and capacity extend to the case of continuous random variables. Let X be a continuous random variable with cumulative distribution function $F(x)$ and probability density function $f(x)$, where

$$\int_S f(x)\mathrm{d}x = 1$$

and where $S = \{x : f(x) > 0\}$ is called the support set of X. The entropy of X is

$$H(X) \triangleq -\int_S f(x)\log f(x)\mathrm{d}x. \tag{7.36}$$

Example 7.7 Consider a continuous random variable X having the uniform distribution

$$f(x) = \begin{cases} \frac{1}{a} , & 0 \le x \le a \\ 0 , & \text{otherwise.} \end{cases}$$

The entropy of X is

$$H(X) = -\int_0^a \frac{1}{a}\log\frac{1}{a}\mathrm{d}x = \log a.$$

Gaussian random variables are particularly of interest in communication systems because the AWGN channel has Gaussian statistics. The capacity C of an AWGN channel is of interest.

Example 7.8 Consider a zero mean normal random variable $X \sim N(0, \sigma^2)$ with pdf

$$f(x) = \frac{1}{\sqrt{2\pi}\sigma}\mathrm{e}^{-x^2/2\sigma^2}, \quad -\infty < x < \infty.$$

(continued)

Example 7.8 (continued)
The entropy of X is

$$
\begin{aligned}
H(X) &= -\int_{-\infty}^{\infty} f(x)\log f(x)\mathrm{d}x \\
&= -\int_{-\infty}^{\infty} f(x)\left[-\frac{x^2}{2\sigma^2} - \log(\sqrt{2\pi}\sigma)\right]\mathrm{d}x \\
&= \tfrac{1}{2}\log(2\pi\sigma^2) + \frac{\mathrm{E}[x^2]}{2\sigma^2} \\
&= \tfrac{1}{2}\left[1 + \log(2\pi\sigma^2)\right] \\
&= \tfrac{1}{2}\log(2\pi \mathrm{e}\sigma^2).
\end{aligned}
$$

Considering a pair of continuous random variables X and Y, the conditional entropy of X given Y is

$$
\begin{aligned}
H(X|Y) &= -\int_S f(x, y)\log f(x|y)\mathrm{d}x\mathrm{d}y \\
&= H(X, Y) - H(Y) \qquad (7.37)
\end{aligned}
$$

and the mutual information is

$$
\begin{aligned}
I(X; Y) &\triangleq \int_S f(x, y)\log\frac{f(x, y)}{q(x)\cdot f(y)}\mathrm{d}x\mathrm{d}y \\
&= \int_S f(y|x)q(x)\log\frac{f(y|x)}{f(y)}\mathrm{d}x\mathrm{d}y \qquad (7.38)
\end{aligned}
$$

where

$$
f(y) = \int f(y|x)q(x)\mathrm{d}x.
$$

Also,

$$
I(X; Y) = H(X) - H(X|Y) = H(Y) - H(Y|X). \qquad (7.39)
$$

Fig. 7.15 Gaussian channel

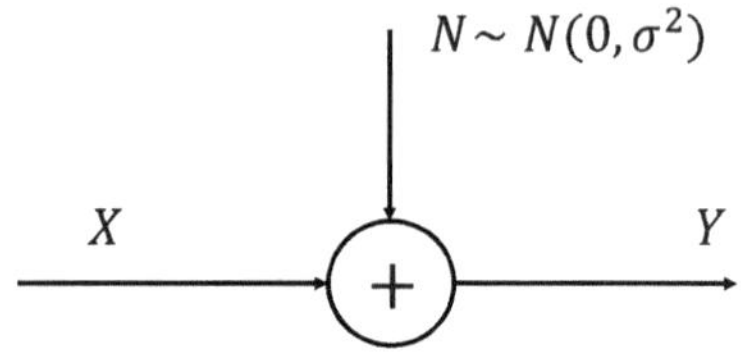

7.5.1 The Gaussian Channel

Consider the Gaussian channel shown in Fig. 7.15. The channel input X is corrupted by additive noise N that is Gaussian distributed $N(0, \sigma^2)$, where σ^2 is the noise power, and the channel output is $Y = X + N$. The power of the input X is constrained, such that

$$\mathrm{E}[X^2] \leq P.$$

To find the capacity, first note that

$$\begin{aligned} I(X:Y) &= H(Y) - H(Y|X) \\ &= H(Y) - H(X+N|X) \\ &= H(Y) - H(N|X)\ , \quad X \text{ and } N \text{ are independent} \\ &= H(Y) - H(N). \end{aligned} \tag{7.40}$$

From Example 7.8

$$H(N) = \tfrac{1}{2}\log(2\pi e \sigma^2). \tag{7.41}$$

Also, since X and N are independent and N has zero mean,

$$\mathrm{E}[Y^2] = \mathrm{E}[(X+N)^2] = \mathrm{E}[X^2] + \mathrm{E}[N^2] = P + \sigma^2. \tag{7.42}$$

Hence, the problem is to maximize $H(Y)$ given that $\mathrm{E}[Y^2] = P + \sigma^2$. This convex optimization problem may be solved in several ways. One method applies variational calculus with the method of Lagrange multipliers. The convex optimization problem finds the density function $f(y)$ to

Example 7.9 Maximize

$$H(Y) = \int_{-\infty}^{\infty} f(y)\log f(y)\mathrm{d}y$$

(continued)

Example 7.9 (continued)
such that

$$\int_{-\infty}^{\infty} y^2 f(y)\mathrm{d}y = P + \sigma^2$$

$$\int_{-\infty}^{\infty} f(y)\mathrm{d}y = 1.$$

From the above constrained optimization problem, to apply the method of Lagrange multipliers, the following auxiliary equation is formed:

$$\mathcal{L} = \int_{-\infty}^{\infty} f(y)\log f(y)\mathrm{d}y + \lambda_1 \left[\int_{-\infty}^{\infty} y^2 f(y)\mathrm{d}y - P - \sigma^2\right] + \lambda_2 \left[\int_{-\infty}^{\infty} f(y)\mathrm{d}y - 1\right].$$

Taking the derivative of $\mathcal{L}$ with respect to $f(y)$ and setting the result equal to zero gives

$$\frac{d\mathcal{L}}{df(y)} = 1 + \log f(y) + \lambda_1 y^2 + \lambda_2 = 0.$$

Hence,

$$f(y) = \mathrm{e}^{-\lambda_1 y^2 - \lambda_2 - 1}.$$

Now apply the constraint

$$\int_{-\infty}^{\infty} f(y)\mathrm{d}y = 1$$

to give

$$\mathrm{e}^{-\lambda_2 - 1} = \sqrt{\frac{\lambda_1}{\pi}}.$$

Therefore,

$$f(y) = \sqrt{\frac{\lambda_1}{\pi}}\mathrm{e}^{-\lambda_1 y^2}.$$

Finally, apply the constraint

$$\int_{-\infty}^{\infty} y^2 p(y)\mathrm{d}y = P + \sigma^2$$

to give

$$\lambda_1 = \frac{1}{2(P + \sigma^2)}$$

and

$$f(y) = \frac{1}{\sqrt{2\pi(P + \sigma^2)}} \mathrm{e}^{-y^2/2(P+\sigma^2)} \sim N(0, P + \sigma^2).$$

Since Y is Gaussian, the entropy of Y is

$$H(Y) = \tfrac{1}{2}\log\left[2\pi\mathrm{e}(P + \sigma^2)\right]. \tag{7.43}$$

Hence,

$$I(X:Y) = H(Y) - H(N) \tag{7.44}$$

$$\leq \tfrac{1}{2}\log(2\pi\mathrm{e}(P + \sigma^2)) - \tfrac{1}{2}\log(2\pi\mathrm{e}\sigma^2)$$

$$= \frac{1}{2}\log\left(1 + \frac{P}{\sigma^2}\right) = C. \tag{7.45}$$

The R.H.S. above represents the capacity C.

Finally, since $Y \sim N(0, P + \sigma^2)$ achieves capacity and $N \sim N(0, \sigma^2)$ is independent of the input X, the capacity achieving input distribution is Gaussian, $X \sim N(0, P)$.

7.5.2 Band-Limited Gaussian Channel

Consider the band-limited AWGN channel shown in Fig. 7.16 below. The ideal low-pass filter has the transfer function

$$H(f) = \begin{cases} 1 \ , & -W \leq f \leq W \\ 0 \ , & \text{elsewhere} \end{cases}$$

The AGWN $w(t)$ has a power spectral density of $N_o/2$ watts/Hz.

To derive the channel capacity, the Shannon sampling theorem is employed. Having highest frequency W, the channel output $y(t)$ can be represented by equally

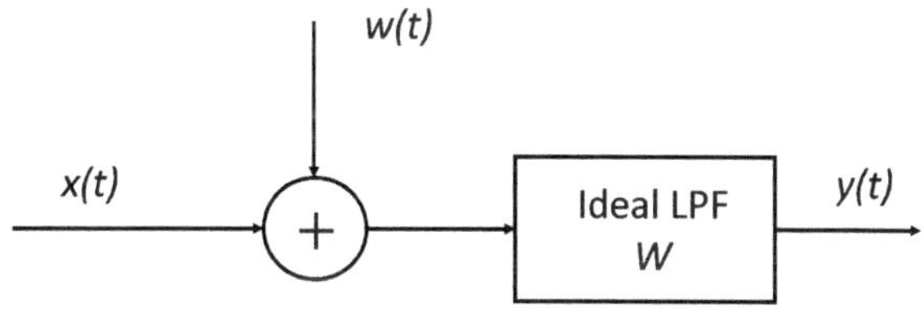

Fig. 7.16 Band-limited Gaussian channel

spaced samples taken $1/2W$ seconds apart. The waveform $y(t)$ can be reconstructed by using the interpolation formula:

$$y(t) = \sum_{n=-\infty}^{\infty} y\left(\frac{n}{2W}\right) \text{sinc}\left(t - \frac{n}{2W}\right). \tag{7.46}$$

Each of the samples $y\left(\frac{n}{2W}\right)$ are corrupted by white Gaussian noise having power spectral density $N_o/2$ watts/Hz and bandwidth W. The noise power in the samples is

$$N = \frac{N_o}{2} \cdot 2W = N_o W \;\; \text{watts} \tag{7.47}$$

and $N \sim N(0, N_o W)$. If data bits are sent with period T and ideal Nyquist pulse shaping is used, then $W = 1/2T$ and the baud rate is $R = 1/T = 2W$ symbols/second.

The capacity per channel use is

$$C = \frac{1}{2}\log\left(1 + \frac{P}{N}\right) \;\; \text{nats/channel use}, \tag{7.48}$$

and with $R = 2W$ channel uses per second, the capacity becomes

$$C = W\log\left(1 + \frac{P}{N}\right) \;\; \text{nats/second}. \tag{7.49}$$

The energy per data bit is equal to $E_b = PT$ so $P = E_b/T = E_b R$, and $N = N_o W$. Hence,

$$C = W\log\left(1 + \frac{E_b R}{N_o W}\right) \;\; \text{nats/second}. \tag{7.50}$$

Dividing both sides by W gives

$$\frac{C}{W} = \log\left(1 + \frac{E_b R}{N_o W}\right) \;\; \text{nats/second/Hz}. \tag{7.51}$$

In the above equation R/W represents the bandwidth efficiency in units of bits/second/Hz.

If $R = C$, i.e., data bits are transmitted at a rate equal to the channel capacity, C, then (using base 2 logarithms)

$$\frac{C}{W} = \log_2\left(1 + \frac{E_b}{N_o} \cdot \frac{C}{W}\right) \text{ bits/second/Hz.} \tag{7.52}$$

Inverting this equation gives E_b/N_o as a function of C/W,

$$\frac{E_b}{N_o} = \frac{2^{C/W} - 1}{C/W}. \tag{7.53}$$

This equation may be used to plot C/W vs. E_b/N_o as shown in Fig. 7.17. All operating points to the right of the C/W curve correspond to $R < C$ and are feasible for reliable communication, while those to the left of the curve correspond to $R > C$ and are impossible. Generally, the objective is to operate as close to the capacity limit as possible. In practice, this is accomplished with error control coding and decoding techniques, the subject of Chap. 8.

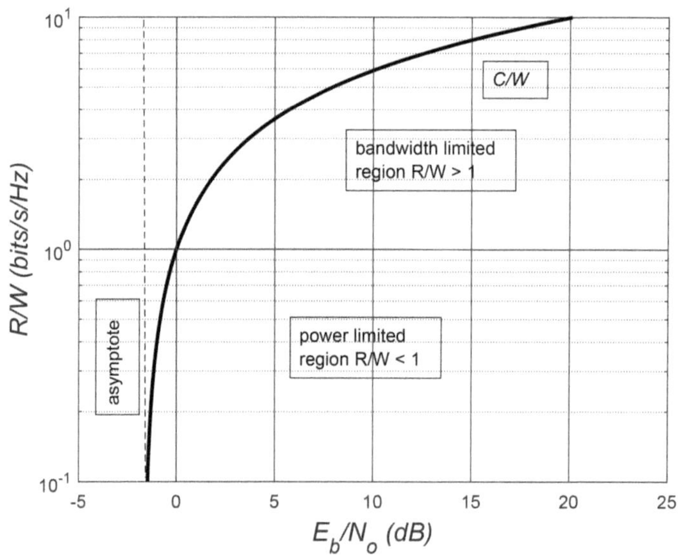

Fig. 7.17 Capacity of band-limited Gaussian channel

Problems

7.1 Find two distinct probability distributions $\{p_i, i = 1, 2, \ldots, N\}$ and $\{q_i, i = 1, 2, \ldots, N\}$ that have the same entropy.

7.2 For the probability distribution $\{p_i, i = 1, 2, \ldots, N\}$ and integer $M,\ 0 < M < N$, define

$$q_M = 1 - \sum_{i=1}^{M} p_i$$

and show that

$$H(p_1, p_2, \ldots, p_N) \leq H(p_1, p_2, \ldots, p_M, q_M) + q_M \log(N - M).$$

7.3 For an arbitrary input probability assignment $\mathbf{q}_o$ to a DMC, let

$$I_o(X = k; Y) = \sum_j p(j|k) \log\left(\frac{p(j|k)}{\sum_\ell q_o(\ell) p(j|\ell)}\right)$$

and show that

$$\sum_k q_o(k) I_o(X = k; Y) \leq C \leq \max_k I_o(X = k; Y).$$

Use this to show that

$$C = \max_{q_o(k)} \max_k I_o(X = k; Y).$$

7.4 A DMC has an input and output alphabet $\{0, 1, \ldots, K - 1\}$. A noise variable Z also takes on values in this set and $Y = X \oplus Z$, where the output is the input plus noise modulo-K.

(a) Show that $I(X; Y) = H(Y) - H(Z)$.
(b) Fine the capacity of this DMC in terms of $H(Z)$ and find the maximizing input probability assignment.

7.5 Find the capacity of the z-channel in Fig. 7.18 and find the maximizing input distribution. Examine these assignments as $\varepsilon \to 1$ and explain your result.

7.6 Let X be a discrete random variable. Show that the entropy of X is greater than or equal to the entropy of any function of X, $g(X)$, by justifying the following steps:

$$\begin{aligned} H(X, g(X)) &= H(X) + H(g(X)|X) \\ &= H(X) \end{aligned}$$

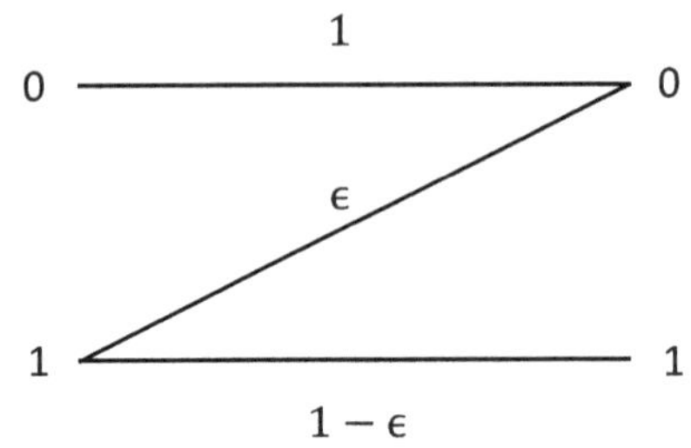

Fig. 7.18 z-channel for Problem 7.5

$$H(X, g(X)) = H(g(X)) + H(X|g(X))$$
$$\geq H(g(X)).$$

Therefore, $H(g(X)) \leq H(X)$.

7.7 If $H(Y|X) = 0$, show that Y is a function of X, That is, for all x with $p(x) > 0$, show there is one and only one possible value of y with $p(x, y) > 0$.

7.8 Consider the joint random variables X, Y, and Z. Show examples that satisfy:

(a) $I(X; Y|Z) < I(X; Y)$
(b) $I(X; Y|Z) > I(X; Y)$

7.9 Let X and Y be a pair of random variables that take on values $x_1, x_2, \ldots, x_R$ and $y_1, y_2, \ldots, y_S$, respectively, and consider the their sum $Z = X + Y$.

(a) Show that $H(Z|X) = H(Y|X)$. Show that if X and Y are independent, then $H(Y) < H(Z)$ and $H(X) < H(Z)$. Hence, the addition of independent random variables increases uncertainty.
(b) Give an example (of dependent random variables) for which $H(Y) > H(Z)$ and $H(X) > H(Z)$.
(c) Under what conditions will $H(Z) = H(X) + H(Y)$?

7.10 Show that the entropy of the distribution $\{p_1, p_2, \ldots, p_i, \ldots, p_j, \ldots, p_m\}$ is smaller than the entropy of the distribution

$$\{p_1, p_2, \ldots, \frac{p_i + p_j}{2}, \ldots, \frac{p_i + p_j}{2}, \ldots, p_m\}.$$

Show that any transfer of probability that makes the distribution more uniform will generally increase the entropy.

7.11 Consider the joint random variables X, Y and Z. Prove the following inequalities and determine the conditions where equality holds:

(a) $H(X, Y|Z) \geq H(X|Z)$
(b) $I(X, Y : Z) \geq I(X : Z)$
(c) $H(X, Y, Z) - H(X, Y) \leq H(X, Z) - H(X)$
(d) $I(X; Z|Y) \geq I(Z; Y|X) - I(Z; Y) + I(X; Z)$

7.12 Let P$[X = i] = p_i, i = 1, 2, \ldots, m$ such that $p_1 \geq p_2, \geq p_3, \geq \cdots \geq p_m$. The minimum error probability predictor of X is $\hat{X} = 1$, with resulting error probability $P_E = 1 - p_1$. Maximize $H(\mathbf{p})$ such that $1 - p_1 = P_E$ to bound P_E in terms of H. This is Fano's equality without any conditioning.

7.13 Find the capacity of the DMC in Fig. 7.19 where the alphabet for X is $\mathcal{X} = \{0, 1\}$ and P$[Z = 0] =$ P$[Z = a] = 1/2$. Assume that Z is independent of X. Note that the channel capacity depends on the parameter a.

7.14 Find the capacity of the binary "errors and erasures" channel shown in Fig. 7.20.

7.15 A binary symmetric channel (BSC) and a binary erasure channel (BEC) are concatenated together as shown in Fig. 7.21. The output of the BSC is applied to the input of the BEC so that the "overall" channel has two inputs and three outputs.

(a) Find the 2×3 channel transition matrix $[p(j|i)]$ (j output, i input) for the "overall channel."

(b) What is the capacity of the overall channel

$$C = \max_{(q(0),q(1))} \sum_j \sum_i p(j|i)q(i)\log_2 \frac{p(j|i)}{\sum_k p(j|k)q(k)}$$

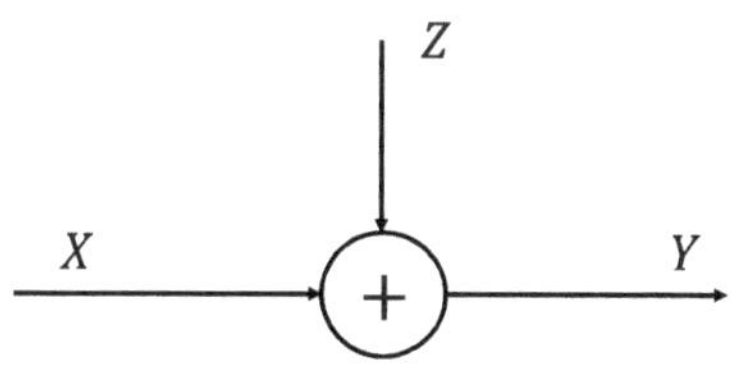

Fig. 7.19 DMC for Problem 7.13

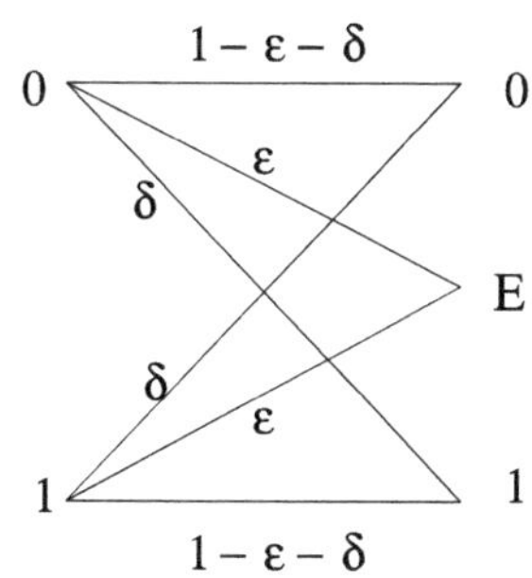

Fig. 7.20 Binary errors and erasures channel for Problem 7.14

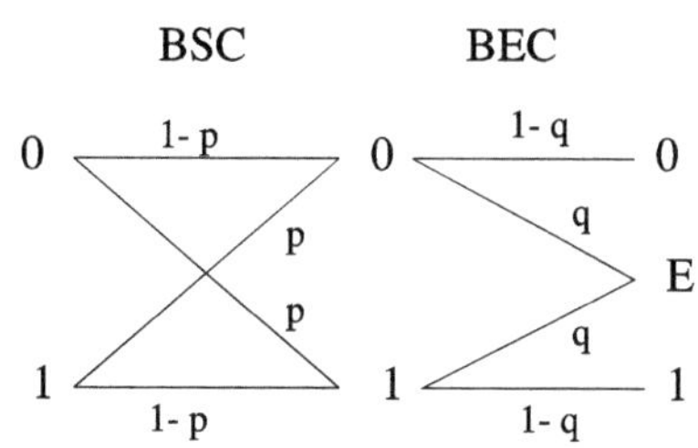

Fig. 7.21 Concatenated channel for Problem 7.15

Chapter 8
Error Control Methods

Error control coding introduces controlled redundancy into the transmitted messages that is exploited at the receiver to correct errors that are introduced by the noisy channel using forward error correction (FEC). An error control code may also be used for error detection in schemes that employ automatic repeat request (ARQ) strategies. ARQ strategies must have a feedback channel to relay the retransmission requests from the receiver back to the transmitter when errors are detected. ARQ schemes require buffering at the transmitter and/or receiver and are, therefore, suitable for data applications but are not suitable for delay sensitive applications such as voice or real-time video. Hybrid ARQ schemes use both error correction and error detection; the code is used to correct the more likely error events and to detect the less likely error events. Upon detection of an error event a retransmission is requested.

There are many different types of error correcting codes, but historically they have been classified into block codes and convolutional codes. To generate a codeword of an (n, k) block code, a block of k data bits is appended by $n - k$ redundant parity bits that are algebraically related to the k data bits, thereby producing a codeword consisting of n code bits. The ratio $R_c = k/n$ is called the code rate, where $0 < R_c < 1$.

Convolutional codes, on the other hand, have codewords that are generated by the discrete-time convolution of an input data sequence with the impulse response of the encoder. While a block encoder operates on k-bit blocks of data bits, a convolutional encoder accepts a continuous sequence of input data bits and outputs a continuous sequenced of code bits that may be terminated to form a fixed-length codeword.

The earliest coded communication systems treated the modulator and encoder as separate entities, leading to disappointing results for bandwidth efficient communications. In 1974, Massey [21] suggested that the performance of a coded communication system could be improved by treating the error control encoder and modulator as a single entity. Ungerboeck later developed the basic principles of trellis-coded modulation (TCM) and identified classes of trellis codes that

G. Stuber, *Introductory Digital Wireless Communications*,
https://doi.org/10.1007/978-3-032-01559-4_8

provide substantial coding gains on bandwidth-limited additive white Gaussian noise (AWGN) channels [29].

The remainder of this chapter is organized as follows. Section 8.1 gives an introduction to block codes and their decoding. Section 8.2 introduces convolutional codes and the Viterbi algorithm for decoding convolutional codes. Section 8.3 introduces trellis coded modulation. This is followed by the performance analysis of convolutional and trellis codes on AWGN channels in Sect. 8.4. Section 8.5 considers block and convolutional interleavers that are useful for error control coding on channels with memory such as land mobile radio channels. Finally, Sect. 8.6 wraps up the chapter with an introduction of automatic repeat request (ARQ) schemes.

8.1 Block Codes

This section introduces the basic concepts of block coding, including the generation of block codes and their decoding. It examines the algebraic properties of block codes and how they are related to error detection and correction.

8.1.1 Binary Block Codes

A binary block encoder accepts a length-k input vector $\mathbf{a} = (a_1, a_2, \ldots, a_k)$, $a_i \in \{0, 1\}$, and generates a length-n codeword $\mathbf{c} = (c_1, c_2, \ldots, c_n)$, $c_i \in \{0, 1\}$, through the linear mapping $\mathbf{c} = \mathbf{aG}$, where $\mathbf{G} = [g_{ij}]_{k\times n}$ is a $k \times n$ *generator matrix*. The matrix $\mathbf{G}$ has full row rank k, and the code C is generated by taking all linear combinations of the rows of the matrix $\mathbf{G}$, where field operations are performed by using modulo-2 arithmetic. The code rate is $R_c = k/n$ and there are 2^k codewords. The task of designing block codes reduces to one of finding generator matrices that produce block codes that are both powerful and easy to decode.

For any block code with generator matrix $\mathbf{G}$, there exists an $(n-k) \times n$ *parity check matrix* $\mathbf{H} = [h_{ij}]_{(n-k)\times n}$ such that $\mathbf{GH}^T = \mathbf{0}_{k\times(n-k)}$. The matrix $\mathbf{H}$ has full row rank $n-k$ and is orthogonal to all codewords, i.e., $\mathbf{cH}^T = \mathbf{0}_{n-k}$. The matrix $\mathbf{H}$ is the generator matrix of a *dual code* C^T, consisting of 2^{n-k} codewords. The parity check matrix of the dual code C^T is the matrix $\mathbf{G}$.

A *systematic block code* is one having a generator matrix of the form

$$\mathbf{G} = [\mathbf{I}_{k\times k}|\mathbf{P}], \tag{8.1}$$

where $\mathbf{P}$ is a $k\times(n-k)$ matrix. For a systematic block code, the first k coordinates of each codeword are equal to the k-bit input vector $\mathbf{a}$, while the last $n-k$ coordinates are the parity check bits. By using elementary row operations, the generator matrix of any linear block code can be put into systematic form. A systematic block code has the parity check matrix

$$\mathbf{H} = [-\mathbf{P}^T | \mathbf{I}_{(n-k)\times(n-k)}]. \tag{8.2}$$

The parity check matrix in (8.2) is a general form that applies to both binary and nonbinary systematic block codes. For binary codes the negative sign in front of the $\mathbf{P}^T$ matrix is not necessary since modulo 2 subtraction is the same as modulo 2 addition. For a binary systematic block code, $\mathbf{G}\mathbf{H}^T = \mathbf{I}_{k\times k}\mathbf{P} \oplus \mathbf{P}\mathbf{I}_{(n-k)\times(n-k)} = \mathbf{P} \oplus \mathbf{P} = \mathbf{0}_{k\times(n-k)}$, where $\oplus$ indicates modulo-2 addition.

Example 8.1 The $(n-k)\times n$ parity check matrix of an (n, k) Hamming code is constructed by listing as columns all nonzero binary $(n-k)$-tuples. There are $n = 2^{n-k} - 1$ such nonzero $(n-k)$-tuples. In general, the (n, k) Hamming code has dimensions

$$n = 2^m - 1$$
$$k = 2^m - m - 1$$

for some integer m.

For example, a systematic (7,4) Hamming code has the parity check matrix

$$\mathbf{H} = \begin{bmatrix} 1\,0\,1\,1 \vdots 1\,0\,0 \\ 1\,1\,1\,0 \vdots 0\,1\,0 \\ 0\,1\,1\,1 \vdots 0\,0\,1 \end{bmatrix}, \tag{8.3}$$

where the columns of **H** consists of all nonzero binary 3-tuples. Note that the last three columns form a 3×3 identity matrix since the code is in systematic form, while the first 4 columns that constitute the $\mathbf{P}^T$ matrix may be placed in any random order with no effect on the performance of the code. Such codes are said to be equivalent.

The generator matrix of this particular (7,4) systematic Hamming code is

$$\mathbf{G} = \begin{bmatrix} 1\,0\,0\,0 \vdots 1\,1\,0 \\ 0\,1\,0\,0 \vdots 0\,1\,1 \\ 0\,0\,1\,0 \vdots 1\,1\,1 \\ 0\,0\,0\,1 \vdots 1\,0\,1 \end{bmatrix} \tag{8.4}$$

(continued)

Example 8.1 (continued)
and it can be verified that $\mathbf{GH}^T = \mathbf{0}_{k\times(n-k)}$. The 16 codewords of the $(7, 4)$ Hamming code are generated by taking all linear combinations of the rows of $\mathbf{G}$ using modulo-2 arithmetic.

8.1.1.1 Minimum Distance

Let $d(\mathbf{c}_1, \mathbf{c}_2)$ denote the Hamming distance between the codewords $\mathbf{c}_1$ and $\mathbf{c}_2$, equal to the number of coordinates in which they differ. For linear block codes $d(\mathbf{c}_1, \mathbf{c}_2) = w(\mathbf{c}_1 \oplus \mathbf{c}_2)$, where $w(\mathbf{c}_1 \oplus \mathbf{c}_2)$ is the weight of $\mathbf{c}_1 \oplus \mathbf{c}_2$, equal to the number of nonzero coordinates of $\mathbf{c}_1 \oplus \mathbf{c}_2$. The free Hamming distance, d_{free}, of a linear block code is the minimum number of coordinates in which any two codewords differ. For a linear code, the sum of any two codewords $\mathbf{c}_1 \oplus \mathbf{c}_2$ is another codeword $\mathbf{c}$ and the all zeroes vector is always a codeword. Hence, the free Hamming distance is

$$d_{\text{free}} = \min_{\mathbf{c}_1,\mathbf{c}_2} d(\mathbf{c}_1, \mathbf{c}_2) = \min_{\mathbf{c}\neq\mathbf{0}} d(\mathbf{c}, \mathbf{0}) = \min_{\mathbf{c}\neq\mathbf{0}} w(\mathbf{c}). \tag{8.5}$$

Therefore, d_{free} is equal to the weight of the minimum weight nonzero codeword.

To derive an upper bound on d_{free}, recall that the generator matrix of any linear block code can be put into systematic form $\mathbf{G} = [\mathbf{I}_{k\times k}|\mathbf{P}]$, where $\mathbf{P}$ is a $k \times (n-k)$ matrix. It is certainly the case that the number of nonzero elements in any row of $\mathbf{P}$ cannot exceed $n-k$. Hence, the number of nonzero elements in any row of $\mathbf{G}$ cannot exceed $n-k+1$. Since the rows of the generator matrix $\mathbf{G}$ are themselves valid codewords, it must be true that

$$d_{\text{free}} \leq n - k + 1, \tag{8.6}$$

a result known as the *Singleton bound*. A code that has $d_{\text{free}} = n-k+1$ is called a maximum distance separable (MDS) code. Intuitively, this means that all codewords are as far apart from each other as possible in Hamming distance, and such codes should perform well.

An example of a simple block code that meets the Singleton bound is the binary repetition code

$$0 \longrightarrow \mathbf{c}_0 = (0, 0, \ldots, 0)_n$$
$$1 \longrightarrow \mathbf{c}_1 = (1, 1, \ldots, 1)_n$$

In this case, $d_{\text{free}} = d(\mathbf{c}_0, \mathbf{c}_1) = n - k + 1$. The repetition code happens to be the only binary MDS code, and no other binary MDS codes exist.

8.1.1.2 Syndromes

Suppose that the codeword $\mathbf{c}$ is transmitted and the vector $\mathbf{y} = \mathbf{c} \oplus \mathbf{e}$ is received, where $\mathbf{e}$ is defined as the error vector. The syndrome of the received vector $\mathbf{y}$ is defined as the length $n - k$ vector:

$$\mathbf{s} = \mathbf{y}\mathbf{H}^T. \tag{8.7}$$

If $\mathbf{s} = \mathbf{0}$, then $\mathbf{y}$ is a codeword; conversely, if $\mathbf{s} \neq \mathbf{0}$, then error(s) must have occurred. Note that if $\mathbf{y}$ is a codeword, then $\mathbf{s} = \mathbf{0}$. Hence, $\mathbf{s} = \mathbf{0}$ does not mean that no errors have occurred. They are just undetectable. Since the sum of any two codewords is another codeword for a linear code, it follows that the number of undetectable error vectors is equal to $2^k - 1$, the number of nonzero codewords. The syndrome only depends upon the error vector $\mathbf{e}$ because

$$\mathbf{s} = \mathbf{y}\mathbf{H}^T = \mathbf{c}\mathbf{H}^T \oplus \mathbf{e}\mathbf{H}^T = \mathbf{0} \oplus \mathbf{e}\mathbf{H}^T = \mathbf{e}\mathbf{H}^T. \tag{8.8}$$

In general, $\mathbf{s} = \mathbf{e}\mathbf{H}^T$ is a system of $n - k$ equations in n variables. Hence, for any given syndrome $\mathbf{s}$, there are 2^k possible solutions having the same error vector $\mathbf{e}$. However, the most likely error vector $\mathbf{e}$ is the one having minimum Hamming weight.

8.1.1.3 Error Detection

A linear block code can detect all error patterns of $d_{\text{free}} - 1$ or fewer errors. If $\mathbf{e} \neq \mathbf{0}$ is a codeword, then no errors are detected. There are $2^k - 1$ undetectable error patterns, but there are $2^n - 1$ possible nonzero error patterns. Hence, the number of detectable error patterns is

$$2^n - 1 - (2^k - 1) = 2^n - 2^k.$$

Usually, $2^k - 1$ is a small fraction of $2^n - 2^k$. For the (7,4) Hamming code considered in Example 8.1, there are $2^4 - 1 = 15$ undetectable error patterns and $2^7 - 2^4 = 112$ detectable error patterns.

8.1.1.4 Weight Distribution

Consider a block code C and let A_i be the number of codewords of weight i. The set $\{A_0, A_1, \ldots, A_n\}$ is called the weight distribution of C. The weight distribution can be expressed as a *weight enumerator polynomial*:

$$A(z) = A_0 z^0 + A_1 z^1 + \cdots + A_n z^n. \tag{8.9}$$

Likewise, the weight distribution of the dual code C^T is $\{B_0, B_1, \ldots, B_n\}$ where B_i is the number of codewords of weight i in the dual code. The weight distribution of the dual code can be expressed as the weight enumerator polynomial:

$$B(z) = B_0 z^0 + B_1 z^1 + \cdots + B_n z^n. \tag{8.10}$$

Example 8.2 For the (7,4) Hamming code in Example 8.1,

$$A_0 = 1, A_2 = 0, A_3 = 7, A_4 = 7, A_5 = 0, A_6 = 0, A_7 = 1.$$

Hence,

$$A(z) = 1 + 7z^3 + 7z^4 + z^7.$$

For the dual to the (7,4) Hamming code

$$B_0 = 1, B_2 = 0, B_3 = 0, B_4 = 7, B_5 = 0, B_6 = 0, B_7 = 0.$$

Hence,

$$B(z) = 1 + 7z^4.$$

Unfortunately, weight enumerator polynomials are generally very difficult to find and are known only for a few classes of codes including the Hamming codes. For a general length-n Hamming code, the weight enumerator polynomial is known to be [3]

$$A(z) = \frac{1}{n+1}\left[(1+z)^n + n(1-z)(1-z^2)^{(n-1)/2}\right] \tag{8.11}$$

and A_i is the coefficient of z^i after expanding the above polynomial.

8.1.1.5 Probability of Undetected Error

The probability of undetected error is

$$\begin{aligned} \mathrm{P}_e[u] &= \mathrm{P}[\mathbf{e} \text{ is a nonzero codeword}] \\ &= \sum_{i=1}^{n} A_i \mathrm{P}[w(\mathbf{e}) = i]. \end{aligned} \tag{8.12}$$

The error probability $\mathrm{P}[w(\mathbf{e}) = i]$ depends on the coding channel, defined as that portion of the communication system that is seen by the coding system. The simplest coding channel is the binary symmetric channel (BSC), where

$$\mathrm{P}[y_i \neq c_i] = p = 1 - \mathrm{P}[y_i = c_i]. \tag{8.13}$$

For a BSC, $\mathrm{P}[w(\mathbf{e}) = i] = p^i(1-p)^{n-i}$ and, hence,

$$\mathrm{P}_e[u] = \sum_{i=1}^{n} A_i p^i (1-p)^{n-i}. \tag{8.14}$$

By using the MacWilliams identity [19], the probability of undetected error for an error control code C may be alternatively expressed in terms of the weight distribution of its dual code C^T as

$$\mathrm{P}_e[u] = 2^{n-k} \sum_{i=0}^{n} B_i (1-2p)^i - (1-p)^n. \tag{8.15}$$

Example 8.3 The (7,4) Hamming code in Example 8.1 has an undetected error probability of

$$\mathrm{P}_e[u] = 7p^3(1-p)^4 + 7p^4(1-p)^3 + p^7. \tag{8.16}$$

For a raw channel error rate of $p = 10^{-2}$, it follows $\mathrm{P}_e[u] = 7 \times 10^{-6}$. Hence, the undetected error rate can be very small even for a fairly simple block code.

Alternatively, the probability of undetected error for the (7,4) Hamming code can be obtained from its (7,3) dual code as

$$\mathrm{P}_e[u] = 2^{-3}\left[1 + 7(1-2p)^4\right] - (1-p)^7. \tag{8.17}$$

Again with $p = 10^{-2}$, it can be verified that $\mathrm{P}_e[u] = 7 \times 10^{-6}$.

8.1.1.6 Error Correction

A linear block code can correct all error patterns of t or fewer errors, where

$$t \leq \lfloor \frac{d_{\mathrm{free}} - 1}{2} \rfloor \tag{8.18}$$

and $\lfloor x \rfloor$ is the largest integer contained in x. A code is usually capable of correcting many error patterns of $t+1$ or more errors. In fact, up to 2^{n-k} error patterns may be corrected, which is equal to the number of syndromes.

For a BSC, the probability of codeword error is

$$\begin{aligned} \mathrm{P}[e] &\leq 1 - \mathrm{P}[t \text{ or fewer errors}] \\ &= 1 - \sum_{i=0}^{t} \binom{n}{i} p^i (1-p)^{n-i}. \end{aligned} \tag{8.19}$$

8.1.1.7 Standard Array Decoding

One conceptually simple method for decoding linear block codes is standard array decoding. The standard array of an (n, k) linear block code is constructed as follows:

1. Write out all 2^k codewords in a row starting with $\mathbf{c}_0 = \mathbf{0}$.
2. From the remaining $2^n - 2^k$ n-tuples, select an error vector $\mathbf{e}_2$ of weight 1 and place it under $\mathbf{c}_0$. Under each codeword put $\mathbf{c}_i \oplus \mathbf{e}_2$, $i = 1, \ldots, 2^k - 1$.
3. Select a minimum weight error vector $\mathbf{e}_3$ from the remaining unused n-tuples and place it under $\mathbf{c}_0 = \mathbf{0}$. Under each codeword put $\mathbf{c}_i \oplus \mathbf{e}_3$, $i = 1, \ldots, 2^k - 1$.
4. Repeat Step 3 until all n-tuples have been used.

Note that every n-tuple appears once and only once in the standard array.

Example 8.4 Consider the (4, 2) code with generator matrix

$$\mathbf{G} = \begin{bmatrix} 1\,1\,0\,0 \\ 0\,1\,0\,1 \end{bmatrix}.$$

The standard array is

$$\begin{bmatrix} \mathbf{e}_1 & 0000 & 1100 & 0101 & 1001 \\ \mathbf{e}_2 & 0001 & 1101 & 0100 & 1000 \\ \mathbf{e}_3 & 0010 & 1110 & 0111 & 1011 \\ \mathbf{e}_4 & 0011 & 1111 & 0110 & 1010 \end{bmatrix}.$$

The standard array consists of 2^{n-k} disjoint rows of 2^k elements. These rows are called cosets and the ith row has the elements

$$F_i = \{\mathbf{e}_i, \mathbf{e}_i \oplus \mathbf{c}_1, \ldots, \mathbf{e}_i \oplus \mathbf{c}_{2^k-1}\}.$$

The first element, $\mathbf{e}_i$, is called the coset leader. The standard array also consists of 2^k disjoint columns. The jth column has the elements

$$D_j = \{\mathbf{c}_j, \mathbf{c}_j \oplus \mathbf{e}_2, \ldots, \mathbf{c}_j \oplus \mathbf{e}_{2^{n-k}}\}.$$

To correct errors, the following procedure is used. When $\mathbf{y}$ is received, find $\mathbf{y}$ in the standard array. If $\mathbf{y}$ is in row i and column j, then the coset leader from row i, $\mathbf{e}_i$, is the most likely error pattern to have occurred and $\mathbf{y}$ is decoded into $\mathbf{y} \oplus \mathbf{e}_i = \mathbf{c}_j$. Hence, $\mathbf{y}$ is simply decoded as the codeword $\mathbf{c}_j$ at the top of its column in the standard array.

A code is capable of correcting all error patterns that are coset leaders. If the error pattern is not a coset leader then erroneous decoding will result. Obviously, standard array decoding is only useful for simple codes, since 2^n vectors must be stored in memory. A somewhat simpler decoding strategy is syndrome decoding.

8.1.1.8 Syndrome Decoding

Syndrome decoding relies on the fact that all 2^k n-tuples in the same coset of the standard array have the same syndrome. This is because the syndrome only depends on the coset leader as shown in (8.8). To perform syndrome decoding:

1. Compute the syndrome $\mathbf{s} = \mathbf{y}\mathbf{H}^T$.
2. Locate the coset leader $\mathbf{e}_\ell$ where $\mathbf{e}_\ell \mathbf{H}^T = \mathbf{s}$.
3. Decode $\mathbf{y}$ into $\mathbf{y} \oplus \mathbf{e}_\ell = \hat{\mathbf{c}}$.

This technique can be used for any linear block code. The calculation in Step 2 can be done by using a simple look-up table. However, for large $n - k$ it becomes impractical because 2^{n-k} syndromes and 2^{n-k} error patterns must be stored.

Cyclic block codes may be construction using one or more primitive polynomials of a Galois extension field $GF(2^m)$. The length $n = 2^m - 1$ Hamming code is one such cyclic code where the columns are arranged in the parity check matrix $\mathbf{H}$ such every row of $\mathbf{H}$ is a cylic shift of every other row, and every row of the corresponding generator matrix $\mathbf{G}$ is a cyclic shift of every other row. Such codes have computationally efficient decoders that can be implemented as simple shift registers and adders [16].

8.2 Convolutional Codes

This section introduces basic convolutional coding and describes the convolutional encoder as a finite-state machine (FSM) with a state and trellis description.

8.2.1 Encoder Description

The encoder for a rate-$1/n$ binary convolutional code can be viewed as an FSM that consists of a ν-stage binary shift register with connections to n modulo-2 adders and a multiplexer that converts the adder outputs to serial codewords. The constraint length of a convolutional code is defined as the number of shifts through the FSM over which a single input data bit can affect the encoder output. For an encoder having a single ν-stage shift register, the constraint length is equal to $K = \nu + 1$. An example rate-1/2, constraint length $K = 3$, binary convolutional encoder is shown in Fig. 8.1.

The above concept can be generalized to rate-k/n binary convolutional code by using k shift registers, n modulo-2 adders, along with input and output multiplexers. For a rate-k/n code, the k-bit information vector $\mathbf{a}_\ell = (a_\ell^{(1)}, \ldots, a_\ell^{(k)})$ is input to the encoder at epoch ℓ to generate the n-bit code vector $\mathbf{c}_\ell = (c_\ell^{(1)}, \ldots, c_\ell^{(n)})$. If K_i denotes the constraint length of the ith shift register, then the overall constraint length is defined as $K = \max_i K_i$. Figure 8.2 shows a simple rate-2/3, constraint length-2 convolutional encoder.

A convolutional encoder can be described by the set of impulse responses, $\{\mathbf{g}_i^{(j)}\}$, where $\mathbf{g}_i^{(j)}$ is the jth output sequence $\mathbf{c}^{(j)}$ that results from the ith input sequence $\mathbf{a}^{(i)} = (1, 0, 0, 0, \ldots)$. The impulse responses can have a duration of at most K and have the form $\mathbf{g}_i^{(j)} = (g_{i,0}^{(j)}, g_{i,1}^{(j)}, \ldots, g_{i,K-1}^{(j)})$. Sometimes the $\{\mathbf{g}_i^{(j)}\}$ are called generator sequences. For the encoder in Fig. 8.1

$$\mathbf{g}^{(1)} = (1,\ 1,\ 1) \qquad \mathbf{g}^{(2)} = (1,\ 0,\ 1) \tag{8.20}$$

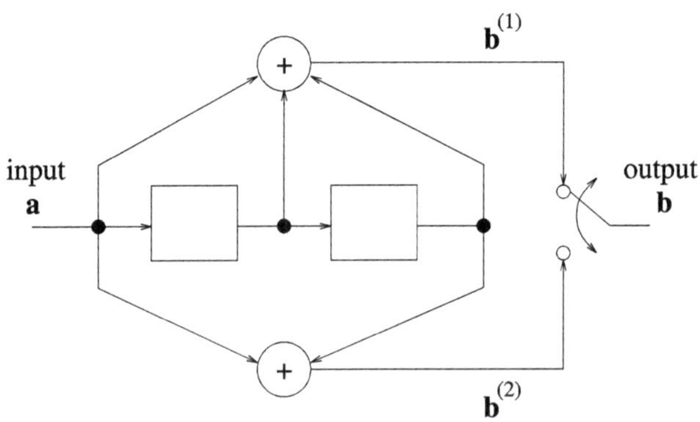

Fig. 8.1 Binary convolutional encoder; $R_c = 1/2$, $K = 3$

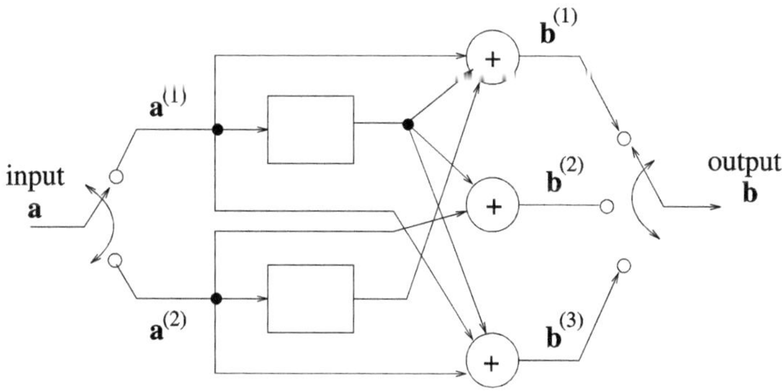

Fig. 8.2 Binary convolutional encoder; $R_c = 2/3$, $K = 2$

and for the encoder in Fig. 8.2

$$\begin{aligned} \mathbf{g}_1^{(1)} &= (1,\ 1),\ \ \mathbf{g}_1^{(2)} = (0,\ 1),\ \ \mathbf{g}_1^{(3)} = (1,\ 1) \\ \mathbf{g}_2^{(1)} &= (0,\ 1),\ \ \mathbf{g}_2^{(2)} = (1,\ 0),\ \ \mathbf{g}_2^{(3)} = (1,\ 0). \end{aligned} \tag{8.21}$$

It follows that the jth output $\mathbf{c}_i^{(j)}$ corresponding to the ith input sequence $\mathbf{a}^{(i)}$ is the discrete convolution $\mathbf{c}_i^{(j)} = \mathbf{a}^{(i)} \circledast \mathbf{g}_i^{(j)}$, where $\circledast$ denotes modulo-2 convolution. The time domain convolutions can be conveniently replaced by polynomial multiplications in a D-transform domain according to

$$\mathbf{c}_i^{(j)}(D) = \mathbf{a}^{(i)}(D)\mathbf{g}_i^{(j)}(D), \tag{8.22}$$

where

$$\mathbf{a}^{(i)}(D) = \sum_{k=0}^{\infty} a_{i,k} D^k \tag{8.23}$$

is the ith input data polynomial,

$$\mathbf{c}_i^{(j)}(D) = \sum_{k=0}^{\infty} b_{i,k}^{(j)} D^k \tag{8.24}$$

is the jth output polynomial corresponding to the ith input, and

$$\mathbf{g}_i^{(j)}(D) = \sum_{k=0}^{K-1} g_{i,k}^{(j)} D^k \tag{8.25}$$

is the associated generator polynomial. It follows that the jth output sequence is

$$\mathbf{c}^{(j)}(D) = \sum_{i=1}^{k} \mathbf{c}_i^{(j)}(D) = \sum_{i=1}^{k} \mathbf{a}^{(i)}(D)\mathbf{g}_i^{(j)}(D). \tag{8.26}$$

The above expression leads to the matrix form

$$\left(\mathbf{c}^{(i)}(D), \ldots, \mathbf{c}^{(n)}(D)\right) = \left(\mathbf{a}^{(1)}(D), \ldots, \mathbf{a}^{(k)}(D)\right) \mathbf{G}(D), \tag{8.27}$$

where

$$\mathbf{G}(D) = \begin{bmatrix} \mathbf{g}_1^{(1)}(D), & \cdots, & \mathbf{g}_1^{(n)}(D) \\ \vdots & & \vdots \\ \mathbf{g}_k^{(1)}(D), & \cdots, & \mathbf{g}_k^{(n)}(D) \end{bmatrix} \tag{8.28}$$

is the generator matrix of the code. For the encoder in Fig. 8.1

$$\mathbf{G}(D) = \left[1 + D + D^2 \ , \ 1 + D^2\right], \tag{8.29}$$

while for the encoder in Fig. 8.2

$$\mathbf{G}(D) = \begin{bmatrix} 1 + D \ , \ D \ , \ 1 + D \\ D \quad\ \ , \ 1 \ , \ 1 \end{bmatrix}. \tag{8.30}$$

After multiplexing the outputs, the final codeword has the polynomial representation

$$\mathbf{c}(D) = \sum_{j=1}^{n} D^{j-1}\mathbf{c}^{(j)}(D^n). \tag{8.31}$$

Systematic convolutional codes are those where first k of the n encoder output sequences, $\mathbf{c}^{(1)}, \ldots, \mathbf{c}^{(k)}$, are equal to the k encoder input sequences $\mathbf{a}^{(1)}, \ldots, \mathbf{a}^{(k)}$.

8.2.2 *State and Trellis Diagrams and Weight Distribution*

The state of a convolutional encoder is defined by its shift register contents. For a rate-k/n code, the ith shift register contains ν_i previous data bits. The state of the encoder at epoch ℓ is defined as

$$\boldsymbol{\sigma}_\ell = \left(a_{\ell-1}^{(1)}, \cdots, a_{\ell-\nu_1}^{(1)} \ ; \cdots ; a_{\ell-1}^{(k)}, \cdots, a_{\ell-\nu_m}^{(k)}\right). \tag{8.32}$$

There are a total of $N_S = 2^{\nu_T}$ encoder states, where $\nu_T \triangleq \sum_{i=1}^{k} \nu_i$ is defined as the total encoder memory. For a rate-$1/n$ code, the encoder state at epoch ℓ is simply $\boldsymbol{\sigma}_\ell = (a_{\ell-1}, \cdots, a_{\ell-\nu})$.

Figures 8.3 and 8.4 show the state diagrams for the codes in Figs. 8.1 and 8.2, respectively. The states are labeled using the convention $\sigma^{(i)}, i = 0, \ldots, \nu_T - 1$, where $\sigma^{(i)}$ represents the encoder state $(c_0, \ldots, c_{\nu_T - 1})$ corresponding to the integer $i = \sum_{j=0}^{\nu_T - 1} c_j 2^j$. In general, for a rate-k/n code there are 2^k branches entering and leaving each state. The branch labels in the state diagram have the convention $\mathbf{a}/\mathbf{b} = (a^{(1)}, a^{(2)}, \ldots, a^{(k)})/(b^{(1)}, b^{(2)}, \ldots, b^{(n)})$. For example, the state transition $\sigma^{(1)} \rightarrow \sigma^{(3)}$ in Fig. 8.3 has the label 1/01. This means that if the encoder in Fig. 8.1 is in state state $\sigma^{(1)} = (01)$ and the input bit is a 1, the encoder will output the code bits 01 and transition to state $\sigma^{(3)} = (11)$.

Convolutional codes are linear codes, meaning that the sum of any two codewords is another codeword and the all-zeroes sequence is a codeword. It follows that the weight distribution and other distance properties of a convolutional code can be obtained from its state diagram. Consider, for example, the encoder in Fig. 8.1 along with its state diagram in Fig. 8.3. Since the self-loop at the zero state $\sigma^{(0)}$ corresponds to the all-zeroes codeword, the zero state $\sigma^{(0)}$ is split into two nodes, representing the input and output of the state diagram. This leads to the modified state diagram shown in Fig. 8.5.

The branches in the modified state diagram have labels of the form $D^i N^j L$, where i is the number of 1s in the encoder output sequence corresponding to a particular state transition, and j is the number of input 1s into the encoder for that transition. Every branch is labeled with the letter L, and the exponent of L is unity because each branch has length one. Each possible path through the modified state diagram corresponds to a non-all-zeroes codeword.

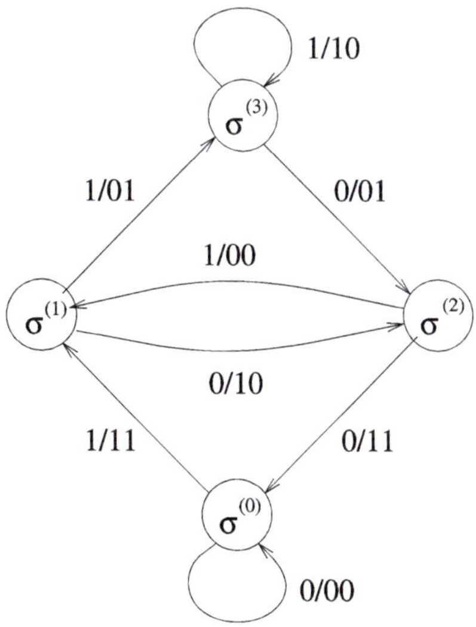

Fig. 8.3 State diagram for the binary convolutional encoder in Fig. 8.1

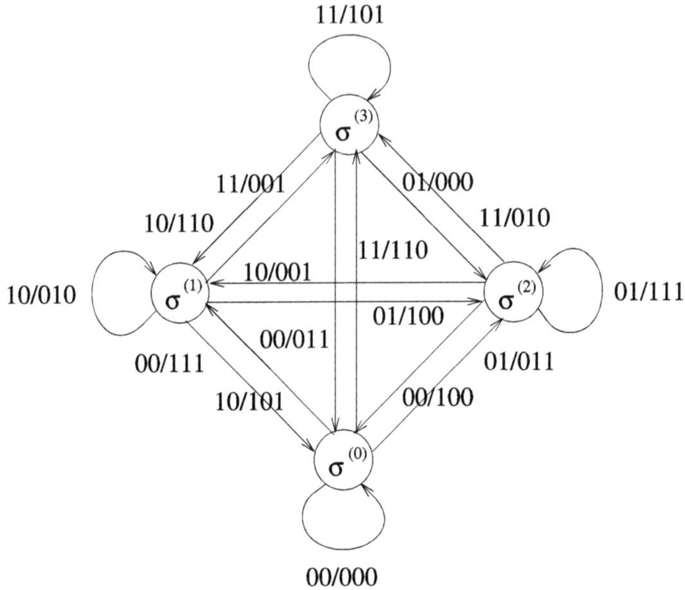

Fig. 8.4 State diagram for the binary convolutional encoder in Fig. 8.2

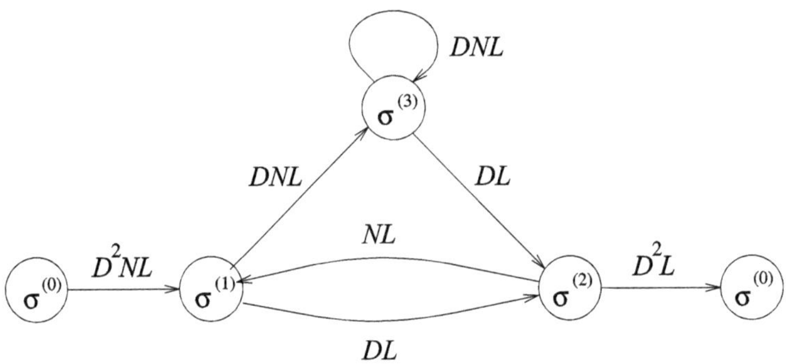

Fig. 8.5 Modified state diagram for the binary convolutional encoder in Fig. 8.1

The weight distribution of a convolutional code can be obtained by computing the transfer function $T(D, N, L)$ from the modified state diagram. Any appropriate technique can be used to obtain the transfer function, including signal flow-graph reduction techniques and Mason's formula [20]. For the example shown in Fig. 8.5, the transfer function is

$$\begin{aligned} T(D,N,L) &= \frac{D^5NL^3}{1-DNL(L+1)} \\ &= D^5L^3N + D^6N^2L^4(L+1) + D^7N^3L^5(L+1)^2 \\ &\quad + \cdots + D^{k+5}N^{k+1}L^{k+3}(L+1)^k + \cdots \end{aligned} \tag{8.33}$$

where the second line was obtained using synthetic division. The terms in the second line of (8.33) enumerate the weight distribution and distance properties of the code. For example, consider the term $D^{k+5}N^{k+1}L^{k+3}(L+1)^k$ appearing in the transfer function. Using the binomial expansion in (6.112), this term can be rewritten as

$$D^{k+5}N^{k+1}\sum_{n=0}^{k}\binom{k}{n}L^{n+k+3}.$$

Hence, there are 2^k paths through the modified state diagram that are at Hamming distance $k+5$ from the all-zeroes path, i.e., have $k+5$ output 1s, and $k+1$ input 1s. Of these 2^k paths, $\binom{k}{n}$ have length $k+n+3$ branches.

Sometimes the transfer function can be simplified if only certain distance properties of the convolutional code are of interest. For example, the output weight distribution of the code can be obtained by setting $N=1$ and $L=1$ in the transfer function. For the particular transfer function in (8.33) this leads to

$$\begin{aligned} T(D) &= \frac{D^5}{1-2D} \\ &= D^5 + 2D^6 + 4D^7 + \cdots + 2^k D^{5+k} + \cdots \end{aligned} \tag{8.34}$$

meaning that there are 2^k codewords at Hamming distance $5+k$ from the all-zeroes codeword. Notice that no nonzero codeword exists with a Hamming distance less than 5 from the all-zeroes codeword. Hence, the free Hamming distance of this particular code is $d_{\text{free}} = 5$. The free Hamming distance can also be seen by inspecting the trellis diagram in Fig. 8.6, where the branches in the trellis

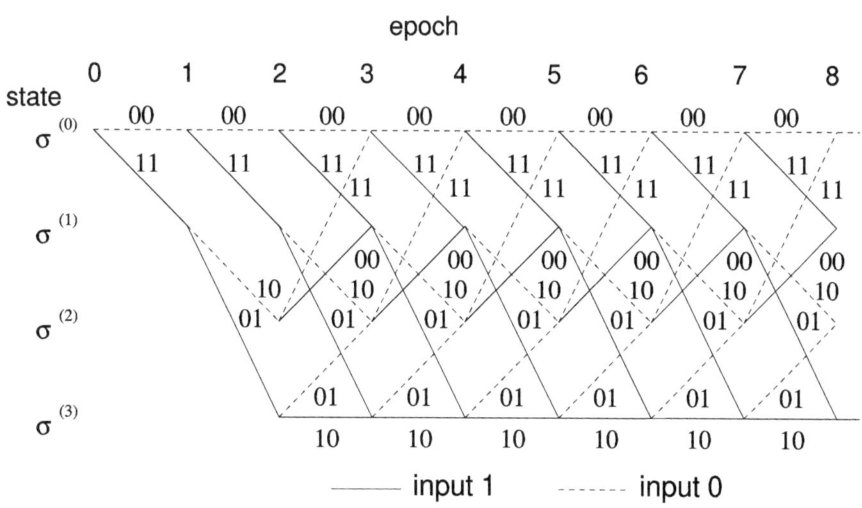

Fig. 8.6 Trellis diagram for the binary convolutional encoder in Fig. 8.1

diagram are labeled with the encoder output bits that correspond to the various state transitions.

Convolutional codes are designed to have the largest possible d_{free} for a given code rate and total encoder memory. Tabulation of convolutional codes that are optimal in this sense have been tabulated, e.g., Proakis [25], Lin and Costello [16], and Clark and Cain [5].

8.2.3 Maximum Likelihood Sequence Estimation

Suppose that a sequence of messages or modulation symbols $\tilde{\mathbf{s}} = \{\tilde{\mathbf{s}}_m\}_{m=1}^{k}$, $\tilde{\mathbf{s}}_m = (\tilde{\mathbf{s}}_{m,1}, \ldots, \tilde{\mathbf{s}}_{m,n})$, corresponding to k branches in a convolutional code trellis are sent on an AWGN channel. After receiving the sequence $\tilde{\mathbf{r}} = \{\tilde{\mathbf{r}}_m\}_{m=1}^{k}$, $\tilde{\mathbf{r}}_m = (\tilde{\mathbf{r}}_{m,1}, \ldots, \tilde{\mathbf{r}}_{m,n})$, the maximum likelihood receiver uses knowledge of the sequence of complex channel gains $\tilde{\mathbf{h}} = \{\tilde{\mathbf{h}}_m\}_{m=1}^{k}$, $\tilde{\mathbf{h}}_m = (\tilde{h}_{m,1}, \ldots, \tilde{h}_{m,n})$ (obtained from a separate channel estimator) to decide in favor of the sequence $\tilde{\mathbf{s}}$ that maximizes the likelihood function:

$$p(\tilde{\mathbf{r}}_k, \ \cdots \ , \ \tilde{\mathbf{r}}_1 | \ \tilde{h}_{k,n}\tilde{\mathbf{s}}_{k,n}, \ \cdots \ , \ \tilde{h}_{1,1}\tilde{\mathbf{s}}_{1,1}) \tag{8.35}$$

or, equivalently, the log-likelihood function

$$\log\{p(\tilde{\mathbf{r}}_k, \ \cdots \ , \ \tilde{\mathbf{r}}_1 | \ \tilde{h}_{k,n}\tilde{\mathbf{s}}_{k,n}, \ \cdots \ , \ \tilde{h}_{1,1}\tilde{\mathbf{s}}_{1,1})\}. \tag{8.36}$$

The log-likelihood function in (8.36) can be rewritten as

$$\begin{aligned}\log\{p(\tilde{\mathbf{r}}_k, \cdots, \tilde{\mathbf{r}}_1 | \tilde{h}_{k,n}\tilde{\mathbf{s}}_{k,n}, \cdots, \tilde{h}_{1,1}\tilde{\mathbf{s}}_{1,1})\} &= \log\{p(\tilde{\mathbf{r}}_k | \tilde{h}_{k,1}\tilde{\mathbf{s}}_{k,1}, \ldots, \tilde{h}_{k,n}\tilde{\mathbf{s}}_{k,n})\} \\ &+ \log\{p(\tilde{\mathbf{r}}_{k-1}, \cdots, \tilde{\mathbf{r}}_1 | \tilde{h}_{k-1,n}\tilde{\mathbf{s}}_{k-1,n}, \cdots, \tilde{h}_{1,1}\tilde{\mathbf{s}}_{1,1})\}.\end{aligned} \tag{8.37}$$

If the second term on the right side of (8.37) has been calculated previously at epoch $k-1$ and stored in memory, then only the first term, called the branch metric, has to be computed for the incoming signal vector $\mathbf{r}_k$ at epoch k. From our treatment in Chap. 6

$$p(\tilde{\mathbf{r}}_k | \tilde{h}_{k,1}\tilde{\mathbf{s}}_{k,1}, \ldots, \tilde{h}_{k,n}\tilde{\mathbf{s}}_{k,n}) = \frac{1}{(2\pi N_o)^N} \exp\left\{ -\frac{1}{2N_o} \sum_{m=1}^{n} \|\tilde{\mathbf{r}}_{k,m} - \tilde{h}_{k,m}\tilde{\mathbf{s}}_{k,m}\|^2 \right\} \tag{8.38}$$

so that the log-likelihood $\log\{p(\tilde{\mathbf{r}}_k|\tilde{h}_{k,1}\tilde{\mathbf{s}}_{k,1},\ldots,\tilde{h}_{k,\ell}\tilde{\mathbf{s}}_{k,\ell})\}$ yields the Euclidean branch metric:

$$\mu_k = \sum_{m=1}^{n} \|\tilde{\mathbf{r}}_{k,m} - \tilde{h}_{k,m}\tilde{\mathbf{s}}_{k,m}\|^2. \tag{8.39}$$

The metric in (8.39) may be readily replaced with the variations in (6.22) and (6.23) or, for equal energy signal sets, in (6.24b). For BPSK, the Euclidean metric reduces the simple form

$$\mu_k = \text{Re}\left\{\sum_{m=1}^{n} \tilde{r}_{k,m}\tilde{h}^*_{k,m}\tilde{s}_{k,m}\right\}. \tag{8.40}$$

If the channel is constant $\tilde{h}^*_{k,m} = \tilde{h}$, then

$$\mu_k = \text{Re}\left\{\tilde{h}^* \sum_{m=1}^{n} \tilde{r}_{k,m}\tilde{s}_{k,m}\right\}. \tag{8.41}$$

A maximum-likelihood sequence estimation (MLSE) receiver generates $\tilde{\mathbf{r}}$ and along with knowledge of the complex channel $\tilde{\mathbf{h}}$ chooses the codeword $\tilde{\mathbf{s}}$ that is closest in Euclidean distance. Since no decisions are made on the individual code symbols $\tilde{\mathbf{s}}_{k,m}$ prior to the sequence decision, this is commonly called *soft decision decoding*.

Hard Decision Decoding
Alternatively, the receiver may process the individual $\tilde{\mathbf{r}}_{k,m}$ with knowledge of the channel $\tilde{h}_{k,m}$ to make hard decisions on the code symbols $\tilde{\mathbf{y}}_{k,m}$ before a sequence decision is made, commonly called *hard decision decoding*. The Euclidean branch metric at epoch k in (8.39) is replaced with the Hamming metric

$$\mu_k = d(\mathbf{y}_k, \mathbf{s}_k). \tag{8.42}$$

In this case, the MLSE receiver finds the codeword $\mathbf{s}$ that is closest in Hamming distance to the received codeword $\mathbf{y}$.

8.2.3.1 The Viterbi Algorithm

The Viterbi algorithm was devised by Viterbi for maximum likelihood decoding of convolutional codes [32, 33]. Based on the recursion in (8.37) and the branch metric in (8.39), the Viterbi algorithm searches through the N_S-state code trellis for the most likely transmitted sequence $\tilde{\mathbf{s}}$ given the sequence of received vectors $\tilde{\mathbf{r}}$ and knowledge of the sequence of complex channel gains $\tilde{\mathbf{h}}$.

At epoch k, the Viterbi algorithm stores N_S survivors $\check{\mathbf{s}}(\sigma_k^{(i)})$ along with their associated path metrics $\Gamma(\sigma_k^{(i)})$ that terminate at state $\sigma_k^{(i)}$, $i = 0, \cdots, N_S - 1$. The path metric is defined as

$$\Gamma(\sigma_k^{(i)}) = \sum_{n=1}^{k} \mu_n^{(i)} , \quad i = 0, \ldots, N_S - 1, \tag{8.43}$$

where $\{\mu_n^{(i)}\}$ is the sequence of branch metrics along the surviving path $\check{\mathbf{s}}(\sigma_k^{(i)})$.

The Viterbi algorithm is initialized at time $k = 0$, by setting all path metrics to zero, i.e., $\Gamma(\sigma_0^{(i)}) = 0, \ i = 1, \ldots, N_S - 1$.

1. After the vector $\tilde{\mathbf{r}}_{k+1}$ has been received, compute the set of path metrics $\Gamma(\sigma_k^{(i)} \rightarrow \sigma_{k+1}^{(j)}) = \Gamma(\sigma_k^{(i)}) + \mu(\sigma_k^{(i)} \rightarrow \sigma_{k+1}^{(j)})$ for all possible paths through the trellis that terminate in each state $\sigma_{k+1}^{(j)}, j = 0, \cdots, N_S - 1$, where $\mu(\sigma_k^{(i)} \rightarrow \sigma_{k+1}^{(j)})$ is the branch metric defined below. For a modulation alphabet of size M, there will be M such paths that terminate in each state $\sigma_{k+1}^{(j)}$.
2. Find $\Gamma(\sigma_{k+1}^{(j)}) = \max_i \Gamma(\sigma_k^{(i)} \rightarrow \sigma_{k+1}^{(j)}), j = 0, \cdots, N_S - 1$ where the maximization is over all M possible paths through the trellis that terminate in state $\sigma_{k+1}^{(j)}$.
3. Store $\Gamma(\sigma_{k+1}^{(j)})$ and its associated surviving sequence $\check{\mathbf{s}}(\sigma_{k+1}^{(j)}), j = 0, \cdots, N_S - 1$. Drop all other paths.
4. Increment the time index k, goto Step 1 and repeat the entire algorithm.

In Step 1 above, $\mu(\sigma_k^{(i)} \rightarrow \sigma_{k+1}^{(j)})$ is the branch metric associated with the state transition $\sigma_k^{(i)} \rightarrow \sigma_{k+1}^{(j)}$ and is computed according to the following variation of (8.39) for soft decision decoding:

$$\mu(\sigma_k^{(i)} \rightarrow \sigma_{k+1}^{(j)}) = \sum_{m=1}^{\ell} \left\| \tilde{\mathbf{r}}_{k,m} - \tilde{h}_{k,m} \tilde{\mathbf{s}}_{k,m}(\sigma_k^{(i)} \rightarrow \sigma_{k+1}^{(j)}) \right\|^2$$

and the following variation of (8.42) for hard decision decoding:

$$\mu(\sigma_k^{(i)} \rightarrow \sigma_{k+1}^{(j)}) = d\left(\mathbf{y}_k, \mathbf{s}_k(\sigma_k^{(i)} \rightarrow \sigma_{k+1}^{(j)})\right)$$

where $\tilde{\mathbf{s}}_{k,m}(\sigma_k^{(i)} \rightarrow \sigma_{k+1}^{(j)})$ is a symbol that is uniquely determined by the state transition $\sigma_k^{(i)} \rightarrow \sigma_{k+1}^{(j)}$ and the symbol mapping being used.

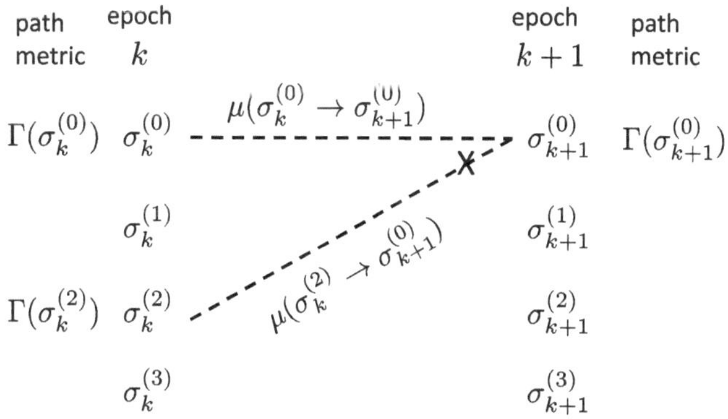

$$\begin{aligned}\Gamma(\sigma_{k+1}^{(0)}) &= \max[\Gamma(\sigma_k^{(0)}) + \mu(\sigma_k^{(0)} \rightarrow \sigma_{k+1}^{(0)}), \Gamma(\sigma_k^{(2)}) + \mu(\sigma_k^{(2)} \rightarrow \sigma_{k+1}^{(2)})] \\ &= \Gamma(\sigma_k^{(0)}) + \mu(\sigma_k^{(0)} \rightarrow \sigma_{k+1}^{(0)})\end{aligned}$$

Fig. 8.7 Path metric update for the code trellis in Fig. 8.6

The calculation of the path metric $\Gamma(\sigma_{k+1}^{(0)})$ for state $\sigma^{(0)}$ at epoch $k+1$ is illustrated in Fig. 8.7 for the code trellis shown in Fig. 8.6. In this case, there are two paths merging into state $\sigma^{(0)}$ at epoch $k+1$. The Viterbi algorithm determines the path metric $\Gamma(\sigma_{k+1}^{(0)})$ in this particular example as

$$\begin{aligned}\Gamma(\sigma_0^{(j)}) &= \max\left\{\Gamma(\sigma_k^{(0)}) + \mu(\sigma_k^{(0)} \rightarrow \sigma_{k+1}^{(0)}), \Gamma(\sigma_k^{(2)}) + \mu(\sigma_k^{(2)} \rightarrow \sigma_{k+1}^{(0)})\right\} \\ &= \Gamma(\sigma_k^{(0)}) + \mu(\sigma_k^{(0)} \rightarrow \sigma_{k+1}^{(0)}).\end{aligned}$$

Hence, the path that includes the state transition $\sigma_k^{(0)} \rightarrow \sigma_{k+1}^{(0)}$ is the survivor and the path that includes the state transition $\sigma_k^{(2)} \rightarrow \sigma_{k+1}^{(0)}$ is "dropped" as indicated by the "X" on the path.

Example 8.5 (Viterbi Algorithm) Consider the $K = 3$, $R = 1/2$ convolutional code in Fig. 8.1, with state diagram in Fig. 8.3 and trellis diagram in Fig. 8.6. The binary code bits are sent using BPSK with hard decision decoding, such that the coding channel appears as a BSC.

Suppose the encoder starts in the all-zeroes state and the 8-bit data sequence 11010011|00 is transmitted, appended by 2 tail bits that return the encoder to the all-zeroes state. The decoder trellis diagram is shown in

(continued)

Example 8.5 (continued)
Fig. 8.8. The data bit **a** is shown across the top along with the corresponding codeword sequence **c**. The sequence **y** beneath the trellis is the received code bit sequence at the output of the BSC, where the asterisks indicate transmission errors.

The branch metrics are computed by calculating the Hamming distance from the corresponding received sequence. This can be done by using the state transition diagram in Fig. 8.8 as an aid, where the branches in the diagram are labeled with the encoder output bits for each state transition. The decoder trellis shows the path metrics stored at each stage in the trellis.

Starting with the third stage in the decoder trellis, there are 2 branches entering each state from 2 different prior states. Using the state transition diagram in Fig. 8.8, branch metrics are calculated for each of these 2 branches using the received sequence **y** and then added to the path metrics of the prior states. The path having the smallest Hamming distance is the survivor and the other path dropped as indicated by the hash marks. This "Add, Compare and Select" operation is illustrated in Fig. 8.8 and repeated for each of the 4 states at each stage of the trellis. Sometimes there is a tie as indicated by the both paths entering a state being circled. In this case, both paths are equally likely and either can be selected as the survivor for that state.

The two tail bits terminate the trellis by forcing the decoder to the all zeroes state. At the end of the trellis there is only one survivor, in this example at Hamming distance 2 from the received sequence due to the 2 transmission errors. The decided sequence is obtained by starting at the end of the trellis and tracing backwards from right to left, following the path that is not blocked by the hash marks. By using the state transition diagram the decided data bits can be determined. In this example, the 2 errors are corrected and the original data sequence is recovered.

8.3 Trellis Coded Modulation

8.3.1 Encoder Description

Conventional convolutional codes realize a coding gain at the expense of the bandwidth efficiency as measured in bits/s/Hz. Although, convolutional codes may be attractive for power-limited applications, they are less suitable for bandwidth-limited applications. One approach for overcoming the loss of bandwidth efficiency is to map the encoder output bits onto a higher-order signal constellation, such as MPSK or QAM. However, the symbol mapping that is used significantly affects

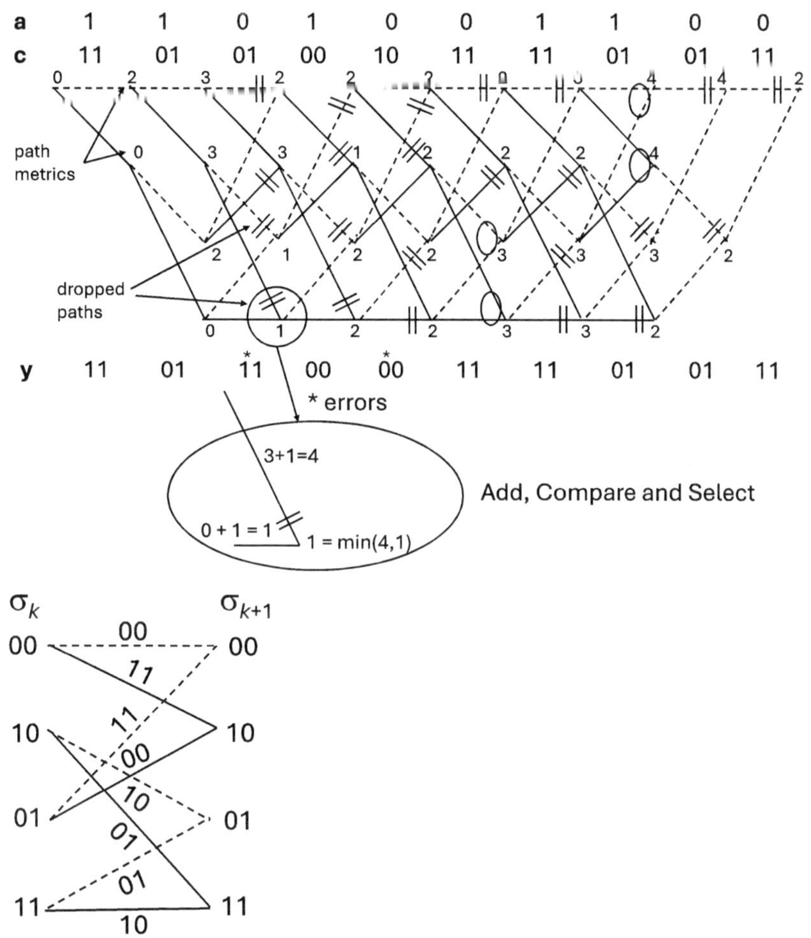

Fig. 8.8 Viterbi decoding example for $R = 1/2$, $K = 3$ convolutional code in Fig. 8.1

the performance. Employing a straightforward mapping, such as Gray mapping or natural mapping, will give disappointing results. The reason is that the decreased Euclidean distance between the signals points in the higher-order constellation offsets the coding gain from the error correcting code.

Ungerboeck showed that a coding gain can be achieved without sacrificing data rate or bandwidth by using a rate-$m/(m + r)$ convolutional encoder and mapping the coded bits onto modulation symbols through a technique called mapping by set partitioning [29]. This combination of coding and modulation, called trellis coded modulation (TCM), has three basic features:

1. An expanded signal constellation is used that is larger than the one necessary for uncoded modulation at the same data rate. The additional constellation points allow redundancy to be inserted without sacrificing data rate or bandwidth.

2. The expanded signal constellation is partitioned such that the intra-subset minimum squared Euclidean distance is maximized at each step in the partition chain.
3. Convolutional encoding and signal mapping is used so that only certain sequences of signal points are allowed.

Figure 8.9 shows the basic encoder structure for Ungerboeck's trellis codes. The n-bit data vector $\mathbf{a}_k = (a_k^{(1)}, \cdots, a_k^{(n)})$ is transmitted at epoch k. At each epoch k, $m \leq n$ data bits are encoded into $m+r$ code bits by using a rate-$m/(m+r)$ linear convolutional encoder. The $m+r$ code bits select one of 2^{m+r} subsets of a 2^{n+r}-point signal constellation. The remaining $n-m$ data bits are used to select one of the 2^{n-m} signal points within the selected subset. This principle is best explained by example, and Fig. 8.10 shows a 4-state 8-PSK Ungerboeck trellis code. The equivalent uncoded system is 4-PSK which has a rate of 2 bits/symbol. The 4-state 8-PSK code uses a rate-1/2 convolutional code along with one uncoded bit to select signal points in an expanded 8-PSK signal constellation. Note that the overall rate is still 2 bits/symbol. Figure 8.11 shows another example of an 8-state 8-PSK Ungerboeck trellis code. The equivalent uncoded system is again 4-PSK with 2 bits/symbol. The 8-state 8-PSK code uses a rate-2/3 convolutional code to select one of the points in an expanded 8-PSK signal constellation so that the overall rate is again 2 bits/symbol.

8.3.2 Mapping by Set Partitioning

The critical step in the design of Ungerboeck trellis codes is the method of mapping by set partitioning. Figure 8.12 shows how the 8-PSK signal constellation is partitioned into subsets such that the intra-subset minimum squared Euclidean distance is maximized for each step in the partition chain. A normalized 8-PSK

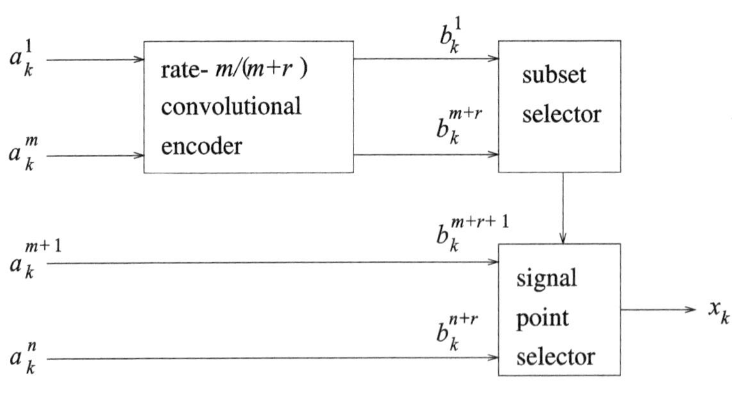

Fig. 8.9 Ungerboeck trellis encoder

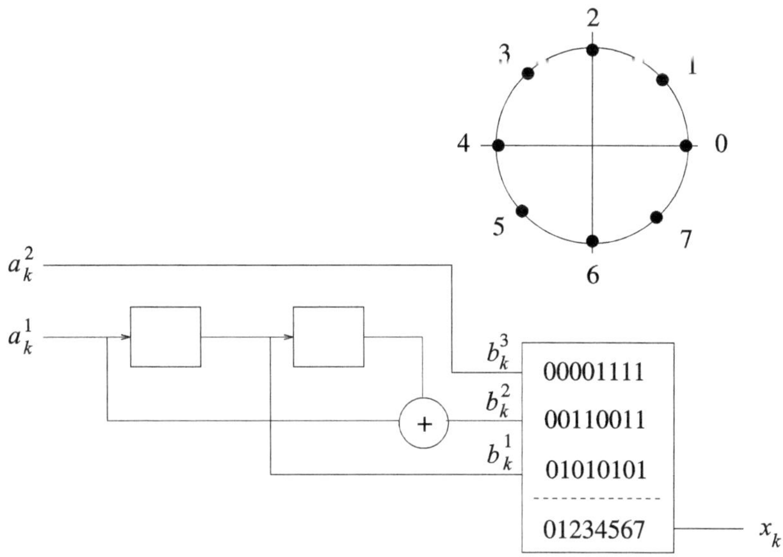

Fig. 8.10 Encoder and signal mapping for the 4-state 8-PSK Ungerboeck trellis code

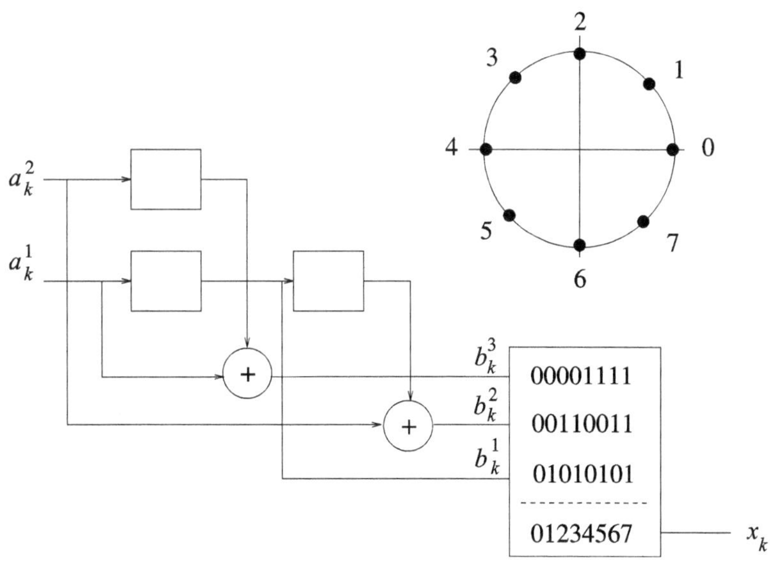

Fig. 8.11 Encoder and signal mapping for the 8-state 8-PSK Ungerboeck trellis code

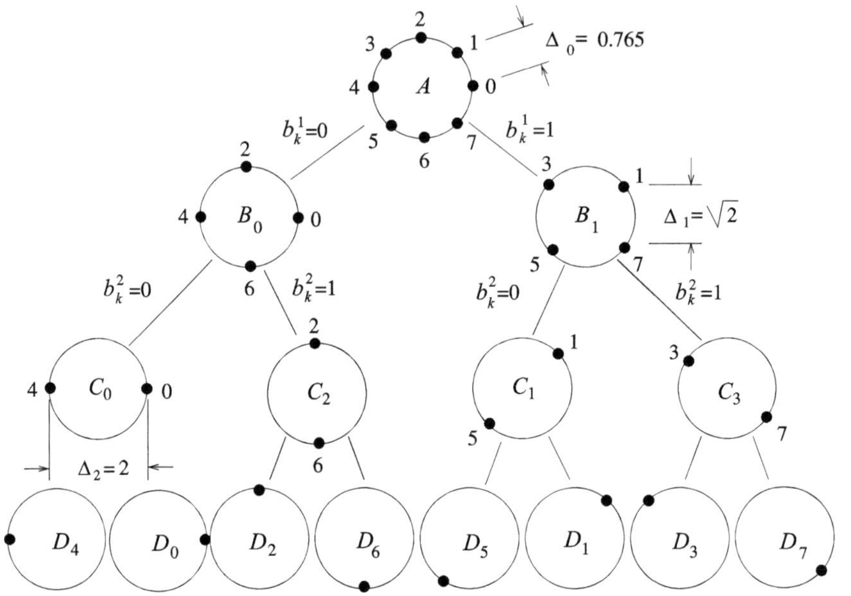

Fig. 8.12 Set partitioning for an 8-PSK signal constellation

signal constellation is assumed having 8 signal points uniformly spaced around a circle of unit radius. Notice that the minimum Euclidean distance between signal points in the normalized 8-PSK signal constellation is $\Delta_0 = 0.765$, while the minimum Euclidean distances between signal points in the first and second level partitions are $\Delta_1 = \sqrt{2}$ and $\Delta_2 = 2$, respectively. Notice that the minimum Euclidean distance increases at each level of partitioning.

The advantages of using TCM can most easily be seen by considering the trellis diagram. For both the 4-state and 8-state 8-PSK trellis codes, the equivalent uncoded system is 4-PSK. The trellis diagram for uncoded 4-PSK is shown in Fig. 8.13. The trellis only has one state and there are 4 parallel transitions between the states. The subsets D_0, D_2, D_4, and D_6 are used as the signal points. The label D_0, D_2, D_4, D_6 means that the branches in the trellis diagram are labeled from top to bottom with signal points taken from the sets D_0, D_2, D_4, D_6. The minimum Euclidean distance between any two paths through the trellis is $d_{\min} = \sqrt{2}$.

The trellis diagram for the 4-state 8-PSK code is shown in Fig. 8.14. Each branch in the 4-state trellis is labeled with one of the four subsets C_0, C_1, C_2, and C_3. Again,

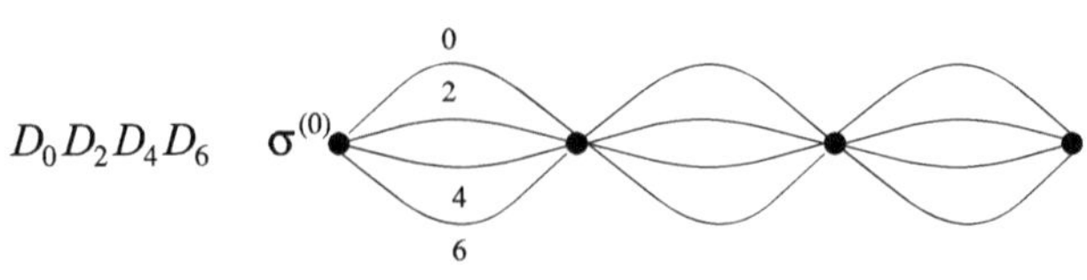

Fig. 8.13 Trellis diagram for uncoded 4-PSK

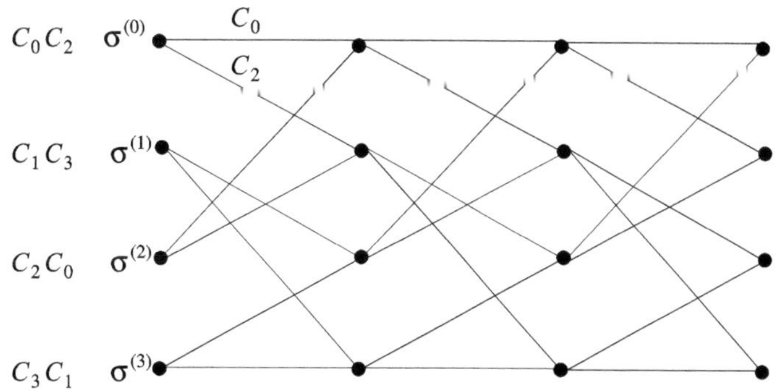

Fig. 8.14 Trellis diagram for 4-state 8-PSK Ungerboeck trellis code

the label C_iC_j associated with a state means that the branches in the trellis diagram originating from that state are labeled from top to bottom with the subsets C_i and C_j. As shown in Fig. 8.12, each subset C_i contains two signal points. Thus, each branch in the trellis diagram actually contains two parallel transitions. For example, branches with the label C_0 have two parallel transitions that are labeled with the signal points 0 and 4. For the 4-state 8-PSK code, it is possible that two coded sequences could differ by just a single parallel transition with a Euclidean distance of $d = 2$. Also, any two signal paths that diverge from a state and merge with the same state after more than one transition have a minimum Euclidean distance of $d = \sqrt{\Delta_1^2 + \Delta_0^2 + \Delta_1^2} = 2.141$. For example, the closest nonparallel code sequence to the all-zeroes sequence $\mathbf{x} = (0, 0, 0)$ is the sequence $\mathbf{x} = (2, 1, 2)$ at Euclidean distance $d = 2.141$. Hence, the minimum Euclidean distance of the code over all parallel and nonparallel pairs of sequences for the 4-state 8-PSK code is $d_{\min} = 2$. This is to be compared with $d_{\min} = \sqrt{2}$ for uncoded 4-PSK.

The concept of mapping by set partitioning was developed by Ungerboeck as a method to maximize the minimum Euclidean distance of a code and consequently to optimize its performance on an AWGN channel. Ungerboeck's construction of the optimum 4-state 8-PSK code was based on the following heuristic rules [30]:

1. Parallel transitions (when they occur) are assigned constellation points having the maximum Euclidean distance between them.
2. The transition starting or ending in any state is assigned the subsets (C_0, C_2) or (C_1, C_3) which have a maximum Euclidean distance between them.
3. All constellation points are used in the trellis diagram with equal frequency.

It is clear that the performance of the 4-state 8-PSK code is limited by the parallel transitions. Larger coding gains can be obtained by introducing more code states so that the parallel transitions are eliminated. For example, the above design rules can be applied to the 8-state 8-PSK code to obtain the code trellis shown in Fig. 8.15. In this case, the minimum Euclidean distance is $d_{\min} = \sqrt{\Delta_1^2 + \Delta_0^2 + \Delta_1^2} = 2.141$.

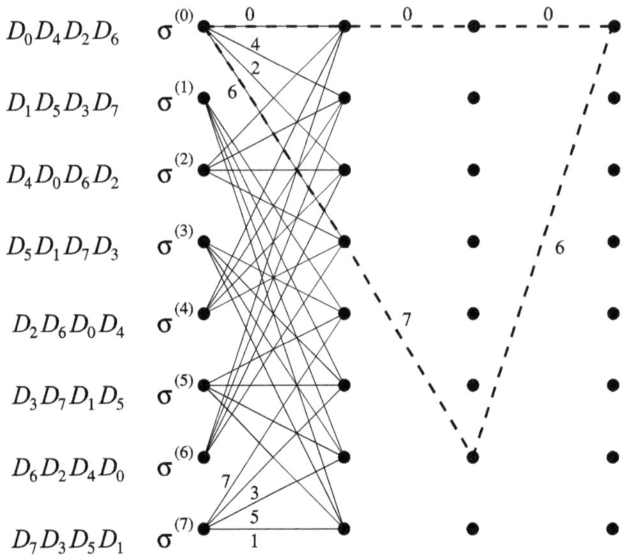

Fig. 8.15 Trellis diagram for 8-state 8-PSK Ungerboeck trellis code. The dashed lines show two minimum distance paths

8.4 Code Performance on AWGN Channels

To develop a union bound on decoded bit error probability, let $\mathbf{a} = \{\mathbf{a}_k\}$ denote the transmitted data sequence. For any other sequence $\hat{\mathbf{a}} \neq \mathbf{a}$, define the corresponding error sequence as $\mathbf{e} = \{e_k\} = \mathbf{a} \oplus \hat{\mathbf{a}}$, where $\oplus$ denotes modulo-2 addition. Since the bit error probability at epoch j_1 is of interest, $e_{j_1} \neq 0$ for all error sequences. An error event occurs between k_1 and k_2 of length $k_2 - k_1$, if $\boldsymbol{\sigma}_{k_1} = \hat{\boldsymbol{\sigma}}_{k_1}$ and $\boldsymbol{\sigma}_{k_2} = \hat{\boldsymbol{\sigma}}_{k_2}$, but $\boldsymbol{\sigma}_j \neq \hat{\boldsymbol{\sigma}}_j$ for $k_1 < j < k_2$, where $k_1 \leq j_1 < k_2$, and $\boldsymbol{\sigma} = \{\sigma_k\}$ and $\hat{\boldsymbol{\sigma}} = \{\hat{\boldsymbol{\sigma}}_k\}$ are the system state sequences associated with $\mathbf{a}$ and $\hat{\mathbf{a}}$, respectively. Let $\mathcal{E}$ be the set of error sequences that have the first nonzero element starting at time j_1. Then, the average bit error probability is bounded by

$$P_b \leq \frac{1}{n} \sum_{\mathbf{e} \in \mathcal{E}} w_b(\mathbf{e}) \sum_{\mathbf{a}} \mathrm{P}[\mathbf{a}] \ \mathrm{P}[\Gamma(\mathbf{a} \oplus \mathbf{e}) \geq \Gamma(\mathbf{a}) \, |\mathbf{a}] \ , \tag{8.44}$$

where $\Gamma(\mathbf{a})$ is the path metric of $\mathbf{a}$, and $w_b(\mathbf{e})$ is the number of bit errors associated with $\mathbf{e}$. The factor $1/n$ appears in front of the first summation, because n information bits are transmitted per epoch (or per branch in the trellis diagram). The second summation is over all possible information sequences, because each sequence $\mathbf{a}$ can have $\mathbf{e}$ as the error sequence. This is necessary for trellis coded modulation because the signal mapping makes the code nonlinear.

Another way of writing the bound on the bit error probability in (8.44) is

$$P_b \leq \sum_{\tilde{\mathbf{s}} \in C} \sum_{\hat{\mathbf{s}} \in C} w_b(\tilde{\mathbf{s}}, \hat{\mathbf{s}}) \mathrm{P}[\tilde{\mathbf{s}}] \mathrm{P}[\tilde{\mathbf{s}} \to \hat{\mathbf{s}}], \tag{8.45}$$

where C is the set of all coded symbol sequences, $w_b(\tilde{\mathbf{s}}, \hat{\mathbf{s}})$ is the number of bit errors that occur when the complex symbol sequence $\tilde{\mathbf{s}} = \{\tilde{\mathbf{s}}_i\}$ is transmitted and the complex symbol sequence $\hat{\mathbf{s}} \neq \tilde{\mathbf{s}}$ is chosen by the decoder, $\mathrm{P}[\tilde{\mathbf{s}}]$ is the *a priori* probability of transmitting $\tilde{\mathbf{s}}$, and $\mathrm{P}[\tilde{\mathbf{s}} \to \hat{\mathbf{s}}]$ is the pairwise error probability.

At high signal-to-noise ratio (SNR), the bit error rate performance on an AWGN channel is dominated by the minimum Euclidean distance error events. The pairwise error probability between two coded symbol sequences $\tilde{\mathbf{s}}$ and $\hat{\mathbf{s}}$ separated by Euclidean distance $d_{\min}$ is

$$\mathrm{P}[\tilde{\mathbf{s}} \to \hat{\mathbf{s}}] = Q\left(\sqrt{\frac{d_{\min}^2}{4N_o}}\right). \tag{8.46}$$

The asymptotic coding gain (at high bit energy-to-noise ratio) is defined by [4]

$$G_a = 10\log_{10} \frac{(d_{\min,\text{ coded}}^2 / E_{\text{coded}})}{(d_{\min,\text{ uncoded}}^2 / E_{\text{uncoded}})} \text{ dB} \tag{8.47}$$

where E is the average energy per symbol in the signal constellation. For the 4-state 8-PSK code shown in Fig. 8.10, the asymptotic coding gain over uncoded 4-PSK is $G_a = 3$ dB. Likewise, the 8-state 8-PSK code in Fig. 8.11 has an asymptotic coding gain of 3.6 dB over uncoded 4-PSK.

8.4.1 Union Bound for Convolutional Codes

For convolutional codes the upper bound in (8.44) simplifies because the codes are linear, meaning that the sum of any two codewords is another codeword and that the all-zeroes sequence is a codeword [16]. Because of this property, the all-zeroes sequence can be assumed to be transmitted, i.e., $\mathbf{a} = \mathbf{0}$, so that the union bound becomes

$$P_b \leq \frac{1}{k} \sum_{\mathbf{e} \in \mathcal{E}} w_b(\mathbf{e}) \, \mathrm{P}\left[\Gamma(\mathbf{e}) \geq \Gamma(\mathbf{0})\right]. \tag{8.48}$$

Note that division is by k rather than n in front of the summation, because a convolutional code transmits k bits per channel use, whereas a trellis code transmits n bits per channel use.

For convolutional codes the set $\mathcal{E}$ in (8.48) consists of all sequences that begin and end at the zero-state in the state diagram. The enumeration of these sequences (or codewords), along with their associated Hamming distances, information weights, and lengths, was obtained earlier by computing the transfer function, $T(D, N, L)$, of the augmented state diagram. When a particular incorrect path through the trellis is selected over the all-zeroes path at a given node in the trellis, the corresponding number of bits errors, $w_b(\mathbf{e})$, is given by the exponent of N in the transfer function. Multiplying $w_b(\mathbf{e})$ by the pairwise error probability $\mathrm{P}[\Gamma(\mathbf{e}) \geq \Gamma(\mathbf{0})]$ for that path and dividing by the number of input bits per branch, k, gives the bit error probability associated with that path. Summing over the set of all possible incorrect sequences $\mathcal{E}$ yields a union bound on the bit error probability.

The pairwise error probability in (8.48) depends on the type of modulation, detection, and decoding that is employed. The code bits are mapped onto symbols taken from a signal constellation and transmitted over the channel. Assuming an AWGN channel having complex gain $\tilde{h}$ and a coherent receiver, the received vector (see Sect. 6.1) at epoch n is

$$\tilde{\mathbf{r}}_n = \tilde{h}\tilde{\mathbf{s}}_n + \tilde{\mathbf{n}}_n, \tag{8.49}$$

where $\tilde{\mathbf{s}}_n$ is the transmitted symbol vector and $\tilde{\mathbf{n}}_n$ is the Gaussian noise vector at epoch n. For convolutional codes, two types of decoding are commonly used, hard decision decoding and soft decision decoding. Soft decision decoders use the sequence of received signal vectors $\mathbf{r} = \{\mathbf{r}_n\}$ to make sequence decisions. For an AWGN channel, the MLSE receiver searches for the sequence of symbol vectors $\tilde{\mathbf{s}} = \{\tilde{\mathbf{s}}_n\}$ that is closest in Euclidean distance to the received sequence of signal vectors $\mathbf{r}$. To do so, the MLSE receiver chooses the sequence $\hat{\mathbf{s}}$ that minimizes the metric

$$\mu(\hat{\mathbf{s}}) = \|\mathbf{r} - \tilde{h}\hat{\mathbf{s}}\|^2. \tag{8.50}$$

The decided sequence $\hat{\mathbf{s}}$ maps one-to-one onto the data bit sequence $\hat{\mathbf{a}}$ that is the final estimate of the transmitted data sequence.

In general, the pairwise error probability for an AWGN channel that is associated with an error event of length ℓ beginning at epoch k_1 is

$$P_2(\ell) = Q\left(\sqrt{\frac{\alpha^2\Delta^2}{4N_o}}\right), \tag{8.51}$$

where

$$\Delta^2 = \sum_{k=k_1}^{k_1+\ell+1} \delta_k^2 \tag{8.52}$$

$$\delta_k^2 = \|\tilde{\mathbf{s}}_k - \hat{\mathbf{s}}_k\|^2 \tag{8.53}$$

and $\tilde{\mathbf{s}} = \{\tilde{\mathbf{s}}_k\}$ and $\hat{\mathbf{s}} = \{\hat{\mathbf{s}}_k\}$ are the symbol sequences corresponding to the data bit sequences $\tilde{\mathbf{a}}$ and $\hat{\mathbf{a}}$, respectively. The parameter δ_k^2 is the squared branch Euclidean distance associated with branch k, and Δ^2 is the squared path Euclidean distance associated with the error event. Clearly, the pairwise error probability depends on the particular mapping between the encoder output bits and the points in the signal constellation. Suppose, for example, the code bits are mapped onto a BPSK signal constellation. Then the pairwise error probability between the two convolutional codewords $\tilde{\mathbf{b}}$ and $\hat{\mathbf{b}}$ that differ in d positions is

$$P_2(d) = Q(\sqrt{2R_c d\gamma_b}) \tag{8.54}$$

where R_c is the code rate and γ_b is the bit energy-to-noise ratio.[1] Note that the pairwise error probability has been explicitly shown to be a function of the Hamming distance between the codewords in (8.54). However, it is important to realize that this property does not always apply. For example, if the outputs of the rate-2/3 convolutional encoder in Fig. 8.2 are mapped onto symbols from an 8-PSK signal constellation, then the pairwise error probability depends not only on the Hamming distance between codewords, but also upon the particular mapping between the 8-PSK symbols and the encoder outputs.

In general, the transfer function $T(D, N)$ for a convolutional code has the form

$$T(D, N) = \sum_{d=d_{\text{free}}}^{\infty} a_d D^d N^{f(d)}, \tag{8.55}$$

where $f(d)$ is the exponent of N as a function of d. For the example in (8.33), $a_d = 2^{d-5}$ and $f(d) = d - 4$. Differentiating $T(D, N)$ with respect to N and setting $N = 1$ gives

$$\frac{\mathrm{d}T(D, N)}{\mathrm{d}N}\Big|_{N=1} = \sum_{d=d_{\text{free}}}^{\infty} a_d f(d) D^d. \tag{8.56}$$

Once again, for the example in (8.33) this leads to

$$\frac{\mathrm{d}T(D, N)}{\mathrm{d}N}\Big|_{N=1} = \sum_{d=d_{\text{free}}}^{\infty} 2^{d-5}(d-4) D^d. \tag{8.57}$$

Using this notation, the union bound on bit error probability for convolutional coding with BPSK signals is

$$P_b \leq \frac{1}{k} \sum_{d=d_{\text{free}}}^{\infty} a_d f(d) P_2(d), \tag{8.58}$$

where $P_2(d)$ is given by (8.54).

[1] The code bit energy-to-noise ratio is $\gamma_s = R_c \gamma_b$.

In contrast to soft decision decoders, hard decision decoders first make symbol-by-symbol decisions on the sequence of received vectors $\tilde{\mathbf{r}} = \{\tilde{\mathbf{r}}_k\}$ to yield the sequence of symbol decisions $\mathbf{y} = \{\mathbf{y}_k\}$. The decoder then operates on the sequence $\mathbf{y}$ to estimate the most likely transmitted data sequence. A minimum distance decoder is one that decides in favor of the symbol sequence $\hat{\mathbf{y}}$ that is closest in Hamming distance to the received symbol sequence $\mathbf{y}$. Again, the pairwise error probability depends on the particular mapping between the encoder outputs and the points in the signal constellation. If BPSK signaling is used, for example, then the pairwise error probability between two codewords at Hamming distance d is

$$P_2(d) = \begin{cases} \sum_{k=(d+1)/2}^{d} \binom{d}{k} p^k (1-p)^{d-k} \,, & d \text{ odd} \\ \sum_{k=d/2+1}^{d} \binom{d}{k} p^k (1-p)^{d-k} + \frac{1}{2}\binom{d}{d/2} p^{d/2}(1-p)^{d/2} \,, & d \text{ even} \end{cases} \tag{8.59}$$

where

$$p = Q(\sqrt{2R_c \gamma_b}) \tag{8.60}$$

is the probability of code bit error. Once again, the pairwise error probability for BPSK is a function of the Hamming distance between the codewords, and the decoded bit error probability is bounded by (8.58).

8.5 Interleaving

Most error control codes are designed for memoryless channels, where successively transmitted code bits experience independent channel realizations. Although this memoryless property holds for AWGN channels, it is not always the case for other coding channels. For example, fading channels exhibit a temporal correlation in their fades. An effective remedy for coding channels with memory is interleaving. There are two basic approaches to interleaving, symbol interleaving and bit interleaving. A symbol interleaver scrambles the modulation symbols *after* the symbol mapper. A bit interleaver scrambles the code bits *before* they are mapped onto a signal constellation.

8.5.1 *Block Interleaver*

A block interleaver can be regarded as a buffer with J rows and M columns, where J represents the interleaving depth and M represents the interleaving span. Such an interleaver is called a (J, M) block interleaver, and the length of the interleaver is JM. The block interleaver can be used to perform either symbol interleaving or bit interleaving. In the case of symbol (bit) interleaving, the code symbols (bits) are input to the buffer row-wise from top to bottom and output from the buffer column-wise from left to right. The deinterleaver simply performs the reverse operation. The block interleaver has the following characteristics [5]:

1. Any burst of symbol (bit) errors of length $j \leq J$ results in single symbol (bit) errors at the deinterleaver output that are each separated by at least M symbols (bits).
2. Any burst of symbol (bit) errors of length $j = kJ, k > 1$ results in bursts of no more than $\lceil k \rceil$ symbol (bit) errors that are separated by no less than $M - \lceil k \rceil$ symbols (bits).
3. A periodic sequence of single symbol (bit) errors that are spaced J code symbols (bits) apart results in a single burst of symbol (bit) errors of length M at the deinterleaver output.
4. The end-to-end interleaving delay is equal to $2JM$ code symbols (bits), and the memory requirements for the interleaver and deinterleaver is equal to JM code symbols (bits).

The interleaver depth J should be chosen so that successive code symbols (bits) that are transmitted J code symbol (bit) durations apart experience independent channel realizations. The choice of the parameter M depends on the type of coding used. For block codes M should be larger than the block length. For convolutional and trellis codes, the interleaver should separate any L_D+1 successive code symbols (bits) as far apart as possible, where L_D is the decoding depth. Hence, the interleaver should have $M \geq L_D + 1$. For convolutional and trellis codes, it is well known that the decoding depth $L_D \approx 5K$, where K is the constraint length [5].

Unfortunately, interleaving introduces communication delay. At the transmitter the $J \times M$ interleaver array must be filled before the symbols (bits) can be sent. At the receiver, the $J \times M$ deinterleaver array must be filled before the symbols (bits) can be decoded. However, delay critical traffic such as real-time voice will impose a limit on the interleaving delay defined as $t_d = 2JMT$ seconds where T is the baud period. One possible solution is to use a code with a smaller decoding depth L_D. Another solution is to use improved interleaving techniques that will reduce the interleaving delay. Once such interleaver is the convolutional interleaver.

8.5.2 Convolutional Interleaver

Convolutional interleavers have been proposed by Ramsey [26] and Forney [10], and the structure proposed by Forney is discussed below. Defining the parameter $M = LJ$, the interleaver is referred to as a (J, M) convolutional interleaver and has properties that are similar to a (J, M) block interleaver.

As shown in Fig. 8.16, the code symbols (bits) are input sequentially into a bank of J registers of increasing lengths. With each new code symbol (bit), the commutator switches to the next register, the code symbol (bit) is input to the register and the oldest code symbol (bit) in that register is shifted out to the channel.

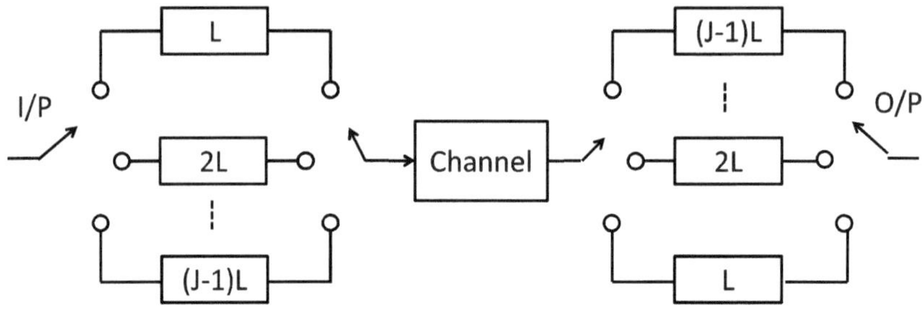

Fig. 8.16 Convolutional interleaver implemented with shift registers

The input and output commutators of the interleaver operate in a synchronous fashion. The deinterleaver performs the reverse operation, and the deinterleaving commutators must be synchronized with the interleaving commutators. The most important properties of the convolutional interleaver are as follows [5]:

1. The minimum separation at the interleaver output is J symbols (bits) for any two symbols that are separated by less than M symbols (bits) at the interleaver input.
2. Any burst of symbol (bit) errors of length $j \leq J$ results in single symbol (bit) errors at the deinterleaver output that are each separated by at least M symbols (bits).
3. A periodic sequence of single symbol (bit) errors that are spaced $M + 1$ code symbols (bits) apart results in a single burst of symbol (bit) errors of length J at the deinterleaver output.
4. The end-to-end interleaving delay is equal to $M(J - 1)$ code symbols (bits), and the memory requirements for both the interleaver and deinterleaver are equal to $M(J - 1)/2$ code symbols (bits). This is half the delay and memory requirement of a (J, M) block interleaver.

The parameters J and M are chosen in the same manner as for a block interleaver.

8.6 Automatic Repeat Request (ARQ) Techniques

As mentioned earlier, there are two methods of employing error control codes. Forward error control (FEC) coding uses the error control code to correct codeword or packet errors as shown in Fig 8.17. Automatic repeat request (ARQ) uses the error control code to detect codeword or packet errors and request retransmissions of the erroneous messages over a reverse channel as shown in Fig 8.18.

Recall that an (n, k) linear block code can detect $2^n - 2^k$ error patterns, but it can only correct 2^{n-k} error patters equal to the number of syndromes. In general, an error control code can be used to detect many more errors than it can correct. Hence, ARQ schemes can achieve much lower packet error probabilities.

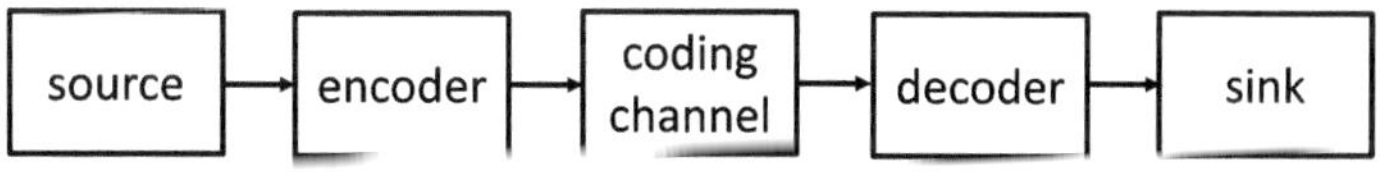

Fig. 8.17 Coding for forward error correction (FEC)

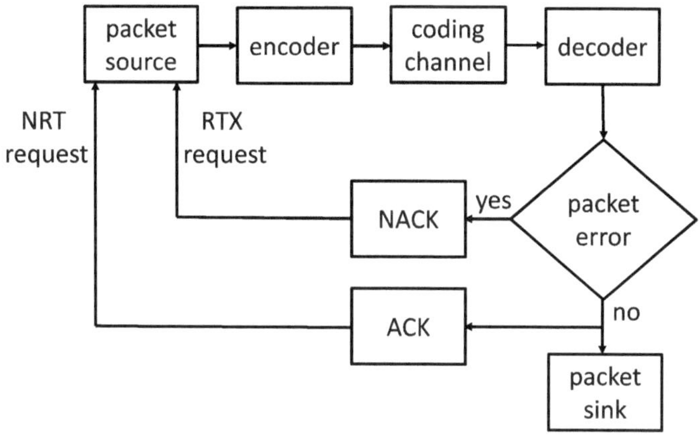

Fig. 8.18 Coding for automatic repeat request (ARQ)

Generally, FEC schemes are use in systems where delay is not tolerable and reliability is not so important, e.g., voice packet transmission. On the other hand, ARQ schemes are used in systems where delay is not so important and reliability is extremely important, e.g., data base transfer. The tradeoff with ARQ systems is increased reliability vs. decreased throughput due to the retransmissions. Hybrid ARQ schemes use both FEC and retransmissions. An error control code is used to detect the mostly likely error patterns, while the same or separate concatenated code, e.g., a cyclic redundancy check (CRC) code, is used to detect infrequent error patterns.

There are three basic types of ARQ systems, (i) stop-and-wait ARQ, selective repeat ARQ, and go-back-*N* ARQ.

8.6.1 Stop-and-Wait (SAW) ARQ

As depicted by the timeline shown in Fig. 8.19, in a stop-and-wait (SAW) ARQ the transmitter sends a packet and waits. The receiver receives the packet, checks for errors, and sends either an acknowledgement (ACK) or a negative acknowledgement (NACK) back to the transmitter over the reverse channel. If an ACK, the transmitter sends a new packet (NTX), and if an NACK, the transmitter retransmits the old packet (RTX). The advantage of SAW ARQ is its simple implementation. The disadvantage is that the idle time reduces throughput.

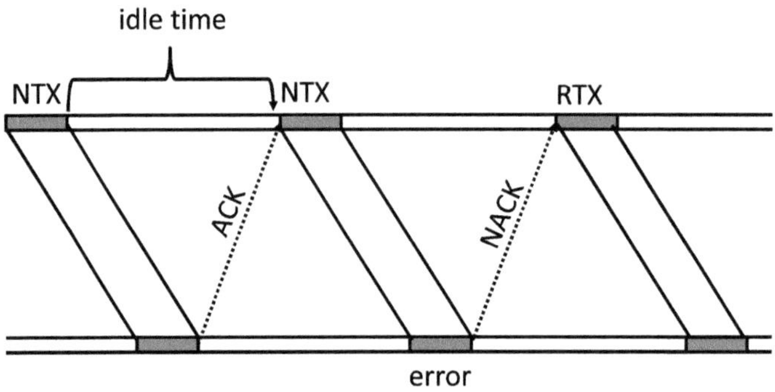

Fig. 8.19 Stop-and-wait (SAW) ARQ

Suppose that an (n, k) linear block code is used for error detection. Define

$$p_c \triangleq \text{P[packet contains no errors]}$$
$$p_d \triangleq \text{P[packet contains detectable errors]}$$
$$p_e \triangleq \text{P[packet contains undetectable errors]}$$
$$p_c + p_d + p_e = 1$$

Then the probability of packet error is

$$P(E) \triangleq \text{P[receiver accepts packets with undetectable errors]}$$
$$= \frac{p_e}{p_c + p_e} = \frac{p_e}{1 - p_d}$$

and the probability of correct packet reception is

$$P(C) \triangleq \text{P[receiver accepts packets with no errors]}$$
$$= \frac{p_c}{p_c + p_e} = 1 - P(E).$$

The above probabilities depend on the type of coding and modulation, and the type of detection and decoding used. For our purpose, assume the coding channel is a BSC with crossover probability p. Consider a general (n, k, d) Hamming code that is used for error detection. For a general Hamming code

$$n = 2^m - 1$$
$$k = 2^m - m - 1$$
$$d = 3.$$

The $(n, k) = (2^m - 1, 2^m - m - 1)$ Hamming code is a dual $(n, k) = (2^m - 1, m)$ linear code called a Simplex code. The Simplex code has a very simple weight distribution, consisting of the all-zeroes code word and $2^m - 1$ code words of weight 2^{m-1}, so its weight enumerator polynomial is

$$B(z) = 1 + (2^m - 1)z^{2^{m-1}}.$$

For a general (n, k, d) Hamming code

$$p_c = (1 - p)^n = (1 - p)^{2^m - 1}.$$

Using the MacWilliams identity the probability of undetected error is, from (8.15),

$$\begin{aligned} p_e &= 2^{-m}\left\{1 + (2^m - 1)(1 - 2p)^{2^m-1}\right\} - (1 - p)^{2^m-1} \\ &= 2^{-m}\left\{1 + (2^m - 1)(1 - 2p)^{2^m-1}\right\} - p_c. \end{aligned}$$

Hence,

$$P(E) = \frac{p_e}{p_c + p_e} = 1 - \frac{(1 - p)^{2^m-1}}{2^{-m}\left\{1 + (2^m - 1)(1 - 2p)^{2^m-1}\right\}}. \tag{8.61}$$

For the generalized Hamming code the rate is

$$R_c = \frac{k}{n} = \frac{2^m - m - 1}{2^m - 1}. \tag{8.62}$$

Example 8.6 Consider error detection with the $(n, k) = (2047, 2035)$ Hamming code that is obtained with $m = 11$. For a code bit error probability $p = 10^{-3}$

$$\begin{aligned} p_c &= 0.12899 \\ p_e &= 0.00016 \\ P(E) &= 0.00124 \\ R_c &= 0.99414 \end{aligned}$$

Without ARQ or coding

$$P(E) = 0.87101.$$

(continued)

Example 8.6 (continued)
Observe that a substantial increase in reliability is obtained with very little redundancy. In this case, there are only $n - k = 11$ parity bits in a packet of length $n = 2047$.

Throughput Efficiency
The throughput efficiency is defined as

$$\eta = \frac{\text{number of data bits successfully sent per unit time}}{\text{number of data bits that could be sent per unit time}}. \tag{8.63}$$

Define

$$\begin{aligned} D &= \text{idle time between two successive transmissions} \\ R_b &= \text{signaling rate in bits/second.} \end{aligned}$$

In one round-trip delay time, the transmitter could have sent

$$n + DR_b \text{ bits}$$

if it did not stay idle. However, a given packet may be sent several times before an ACK is received. The probability that a received packet will be accepted by the receiver is

$$\text{P}[ACK] = p_c + p_e = P. \tag{8.64}$$

The average number of bits that could have been sent until an ACK is received is

$$\begin{aligned} T_{SW} &= (n + DR_b)P + 2((n + DR_b)P(1 - P) \\ &\quad +3((n + DR_b)P(1 - P)^2 \\ &\quad +\cdots \\ &\quad +k((n + DR_b)P(1 - P)^{k-1} + \cdots \\ &= (n + DR_b)P \left\{ \sum_{k=1}^{\infty} k(1 - P)^{k-1} \right\}. \end{aligned}$$

However,

$$\sum_{k=1}^{\infty} kx^{k-1} = \frac{\mathrm{d}}{\mathrm{d}x} \sum_{k=0}^{\infty} x^k = \frac{\mathrm{d}}{\mathrm{d}x} \frac{1}{1 - x} = \frac{1}{(1 - x)^2}. \tag{8.65}$$

Hence, letting $x = (1 - P)$ yields

$$T_{SW} = \frac{(n + DR_b)P}{P^2} = \frac{n + DR_b}{P}. \tag{8.66}$$

The number of bits successfully sent until an ACK is received is

$$kP(C) = k\frac{p_c}{p_c + p_e}.$$

Finally, taking the ratio of the two and using (8.64)

$$\begin{aligned}\eta_{SW} &= \frac{k\frac{p_c}{p_c+p_e}}{\frac{n+DR_b}{P}} \\ &= k\frac{p_c}{n + DR_b} \\ &= \left(\frac{k}{n}\right)p_c\frac{1}{1 + DR_b/n}.\end{aligned} \tag{8.67}$$

The factor DR_b/n can be interpreted as the number of packets that could be transmitted during the idle time. The throughput efficiency with SAW ARQ is seen to degrade with increased propagation delay, and it is desirable that $DR_b \ll n$. In fact, the throughput can never reach the maximum value k/n even if the channel is noiseless ($P = 1$) due to the idle time. Therefore, SAW ARQ may be acceptable for local area networks with small D, but performs poorly for satellite networks with large D.

8.6.2 Go-Back-*N* (GBN) ARQ

As depicted by the timeline shown in Fig. 8.20, with go-back-N (GBN) ARQ the transmitter sends packets continuously. The receiver sends an ACK or NACK for each packet which arrives at the transmitter after the round-trip delay. If the transmitter receives a NACK, it goes-back-N to the corresponding packet and begins sending packets continuously from there. The $N - 1$ packets that were transmitted previously are sent again even though they may have been received correctly. At the receiver, these $N - 1$ packets are discarded regardless of whether or not they are error free. The transmitter must buffer these N packets for retransmission; however, the receiver only needs to store one packet at a time. The advantage of GBN ARQ is better throughput than SAW ARQ due to the continuous transmission and no requirement for receiver buffering. The disadvantage is sensitivity to round-trip delay and the requirement for transmitter buffering.

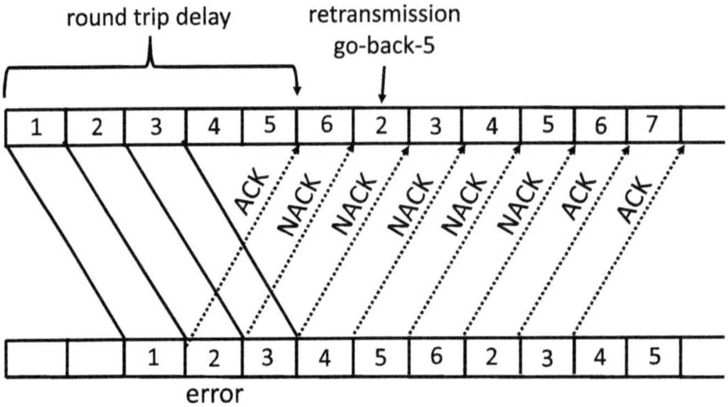

Fig. 8.20 Go-back-N (GBN) ARQ

For a packet to be accepted by the receiver, the average number of packet retransmissions, including the original transmission, is

$$\begin{aligned}
T_{GBN} &= 1 \cdot P + (N+1)P(1-P) + (2N+1)P(1-P)^2 + \cdots \\
&\quad + (kN+1)P(1-P)^k + \cdots \\
&= NP(1-P)\left\{1 + 2(1-P) + 3(1-P)^2 + \cdots\right\} \\
&\quad + P\left\{1 + (1-P) + (1-P)^2 + (1-P)^3 + \cdots\right\} \\
&= NP(1-P)\sum_{k=0}^{\infty} k(1-P)^{k-1} + P\sum_{k=0}^{\infty}(1-P)^k \\
&= \frac{NP(1-P)}{P^2} + \frac{P}{P} \\
&= \frac{N(1-P)}{P} + 1
\end{aligned} \tag{8.68}$$

where the first sum was reduced using the identity in (8.65). During this time

$$\left(\frac{N(1-P)}{P} + 1\right) n$$

bits could have been sent. Therefore, using (8.64) the throughput efficiency becomes

$$\begin{aligned}
\eta_{GBN} &= \frac{k \frac{p_c}{p_c + p_e}}{\left(\frac{N(1-P)}{P} + 1\right) n} \\
&= \left(\frac{k}{n}\right) \frac{p_c}{N(1-P) + P}.
\end{aligned} \tag{8.69}$$

With GBN ARQ, the value of N is determined by the round-trip delay. To obtain good throughput efficiency, it is required that $N(1 - P)$ be small

8.6.3 *Selective Repeat (SR) ARQ*

As depicted by the timeline shown in Fig. 8.21, with selective repeat (SR) ARQ packets are sent continuously. However, the transmitter only retransmits the packets that are NACK'd. After transmitting the NACK'd packet the transmitter goes back to where it left off. The key advantage of SR ARQ is that the throughput efficiency is independent of the round-trip delay and, therefore, excellent throughput efficiency is obtained. The disadvantage is that extensive buffering is required at both the transmitter and the receiver. Receiver buffering is required, since the packets must be delivered to the user in proper order. When a NACK'd packet is successfully received, the receiver releases the ACK'd packets in consecutive order until the next erroneously received packet is encountered at the receiver.

For a packet to be accepted by the receiver, the average number of packet retransmissions, including the original transmission, is

$$\begin{aligned} T_{SR} &= 1 \cdot P + 2P(1-P) + 3P(1-P)^2 + \cdots \\ &= \frac{1}{P} \end{aligned} \tag{8.70}$$

and during this time n/P bits could have been sent. Hence, the throughput efficiency with SR ARQ is

$$\eta_{SR} = \frac{k\frac{p_c}{p_c+p_e}}{n/P} = \left(\frac{k}{n}\right) p_c. \tag{8.71}$$

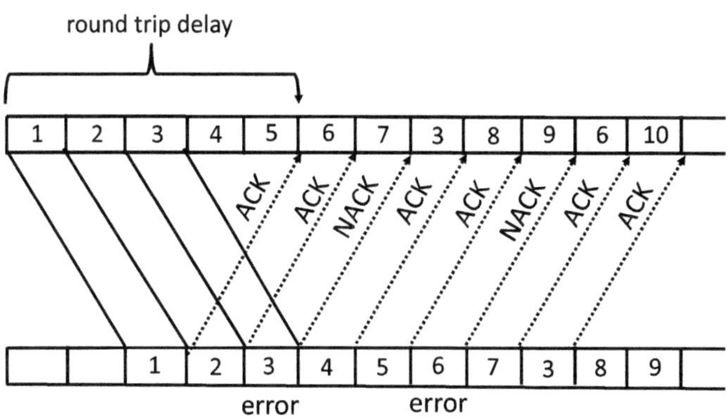

Fig. 8.21 Selective repeat (SR) ARQ

Problems

8.1 Consider a simple repetition code that generates codewords by simply repeating each information symbol L times.

(a) What is the rate of the code?
(b) How many codewords are there in the code?
(c) What is the minimum distance of the code?
(d) How many channel errors would have to occur to confuse one codeword with another?

8.2 Consider a systematic (8,4) code whose parity check equations are

$$p_0 = a_1 + a_2 + a_3$$

$$p_1 = a_0 + a_1 + a_2$$

$$p_2 = a_0 + a_1 + a_3$$

$$p_3 = a_0 + a_2 + a_3$$

where a_0, a_1, a_2, a_3 are the message digits and p_0, p_1, p_2, p_3 are parity-check digits, i.e., the codewords are $(a_0, a_1, a_2, a_3, p_0, p_1, p_2, p_3)$.

(a) Find the generator and parity-check matrices for this code.
(b) Show that the minimum distance of this code is $d_{\min} = 4$.

8.3

(a) Determine the weight distribution of the (8,4) code linear code in Problem 8.2.
(b) Let the transition probability of a binary symmetric channel be $p = 10^{-2}$. Compute the probability of undetected error of this code.

8.4 A systematic (5,3) code has the generator matrix

$$\mathbf{G} = \begin{bmatrix} 1\,0\,0\,1\,1 \\ 0\,1\,0\,0\,1 \\ 0\,0\,1\,1\,0 \end{bmatrix}$$

(a) What is the corresponding parity check matrix $\mathbf{H}$?
(b) Construct the standard array along with the correctable error patterns.

8.5 The generator matrix for a (6,3) systematic binary linear block code is

$$\mathbf{G} = \begin{bmatrix} 0\,1\,1\,1\,0\,0 \\ 1\,0\,1\,0\,1\,0 \\ 1\,1\,0\,0\,0\,1 \end{bmatrix}$$

(a) What is the party check matrix **H** for this code?
(b) Generate the standard array for this code.
(c) Calculate the syndrome vector for all of the correctable error patterns.
(d) Decode the received sequence $\mathbf{y} = 101101$, i.e., find the message vector.

8.6 The message $\mathbf{a} = 101$ is transmitted over a communications link using BPSK modulation. The system achieves E_b/N_o =9.0 dB.

(a) What is the probability of correctly receiving this message without forward error correction?
(b) What is the probability of correctly receiving this message if the (6,3) code of Problem 8.5 is used?

8.7 The parity check matrix **H** for a linear block code is given as follows (note that the identity matrix is at the beginning of the **H** matrix rather than the end):

$$\mathbf{H} = \begin{bmatrix} 1\,0\,0\,1\,1 \\ 0\,1\,0\,1\,1 \\ 0\,0\,1\,0\,1 \end{bmatrix}$$

(a) Construct a standard array decoding table for this code.
(b) How many error patterns can this code correct.
(c) If this code is used for error detection on a binary symmetric channel with crossover probability p, what is the probability of undetected error?

8.8 A systematic (5,3) linear block code has the generator matrix

$$\mathbf{G} = \begin{bmatrix} 1\,0\,0\,1\,1 \\ 0\,1\,0\,0\,1 \\ 0\,0\,1\,1\,0 \end{bmatrix}$$

(a) Write down the weight enumerator polynomial of the code. What is the minimum distance of the code?
(b) Construct the parity check matrix **H** of the code.
(c) Construct the standard array, and syndrome decoding table.
(d) If the code is used for error correction using standard array decoding and the received vector is $\mathbf{y} = (1, 1, 1, 1, 1)$ what is the decoded codeword?
(e) If the code is used for error detection on a binary symmetric channel with crossover probability $p = 0.1$, what is the probability of undetected error?

8.9 Consider a systematic (15,11) Hamming code.

(a) Construct a parity check matrix for a systematic (15,11) Hamming code.
(b) Construct a syndrome table for the code defined by the parity check matrix.
(c) If the (15,11) Hamming code is used for error detection on a binary symmetric channel with crossover probability p, what is the probability of undetected error?

8.10 The generator matrix of a systematic (7,4) Hamming code is

$$\mathbf{G} = \begin{bmatrix} 1\,0\,0\,0\,1\,1\,0 \\ 0\,1\,0\,0\,0\,1\,1 \\ 0\,0\,1\,0\,1\,1\,1 \\ 0\,0\,0\,1\,1\,0\,1 \end{bmatrix}$$

The corresponding parity check matrix is

$$\mathbf{H} = \begin{bmatrix} 1\,0\,1\,1\,1\,0\,0 \\ 1\,1\,1\,0\,0\,1\,0 \\ 0\,1\,1\,1\,0\,0\,1 \end{bmatrix}$$

The dual to a (7,4) Hamming code can be obtained by exchanging the generator and parity check matrices, i.e., let $\mathbf{H}$ be the generator matrix and let $\mathbf{G}$ be the parity check matrix. The dual to a (7,4) Hamming code is called a (7,3) Simplex code.

(a) List out the codewords of the (7,3) Simplex code.
(b) Find the weight distribution of the code.
(c) The (7,3) Simplex code is capable of either detecting all patterns of s or fewer errors, or correcting all patterns of t or fewer errors. Find s and t.
(d) Suppose that the (7,3) Simplex code is used for error detection. The code bits are transmitted using BPSK over an additive white Gaussian noise channel and $E_b/N_o = 6$ dB. What is the probability of undetected error?

8.11 Consider the simple 2-state convolutional code shown in Fig. 8.22.

(a) Draw the state diagram.
(b) Draw the augmented state diagram.
(c) Determined the transfer function $T(D, N, L)$ of the code.
(d) What is the minimum distance of the code, d_{free}?

8.12 Consider a rate-1/3 convolutional code with generators $\mathbf{g}^{(1)} = (111)$, $\mathbf{g}^{(2)} = (111)$, and $\mathbf{g}^{(3)} = (101)$.

(a) Draw a block diagram of the encoder structure.

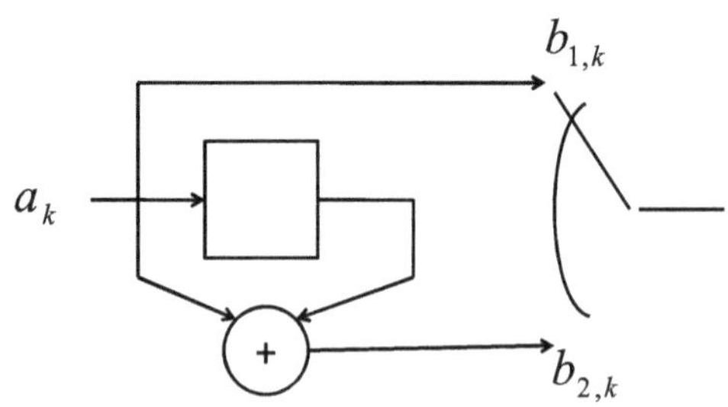

Fig. 8.22 Two-state convolutional code for Problem 8.11

(b) Draw the state diagram and trellis diagram.
(c) Determine the output sequence corresponding to the input sequence 1110101.

8.13 A rate-1/3 convolutional code is defined by the generator sequences

$$\mathbf{g}^{(1)} = (1, 1, 0)$$
$$\mathbf{g}^{(2)} = (1, 1, 1)$$
$$\mathbf{g}^{(3)} = (1, 0, 1)$$

(a) Draw the shift-register convolutional encoder.
(b) Draw the state diagram for this encoder.
(c) Draw the trellis diagram for this encoder.
(d) Draw the augmented or modified state diagram for this encoder.
(e) Find the transfer function $T(D, N, L)$ for this encoder.
(f) What is the free Hamming distance of the code, d_{free}?

8.14 For the code in Problem 8.13 and hard decision decoding, determine:

(a) A union bound on the probability of decoding error
(b) An upper bound on the probability of decoded bit error

8.15 A rate-1/3 convolutional code is defined by the generator sequences

$$\mathbf{g}^{(1)} = (1, 0, 1)$$
$$\mathbf{g}^{(2)} = (1, 1, 1)$$
$$\mathbf{g}^{(3)} = (1, 1, 1)$$

(a) Draw the shift-register convolutional encoder.
(b) Draw the state diagram for this encoder.
(c) Draw the augmented or modified state diagram for this encoder.
(d) Find the transfer function $T(D, N, L)$ for this encoder.

8.16 The convolutional code in Problem 8.15 is used for transmission over a binary symmetric channel. The output of the demodulator detector is (101001011110111....). Using the Viterbi algorithm, find the data sequence, assuming that the convolutional code begins in the zero state and is terminated at the zero state.

8.17 The output of a rate-1/3 convolutional encoder with constraint length 3 to the input $\mathbf{a} = (1, 1, 0, \ldots)$ is $\mathbf{b} = (111, 110, 010, \ldots)$

(a) Determine the transfer function $T(D, N, L)$.
(b) Determine the number of paths through the state diagram or trellis that diverge from the all-zeroes state and merge with the all-zeroes state 7 branches later.
(c) Determine the number of paths of Hamming distance 20 from the all zeroes sequence.

8.18 A rate 1/2 convolutional code has the generator sequences

$$\mathbf{g}^{(1)} = (1, 1, 1)$$
$$\mathbf{g}^{(2)} = (0, 1, 1)$$

(a) Draw the shift register convolutional encoder.
(b) Draw the state transition diagram.
(c) Draw the modified or augmented state diagram.
(d) Determined the transfer function $T(D, N, L)$ of the code.
(e) What is the minimum distance d_{free} (or d_{min}) of the code?

8.19 Consider the 8-PAM and 32-CROSS signal constellations in Fig. 8.23.

(a) Construct the partition chain as in Fig. 8.12 and compute the minimum Euclidean distance between signal points at each step in the partition chain.
(b) What is the average symbol energy for each of the signal constellations?

8.20 Consider the 2-state, rate-1/2, trellis encoder shown in Fig. 8.24. By using this encoder with a 4-PAM and 8-PAM signal constellation TCM systems can be constructed having bandwidth efficiencies of 1 bit/s/Hz and 2 bits/s/Hz, respectively.

(a) Determine the appropriate partitions for the signal constellation for the 2-state, 4-PAM and 8-PAM trellis codes.

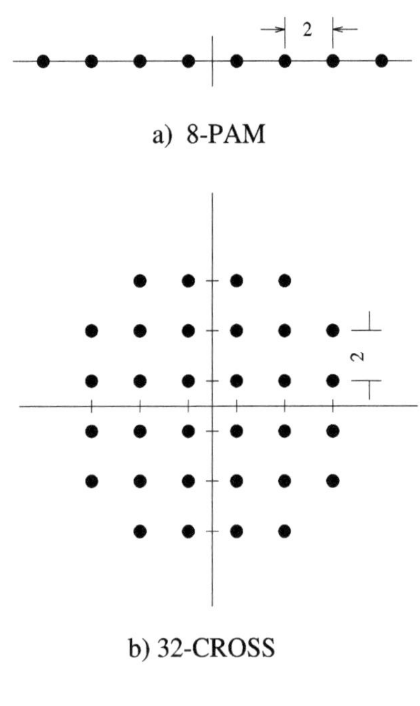

Fig. 8.23 Signal constellations for Problem 8.19

Fig. 8.24 Trellis encoder for Problem 8.20

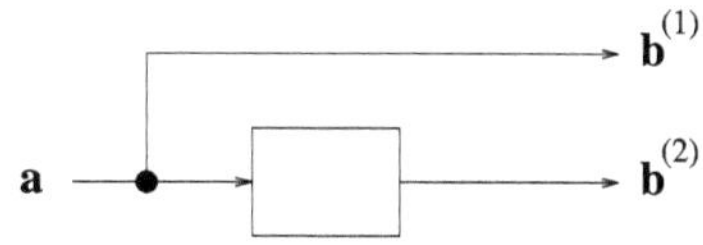

(b) Construct and label the trellis diagrams for the 2-state 4-PAM and 8-PAM trellis codes.

(c) Determine the minimum Euclidean distance for each trellis code and the asymptotic coding gain on an AWGN channel relative to the equivalent uncoded systems.

Chapter 9
Intersymbol Interference, Equalization, and Partial Response Signals

Communication channels are often modeled as time-invariant or time-varying linear filters. For anything other than an ideal channel, $\tilde{h}(t) = \alpha\delta(t - t_o)$, some degree of amplitude and phase distortion will be introduced into the message waveform. The overall pulse consisting of the concatenation of the transmitter pulse shaping filter, communication channel, and receiver matched filter, $p(t)$, will no longer be a Nyquist pulse and *intersymbol interference* (ISI) will be introduced.

ISI will degrade the bit error rate performance, and if the ISI is severe enough, errors will occur even in the absence of noise. Consequently, it is necessary to compensate for amplitude and phase distortion introduced by the channel at the receiver and this compensation is accomplished by using an *equalizer* to correct for the amplitude and phase distortion introduced by the channel.

The need for equalization will depend on the delay spread or length of the waveform channel $\tilde{h}(t)$ relative to the baud period T. Generally, as the baud rate $R = 1/T$ increases, the ISI will span a larger number of modulation symbols. However, depending on the shaping pulse $h_a(t)$ used, even a relatively small delay spread can result in an overall sampled pulse $p_k = p(kT)$ that spans a significant number of modulation symbols. Thus, for high data rate or broadband communication, the ISI will become more severe and a more powerful equalizer is needed.

Equalization may be accomplished either in the time domain or frequency domain. This chapter introduces the basic concepts of time-domain equalization, while frequency-domain equalization is considered in Chap. 10. There are two categories of time-domain equalizers. The first category is the *symbol-by-symbol equalizers*, where data symbol decisions are made on a symbol-by-symbol basis after equalization. The second category is the *sequence estimators*, where sequence decisions on the data symbol decisions are made using maximum likelihood sequence estimation (MLSE) or other sequence estimator. Similar to convolutional codes, the finite-length ISI channel can be treated as a finite-state machine and the

G. Stuber, *Introductory Digital Wireless Communications*,
https://doi.org/10.1007/978-3-032-01559-4_9

Viterbi algorithm applied to find the most likely symbol (or data) sequence. Many different types of equalizers exist in each category.

This chapter first considers the receiver front-end processing for channels with ISI in Sect. 9.1. Afterwards, a simple linear symbol-by-symbol time-domain equalizer called a *zero-forcing equalizer* is introduced. Finally, partial response signals are considered in Sect. 9.3, where controlled ISI is introduced at the transmitter to achieve bandwidth efficient communication within the Nyquist bandwidth while employing realizable filters.

9.1 Front-End Processing

The cascade of the transmit pulse $h_a(t)$ and the waveform channel $\tilde{h}(t)$ yields a received pulse $p_r(t)$ that is the convolution of the two pulses:

$$p_r(t) = h_a(t) * \tilde{h}(t). \tag{9.1}$$

Assume that the signal constellation has dimension $N = 1$, such that the messages are complex scalars. The received complex envelope is

$$\tilde{r}(t) = \sum_n x_n p_r(t - nT) + \tilde{n}(t), \tag{9.2}$$

where n is the discrete-time index and x_n represents the nth message symbol, and $\tilde{n}(t)$ is the AWGN.

Section 6.2 showed that, for an ideal channel $\tilde{h}(t) = \tilde{h}\delta(t - t_o)$, the optimum receiver employs a receiver filter that is matched to the transmitted pulse $h_a(t)$ followed by a baud-rate sampler as shown in Fig. 9.1. The receiver filter $h_a(T_o - t)$ is optimal in this case because the ideal channel does not alter the transmitted pulse $h_a(t)$ except to delay it by t_o seconds and scale it by the complex gain $\tilde{h}$. However, if the channel impulse response $\tilde{h}(t)$ in (9.1) is nonideal and unknown to the receiver, then so is the received pulse $p_r(t)$. Therefore, the matched filter $p_r(-t)$ cannot be implemented at the receiver.

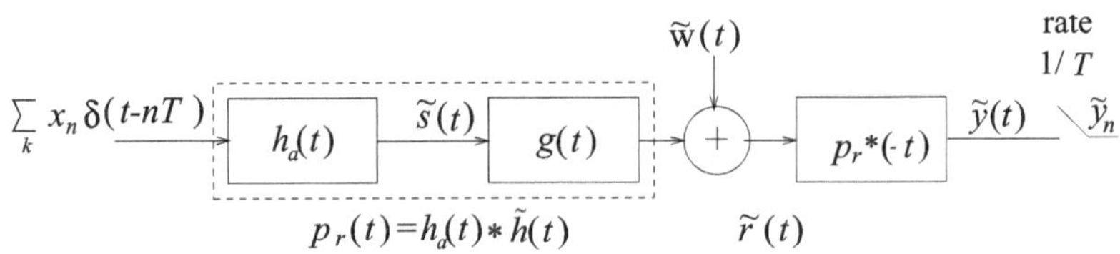

Fig. 9.1 Digital signaling on an ISI channel. The optimum front-end processor implements a filter that is matched to the received pulse $p_r(t)$ followed by a baud-rate sampler

Various practical front-end solutions have been suggested in the literature to overcome the above limitation. For pulses with less than 100% excess bandwidth, such as the root raised cosine pulse in Sect. 5.4.2, the pulse bandwidth is less than the baud rate $1/T$. One front-end processing solution that is optimal in a maximum likelihood sense passes the received complex envelope $\tilde{r}(t)$ through an "ideal" low-pass filter having a cutoff frequency of $1/T$ and sampling the output at rate $2/T$ (two samples per baud). Since the bandwidth of $\tilde{r}(t)$ is limited to $1/T$, there is no aliasing. Meanwhile, the ideal low-pass filter that bandlimits the noise also serves as an ideal anti-aliasing filter. The $T/2$-spaced noise samples taken at the output of the ideal low-pass filter with edge frequency $1/T$ are uncorrelated as shown in Sect. 4.4.2.

Moreover, since the signal bandwidth is less than $1/T$ and the noise is white, no information is lost by rate-$2/T$ sampling and the $T/2$-spaced samples provide sufficient statistics for maximum likelihood detection. The operations of matched filtering and maximum-likelihood detection are subsequently combined into one discrete-time subsystem. This approach, however, requires a *fractionally spaced* equalizer that operates on the $T/2$-spaced samples. Vachula and Hill [31] showed that such a receiver is optimal in a maximum likelihood sense. However, it may not be suitable for bandwidth efficient multi-access systems that use frequency division multiplexing, because the cutoff frequency of the low-pass filter, $1/T$ Hz, is excessive when the excess bandwidth of the shaping pulse $h_a(t)$ beyond the Nyquist bandwidth $1/2T$ Hz is small.

Another approach that is optimal in a maximum-likelihood sense uses a receiver filter $h_r(t)$ that is matched to the transmitted pulse $h_r(t) = h_a(T_o - t)$, followed by a rate-$2/T$ sampler, and a $T/2$-spaced noise whitening filter. Rate $2/T$ sampling satisfies the Shannon sampling theorem for shaping pulses $h_a(t)$ having up to 100% excess bandwidth. The matched filter $h_a(T_o - t)$ is a root-Nyquist pulse so T-spaced noise samples of the filtered AWGN noise taken at the output of the filter are uncorrelated. However, with a rate-$2/T$ sampler the $kT/2$-spaced noise samples taken at the output of the matched filter $h_r(t) = h_a(T_o - t)$ are correlated for odd k. Such noise correlation greatly complicates any further signal processing. The remedy is to whiten the colored noise samples prior to further processing with a fractionally spaced discrete-time noise whitening filter $G_h^{(2)^*}(1/z^*)$ as shown by the arrangement in Fig. 9.2 [28]. The $T/2$-spaced samples at the output of the noise whitening filter may next be applied to a *fractionally spaced* equalizer, such as a maximum likelihood sequence estimator as implemented with the Viterbi algorithm that operates on 2 samples per symbol to recover the data sequence.

In the special case of zero excess bandwidth, both the transmitted pulse and the receiver matched filter are the ideal root Nyquist pulse $h_a(t) = h_r(t) = h_N(t) = (1/\sqrt{T})\text{sinc}(t/T)$. In this case, T-spaced samples taken at the output of receiver matched filter will satisfy the Shannon sampling theorem and provide sufficient statistics with uncorrelated noise components. The cascade of the transmit filter

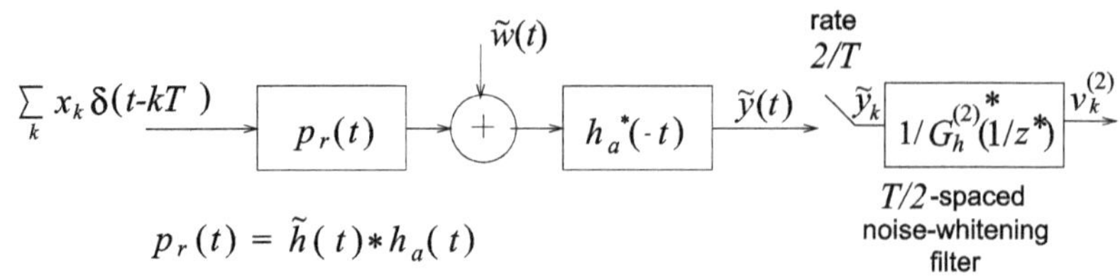

Fig. 9.2 Block diagram of system that implements a filter matched to $h_a(t)$ followed by a $T/2$-spaced noise whitening filter $G_h^{(2)^*}(1/z^*)$. The structure of the noise whitening filter depends only on the known pulse $h_a(t)$

$h_a(t)$, channel $\tilde{h}(t)$, and receiver matched filter $h_r(t)$ yields the overall pulse

$$p(t) = h_a(t) * \tilde{h}(t) * h_r(t). \tag{9.3}$$

The complex envelope at the output of the receiver matched filter is

$$\tilde{y}(t) = \sum_n x_n p(t - nT) + \tilde{n}(t), \tag{9.4}$$

where n is the discrete-time index, x_n represents the nth modulation symbol, and

$$\tilde{n}(t) = \int_{-\infty}^{\infty} \tilde{w}(\tau) h_r(t - \tau) \mathrm{d}\tau \tag{9.5}$$

is the noise at the output of the matched filter. If T-spaced samples are taken at the output of the receiver matched filter at the correct sampling instants, the kth output sample is

$$\tilde{y}_k = \tilde{y}(kT) = \sum_n x_n p_{k-n} + \tilde{n}_n, \tag{9.6}$$

where $p_k = p(kT)$ and $\tilde{n}_n = \tilde{n}(nT)$.

In practice, the transmitter and receiver matched filters are all approximated with causal finite-length filters having length T_o. Their length will increase as the excess bandwidth decreases, with the ideal root Nyquist pulses requiring the largest T_o for useful approximation. Hence, the overall pulse $p(t)$ is modeled here as a finite-length causal pulse. Thus, with T-spaced sampling, the overall discrete-time channel impulse response $\{p_n\}$ can be represented by the overall channel vector

$$\mathbf{p} = (p_0, p_1, \ldots, p_L), \tag{9.7}$$

where L is the length of the discrete-time channel. The sample sequence $\{\tilde{y}_k\}$ is next applied to the equalizer. While a variety of equalizers may be considered, our introductory treatment considers a simple linear symbol-by-symbol equalizer.

9.2 Linear Time-Domain Equalization

Figure 9.3 shows a linear symbol-by-symbol equalizer, consisting of a finite impulse response (FIR) digital filter. The tap coefficients of the symbol-by-symbol equalizer are denoted by the vector

$$\mathbf{w} \stackrel{\triangle}{=} (w_0,\ w_1,\ \cdots,\ w_{N-1}) \tag{9.8}$$

where N is the number of equalizer taps. Assuming that the equalizer inputs the sequence $\{\tilde{y}_k\}$ with samples in (9.6), the nth equalizer output sample is

$$\tilde{x}_n = \sum_{j=0}^{N-1} w_j \tilde{y}_{n-j}, \tag{9.9}$$

which is then applied to a symbol-by-symbol decision device to produce the decided sequence $\hat{x}_n$. If error control coding is used, then the soft outputs $\tilde{x}_n$ may be applied directly to the decoder in the case of soft decision decoding.

The concatenation of the overall discrete-time channel $\mathbf{p} = \{p_n\}$ and equalizer $\mathbf{w} = \{w_n\}$ can be represented by an overall discrete-time system having the impulse response

$$\mathbf{q} = (q_0, q_1,\ \ldots,\ q_{N+L-1}) \tag{9.10}$$

where

$$\begin{aligned} q_n &= \sum_{j=0}^{N-1} w_j p_{n-j} \\ &= \mathbf{w}\mathbf{p}(n)^T \end{aligned} \tag{9.11}$$

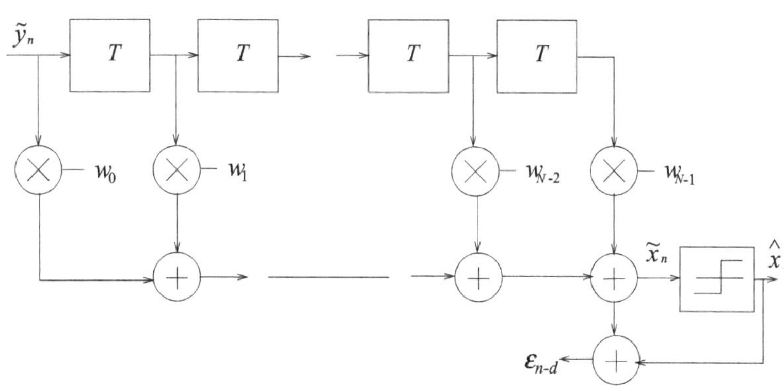

Fig. 9.3 Linear transversal equalizer

with

$$\mathbf{p}(n) = (p_n, p_{n-1}, p_{n-2}, \ldots, p_{n-N+1}) \tag{9.12}$$

and $p_i = 0, i < 0, i > L$. That is, $\mathbf{q}$ is the discrete-time linear convolution of $\mathbf{p}$ and $\mathbf{w}$.

Let the component of $\mathbf{p}$ having the greatest magnitude be denoted by p_{d_1}. Let the number of equalizer taps be equal to $N = 2d_2 + 1$ where d_2 is an integer. Perfect equalization means that

$$\mathbf{q} = \mathbf{e}_d = (\underbrace{0,\ 0,\ \ldots,\ 0}_{d-1 \text{ zeroes}},\ 1,\ 0,\ \ldots,\ 0,\ 0) \tag{9.13}$$

where $d-1$ zeroes precede the "1" and d is an integer representing the overall delay, a parameter to be optimized. Unfortunately, perfect equalization is not always possible with a finite-length equalizer and does not always yield the best performance.

9.2.1 Zero-Forcing Equalizer

Various approaches exist for determining the equalizer tap coefficients $\mathbf{w}$ that differ in their performance. The first solution was proposed by the pioneer Lucky [17, 18] that yields a *zero-forcing* (ZF) equalizer. With a zero-forcing (ZF) equalizer, $\mathbf{w}$ is chosen to minimize the peak distortion of the equalized channel, defined as

$$D_{\mathrm{p}} = \frac{1}{|q_d|} \sum_{\substack{n=0 \\ n \neq d}}^{N+L-1} |q_n - \hat{q}_n| \tag{9.14}$$

where $\hat{\mathbf{q}} = (\hat{q}_0, \ \ldots, \ \hat{q}_{N+L-1})$ is the *desired equalized channel* and the delay d is a positive integer optimized to have the value $d = d_1 + d_2$.

Lucky showed [17, 18] that if the initial distortion without equalization is less than unity, i.e.,

$$D = \frac{1}{|p_{d_1}|} \sum_{\substack{n=0 \\ n \neq d_1}}^{L} |p_n| < 1 , \tag{9.15}$$

then D_p is minimized by those N tap values which simultaneously cause $q_j = \hat{q}_j$ for $d - d_2 \leq j \leq d + d_2$. However, if the initial distortion before equalization is greater than unity, then the ZF criterion is not guaranteed to minimize the peak distortion D_{p}.

For the case when $\hat{\mathbf{q}} = \mathbf{e}_d^T$ the equalized channel is given by

$$\mathbf{q} = (q_0, \ldots, q_{d_1-1}, 0, \ldots, 0, 1, 0, \ldots, 0, q_{d_1+N}, \ldots, q_{N+L-1})^T. \tag{9.16}$$

In this case the equalizer forces zeroes into the equalized discrete-time channel, hence, the name "zero-forcing equalizer."

If the ZF equalizer has an infinite number of taps, the tap weights may be selected so that $D_\mathrm{p} = 0$, i.e., $\boldsymbol{q} = \hat{\boldsymbol{q}}$. Assuming that $\hat{q}_n = \delta_{n0}$ and taking the z-transform of both sides of (9.11), this means that

$$Q(z) = 1 = W(z)P(z). \tag{9.17}$$

Therefore,

$$W(z) = \frac{1}{P(z)} \tag{9.18}$$

and the ideal ZF equalizer has a system function $W(z)$ that is the inverse of channel system function $P(z)$.

9.2.2 Zero-Forcing Solution

For a known channel impulse response, $\mathbf{p}$, the tap gains of the ZF equalizer can be found by the direct solution of a simple set of linear equations. This is done by forming the matrix

$$\mathbf{P} = \begin{bmatrix} \mathbf{p}(d_1) \\ \vdots \\ \mathbf{p}(d) \\ \vdots \\ \mathbf{p}(N + d_1 - 1) \end{bmatrix} \tag{9.19}$$

and the vector

$$\tilde{\mathbf{q}} = \left(\hat{q}_{d_1}, \ldots, \hat{q}_d, \ldots, \hat{q}_{N+d_1-1}\right). \tag{9.20}$$

Then the vector of optimal tap gains, $\mathbf{w}_\mathrm{op}$, satisfies

$$\mathbf{P}\mathbf{w}_\mathrm{op} = \tilde{\mathbf{q}} \longrightarrow \mathbf{w}_\mathrm{op}^T = \mathbf{P}^{-1}\tilde{\mathbf{q}}. \tag{9.21}$$

Example 9.1 Suppose that the overall discrete-time channel impulse response is

$$\mathbf{p} = (0.1, -0.2, 0.9, -0.3, 0.1, 0, \ldots, 0),$$

where $p_i = 0, i < 0, i > 4$. Hence, $d_1 = 2$ and the distortion before equalization is

$$D = \frac{1}{|p_{d_2}|} \sum_{\substack{n=0 \\ n \neq d_2}}^{4} |p_n| = 0.777778 < 1.$$

Since $D < 1$, the eye is initially open and the distortion is minimized by the ZF solution.

Consider a 3-tap ZF equalizer. Since p_2 is the component of $\mathbf{p}$ having the largest magnitude, $d_1 = 2$, and the equalizer delay is chosen as $d = d_1 + d_2 = 3$. The desired response is

$$\mathbf{q} = \mathbf{e}_3 = (0,\ 0,\ 0,\ 1,\ 0,\ 0,\ 0,\ \ldots,\ 0)$$

and

$$\tilde{\mathbf{q}} = (0,\ 1,\ 0).$$

Next ,construct the matrix

$$\mathbf{P} = \begin{bmatrix} \mathbf{p}(2) \\ \mathbf{p}(3) \\ \mathbf{p}(4) \end{bmatrix} = \begin{bmatrix} 0.9 & -0.2 & 0.1 \\ -0.3 & 0.9 & -0.2 \\ 0.1 & -0.3 & 0.90 \end{bmatrix}$$

and obtain the optimal tap solution:

$$\mathbf{w}_{\text{op}} = (\mathbf{P}^{-1})\tilde{\mathbf{q}} = (0.24,\ 1.28,\ 0.4).$$

The overall discrete-time impulse response of the channel and equalizer is

$$\mathbf{q} = (\,0.024,\ 0.08,\ 0.0,\ 1.0,\ 0.0,\ 0.008,\ 0.04,\ 0, \ldots).$$

(continued)

Example 9.1 (continued)
The distortion after equalization is

$$D_p = \frac{1}{|q_3|} \sum_{n=0}^{6} |q_n - \hat{q}_n| = 0.152,$$

which is considerably less than the distortion before equalization. A smaller D_p can be obtained by using a longer equalizer, i.e., a larger value of N.

9.2.3 *Adaptive Zero-Forcing Equalizer*

In practice, the channel vector **p** is unknown to the receiver so it is not possible to obtain the equalizer **w** by direct solution. Usually, a known finite-length training sequence **a** is sent to train the equalizer that is set with some initial value for **w**. After sending the training sequence, the transmitter sends the data sequence **x** that is processed with the trained equalizer. The equalizer may use decided elements of the data sequence, $\hat{\mathbf{x}}$, as a training sequence to further update the equalizer taps.

During the *training mode*, the equalizer taps can be obtained by using the steepest-descent recursive algorithm:

$$w_j^{n+1} = w_j^n + \alpha \epsilon_n a_{n-j-d_1}\,, \qquad j = 0, \ldots, N-1, \tag{9.22}$$

where the training sequence $\mathbf{a} = \{a_n\}$ is assumed to be real-valued and known,

$$\begin{aligned} \epsilon_n &= a_{n-d} - \tilde{a}_n \\ &= a_{n-d} - \sum_{i=0}^{N-1} w_i^n y_{n-i} \end{aligned} \tag{9.23}$$

is the error sequence, and $\{w_j^n\}$ is the set of equalizer tap at epoch n. The parameter α is an adaptation step-size that can be optimized to trade off convergence rate and steady-state performance.

The adaptation rule in (9.22) attempts to force the crosscorrelations $\epsilon_n a_{n-j-d_1}$, $j = 0, \ldots, N-1$, to zero. To see that this is the case, consider

$$\mathrm{E}[\epsilon_n a_{n-j-d_1}] = \mathrm{E}[a_{n-d} a_{n-j-d_1}] - \sum_{i=0}^{N-1} \sum_{\ell=0}^{L} w_i^n p_\ell \mathrm{E}[a_{n-i-\ell} a_{n-j-d_1}]$$

$$= \sigma_a^2 \left(\delta_{d_2-j} - \sum_{i=0}^{N-1} w_i^n p_{j+d_1-i} \right)$$

$$= \sigma_a^2 (\delta_{d_2-j} - q_{j+d_1}) \,, \quad j = 0, \ 1, \ \ldots, \ N-1, \tag{9.24}$$

where $\sigma_a^2 = \mathrm{E}[|a_k|^2]$.

The conditions $\mathrm{E}[\epsilon_n a_{n-j-d_1}] = 0$ are satisfied when $q_d = 1$ and $q_i = 0$ for $d - d_2 \leq i < d$ and $d < i \leq d + d_2$, which is the zero-forcing solution. Note the ensemble average is over both the noise and the training symbol alphabet.

After training the equalizer, a decision-feedback mechanism may be employed where the decided symbol sequence $\hat{\mathbf{x}}$ is used as a training sequence to update the equalizer tap coefficients. This mode is called the *data mode* and allows the equalizer to track variations in the channel vector $\mathbf{p}$. In the data mode,

$$w_j^{n+1} = w_j^n + \alpha \epsilon_n \hat{x}_{n-j-d_1} \,, \qquad j = 0, \ldots, N-1, \tag{9.25}$$

where the error term ϵ_n in (9.23) becomes

$$\epsilon_n = \hat{x}_{n-d} - \sum_{i=0}^{N-1} w_i^n \tilde{y}_{n-i} \tag{9.26}$$

and, again, $\hat{x}_{n-d}$ is the decision on the equalizer output $\tilde{x}_n$ delayed by d samples.

9.3 Partial Response Signaling

The minimum required bandwidth for ISI-free transmission at a baud rate of R symbols/second is $W = R/2 = 1/2T$ Hz, referred to as the Nyquist bandwith. However, the corresponding overall pulse $P(f)$ is an ideal low-pass filter which is noncausal and not realizable. Nyquist pulse shaping in Sect. 5.4 was suggested as one solution. The excess bandwidth in the overall pulse $P(f)$ beyond the Nyquist bandwidth allows for a gentler roll-off of the overall pulse $P(f)$ in the frequency domain and gives a faster roll-off of sidelobes of the pulse $p(t)$ in the time domain.

Partial response modulation schemes are another solution that can offer a rate of R symbols/second in the Nyquist bandwidth with realizable shaping filters. This is accomplished by introducing controlled ISI whose effects can be completely eliminated at the receiver.

Assume again an ideal channel $\tilde{h}(t) = \delta(t)$, such that the transmitter uses shaping pulse $h_a(t)$ and the receiver employs a matched filter $h_a(T_o - t)$. As with Nyquist pulse shaping, the delay T_o required to make these filters causal is ignored when designing the overall pulse $p(t) \leftrightarrow P(f)$ and

$$p(t) = h_a(t) * h_a(-t) \tag{9.27a}$$

$$P(f) = |H_a(f)|^2. \tag{9.27b}$$

9.3.1 Duobinary Signaling

With *duobinary* partial response signaling, the overall frequency domain pulse $P(f)$ can be described by the system shown in Fig. 9.4. The filter $H_N(f)$ is the ideal Nyquist pulse:

$$H_N(f) = \frac{1}{2W}\text{rect}\left(\frac{f}{2W}\right) = \begin{cases} 1 & |f| < W \\ 0 & \text{else} \end{cases} \tag{9.28}$$

From the system shown in Fig. 9.4, the overall frequency domain pulse is

$$\begin{aligned} P(f) &= \left(1 + e^{-j2\pi fT}\right) H_N(f) \\ &= 2e^{-j\pi fT}\left(\frac{e^{j\pi fT} + e^{-j\pi fT}}{2}\right) H_N(f) \\ &= 2\cos(\pi fT)e^{-j\pi fT} H_N(f) \\ &= 2\cos(\pi fT)e^{-j\pi fT}\frac{1}{2W}\text{rect}\left(\frac{f}{2W}\right) \\ &= \begin{cases} 2T\cos(\pi fT)e^{-j\pi fT} & |f| < W \\ 0 & \text{else} \end{cases} \end{aligned} \tag{9.29}$$

The magnitude and phase response of $P(f)$ are sketched in Fig. 9.5. To obtain $p(t)$, the time-shift property of the Fourier transform may be used

$$\left\{P(f) = H_N(f) + H_N(f)e^{j2\pi fT}\right\} \leftrightarrow \left\{p(t) = \text{sinc}\left(\frac{t}{T}\right) + \text{sinc}\left(\frac{t-T}{T}\right)\right\}. \tag{9.30}$$

The corresponding time domain pulse $p(t)$ is plotted in Fig. 9.6.

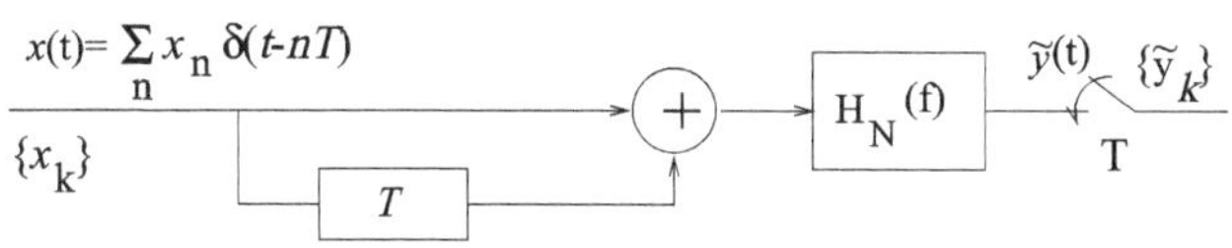

Fig. 9.4 Duobinary partial response system

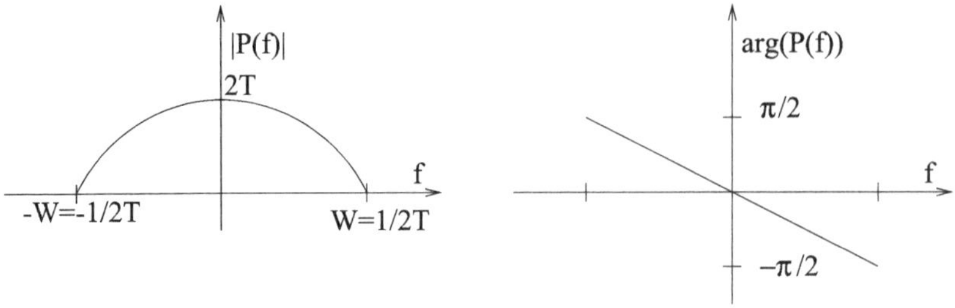

Fig. 9.5 Magnitude and phase response of duobinary partial response system

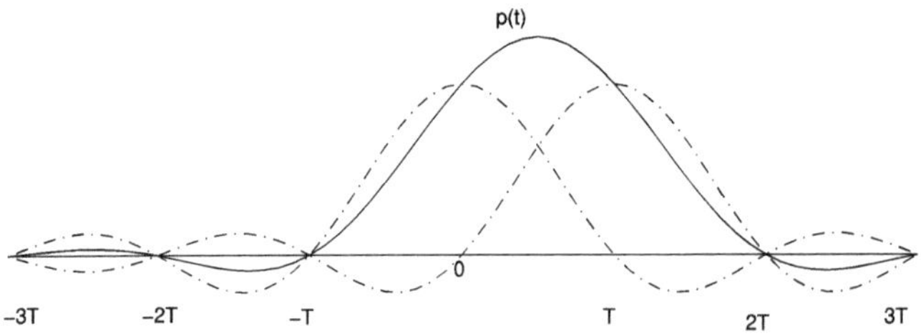

Fig. 9.6 Time domain response of duobinary partial response system

Referring to Fig. 9.4, the noiseless waveform at the output of the receiver matched filter is

$$\tilde{y}(t) = \sum_n x_n p(t - nT)$$

and the perfectly synchronized sampled output is

$$\tilde{y}_k = \tilde{y}(kT) = \sum_n x_n p((k-n)T) = \sum_n x_n p_{k-n}.$$

The sampled duobinary pulse is

$$p_n = p(nT) = \begin{cases} 1 \ , & n = 0, 1 \\ 0 \ , & n \text{ else} \end{cases} \tag{9.31}$$

Therefore,

$$\tilde{y}_k = x_k + x_{k-1}$$

Clearly, the duobinary system introduces ISI. For simplicity, assume that $x_n \in \{-1, +1\}$, in which case $\tilde{y}_k \in \{-2, 0, 2\}$.

The data sequence $\{x_k\}$ may be recovered from the sequence $\{\tilde{y}_k\}$ by

$$x_k = \tilde{y}_k - x_{k-1}$$

assuming an initial value, e.g., $x_0 = -1$ or $+1$. This mechanism is called *decision feedback* detection. Unfortunately, decision feedback detection results in error propagation when noise is present. To see this, let the decided data sequence be $\hat{x}_k = \tilde{y}_k - \hat{x}_{k-1}$. If $\hat{x}_{k-1}$ happens to be in error, then $\hat{x}_k$ is likely to be in error as well, and the errors will tend to propagate.

Error propagation can be avoided by precoding the data sequence as shown in Fig. 9.7. Precoding starts with the logical data sequence $\{b_k\}$ where and $b_k \in \{0, 1\}$. The duobinary precoded sequence is formed as

$$d_k = b_k \oplus d_{k-1} \tag{9.32}$$

where $\oplus$ indicates modulo-2 addition or a logical *exclusive or* operation. From the precoded logical sequence $\{b_k\}$, the modulation symbols are constructed as

$$x_k = 2d_k - 1$$

so that $0 \to -1$ and $1 \to +1$.

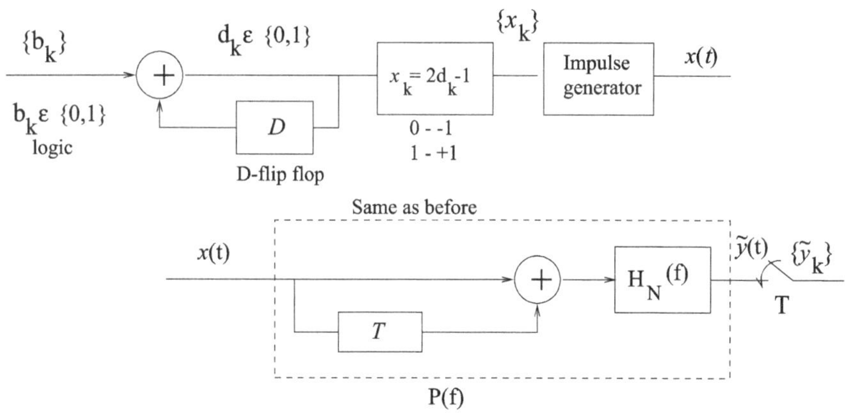

Fig. 9.7 Precoded duobinary partial response system

Example 9.2 Consider the following example of a precoded duobinary system operating on a noiseless channel, where the various sequences are shown.

$\{b_k\}$		0	0	1	1	1	0	1	0	0	0
$\{d_k\}$	1	1	1	0	1	0	0	1	1	1	1
$\{x_k\}$	+1	+1	+1	−1	+1	−1	−1	+1	+1	+1	+1
$\{\tilde{y}_k\}$		+2	+2	0	0	0	−2	0	+2	+2	+2

With precoding, once again $\tilde{y}_k \in \{-2, 0, 2\}$. By comparing the sequences $\{b_k\}$ and $\{\tilde{y}_k\}$ in the above example, the data sequence $\{b_k\}$ may be recovered from $\{\tilde{y}_k\}$ by using the simple decision rule:

$$b_k = \begin{cases} 0 \ , & \tilde{y}_k = \pm 2 \\ 1 \ , & \tilde{y}_k = 0 \end{cases}$$

Therefore, the decided data sequence $\{\hat{b}_k\}$ may be recovered from the decided received sequence $\{\tilde{y}_k\}$ by using simple *symbol-by-symbol* detection. With symbol-by-symbol detection, there is no longer any error propagation. However, the decided received sequence $\{\tilde{y}_k\}$ is ternary-valued rather than binary-valued. Hence, there is a power penalty in terms of the required E_b/N_o as compared to binary antipodal signals.

9.3.1.1 Error Probability with Duobinary Signaling

Consider the error probability of precoded duobinary signaling with symbol-by-symbol detection. The transmit and receiver filters are implemented as root-duobinary pulses, such that

$$|H_a(f)| = \sqrt{P(f)}.$$

This ensures that the receiver filter will be a matched filter to the transmitted pulse $h_a(t)$.

With AWGN, the noise process at the output of the receiver matched filter has psd

$$\Phi_{\tilde{n}\tilde{n}}(f) = N_o|H_a(f)|^2 = N_o P(f)$$

and autocorrelation function

$$\psi_{\tilde{n}\tilde{n}}(\tau) = N_o p(\tau).$$

The overall pulse $p(t)$ is not a Nyquist pulse, i.e., $p_k = p(kT) \neq \delta_{k0}$. Since $p_0 = p_1 = 1$, adjacent noise samples are correlated and, therefore, symbol-by-symbol decisions are suboptimal and will result in a loss in E_b/N_o performance.

The complex Gaussian noise samples at the output of the receiver matched filter have zero mean and variance:

$$\sigma_n^2 = N_o \int_{-\infty}^{\infty} P(f)\mathrm{d}f = N_o \int_{-1/2T}^{1/2T} 2\cos(\pi f T)\mathrm{d}f = \frac{4N_o}{\pi}.$$

Hence, the sampled outputs of the matched filter have the complex Gaussian probability density function:

$$\tilde{y}_k \sim \begin{cases} CN(\pm 2, 4N_o/\pi)\ , & b_k = 0 \\ CN(0, 4N_o/\pi) & , b_k = 1 \end{cases}$$

where the nonzero means ± 2 each occur with probability 1/4 and the zero mean 0 occurs with probability 1/2.

Assume the receiver makes minimum distance decisions according to

$$\hat{b}_k = \begin{cases} 1, & |\mathrm{Re}\{\tilde{y}_k\}| < 1 \\ 0, & |\mathrm{Re}\{\tilde{y}_k\}| > 1 \end{cases}$$

The probability of bit error is

$$P_b = \tfrac{1}{4} \cdot Q + \tfrac{1}{4} \cdot Q + \tfrac{1}{2} \cdot 2Q = \tfrac{3}{2} Q$$

where

$$Q = Q\left(\sqrt{\frac{\pi}{4N_o}}\right).$$

With $x_n \in \{-1, +1\}$ and assuming bandpass modulation, the energy per data bit is

$$E_b = \frac{1}{2}\int_{-\infty}^{\infty} |H_a(f)|^2 \mathrm{d}f = \frac{1}{2}\int_{-\infty}^{\infty} P(f)\mathrm{d}f = \frac{2}{\pi}.$$

Hence, $\frac{\pi}{2} E_b = 1$ and Q becomes

$$Q = Q\left(\sqrt{\frac{\pi}{4N_o} \cdot \frac{\pi}{2} E_b}\right) = Q\left(\sqrt{\left(\frac{\pi}{4}\right)^2 \frac{2E_b}{N_o}}\right)$$

and, finally,

$$P_b = \frac{3}{2} Q\left(\sqrt{\left(\frac{\pi}{4}\right)^2 \frac{2E_b}{N_o}}\right) \tag{9.33}$$

Recall that with binary antipodal signaling, e.g., BPSK with root raised cosine pulse shaping

$$P_b = Q\left(\sqrt{\frac{2E_b}{N_o}}\right).$$

At larger values of E_b/N_o, the factor of 3/2 multiplying the Q function in (9.33) has a negligible effect on the bit error rate performance and only the argument of the Q function is important. The extra factor of $(\pi/4)^2$ that multiplies E_b/N_o indicates a performance loss of $-10\log_{10}(\pi/4)^2 = 2.1$ dB compared to binary antipodal signaling to achieve the same bit error probability.

The advantage of duobinary signaling is that the pulse magnitude response $|H_a(f)| = \sqrt{2T \cos(\pi f T)}$ has a bandwidth of $W = R/2$ Hz, equal to that of the ideal Nyquist pulse $H_N(f)$, while having a much more gradual roll-off in the frequency domain. As before, the pulse $|H_a(f)|$ is strictly bandlimited and, hence, it is noncausal. However, the gradual roll-off of $|H_a(f)|$ makes it easier to approximate the infinite-length noncausal pulse $h_a(t)$ with a finite length-LT pulse $\hat{h}_a(t)$ as was done for the root-raised cosine pulse in Sect. 5.4.2.

9.3.2 Modified Duobinary Signaling

With a *modified duobinary* partial response system, the overall frequency domain pulse $P(f)$ can be described by the system shown in Fig. 9.8, which also shows the modified duobinary precoder. The overall frequency domain pulse is

$$\begin{aligned} P(f) &= (1 - e^{-j4\pi f T}) H_N(f) \\ &= j2e^{-j2\pi f T}\left(\frac{e^{j2\pi f T} - e^{-j2\pi f T}}{j2}\right) H_N(f) \\ &= j2H_N(f)\sin 2\pi f T e^{-j2\pi f T} \\ &= \begin{cases} 2T \sin 2\pi f T e^{j(\pi/2 - 2\pi f T)} & |f| < 1/2T \\ 0 & |f| > 1/2T \end{cases} \end{aligned} \tag{9.34}$$

The magnitude and phase response of $P(f)$ is sketched in Fig. 9.9.

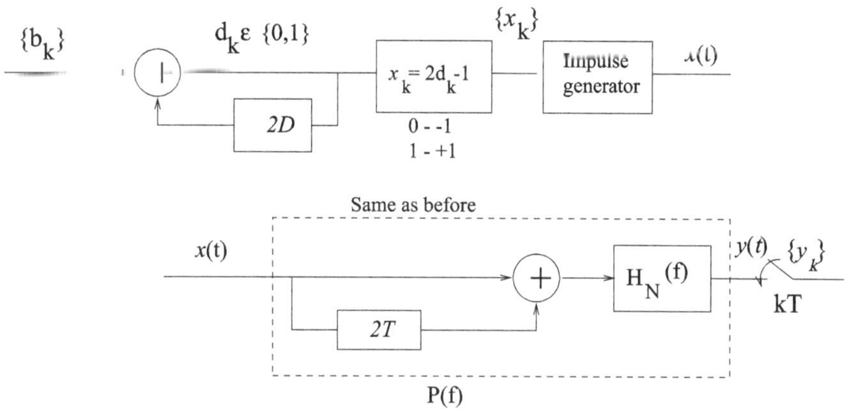

Fig. 9.8 Modified duobinary partial response system

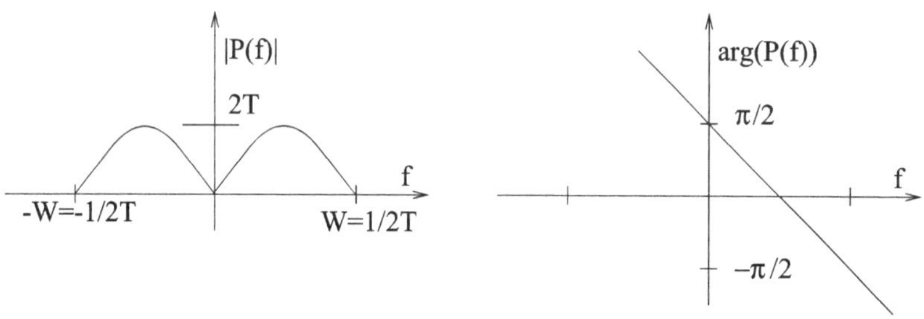

Fig. 9.9 Magnitude and phase response of modified duobinary partial response system

Using the time-shift property of the Fourier transform,

$$\left\{P(f) = H_N(f) - e^{-j4\pi f T} H_N(f)\right\} \leftrightarrow \left\{p(t) = \operatorname{sinc}\left(\frac{t}{T}\right) - \operatorname{sinc}\left(\frac{t-2T}{T}\right)\right\}. \tag{9.35}$$

The corresponding time domain pulse $p(t)$ is plotted in Fig. 9.10.

The modified duobinary precoder operates on the logical data sequence $\{b_n\}$, $b_n \in \{0, 1\}$, to produce the modified duobinary precoded sequence as

$$d_k = b_k \oplus d_{k-2}. \tag{9.36}$$

From the precoded logical sequence $\{b_k\}$, the modulation symbols are constructed as

$$x_k = 2d_k - 1$$

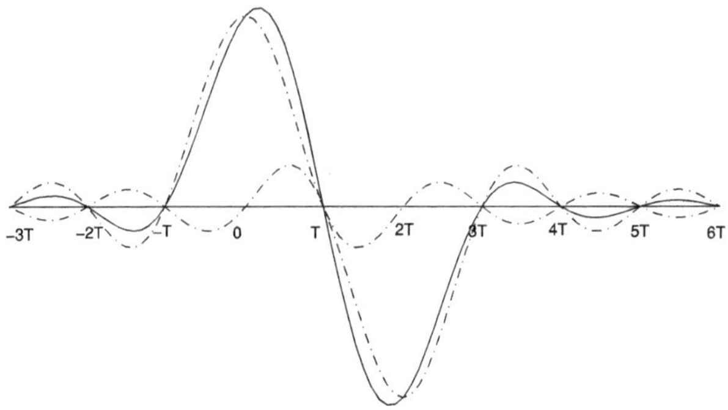

Fig. 9.10 Time domain response of modified duobinary partial response system

so that $0 \to -1$ and $1 \to +1$. At the receiver, in the absence of noise

$$\tilde{y}_k = x_k - x_{k-2}.$$

Example 9.3 Consider the following example of a modified precoded duobinary system operating on a noiseless channel, where the various sequences are shown.

$\{b_k\}$			0	0	1	1	1	0	1	0	0	0
$\{d_k\}$	1	1	1	1	0	0	1	0	0	0	0	0
$\{x_k\}$	+1	+1	+1	+1	−1	−1	+1	−1	−1	−1	−1	−1
$\{\tilde{y}_k\}$			0	0	−2	−2	2	0	−2	0	0	0

From the above example, the symbol-by-symbol decoding rule for modified duobinary is

$$b_k = \begin{cases} 1 \ , & \tilde{y}_k = \pm 2 \\ 0 \ , & \tilde{y}_k = 0 \end{cases}$$

When noise is present decisions on the $\{\hat{b}_k\}$ may be recovered from the decided received sequence $\{\hat{y}_k\}$ according to minimum distance decisions:

$$\hat{b}_k = \begin{cases} 1, & |\text{Re}\{\tilde{y}_k\}| > 1 \\ 0, & |\text{Re}\{\tilde{y}_k\}| < 1 \end{cases}$$

It is left as an exercise for the reader to show that the bit error rate performance of precoded duobinary signaling and precoded modified duobinary signaling with symbol by symbol decoding is the same.

Finally, the precoder for duorbinary and modified duobinary signaling can be viewed as a finite state machine having 2 states. In fact, a rate-1 recursive convolutional code. By using MLSE on the 2-state trellis as implemented by the Viterbi algorithm as described in Sect. 8.2.3, the 2.1 dB loss in E_b/N_o performance can be completely recovered.

Problems

9.1 Consider a binary communication system that transmits binary data bits, x_n, where $x_n \in \{-1, +1\}$ with equal probability. The overall pulse response $p(t)$ is such that

$$p_k = p(kT) = \begin{cases} 1, & k = 0 \\ 0.25, & k = 1,\ 2 \\ 0, & \text{elsewhere} \end{cases}$$

The samples at the output of the receiver matched filter are

$$\tilde{y}_n = p_0 x_n + \sum_{k \neq n} x_k p_{n-k} + \tilde{n}_n$$

where the $\tilde{n}_n$ are independent zero-mean Gaussian random variables with variance σ^2. Data decisions are made such that

$$\hat{x}_n = \begin{cases} -1\ , & \text{if } \tilde{y}_n < 0 \\ +1\ , & \text{if } \tilde{y}_n > 0 \end{cases}$$

(a) What are the possible values for the ISI term and their probabilities?
(b) Derive an expression for the probability of bit error, $P_b = P[\hat{x}_n \neq x_n]$.

9.2 Suppose that the frequency response of a channel is given by

$$\tilde{H}(f) = 1 + \alpha \cos(2\pi f t_o)$$

where $|\alpha| < 1$. An input pulse $h_a(t) = (1/\sqrt{T})u_T(t)$ is passed through the channel, yielding the output $\tilde{y}(t) = h_a(t) * \tilde{h}(t)$.

(a) In the absence of noise, show that

$$\tilde{y}(t) = h_a(t) + \frac{\alpha}{2}[h_a(t - t_o) + h_a(t + t_o)].$$

Thus, the channel produces a pair of echoes.

(b) Suppose the noiseless received signal $\tilde{y}(t)$ is passed through a filter $h_r(t)$ that is matched to $h_a(t)$. Determine the output of the matched filter at $t = kT, k = 0, \pm 1, \pm 2, \ldots$ where T is the symbol duration.

(c) What is the sampled pulse $p_k = p(kT)$, where $p(t) = h_a(t) * \tilde{h}(t) * h_r(t)$ if $t_o = T$?

9.3 Consider a binary antipodal communication system using the amplitude shaping pulse $h_a(t) = (1/\sqrt{T})u_T(t)$. The channel has the impulse response

$$\tilde{h}(t) = 0.5\delta(t - T/2) + 0.25\delta(t - T).$$

The receiver employs a filter $h_r(t)$ that is matched to the pulse $h_a(t)$. The filter output is $\tilde{y}(t)$.

(a) Determine and sketch the overall pulse $p(t)$.

(b) Does the sampled pulse $p_k = p(kT)$ result in zero ISI?

(c) What is the output of the receiver filter at time kT, i.e., what is $\tilde{y}_k = \tilde{y}(kT)$?

9.4 Consider a communication system that transmits the binary messages $\{\tilde{s}_k\}$, where $\tilde{s}_k \in \{-\sqrt{2E_b}, +\sqrt{2E_b}\}$ with equal probability using the rectangular pulse $h_a(t) = (1/\sqrt{T})u_T(t)$, where E_b is the energy per data bit. The received complex envelope is applied to a matched filter $h_r(t)$ whose output is

$$\tilde{y}(t) = \sum_n \tilde{s}_n p(t - nT) + \tilde{n}(t)$$

where $p(t) = h_a(t) * h_r(t)$, $\tilde{n}(t) = \tilde{w}(t) * h_r(t)$, and where $\tilde{w}(t)$ is complex AWGN with power spectral density N_o watts/Hz. The matched filter output $\tilde{y}(t)$ is sampled every T seconds to yield the sample sequence $\{\tilde{y}_k\}$, such that

$$\tilde{y}_k = \tilde{y}(kT + \varepsilon T) = \sum_n x_n p(kT + \varepsilon T - nT) + \tilde{n}_k$$

where εT is the *timing error* with $|\varepsilon| < 0.5$, and $\tilde{n}_k = \tilde{n}(kT + \varepsilon T)$.

(a) Sketch the filter, $h_r(t)$, matched to the transmitted pulse $h_a(t)$.

(b) What is the overall pulse $p(t) = h_a(t) * h_r(t)$?

(c) Due to the presence of a timing offset, ε, intersymbol interference (ISI) is introduced into the samples $\tilde{y}_k$. Obtain an expression for the samples $\tilde{y}_k$ in terms of the sampled pulse $p_k = p(kT + \varepsilon T)$ and the noise samples $\tilde{n}_k$.

(d) Using your results from part (c) derive an expression for the probability of bit error in terms of the bit energy-to-noise ratio, E_b/N_0.

9.5 Suppose that the overall channel vector $\mathbf{p} = \{p_k, k = 0, 1, \ldots\}$ is

$$\mathbf{p} = \{0.1, -0.5, 2.1, 0.8, 0.5 - .1\}$$

and the remaining values of $p_k = 0$.

(a) Design a 3-tap zero-forcing equalizer.
(b) What is the overall pulse $\mathbf{q}$?
(c) What is the distortion before equalization D_p and after equalization D?

9.6 A communication channel is tested for ISI. The overall pulse is observed to be

$$p(t) = \begin{cases} 1 - \frac{|t-2T|^2}{2T} & , \quad 0 \le t \le 4T \\ 0 & , \quad \text{elsewhere} \end{cases}$$

Assuming that the zero-forcing solution minimizes the peak distortion, design a 3-tap zero-forcing equalizer to equalize the channel response.

9.7 Consider the partial response system shown in Fig. 9.11.

(a) What is the overall pulse response $P(f)$? Sketch the magnitude and phase response of $P(f)$ showing all important values.
(b) If matched filtering is used at the receiver, what is $|H_a(f)|$ and $|H_r(f)|$, the magnitude responses of the transmitter and receiver filter, respectively?

9.8 Consider the partial response system shown in Fig. 9.12.

(a) What is the overall pulse response $P(f)$? Sketch the magnitude and phase response of $P(f)$ showing all important values.

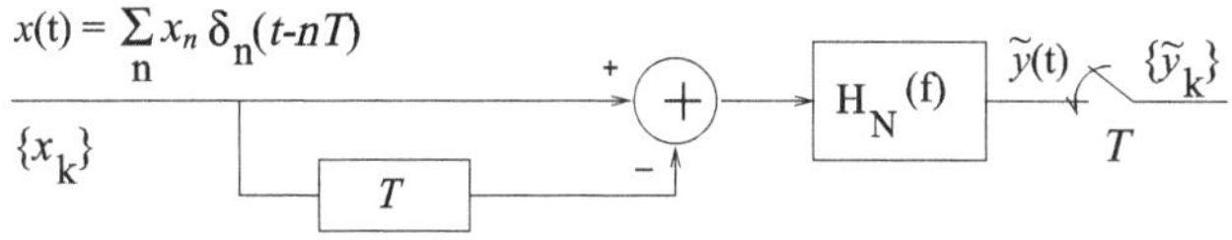

Fig. 9.11 Partial response system for Problem 9.7

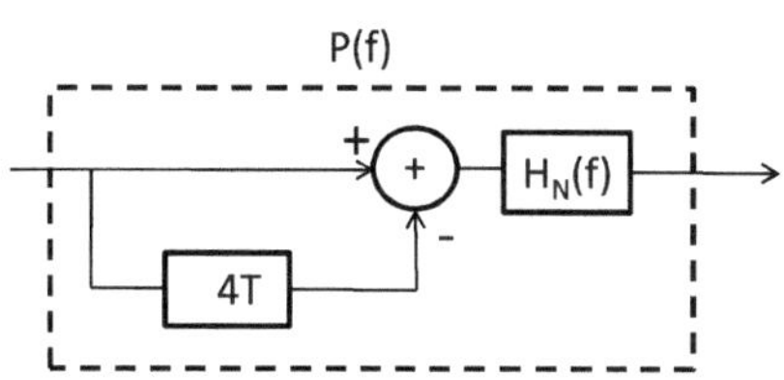

Fig. 9.12 Partial response system for Problem 9.8

(b) What is the corresponding time domain pulse $p(t)$? Sketch it showing all important values.

(c) Using $P(f)$ from part (a) obtain the frequency response of the transmit filter $H_a(f)$ and receiver matched filter $H_r(f)$.

Chapter 10
Orthogonal Frequency Division Multiplexing

Modulation techniques may be classified as either *single-carrier* that employ a single RF carrier or *multicarrier* where information is transmitted in parallel by using multiple subcarriers. Orthogonal frequency division multiplexing (OFDM) is a well-known multicarrier modulation technique. The great advantage of OFDM and other multicarrier schemes is their ease of implementation with common DSP processing functions such as the Fast Fourier Transform (FFT) and its inverse (IFFT). Moreover, equalization is simple when the channel exhibits ISI, since equalization is done in the frequency domain rather than in the time domain. Finally, the maximum likelihood receiver is straightforward to implement in digital baseband with simple processing.

This chapter begins with the basic description of OFDM signals in Sect. 10.1. Section 10.2 shows how the OFDM baseband modulator can be efficiently implemented by using an inverse discrete-time Fourier transform (IDFT). Likewise, Sect. 10.3 shows that the OFDM baseband demodulator can be implemented as a discrete Fourier transform (DFT). Section 10.4 considers the error probability analysis of OFDM waveforms on flat fading channels with AWGN. Section 10.5 considers the performance of OFDM on frequency selective channels with AWGN.

10.1 OFDM Waveforms

OFDM is a block modulation scheme where data symbols are transmitted in parallel on orthogonal subcarriers. A serial data stream of data symbols with rate $R_s = 1/T_s$ is passed through a serial-to-parallel converter to create blocks of N data symbols, called OFDM symbols, that have duration $T = NT_s$. The N parallel data symbols

G. Stuber, *Introductory Digital Wireless Communications*,
https://doi.org/10.1007/978-3-032-01559-4_10

modulate N subcarriers that are spaced in frequency $1/T$ Hz apart. The OFDM complex envelope is given by

$$\tilde{s}(t) = \sum_n b(t - nT, \mathbf{x}_n), \tag{10.1}$$

where

$$b(t, \mathbf{x}_n) = \sum_{k=0}^{N-1} x_{n,k} \exp\left\{ j\frac{2\pi kt}{T} \right\} u_T(t) \tag{10.2}$$

is the nth OFDM symbol, n is the block or OFDM symbol index, k is the subcarrier index, N is the number of subcarriers, and $\mathbf{x}_n = \{x_{n,0},\ x_{n,1},\ \ldots,\ x_{n,N-1}\}$ is the nth data symbol block. The amplitude shaping on each of the subcarriers is the rectangular pulse $u_T(t)$.

With OFDM, the data symbols $x_{n,k}$ are often chosen from a QAM signal constellation, although other signal constellations such as PSK may be used. In the case of QAM, the data symbols may be defined as

$$x_{n,k} = \sqrt{\frac{2E_o}{T}} A_{n,k}.$$

where

$$A_m \triangleq A_m^c + jA_m^s$$

and with a square constellation

$$A_m^{\{c,s\}} \in \{\pm 1, \pm 3, \pm 5, \ldots \pm (\sqrt{M} - 1)\}$$

where $2E_o$ is the bandpass energy of the message with the lowest amplitude, i.e., when $A_m^c, A_m^s = \pm 1$.

The subcarriers are spaced $1/T$ Hz apart so the individual complex envelopes transmitted on the N subcarriers are mutually orthogonal regardless of the subcarrier phases that are imparted by the random data symbols $x_{n,k}$ as shown in Problem 5.8. If it happens that the data symbols are all real-valued or all imaginary-valued, then the subcarriers may be spaced $1/2T$ Hz apart.

A cyclic extension (or guard interval) is added to the OFDM waveform in (10.1) and (10.2) to combat intersymbol interference (ISI) as described in the sequel. The cyclic extension can be implemented as either a cyclic prefix or a cyclic suffix or both. With a cyclic suffix, the first $\alpha_g T$ seconds of each T-second OFDM symbol is appended to the end of each OFDM symbol, such that the OFDM complex envelope becomes

$$\tilde{s}_g(t) = \begin{cases} \tilde{s}(t) & , \quad 0 \le t < T \\ \tilde{s}(t-T) & , \quad T \le t \le T + \alpha_g T \end{cases} . \tag{10.3}$$

Hence, with a cyclic suffix

$$\tilde{s}_g(t) = \sum_n b(t - nT_g, \mathbf{x}_n), \tag{10.4}$$

where

$$b(t, \mathbf{x}_n) = \sum_{k=0}^{N-1} x_{n,k} \exp\left\{ j\frac{2\pi kt}{T} \right\} u_T(t) + \sum_{k=0}^{N-1} x_{n,k} \exp\left\{ j\frac{2\pi k(t-T)}{T} \right\} u_{\alpha_g T}(t-T), \tag{10.5}$$

and $T_g = (1+\alpha_g)T$ is the OFDM symbol period with addition of the guard interval of length $\alpha_g T$. With a cyclic suffix, the last $\alpha_g T$ seconds of each OFDM symbol waveform $b(t, \mathbf{x}_n)$ is exactly the same as the first $\alpha_g T$ seconds.

With a cyclic prefix, the last $\alpha_g T$ seconds of each T-second OFDM symbol is appended to the beginning of each OFDM symbol, such that the OFDM complex envelope becomes

$$\tilde{s}_g(t) = \begin{cases} \tilde{s}(t+T) & , \quad -\alpha_g T \le t < 0 \\ \tilde{s}(t) & , \quad 0 \le t < T \end{cases} . \tag{10.6}$$

Hence, with a cyclic prefix, the OFDM complex envelope is given by (10.4) with

$$\begin{aligned} b(t, \mathbf{x}_n) = & \sum_{k=0}^{N-1} x_{n,k} \exp\left\{ j\frac{2\pi k(t+T)}{T} \right\} u_{\alpha_g T}(t + \alpha_g T) \\ & + \sum_{k=0}^{N-1} x_{n,k} \exp\left\{ j\frac{2\pi kt}{T} \right\} u_T(t). \end{aligned} \tag{10.7}$$

With a cyclic prefix, the first $\alpha_g T$ seconds of each OFDM symbol waveform $b(t, \mathbf{x}_n)$ is exactly the same as the last $\alpha_g T$ seconds.

10.2 DFT-Based OFDM System

A key advantage of using OFDM is that the baseband modulator and demodulator can be implemented by using an inverse discrete-time Fourier transform (IDFT) and discrete Fourier transform (DFT), respectively. When the number of subcarriers N

is a power of 2, the computationally efficient fast Fourier transform (FFT) and its inverse (IFFT) may be used.

Consider the OFDM complex envelope defined by (10.1) and (10.2). The nth OFDM symbol, transmitted during time interval $nT \le t < (n+1)T$, has complex envelope

$$\tilde{s}(t) = \sum_{k=0}^{N-1} x_{n,k} \exp\left\{\frac{j2\pi k(t-nT)}{T}\right\} u_T(t-nT), \quad nT \le t < (n+1)T. \tag{10.8}$$

Now suppose that the complex envelope in (10.8) is sampled at T_s second intervals beginning at time nT to yield the sample sequence

$$X_{n,m} = \tilde{s}(nT + mT_s) = \sum_{k=0}^{N-1} x_{n,k} \exp\left\{\frac{j2\pi km}{N}\right\}, \quad m = 0,\ 1,\ \ldots,\ N-1. \tag{10.9}$$

Observe that the vector $\mathbf{X}_n = \{X_{n,m}\}_{m=0}^{N-1}$ is the IDFT of the data symbol vector $\mathbf{x}_n = \{x_{n,k}\}_{k=0}^{N-1}$. Contrary to conventional notation, however, the lower case vector of data symbols $\mathbf{x}_n$ represents the frequency-domain coefficients, while the upper case vector $\mathbf{X}_n$ represents the time-domain coefficients.

Cyclic Guard Insertion

A cyclic prefix or suffix extension is added to each $\mathbf{X}_n$. With the addition of a cyclic suffix, the nth time-domain OFDM symbol $\mathbf{X}_n^g = \{X_{n,m}^g, m = 0, \ldots N+G-1\}$ becomes

$$X_{n,m}^g = X_{n,(m)_N} \tag{10.10a}$$

$$= \sum_{k=0}^{N-1} x_{n,k} \exp\left\{\frac{j2\pi km}{N}\right\}, \quad m = 0,\ 1,\ \ldots,\ N+G-1, \tag{10.10b}$$

where G is the length of the guard interval, and $(m)_N$ is the residue of m modulo N, equal to the integer remainder after dividing m by N. This yields the length $N+G$ vector $\mathbf{X}_n^g = \{X_{n,m}^g\}_{m=0}^{N+G-1}$ where the first and last G coordinates are the same.

With the addition of a cyclic prefix, the nth time-domain OFDM symbol becomes

$$X_{n,m}^g = X_{n,(m)_N} \tag{10.11a}$$

$$= \sum_{k=0}^{N-1} x_{n,k} \exp\left\{\frac{j2\pi km}{N}\right\}, \quad m = -G,\ \ldots, -1,\ 0,\ 1,\ \ldots,\ N-1. \tag{10.11b}$$

This yields the vector $\mathbf{X}_n^g = \{X_{n,m}^g\}_{m=-G}^{N-1}$ where, again, the first and last G coordinates of the vector $\mathbf{X}_n^g$ are the same.

The OFDM complex envelope $\tilde{s}(t) = \tilde{s}_I(t) + j\tilde{s}_Q(t)$ can be generated by splitting the complex-valued vector $\mathbf{X}_n^g$ into its real and imaginary parts, $\mathrm{Re}(\mathbf{X}_n^g)$ and $\mathrm{Im}(\mathbf{X}_n^g)$, respectively. The sequences $\{\mathrm{Re}(X_{n,m})\}$ and $\{\mathrm{Im}(X_{n,m})\}$ are then input to a pair of digital-to-analog (D/A) converters to generate the real and imaginary components of the complex envelope $\tilde{s}_I(t)$ and $\tilde{s}_Q(t)$, respectively, during the time interval $nT \leq t < (n+1)T$. So that the OFDM modulator will keep up with the digital source that produces message symbols with period T_s, the clock period of the D/A converter pair after insertion of the guard interval, T_s^g, must be reduced such that

$$T_s^g = \frac{N}{N+G} T_s . \tag{10.12}$$

The overall OFDM baseband modulator is shown in Fig. 10.1.

10.2.1 Discrete-Time Continuous-Time Equivalence?

The complex envelope that is generated with the IDFT implementation of the OFDM baseband modulator will not be *not exactly* the same as the continuous-time OFDM complex envelope that is sampled. Consider the OFDM waveform without a cyclic guard in (10.1) and (10.2). The amplitude shaping used on each subcarrier is strictly time-limited to T seconds, here the rectangular pulse $h_a(t) = u_T(t)$. Hence, the per subcarrier psd is not band-limited and no sampling rate will satisfy the Shannon sampling theorem, and some aliasing will exist. With rectangular pulse shaping the persubcarrier psd $\propto |H_a(f)|^2 = T^2 \mathrm{sinc}^2(fT)$ which has high sidelobes. The out-of-band emissions can be reduced by time-domain windowing

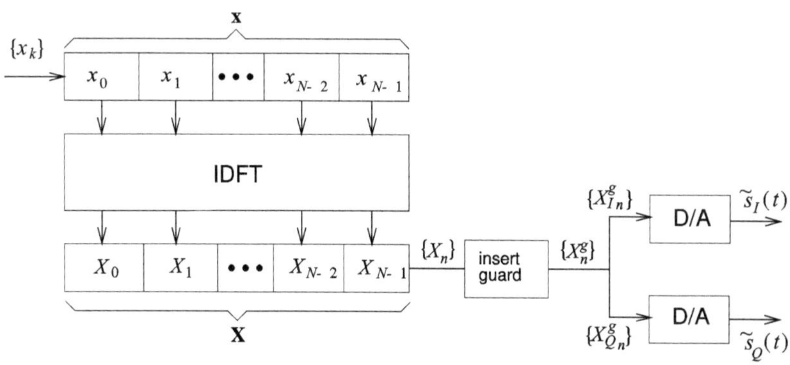

Fig. 10.1 Block diagram of IDFT-based baseband OFDM modulator with guard interval insertion and digital-to-analog conversion

the OFDM symbols with something other than a rectangular window, for example, with a time-domain raised cosine [28]. However, side lobes are inherent in the continuous-time OFDM waveform due to the subcarrier pulse shaping with a time-limited pulse.

The IDFT OFDM baseband modulator applies a sequence of complex-valued time-domain samples to a pair of D/A converters as in Fig. 10.1. The perfect digital-to-analog (D/A) converter is an ideal low-pass filter having cutoff frequency equal to the Nyquist frequency $1/(2T_s)$ and having non-causal impulse response $h(t) = \text{sinc}(t/T_s)$. Without a guard interval ($G = 0$), the time-domain samples $X_{n,m}$ are all uncorrelated and the psd of the OFDM complex envelope with a perfect reconstruction filter $\propto |H(f)|^2 = T_s^2\text{rect}(fT_s)$. This is identical to the power spectrum of single-carrier modulation with an ideal Nyquist shaping pulse $h_N(t)$ and baud rate $R_s = 1/T_s$. Since the perfect (D/A) converter is non-causal, a practical a reconstruction filter must be used instead. In this case, out-of-band or side lobes emissions are produced by the non-ideal (D/A) converter.

10.3 DFT-Based OFDM Baseband Demodulator

Suppose that the discrete-time sequence $\mathbf{X}_n^g = \{X_{n,m}^g\}_{m=0}^{N+G-1}$ is passed through a balanced pair of digital-to-analog (D/A) converters, as shown in Fig. 10.1, and the resulting complex envelope is transmitted over an ideal channel with complex gain $\tilde{h} = \alpha e^{j\phi}$. For ideal channels, a cyclic guard interval is unnecessary, but is included here anyway to handle the more general case of ISI channels later.

The receiver first uses a quadrature demodulator to extract the received complex envelope $\tilde{r}(t) = \tilde{r}_I(t) + j\tilde{r}_Q(t)$. The quadrature components $\tilde{r}_I(t)$ and $\tilde{r}_Q(t)$ are each passed through an anti-aliasing filter. For a sampling rate $R_s^g = 1/T_s^g$, the perfect anti-aliasing filter is an ideal low-pass filter having a cutoff frequency $1/(2T_s^g)$. The anti-aliasing filter outputs are input to a pair of analog-to-digital (A/D) converters that are clocked with rate R_s^g as shown in Fig. 10.2. This produces the received complex-valued discrete-time sample sequence $\mathbf{R}_n^g = \{R_{n,m}^g\}_{m=0}^{N+G-1}$, where

$$R_{n,m}^g = \tilde{h}X_{n,m}^g + \tilde{n}_{n,m}, \tag{10.13}$$

and the $\tilde{n}_{n,m}$ are the complex-valued Gaussian noise samples. For an ideal anti-aliasing filter having a cutoff frequency $1/(2T_s^g)$, Sect. 4.4.2 shows that the $\tilde{n}_{n,m}$ are i.i.d. zero-mean complex Gaussian random variables with variance

$$\sigma_{\tilde{n}}^2 = \frac{1}{2}\text{E}[|\tilde{n}_{n,m}|^2] = \frac{N_o}{T_s^g} = \frac{N+G}{N}\frac{N_o}{T_s}. \tag{10.14}$$

Once the nth received OFDM symbol $\mathbf{R}_n^g$ has been obtained, the next stage is guard interval removal.

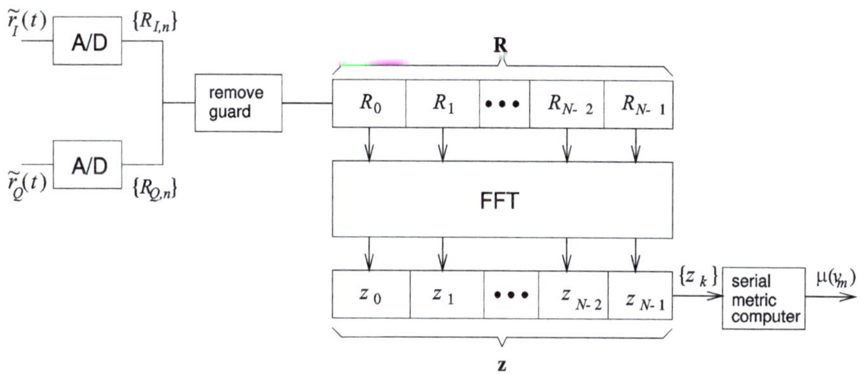

Fig. 10.2 Block diagram of OFDM receiver

Cyclic Guard Removal

Addition and removal of a cyclic guard is essential for OFDM on ISI channels, so a length-G cyclic suffix is assumed as discussed in Sect. 10.2. The cyclic guard is removed from the received OFDM symbol according to

$$R_{n,m} = R^g_{n,G+(m-G)_N}, \; 0 \le m \le N-1, \tag{10.15}$$

where $(m)_N$ is the residue of m modulo N. This replaces the first G coordinates with the last G coordinates in each received OFDM symbol $\mathbf{R}^g_n = \{R^g_{n,m}\}_{m=0}^{N+G-1}$. For an ideal channel, the first and last G coordinates of $\mathbf{R}^g_n$ are identical, so the cyclic guard interval serves no useful purpose.

In case of a cyclic prefix, the receiver removes the guard interval by simply deleting the first G samples in each received OFDM symbol $\mathbf{R}^g_n$ according to

$$R_{n,m} = R^g_{n,m}, \; 0 \le m \le N-1. \tag{10.16}$$

DFT Baseband Demodulation

After guard interval removal, baseband demodulation is performed by taking the DFT or FFT of the time-domain vector $\mathbf{R}_n = \{R_{n,m}\}_{m=0}^{N-1}$ to yield the frequency-domain vector $\mathbf{z}_n = \{z_{n,k}\}_{k=0}^{N-1}$

$$\begin{aligned} z_{n,k} &= \frac{1}{N}\sum_{m=0}^{N-1} R_{n,m}\exp\left\{-\frac{j2\pi km}{N}\right\} \\ &= \tilde{h}x_{n,k} + \nu_{n,k}\,, \quad k = 0,\ldots,N-1, \end{aligned} \tag{10.17}$$

where

$$\nu_{n,k} = \frac{1}{N}\sum_{m=0}^{N-1} \tilde{n}_{n,m}\exp\left\{-\frac{j2\pi km}{N}\right\}, \quad k = 0, \ldots, N-1. \tag{10.18}$$

10.4 Error Probability of OFDM

Exploiting the circular symmetry of the complex Gaussian distribution and expressing the DFT and IDFT operations as unitary matrices, the $\nu_{n,k}$ are zero-mean complex Gaussian random variables with covariance

$$\phi_{j,k} = \tfrac{1}{2}\mathrm{E}[\nu_{n,j}\nu_{n,k}^*] = \frac{N_o}{NT_s^g}\delta_{jk}. \tag{10.19}$$

Hence, the $z_{n,k}$ are independent complex Gaussian random variables with mean $\tilde{h}x_{n,k}$ and variance N_o/NT_s^g. To be meaningful with our earlier error probability results in Sect. 6.3.3, the $z_{n,k}$ may be multiplied by the scalar constant $\sqrt{NT_s^g}$. Such scaling gives

$$\tilde{z}_{n,k} = \tilde{h}\sqrt{N/(N+G)}x_{n,k} + \tilde{\nu}_{n,k}, \tag{10.20}$$

where the $\tilde{\nu}_{n,k}$ are i.i.d. zero-mean complex Gaussian random variables with variance N_o. Comparing with the results in Sect. 6.3.3 the extra term $\sqrt{N/(N+G)}$ in front of the message symbol means a loss in effective symbol energy or symbol energy-to-noise ratio by a factor equal to $N/(N+G)$, due to the cyclic guard interval, which, on an ideal channel, serves no purpose. The error rate performance is otherwise identical to that with single-carrier modulation.

For *each* of the $\tilde{z}_{n,k}$, the OFDM receiver decides in favor of the message $x_{n,k}$ that minimizes the squared Euclidean distance

$$\mu_1(x_{n,k}) = \|\tilde{z}_{n,k} - \tilde{h}\sqrt{N/(N+G)}x_{n,k}\|^2, \quad k = 0, \ldots, N-1, \tag{10.21}$$

or alternatively maximizes the metric $\mu_2(x_{n,k})$ in (6.22). Thus, for each OFDM block, N symbol decisions are made, one for each of the N subcarriers. It is apparent from (10.20) that the probability of symbol error is identical to that achieved with independent modulation on each of the subcarriers, except for the loss of signal energy $N/(N+G)$ due to the cyclic guard interval. This is expected, because the waveforms sent on the OFDM subcarriers are mutually orthogonal.

10.5 Frequency-Domain Equalization

Consider a linear time-invariant channel having impulse response $\tilde{h}(t)$ that is a nonideal channel. Being a nonideal channel, $\tilde{h}(t)$ will have a nonzero finite length, called the *delay spread*. As shown in Fig. 10.3, the serial concatenation of the D/A converter, waveform channel $\tilde{h}(t)$, anti-aliasing filter, and A/D converter constitutes an overall discrete-time channel having a sampled impulse response $\tilde{\mathbf{h}} = \{\tilde{h}_m\}_{m=0}^{L}$, where L is the length of the discrete-time channel impulse response. Note that channels $\tilde{h}(t)$ with a small delay spread can have a surprisingly large value of L due to the length of the reconstruction pulse.

If the number of subcarriers N is chosen such that $N \gg L$, the ISI will span only a small fraction of each time-domain OFDM symbol. Viewed in the frequency domain, the subcarriers are spaced $1/T = 1/(NT_s)$ Hz apart. Hence, as N becomes large, the channel transfer function $\tilde{H}(f)$ can be modeled as being piecewise flat across sub-bands of width $1/T$ Hz that become increasingly narrow with N. That is,

$$\tilde{H}(f) \doteq \tilde{H}(k\Delta_f), \quad k\Delta_f - 1/(2T) \leq f \leq k\Delta_f + (1/(2T), \tag{10.22}$$

Weinstein and Ebert [34] proposed an ingenious solution to mitigate the effects of ISI, whereby a cyclic guard interval is appended in the form of a length-G cyclic prefix or suffix to each IDFT output vector $\mathbf{X}_n$ to yield the vector $\mathbf{X}_n^g$ as described earlier. If the length of the guard interval G is chosen such that $G \geq L$, then the ISI can be completely removed in a very efficient manner as is now described.

Suppose the OFDM symbols $\mathbf{X}_n^g$ have a length-G cyclic suffix. The discrete-time *linear* convolution of the transmitted time-domain sequence $\{\mathbf{X}_n^g\}$ with the discrete-time channel $\tilde{\mathbf{h}}$ produces the discrete-time received sequence $\{\mathbf{R}_n^g\}$, where

$$R_{n,m}^g = \begin{cases} \sum_{i=0}^{m} \tilde{h}_i X_{n,m-i}^g \\ \quad + \sum_{i=m+1}^{L} \tilde{h}_i X_{n-1,N+G+m-i}^g + \tilde{n}_{n,m} & , \; 0 \leq m < L \\ \sum_{i=0}^{L} \tilde{h}_i X_{n,m-i}^g + \tilde{n}_{n,m} & , \; \leq m \leq N+G-1 \end{cases} . \tag{10.23}$$

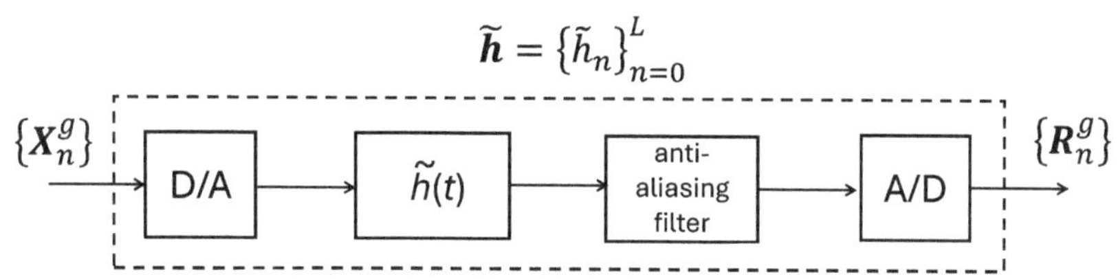

Fig. 10.3 The discrete-time channel

The sample range $0 \leq m < L$ represents the coordinates where the nth received OFDM symbol $\mathbf{R}_n^g$ is corrupted with ISI from the $(n-1)$th OFDM symbol, as given by the second summation in (10.23). To remove the ISI introduced by the channel, the first G received samples of each received OFDM symbol $\mathbf{R}_n^g$ are discarded and replaced with the last G received samples of $\mathbf{R}_n^g$, as shown in Fig. 10.4. If the guard interval length satisfies $G \geq L$, this can be shown to give the received sequence

$$\begin{aligned} R_{n,m} &= R^g_{n,G+(m-G)_N} \\ &= \sum_{i=0}^{L} \tilde{h}_i X_{n,(m-i)_N} + \tilde{n}_{n,(m-i)_N} \; , \quad 0 \leq m \leq N-1. \end{aligned} \tag{10.24}$$

The first term in (10.24) is the *circular* convolution of the transmitted sequence $\mathbf{X}_n$ with the discrete-time channel $\tilde{\mathbf{h}}$. The addition and removal of the cyclic guard interval are seen to convert a linear convolution of the transmitted discrete-time OFDM symbol with the discrete-time channel into a circular convolution. For the discrete Fourier transform (DFT), a circular convolution in the time domain corresponds to multiplication in the frequency domain and vice versa.

As shown in Fig. 10.2, the OFDM baseband demodulator computes the DFT of the vector $\mathbf{R}_n$. The DFT output vector is

$$\begin{aligned} z_{n,i} &= \frac{1}{N} \sum_{m=0}^{N-1} R_{n,m} \exp\left\{-j\frac{2\pi mi}{N}\right\} \\ &= \tilde{H}_i x_{n,i} + \nu_{n,i} \; , \quad 0 \leq i \leq N-1, \end{aligned} \tag{10.25}$$

where

$$\tilde{H}_i = \sum_{m=0}^{L} \tilde{h}_m \exp\left\{-j\frac{2\pi mi}{N}\right\} \tag{10.26}$$

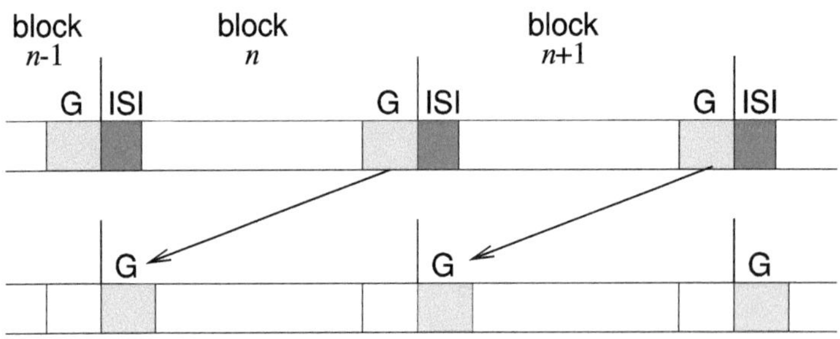

Fig. 10.4 Removal of ISI by using the cyclic suffix

and the noise samples $\{\nu_{n,i}\}$ are i.i.d circularly symmetric complex Gaussian random variables having zero-mean and variance $N_o/(NT_s^g)$.

The vector $\mathbf{H} = \{H_i\}_{i=0}^{N-1}$ is the sampled frequency response of the channel and may be obtained by zero padding the length-L channel vector $\tilde{\mathbf{h}}$ to length N and taking its N-point DFT. Again, to be consistent with our earlier results in Chap. 6, the $z_{n,i}$ may be multiplied by the scalar $\sqrt{NT_s^g}$. Following the same argument used in Sect. 10.4, such scaling gives

$$\tilde{z}_{n,i} = \tilde{H}_i \sqrt{N/(N+G)} x_{n,i} + \tilde{\nu}_{n,i} \quad i = 0, \ldots, N-1, \tag{10.27}$$

where the $\tilde{\nu}_{n,i}$ are i.i.d. zero-mean circularly symmetric complex Gaussian random variables with variance N_o. Observe that each $\tilde{z}_{n,i}$ depends only on the corresponding data symbol $x_{n,i}$ sent on that subcarrier and, therefore, the ISI has been completely removed. Compared to the $\tilde{z}_{n,i}$ for an ideal channel with complex gain $\tilde{h}$ in (10.20) the only difference is that each subcarrier experiences a different complex gain $\tilde{H}_i$. Finally, for *each* of the $\tilde{z}_{n,i}$, the receiver decides in favor of the message $x_{n,i}$ that minimizes the squared Euclidean distance

$$\mu_1(x_{n,i}) = \|\tilde{z}_{n,i} - \tilde{H}_i \sqrt{N/(N+G)} x_{n,i}\|^2 \tag{10.28}$$

or alternatively maximizes the metric $\mu_2(x_{n,i})$ in (6.22). The symbol error rate performance is otherwise the same as discussed in Sect. 6.3.3.

Problems

10.1 Consider the time-domain sample sequence for the nth OFDM block

$$X_{n,m} = \sum_{k=0}^{N-1} x_{n,k} \exp\left\{ j \frac{2\pi km}{N} \right\}.$$

The data symbols $x_{n,k},\ k = 0, \ldots, N-1$, are independent and each is chosen with equal probability from a BPSK symbol alphabet, such that $x_{n,k} \in \{-1, +1\}$ with equal probability.

OFDM waveforms are highly susceptible to envelope clipping caused by amplifier saturation. The peak-to-average power ratio (PAPR) is the required backoff to keep power amplifier operation in its linear range and avoid saturation. The PAPR of the sample sequence for block n can be defined as follows:

$$\text{PAPR} = \frac{\max_m |X_{n,m}|^2}{N^{-1} \sum_{m=0}^{N-1} |X_{n,m}|^2}.$$

By using Parseval's theorem and the triangle inequality, show that PAPR $\leq N$.

10.2 OFDM systems are known to be resilient to timing errors. Suppose that the OFDM complex envelope

$$\tilde{s}(t) = \sum_{k=0}^{N-1} x_k \exp\left\{ j\frac{2\pi kt}{T} \right\} u_t(t)$$

is sampled at time instants $t = nT_s + \Delta_t$, where Δ_t is a timing offset, to yield the samples

$$R_n = \tilde{s}(nT_s + \Delta_t) \ , \quad n = 0, \ldots, N-1$$

and an N-point DFT is taken on the samples $\{R_n\}_{n=0}^{N-1}$ to yield the coefficients $\{z_k\}_{k=0}^{N-1}$.

Assume that the timing offset Δ_t lies somewhere in the OFDM guard interval such that all N DFT coefficients belong to the same OFDM symbol. Determine the FFT coefficients $\{z_k\}_{k=0}^{N-1}$.

10.3 Consider the transmission of binary antipodal symbols using OFDM with $N = 4$ sub-carriers. The the data vector for the nth OFDM symbol is

$$\mathbf{x}_n = \{x_{n0}, x_{n1}, x_{n2}, x_{n3}\} = \{+1, +1, +1, -1\}$$

(a) Determine the corresponding time-domain coefficients

$$\mathbf{X}_n = \{X_{n0}, X_{n1}, X_{n2}, X_{n3}\}$$

(b) If a length $G = 2$ cyclic prefix is used, what discrete time-domain sequence is sent during the nth OFDM symbol?

(c) Determine the peak-to-average power ratio for the nth block, given by

$$\text{PAPR}_n = \frac{\max_m |X_{n,m}|^2}{\frac{1}{N+G}\sum_{m=0}^{N+G-1} |X_{n,m}|^2}.$$

Express your result in dB units.

10.4 Consider the transmission of binary antipodal symbols using OFDM with $N = 4$ subcarriers. The data vector for the nth OFDM symbol is

$$\mathbf{x}_n = \{x_{n0}, x_{n1}, x_{n2}, x_{n3}\} = \{-1, +1, +1, -1\}$$

(a) Determine the corresponding time-domain coefficients

$$\mathbf{X}_n = \{X_{n0}, X_{n1}, X_{n2}, X_{n3}\}$$

(b) If a length $G = 4$ cyclic guard is used, what time-domain sequence is sent during the nth OFDM symbol?

(c) Following part (b), determine the peak-to-average power ratio for the nth block, PAPR_n given by

$$\text{PAPR}_n = \frac{\max_m |X_{n,m}|^2}{\frac{1}{N+G}\sum_{m=0}^{N+G-1} |X_{n,m}|^2}.$$

Express your result in dB units.

(d) Assuming $N = G = 4$, specify a binary data vector that will maximize PAPR_n, and what is the maximum PAPR_n in dB units? In case of multiple solutions, one solution will do.

(e) Assuming $N = G = 4$, specify a binary data vector that will minimize PAPR_n, and what is the minimum PAPR_n in dB units? In case of multiple solutions, one solution will do.

10.5 Consider the transmission of independent and equally likely binary data bits drawn from the alphabet $\{-1, +1\}$, using OFDM with $N = 4$ sub-carriers and a length $G = 2$ cyclic suffix. The data symbol vector for the nth OFDM symbol is equal to

$$\mathbf{x}_n = \{x_{n0}, x_{n1}, x_{n2}, x_{n3}\} = \{+1, +1, -1, -1\}$$

(a) What time-domain sequence will be transmitted for the nth OFDM symbol?

(b) The discrete-time sequence in part a) is sent over a length $L = 2$ linear time-invariant ISI channel having discrete-time impulse response $\tilde{\mathbf{h}} = \{\tilde{h}_m\}_{m=0}^{L} = (\sqrt{3}/2,\ 0,\ 1/2)$. What are the frequency-domain coefficients of the channel $\tilde{H}_i, i = 0, \ldots, 3$?

(c) The receiver makes minimum distance decisions on the data bits. What is the probability of bit error as a function of the bit energy-to-noise ratio E_b/N_o when averaged over the four subcarriers?

Chapter 11
Land Mobile Radio Channels

The design of spectrally efficient wireless communication systems requires a thorough understanding of the radio propagation channel. This chapter emphasizes land mobile radio channels, particularly those found in cellular land mobile radio systems. The characteristics of the radio channel will vary greatly depending on the operating frequency, and the propagation environment, e.g., line-of-sight (LoS) versus non-line-of-sight (NLoS), stationary versus mobile transmitters, receivers and scatterers, and other factors. However, many of the concepts are of a fundamental nature and will apply, with appropriate modification, to other types of mobile radio channels as well, such as vehicle-to-vehicle and drone communication channels.

A cellular land mobile radio network consists of a collection of fixed base stations (BSs) that define radio coverage areas known as cells. The height and placement of the BS antennas affect the proximity of local scatterers at the BSs. In a macrocellular environment where the cell radii are large, the BS antennas are well elevated above the local terrain and are free of local scatterers. Mobile stations (MSs), on the other hand, are typically surrounded by local scatterers due to their low elevation antennas. Sometimes, a line-of-sight (LoS) condition will exist between a BS and an MS, for example, in a rural (or open) environment. However, in a typical urban or suburban environment, the BSs and MSs are usually in a non-line-of-sight (NLoS) condition, and the radio waves propagate between the BSs and MSs via reflections and diffractions. Due to the typically large distance between the BSs and MSs in cellular systems, radio propagation is often modeled in a two-dimensional (2-D) plane. As shown in Fig. 11.1, multiple plane waves will arrive at the MS (or BS) receiver antenna(s) from different directions, with each having a distinct amplitude, phase, delay, and Doppler shift. This phenomenon is called multipath propagation. The multipath components combine vectorially at the target MS (or BS) receiver antenna to produce a composite received signal.

G. Stuber, *Introductory Digital Wireless Communications*,
https://doi.org/10.1007/978-3-032-01559-4_11

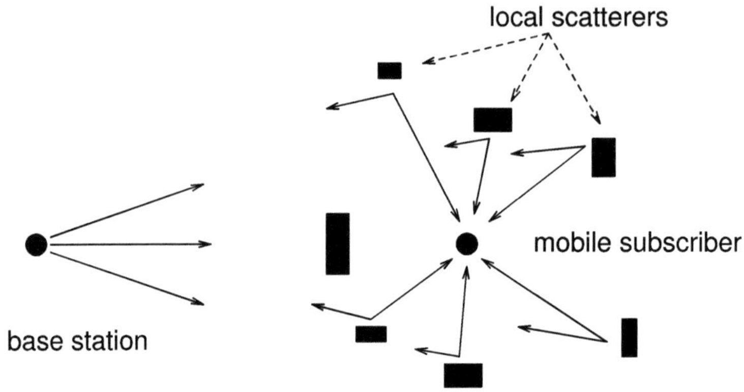

Fig. 11.1 Typical cellular land mobile radio propagation environment

Cellular telephone systems operate predominately at microwave frequencies where the carrier wavelength, λ_c, is in the centimeter range. For example, at frequency $f_c = 1$ GHz, $\lambda_c = 29.98$ cm or roughly 1 foot, where $c = f_c\lambda_c$ is the speed of light. Small changes in the propagation delays of the individual multipath components due to changes in MS location on the order of mere centimeters will cause large changes in the phases of the multipath components arriving at the receiver antenna(s). The phenomenon is even greater at millimeter wave frequencies where the wavelength is in the millimeter range. The multiple plane waves arriving at the receiver antenna will combine vectorially and experience constructive and destructive addition depending on their relative phases that will depend on spatial location. If all arriving components have the same frequency, the phasor addition rule can be applied to combine the sinusoids, a favorite technique used in introductory signal processing. If the MS is moving, however, the arriving sinusoids will not have the same frequency and the phasor addition rule cannot be applied. There will be time variations in the amplitude and phase of the composite signal received at each MS (or BS) antenna, a phenomenon known as envelope fading.

Radio channels are reciprocal, meaning if a propagation path exists, it will carry energy equally well in either direction. An MS in a typical NLoS macrocellular environment is usually surrounded by local scatterers, so that the plane waves on the forward link from BS to MS will arrive at the MS receiver antenna from many different azimuth angles, as shown in Fig. 11.1. The BSs on the other hand are relatively free from local scatterers. Hence, the plane waves on the reverse link from MS to BS tend to arrive at the BS receiver antenna with a small azimuth angle of arrival spread in the direction of the MS as shown in Fig. 11.1.

The remainder of this chapter begins with the mathematical modeling of multipath fading channels in Sect. 11.1. The received envelope and phase distribution is derived in Sect. 11.2 including Rayleigh and Rician fading and its effect on error probability. Section 11.3 considers the received envelope time correlation and Doppler spectrum. The simulation of fading channels is treated in Sect. 11.4 including filtered white noise and sum-of-sinusoids methods. This is extended to frequency-selective fading channels in Sect. 11.5. Shadowing models are considered in Sect. 11.6. The chapter wraps up with a treatment of path loss models in Sect. 11.7.

11.1 Multipath Fading

Figure 11.2 depicts radio wave propagation in an $x - y$ plane, where an MS is moving in the direction of the positive x-axis with velocity v. The BS is assumed to be stationary and located somewhere in the $x - y$ plane. Suppose the BS transmits a continuous wave sinusoidal tone with frequency f_c. At the MS, the nth plane wave arrives at the MS antenna with an angle of incidence θ_n with respect to the x-axis. The MS movement imparts a Doppler shift, or frequency shift, into each incident plane wave. The Doppler frequency or Doppler shift associated with the nth arriving plane wave is given by

$$\nu_{D,n} = \nu_m \cos(\theta_n) \text{ Hz}, \tag{11.1}$$

where $\nu_m = v/\lambda_c$, and ν_m is the maximum positive Doppler frequency occurring when $\theta_n = 0$. Plane waves arriving from the right half $x - y$ plane will experience a positive Doppler shift, while those arriving from the left half $x - y$ plane will experience a negative Doppler shift.

Now consider the propagation of a band-pass waveform

$$s(t) = \text{Re}\left\{\tilde{s}(t)\mathrm{e}^{j2\pi f_c t}\right\}. \tag{11.2}$$

where $\tilde{s}(t)$ is the complex envelope of the band-pass waveform, f_c is the carrier frequency. If the channel is comprised of N propagation paths, then the received

Fig. 11.2 A typical plane wave incident on an MS receiver moving along the positive x-axis with speed v

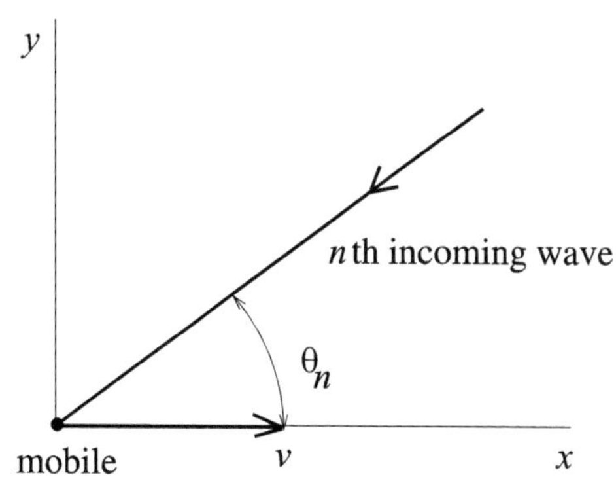

band-pass waveform is

$$r(t) = \text{Re}\left\{\sum_{n=1}^{N} C_n e^{j\phi_n - j2\pi c\tau_n/\lambda_c + j2\pi(f_c+\nu_{D,n})t}\tilde{s}(t-\tau_n)\right\}, \tag{11.3}$$

where C_n, ϕ_n, $\nu_{D,n}$, and τ_n are the amplitude, phase change, Doppler shift and time delay, respectively, associated with the nth propagation path. The amplitude C_n is directly proportional to the cross sectional area of the nth reflecting surface or the length of the nth diffracting edge. The phase ϕ_n is randomly introduced by interaction of the propagating wave and the nth scatterer and is assumed uniformly distributed on the interval $[-\pi, \pi)$. The delay $\tau_n = d_n/c$ is the propagation delay associated with the nth propagation path, where d_n is the length of the path. The set of path lengths, $\{d_n\}$, will depend on the physical scattering geometry which has not been specified at this point. The Doppler shift $\nu_{D,n}$ of the nth propagation path is given by (11.1).

Similar to (11.2), the received band-pass waveform $r(t)$ has the complex envelope representation

$$r(t) = \text{Re}\left\{\tilde{r}(t)e^{j2\pi f_c t}\right\}, \tag{11.4}$$

where the received complex envelope is

$$\tilde{r}(t) = \sum_{n=1}^{N} C_n e^{j\phi_n(t)}\tilde{s}(t-\tau_n) \tag{11.5}$$

and

$$\phi_n(t) = \phi_n - 2\pi c\tau_n/\lambda_c + 2\pi\nu_{D,n}t \tag{11.6}$$

is the time-variant phase associated with the nth propagation path. The time variant phases $\{\phi_n(t)\}$ make the channel a time-varying linear system with impulse response $\tilde{h}(t,\tau)$. Sect. 4.2 showed that $\tilde{s}(t)$, $\tilde{h}(t,\tau)$, and $\tilde{r}(t)$ are related through the convolution integral

$$\tilde{r}(t) = \int_0^t \tilde{h}(t,\tau)\tilde{s}(t-\tau)\text{d}\tau. \tag{11.7}$$

Deconvolution can be performed on (11.5), and it follows that

$$\tilde{h}(t,\tau) = \sum_{n=1}^{N} C_n e^{j\phi_n(t)}\delta(\tau-\tau_n), \tag{11.8}$$

where $\delta(\,\cdot\,)$ is the Dirac delta function or unit impulse function. To verify the above result, (11.8) may be back substituted into (11.7) to yield (11.5).

From (11.5) and (11.6), several interesting observations can be made. Since the carrier wavelength λ_c is small ($\lambda_c = 29.98$ cm or almost 1 foot at $f_c = 1$ GHz), even small changes in the path lengths $d_n = c\tau_n$ will cause large changes in the phases $\phi_n(t)$, due to the term $2\pi c\tau_n/\lambda_c$. Also, due to the Doppler frequency $\nu_{D,n}$, the phases $\phi_n(t)$ vary with time. Hence, at any given point in the space-time continuum, the random phases $\phi_n(t)$ will result in the constructive or destructive addition of the N multipath components, a phenomenon known as fading. Once again, due to the different angles of arrival θ_n, the multipath components have different Doppler frequencies so the phasor addition rule cannot be applied to obtain the received complex envelope.

If the differential path delays $\tau_i - \tau_j$ for all i, j are very small compared to the modulation symbol period, $T = 1/R$, where R is the baud rate, then the $\{\tau_n\}$ that appear in the argument of Dirac delta function in (11.8) may be replaced by the average delay, defined as

$$\bar{\tau} \triangleq \left(\sum_{n=1}^{N} C_n^2 \tau_n\right) \Big/ \left(\sum_{n=1}^{N} C_n^2\right). \tag{11.9}$$

In this case, the complex channel impulse response reduces to the simplified form

$$\tilde{h}(t,\tau) = \tilde{h}(t)\delta(\tau - \bar{\tau}), \tag{11.10}$$

where

$$\tilde{h}(t) = \sum_{n=1}^{N} C_n e^{j\phi_n(t)}. \tag{11.11}$$

Note that $\phi_n(t)$ remains as defined with the τ_n in (11.6), since the approximation $f_c\tau_n \approx f_c\bar{\tau}$ *cannot* be made because f_c is large in the microwave or millimeter wave frequency bands. It follows that the received complex envelope is

$$\tilde{r}(t) = \tilde{h}(t)\tilde{s}(t - \bar{\tau}) \tag{11.12}$$

which experiences fading due to the time-varying complex channel gain $\tilde{h}(t)$.

Example 2.1 (Two-Ray Fading Channel) Suppose the bandpass waveform $s(t) = \cos(2\pi f_c t)$ is transmitted and the channel consists of two rays such that the received waveform is

$$r(t) = \alpha\cos(2\pi f_c t) + \beta\cos(2\pi(f_c + \nu_D)t)\ ,$$

(continued)

Example 2.1 (continued)
where ν_D is a "Doppler" shift. Using the complex phaser representation of sinusoids,

$$r(t) = A(t)\cos(2\pi f_c t + \phi(t))$$

where

$$A(t) = \sqrt{\alpha^2 + \beta^2 + 2\alpha\beta\cos(2\pi\nu_D t)}$$

$$\phi(t) = -\text{Tan}^{-1}\frac{\beta\sin(2\pi\nu_D t)}{\alpha + \beta\cos(2\pi\nu_D t)}$$

Suppose $\alpha = \beta = 1$. Then

$$A(t) = \sqrt{2 + 2\cos(2\pi\nu_D t)}$$

$$\phi(t) = -\text{Tan}^{-1}\frac{\sin(2\pi\nu_D t)}{1 + \cos(2\pi\nu_D t)}$$

The magnitude $|A(t)|$ and phase $\phi(t)$ are plotted in Figs. 11.3 and 11.4, respectively. Observe that the fading channel is *time varying* and exhibits deep fades where the received amplitude $A(t) = 0$.

In the frequency domain, the received complex envelope is

$$\tilde{R}(f) = \tilde{H}(f) * \left(\tilde{S}(f)\mathrm{e}^{-j2\pi f\bar{\tau}}\right). \tag{11.13}$$

Fig. 11.3 Two-ray fading channel magnitude $|A(t)|$

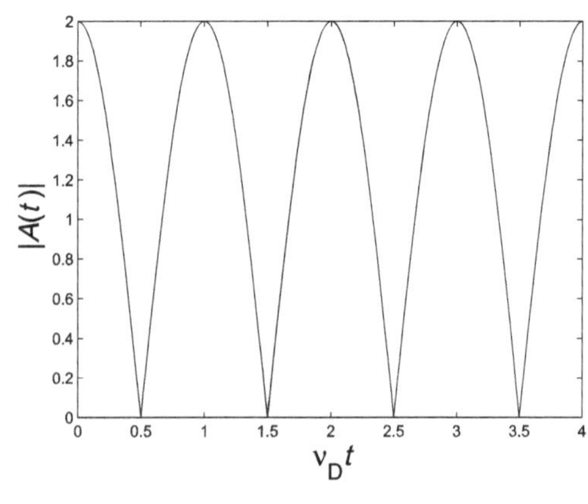

Fig. 11.4 Two-ray fading channel phase, $\phi(t)$

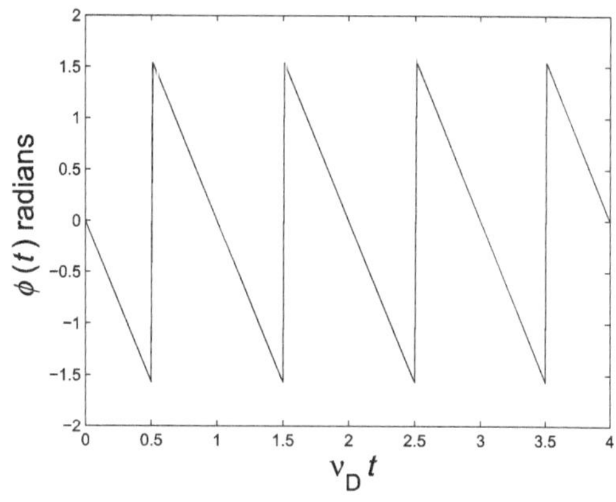

Since the channel varies with time, $\tilde{H}(f)$ is not an impulse function and has a finite nonzero width in the frequency domain that is bounded by the maximum Doppler frequency $\nu_m = v/\lambda_c$. Due to the convolution operation, the output spectrum $\tilde{R}(f)$ will be wider than the input spectrum $\tilde{S}(f)$. This broadening of the transmitted signal spectrum is caused by the channel time variations and is called frequency spreading or Doppler spreading. If the maximum Doppler frequency ν_m is much less than the width of the signal spectrum $\tilde{S}(f)$, then the Doppler spreading will not significantly distort $\tilde{S}(f)$. Fortunately, this is often the case for land mobile radio. For example, if $f_c = 1$ GHz and $v = 60$ mph, then $\nu_m \approx 88$ Hz.

The time-variant channel transfer function $\tilde{H}(t, f)$ is a Fourier transform pair with time-variant impulse response $\tilde{h}(t, \tau)$ with respect to the time-delay variable τ and frequency variable f

$$\tilde{H}(t, f) = \int_{-\infty}^{\infty} \tilde{h}(t, \tau) \mathrm{e}^{-j2\pi\nu\tau} \mathrm{d}\tau \tag{11.14a}$$

$$\tilde{h}(t, \tau) = \int_{-\infty}^{\infty} \tilde{H}(t, f) \mathrm{e}^{j2\pi f\tau} \mathrm{d}f \tag{11.14b}$$

For the channel in (11.10),

$$\tilde{H}(t, f) = \mathcal{F}_\tau\{\tilde{h}(t, \tau)\} = \tilde{h}(t) \mathrm{e}^{-j2\pi f\bar{\tau}}. \tag{11.15}$$

The channel magnitude response $|\tilde{H}(t, f)| = |\tilde{h}(t)|, \ \forall f$ is flat in its frequency variable f, but it has a time-variant gain $|\tilde{h}(t)|$. Such a channel is said to exhibit flat fading.

11.2 Received Envelope and Phase Distribution

For a fading channel, the complex envelope $\tilde{h}(t) = \tilde{h}_I(t) + j\tilde{h}_Q(t) = \alpha(t)\mathrm{e}^{j\phi(t)}$ is a complex-valued random process. Sampling $\tilde{h}(t)$ at any time t yields a complex-valued random variable $\tilde{h} = \tilde{h}_I + j\tilde{h}_Q = \alpha\mathrm{e}^{j\phi}$. This section considers the pdfs of α, α^2, and ϕ.

In the limit as $N \rightarrow \infty$ in (11.11), which is typical for dense multipath environments, the *central limit theorem* may be invoked and the real and imaginary components of the flat faded envelope

$$\tilde{h}_I(t) = \sum_{n=1}^{N\rightarrow\infty} C_n \cos\phi_n(t) \qquad \tilde{h}_Q(t) = \sum_{n=1}^{N\rightarrow\infty} C_n \sin\phi_n(t) \tag{11.16}$$

can be treated as *Gaussian random processes* due to the random phases $\phi_n(t)$, meaning that at any time t_1, $\tilde{h}_I = \tilde{h}_I(t_1)$, and $\tilde{h}_Q = \tilde{h}_Q(t_1)$ are a pair of Gaussian random variables that generally correlated.

11.2.1 Rayleigh Fading

For some types of scattering environments, which will be specified in the sequel, $\tilde{h}_I$ and $\tilde{h}_Q$ are a pair of i.i.d. Gaussian random variables with zero mean and variance $\sigma^2 = \mathrm{E}[\tilde{h}_I^2] = \mathrm{E}[\tilde{h}_Q^2]$. Under these conditions, the magnitude of the complex envelope

$$\alpha \triangleq |\tilde{h}| = \sqrt{\tilde{h}_I^2 + \tilde{h}_Q^2} \tag{11.17}$$

has a Rayleigh pdf, as shown in Sect. 2.7.4,

$$p_\alpha(x) = \frac{2x}{\Omega_p} \exp\left\{-\frac{x^2}{\Omega_p}\right\}, \qquad x \geq 0. \tag{11.18}$$

where the second moment of α

$$\Omega_p = \mathrm{E}[\alpha^2] = \mathrm{E}[\tilde{h}_I^2] + \mathrm{E}[\tilde{h}_Q^2] = 2\sigma^2 \tag{11.19}$$

is the envelope power. This type of fading is called Rayleigh fading. .

The corresponding squared envelope $\alpha^2 = |\tilde{h}|^2$ is shown in Sect. 2.7.5 to be exponentially distributed at any time t with pdf

$$p_{\alpha^2}(x) = \frac{1}{\Omega_p} \exp\left\{-\frac{x}{\Omega_p}\right\}, \qquad x \geq 0. \tag{11.20}$$

The squared envelope is important for the performance analysis of digital communication systems because it is proportional to the received signal power and, hence, the received bit energy-to-noise ratio $\gamma_b = \alpha^2 E_b/N_o$.

11.2.2 Ricean Fading

Some types of cellular propagation environments have a dominant specular or line-of-sight (LoS) multipath component. In this case, the complex channel gain becomes

$$\tilde{h}(t) = \sum_{n=0}^{N} C_n e^{j\phi_n(t)} = C_0 e^{j\phi_0(t)} + \sum_{n=1}^{N} C_n e^{j\phi_n(t)} \tag{11.21}$$

and

$$\tilde{h}_I(t) = C_0 \cos(\phi_0(t)) + \sum_{n=1}^{N} C_n \cos(\phi_n(t)) \tag{11.22a}$$

$$\tilde{h}_Q(t) = C_0 \sin(\phi_0(t)) + \sum_{n=1}^{N} C_n \sin(\phi_n(t)) \tag{11.22b}$$

where C_0 and $\phi_0(t)$ are the gain and time-variant phase, respectively, associated with the dominant specular or LoS component.

Aulin's Ricean fading model [2] follows directly from the above, such that $\tilde{h}_I(t)$ and $\tilde{h}_Q(t)$ have time-variant means

$$m_I(t) = \mathrm{E}[\tilde{h}_I(t)] = s \cdot \cos(2\pi \nu_m \cos(\theta_0)t + \phi_0) \tag{11.23a}$$

$$m_Q(t) = \mathrm{E}[\tilde{h}_Q(t)] = s \cdot \sin(2\pi \nu_m \cos(\theta_0)t + \phi_0) \tag{11.23b}$$

where $s = C_0 = \sqrt{m_I^2(t) + m_Q^2(t)}$ is the amplitude, $\nu_m \cos(\theta_0)$ is the Doppler shift, and ϕ_0 is the random phase associated with the LoS or specular component arriving at angle θ_0. Note that the phase term $2\pi c\tau_0/\lambda_c$ is ignored in the means $m_I(t)$ and $m_Q(t)$ because it can be readily absorbed into the uniformly distributed random phase ϕ_0.

By invoking the central limit theorem, $\tilde{h}_I = \tilde{h}_I(t_1)$ and $\tilde{h}_Q = \tilde{h}_Q(t_1)$ at any time t_1 may once again be treated as a pair of Gaussian random variables with means $m_I = m_I(t_1)$ and $m_Q = m_Q(t_1)$, respectively, and variance $\sigma^2 = \mathrm{E}[(\tilde{h}_I - m_I)^2] = \mathrm{E}[(\tilde{h}_Q - m_Q))^2]$. If it is again assumed that $\tilde{h}_I$ and $\tilde{h}_Q$ are uncorrelated (and because they are Gaussian independent), then the magnitude of the received complex envelope $\alpha = |\tilde{h}|$ has a Rice distribution, derived in Sect. 2.7.6 to have the form

$$p_\alpha(x) = \frac{2(K+1)x}{\Omega_p} \exp\left\{-K - \frac{(K+1)x^2}{\Omega_p}\right\} I_o\left(2x\sqrt{\frac{K(K+1)}{\Omega_p}}\right), \quad x \geq 0, \tag{11.24}$$

where $s^2 = m_I^2 + m_Q^2$ is the specular power, $2\sigma^2$ is the scattered power, the Rice factor $K = s^2/2\sigma^2$ is their ratio, and the envelope power $\Omega_p = s^2 + 2\sigma^2$ is their sum. When $K = 0$, there is no LoS or specular component and the envelope exhibits Rayleigh fading. When $K = \infty$, there is no scatter component and the channel does not exhibit any fading at all. Figure 2.7 plots the Rice pdf for several values of K with $\Omega_p = 1$. The curve for $K = 0$ is the Rayleigh pdf in (11.18).

The Rice pdf in (11.24) only requires that the means of h_I and h_Q, m_I and m_Q, be nonzero, particularly they could be constant values. However, Aulin's model in (11.23a) and (11.23b) shows that the means $m_I(t)$ and $m_Q(t)$ in a land mobile radio channel will vary with time t and that a simple model with constant means has no basis in reality.

Finally, the squared envelope $\alpha^2 = |\tilde{h}|^2$ has the following non-central Chi-square distribution with two degrees of freedom

$$p_{\alpha^2}(x) = \frac{(K+1)}{\Omega_p} \exp\left\{-K - \frac{(K+1)x}{\Omega_p}\right\} I_o\left(2\sqrt{\frac{K(K+1)x}{\Omega_p}}\right), \quad x \geq 0. \tag{11.25}$$

11.2.3 Envelope Phase

The phase of the channel complex envelope $\tilde{h} = \tilde{h}_I + j\tilde{h}_Q$ is

$$\phi = \mathrm{Tan}^{-1}\left(\frac{\tilde{h}_Q}{\tilde{h}_I}\right). \tag{11.26}$$

For Rayleigh fading, $\tilde{h}_I$ and $\tilde{h}_Q$ are i.i.d. zero-mean Gaussian random variables. Sect. 2.7.4 shows the phase ϕ to be uniformly distributed on the interval $[-\pi, \pi)$,

$$p_\phi(x) = \frac{1}{2\pi}, \quad -\pi \leq x \leq \pi. \tag{11.27}$$

For Ricean fading channels, the phase ϕ is not uniformly distributed and its pdf takes on a more complicated integral form that can be obtained from the joint pdf in (2.99).

11.2.4 Error Probability with Rayleigh Fading

When the channel experiences fading, the error probability must be averaged over the fading distribution. For example, if the channel is Rayleigh faded, then α is a Rayleigh random variable and the squared envelope α^2 is an exponential random variable at any given time, as discussed in Sect. 11.2.1. It follows that the symbol energy-to-noise γ_s in (6.58) has the exponential pdf

$$p_{\gamma_s}(x) = \frac{1}{\bar{\gamma}_s} e^{-x/\bar{\gamma}_s} , \quad x \geq 0, \tag{11.28}$$

where $\bar{\gamma}_s$ is the average symbol energy-to-noise ratio. Since there are $\log_2 M$ bits per message, $\bar{\gamma}_s = \bar{\gamma}_b \log_2 M$, where $\bar{\gamma}_b$ is the average bit energy-to-noise ratio. A corresponding expression for the distribution of γ_s may be obtained in the case of Ricean fading.

For BPSK and QPSK and Rayleigh fading, the probability of bit error averaged over the distribution of the bit energy-to-noise ratio γ_b is

$$\begin{aligned} P_b &= \int_0^{\infty} Q(\sqrt{2x}) p_{\gamma_b}(x) \mathrm{d}x \\ &= \frac{1}{2}\left(1 - \sqrt{\frac{\bar{\gamma}_b}{1+\bar{\gamma}_b}}\right) \\ &\approx \frac{1}{4\bar{\gamma}_b} \qquad \text{for } \bar{\gamma}_b \gg 1. \end{aligned} \tag{11.29}$$

The BPSK and QPSK bit error probability is plotted in Fig. 11.5 for an AWGN channel and a Rayleigh fading channel. Observe that Rayleigh fading converts an exponential dependency of the bit error probability on the average received bit energy-to-noise ratio into an inverse linear one.

For non-binary modulation schemes, the symbol error probability averaged over the envelope fading is

$$P_e = \int_0^{\infty} P_e(x) p_{\gamma_s}(x) \mathrm{d}x. \tag{11.30}$$

In the case of QAM, Fig. 11.6 plots the (approximate) bit error probability $P_b \approx P_e/\log_2 M$ against the average received *bit* energy-to-noise ratio, $\bar{\gamma}_b = \bar{\gamma}_s/\log_2 M$, for several values of M. Once again, Rayleigh fading converts an exponential dependency of the bit error probability on the average received bit energy-to-noise ratio into an inverse linear one.

The inverse linear dependency of P_b on $\bar{\gamma}_b$ will be observed for all types of modulation and detectors with flat Rayleigh fading, and results in a huge loss in power efficiency unless appropriate countermeasures such as diversity and coding

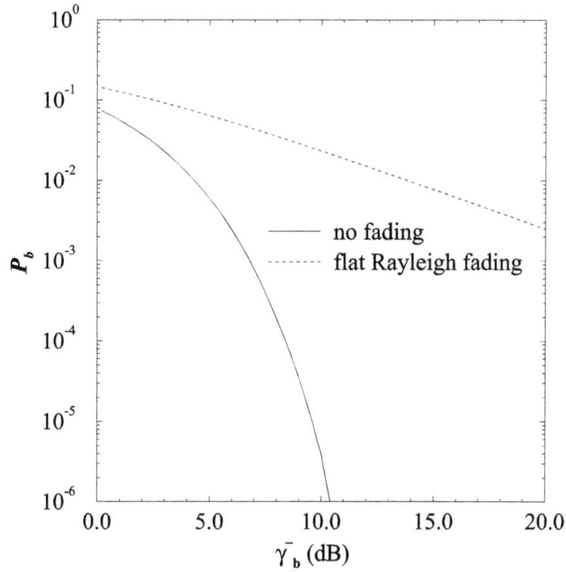

Fig. 11.5 Bit error probability for BPSK and QPSK for a slow flat Rayleigh fading channel with AWGN

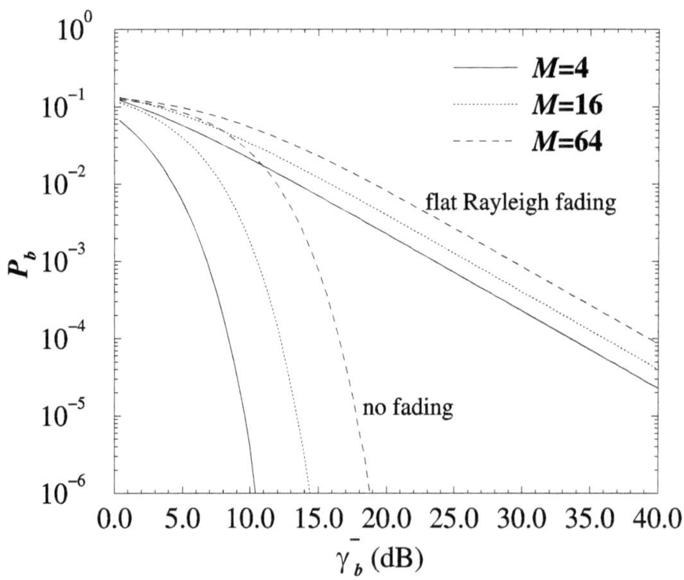

Fig. 11.6 Bit error probability for M-QAM on an AWGN channel and a Rayleigh fading channel with AWGN

are used. For example, for orthogonal MFSK with non-coherent detection on Rayleigh fading channels, the bit error probability in (6.119) can be averaged over the distribution of γ_s in (11.28). This gives the following simple closed form for the bit error probability

$$P_b = \frac{M}{2(M-1)} \sum_{k=1}^{M-1} \frac{(-1)^{k+1}\binom{M-1}{k}}{1+k+k\bar{\gamma}_b \log_2 M}. \tag{11.31}$$

Once again, the bit error probability has an inverse linear dependency on $\bar{\gamma}_b$.

11.3 Envelope Time Correlation

A flat fading channel can be characterized using a continuous wave sinusoid at the carrier frequency f_c, because its magnitude response is flat. For convenience, let the complex envelope in (11.5) be $\tilde{s}(t) = 1$. From (11.4) and (11.5), the corresponding received band-pass signal has the quadrature form

$$r(t) = \tilde{h}_I(t)\cos(2\pi f_c t) - \tilde{h}_Q(t)\sin(2\pi f_c t), \tag{11.32}$$

where

$$\tilde{h}_I(t) = \sum_{n=1}^{N} C_n \cos(\phi_n(t)) \tag{11.33a}$$

$$\tilde{h}_Q(t) = \sum_{n=1}^{N} C_n \sin(\phi_n(t)) \tag{11.33b}$$

The $\{C_n\}$ will vary from one locality to the next due to changes in the site specific scattering environment. However, over a local area on the dimension of several tens of wavelengths, the $\{C_n\}$ may be treated as being constant. Such a channel model is said to be locally wide-sense stationarity (WSS).

Following the results in Sect. 4.3 for bandpass random signals, the auto-correlation function of the bandpass waveform $r(t)$ in (11.32) is

$$\phi_{rr}(\tau) = \phi_{\tilde{h}_I\tilde{h}_I}(\tau)\cos 2\pi f_c\tau - \phi_{\tilde{h}_I\tilde{h}_Q}(\tau)\sin 2\pi f_c\tau \tag{11.34}$$

where

$$\phi_{\tilde{h}_I\tilde{h}_I}(\tau) \triangleq \mathrm{E}[\tilde{h}_I(t)\tilde{h}_I(t+\tau)] \tag{11.35a}$$

$$\phi_{\tilde{h}_I\tilde{h}_Q}(\tau) \triangleq \mathrm{E}[\tilde{h}_I(t)\tilde{h}_Q(t+\tau)]. \tag{11.35b}$$

Considering again the flat fading model in (11.12) and ignoring the average time delay $\bar{\tau}$, the auto-correlation of the received complex envelope $\tilde{r}(t) = \tilde{h}(t)\tilde{s}(t)$ is

$$\begin{aligned}\phi_{\tilde{r}\tilde{r}}(\tau) &= \tfrac{1}{2}\mathrm{E}[\tilde{h}^*(t)\tilde{s}^*(t)\tilde{h}(t+\tau)\tilde{s}(t+\tau)] \\ &= \mathrm{E}[\tilde{h}^*(t)\tilde{h}(t+\tau)]\tfrac{1}{2}\mathrm{E}[\tilde{s}^*(t)\tilde{s}(t+\tau)] \\ &= \phi_{\tilde{h}\tilde{h}}(\tau)\phi_{\tilde{s}\tilde{s}}(\tau) \end{aligned} \tag{11.36}$$

where the second line makes the reasonable assumption that the message waveform $\tilde{s}(t)$ and channel $\tilde{h}(t)$ are uncorrelated and the channel correlation function is defined as

$$\phi_{\tilde{h}\tilde{h}}(\tau) \triangleq \mathrm{E}[\tilde{h}^*(t)\tilde{h}(t+\tau)]. \tag{11.37}$$

Note the absence of the factor of 1/2 in the definition of the channel correlation function in (11.37), a result that follows from (11.36). The channel correlation function may be expressed as

$$\begin{aligned}\phi_{\tilde{h}\tilde{h}}(\tau) &= \mathrm{E}\left[\tilde{h}^*(t)\tilde{h}(t+\tau)\right] \\ &= \mathrm{E}\left[(\tilde{h}_I(t) - j\tilde{h}_Q(t))(\tilde{h}_I(t+\tau) + j\tilde{h}_Q(t+\tau))\right] \\ &= \mathrm{E}\left[(\tilde{h}_I(t)\tilde{h}_I(t+\tau)\right] + \mathrm{E}\left[(\tilde{h}_Q(t)\tilde{h}_Q(t+\tau)\right] \\ &\quad + j\mathrm{E}\left[(\tilde{h}_I(t)\tilde{h}_Q(t+\tau)\right] - j\mathrm{E}\left[(\tilde{h}_Q(t)\tilde{h}_I(t+\tau)\right] \\ &= 2\phi_{\tilde{h}_I\tilde{h}_I}(\tau) + j2\phi_{\tilde{h}_I\tilde{h}_Q}(\tau) \end{aligned} \tag{11.38}$$

From (11.34)

$$\phi_{rr}(\tau) = \tfrac{1}{2}\mathrm{Re}\left\{\phi_{\tilde{h}\tilde{h}}(\tau)\mathrm{e}^{j2\pi f_c\tau}\right\}. \tag{11.39}$$

The received *envelope power* is

$$\Omega_p = \phi_{\tilde{h}\tilde{h}}(0) = \mathrm{E}[|\tilde{h}(t)|^2] = \mathrm{E}[\tilde{h}_I^2(t)] + \mathrm{E}[\tilde{h}_Q^2(t)]. \tag{11.40}$$

Evaluating both sides of (11.39) once again shows that when $\tilde{r}(t) = \tilde{h}(t)$, i.e., $\tilde{s}(t) = 1$, the received bandpass power is $P = \Omega_p/2$ equal to one-half the received envelope power. Finally, from (11.36), $\phi_{\tilde{r}\tilde{r}}(0) = \phi_{\tilde{h}\tilde{h}}(0)\phi_{\tilde{s}\tilde{s}}(0)$, meaning that the received envelope power is the product of the channel envelope power and the signal envelope power.

The correlations $\phi_{\tilde{h}_I\tilde{h}_I}(\tau)$ and $\phi_{\tilde{h}_I\tilde{h}_Q}(\tau)$ in (11.35a) and (11.35b), are now evaluated. At any time t, the phases $\phi_n(t)$ in (11.6) can be treated as being independent and uniformly distributed over the interval $[-\pi, \pi)$, because the random phases ϕ_n

are independent and uniform on the interval $[-\pi, \pi)$. The azimuth angles of arrival, θ_n, are all independent due to the random placement of scatterers. In the limit as $N \to \infty$, the discrete azimuth angles of arrival θ_n can be replaced by a continuous random variable θ having the pdf $p(\theta)$ known as the power angle distribution.

By using the above properties, the auto-correlation $\phi_{\tilde{h}_I\tilde{h}_I}(\tau)$ can be obtained as follows:

$$
\begin{aligned}
\phi_{\tilde{h}_I\tilde{h}_I}(\tau) &= \mathrm{E}[\tilde{h}_I(t)\tilde{h}_I(t+\tau)] \\
&= \mathrm{E}\left[\sum_{n=1}^{N} C_n \cos(\phi_n(t)) \sum_{m=1}^{N} C_m \cos(\phi_m(t+\tau))\right] \\
&= \mathrm{E}\left[\sum_{n=1}^{N}\sum_{m=1}^{N} C_n C_m \cos(\phi_n(t)) \cos(\phi_m(t+\tau))\right] \\
&= \frac{1}{2}\mathrm{E}\left[\sum_{n=1}^{N}\sum_{m=1}^{N} C_n C_m \Big(\cos(\phi_n(t)+\phi_m(t+\tau)) \right. \\
&\qquad\qquad \left. + \cos(\phi_m(t+\tau)-\phi_n(t)) \Big)\right] \\
&= \frac{1}{2}\mathrm{E}\left[\sum_{n=1}^{N} C_n^2 \Big(\cos(\phi_n(t)+\phi_n(t+\tau)) \right. \\
&\qquad\qquad \left. + \cos(\phi_n(t+\tau)-\phi_n(t)) \Big)\right] \\
&\quad + \frac{1}{2}\mathrm{E}\left[\sum_{n=1}^{N}\sum_{\substack{m=1\\ m\neq n}}^{N} C_n C_m \Big(\cos(\phi_n(t)+\phi_m(t+\tau)) \right. \\
&\qquad\qquad \left. + \cos(\phi_m(t+\tau)-\phi_n(t)) \Big)\right]
\end{aligned}
$$

where, once again,

$$
\phi_n(t) = \phi_n - 2\pi d_n/\lambda_c + 2\pi \nu_m \cos(\theta_n)t \tag{11.41}
$$

In the phase terms $\phi_n(t)$, the variables ϕ_n, d_n, and θ_n are all independent random variables.

For the double sum term,

$$
\frac{1}{2}\mathrm{E}\left[\sum_{n=1}^{N}\sum_{\substack{m=1\\ m\neq n}}^{N} C_n C_m \Big(\cos(\phi_n(t)+\phi_m(t+\tau)) + \cos(\phi_m(t+\tau)-\phi_n(t))\Big)\right]
$$

$$= \frac{1}{2}\sum_{n=1}^{N}\sum_{\substack{m=1\\m\neq n}}^{N} C_n C_m \Big(\mathrm{E}\,[\cos(\phi_n + \phi_m - 2\pi(d_n + d_m)/\lambda_c$$

$$+2\pi\nu_m\cos(\theta_n)t + 2\pi\nu_m\cos(\theta_m)(t+\tau))]$$

$$+\mathrm{E}\,[\cos(\phi_m - \phi_n - 2\pi(d_m - d_n)/\lambda_c$$

$$+2\pi\nu_m\cos(\theta_m)(t+\tau) - 2\pi\nu_m\cos(\theta_n)t)]\Big) = 0.$$

The above result is due to the expectation E[·] being taken over the random phases ϕ_n that are independent and uniformly distributed on the interval $[-\pi, \pi)$.

For the single sum term,

$$\frac{1}{2}\mathrm{E}\left[\sum_{n=1}^{N} C_n^2 \Big(\cos(\phi_n(t) + \phi_n(t+\tau)) + \cos(\phi_n(t+\tau) - \phi_n(t))\Big)\right]$$

$$= \frac{1}{2}\sum_{n=1}^{N} C_n^2 \Big(\mathrm{E}\,[\cos(2\phi_n - 4\pi d_n/\lambda_c + 2\pi\nu_m\cos(\theta_n)t$$

$$+2\pi\nu_m\cos(\theta_n)(t+\tau))]$$

$$+\mathrm{E}\,[\cos(2\pi\nu_m\cos(\theta_n)\tau)]\Big)$$

$$= \frac{1}{2}\sum_{n=1}^{N} C_n^2 \mathrm{E}\,[\cos(2\pi\nu_m\cos(\theta_n)\tau)]$$

where the first expectation above is zero when the ensemble average is taken over the random phases ϕ_n. Hence,

$$\begin{aligned}\phi_{\tilde{h}_I\tilde{h}_I}(\tau) &= \frac{1}{2}\sum_{n=1}^{N} C_n^2 \mathrm{E}\,[\cos(2\pi\nu_m\cos(\theta_n)\tau)]\\ &= \frac{\Omega_p}{2}\mathrm{E}_\theta\,[\cos(2\pi\nu_m\cos(\theta)\tau)]\end{aligned} \tag{11.42}$$

where the second line follows because the azimuth arrival angles θ_n are i.i.d. random variables, and

$$\Omega_p = \mathrm{E}[|\tilde{h}(t)|^2] = \sum_{n=1}^{N} C_n^2 \tag{11.43}$$

is the envelope power.

Similarly, the cross-correlation $\phi_{\tilde{h}_I \tilde{h}_Q}(\tau)$ may be obtained as follows:

$$\begin{aligned}
\phi_{\tilde{h}_I \tilde{h}_Q}(\tau) &= \mathrm{E}[\tilde{h}_I(t)\tilde{h}_Q(t+\tau)] \\
&= \mathrm{E}\left[\sum_{n=1}^{N} C_n \cos(\phi_n(t)) \sum_{m=1}^{N} C_m \sin(\phi_m(t+\tau))\right] \\
&= \mathrm{E}\left[\sum_{n=1}^{N}\sum_{m=1}^{N} C_n C_m \cos(\phi_n(t)) \sin(\phi_m(t+\tau))\right] \\
&= \frac{1}{2}\mathrm{E}\left[\sum_{n=1}^{N}\sum_{m=1}^{N} C_n C_m \left(\sin(\phi_n(t)+\phi_m(t+\tau)) \right.\right. \\
&\quad \left.\left. - \sin(\phi_n(t)-\phi_m(t+\tau))\right)\right] \\
&= \frac{\Omega_p}{2}\mathrm{E}_\theta\left[\sin(2\pi\nu_m \cos(\theta)\tau)\right]
\end{aligned} \tag{11.44}$$

Evaluation of the pair of the expectations in (11.42) and (11.44) requires the power angle distribution $p(\theta)$. One simple and useful model assumes that $p(\theta)$ is uniformly distributed, such that $p(\theta) = 1/(2\pi), \theta \in [-\pi, \pi)$. This model was first suggested by Clarke [6] and is commonly referred to as Clarke's 2-D isotropic scattering model. With 2-D isotropic scattering, the expectation in (11.42) becomes

$$\begin{aligned}
\phi_{\tilde{h}_I \tilde{h}_I}(\tau) &= \frac{\Omega_p}{2}\int_{-\pi}^{\pi} \cos\left(2\pi\nu_m\tau\cos(\theta)\right) p(\theta)\mathrm{d}\theta \\
&= \frac{\Omega_p}{2\pi}\int_{0}^{\pi} \cos\left(2\pi\nu_m\tau\cos(\theta)\right) \mathrm{d}\theta \\
&= \frac{\Omega_p}{2} J_0(2\pi\nu_m\tau),
\end{aligned} \tag{11.45}$$

where

$$J_0(x) = \frac{1}{\pi}\int_0^{\pi} \cos\left(x\cos(\theta)\right)\mathrm{d}\theta \tag{11.46}$$

is the zero-order Bessel function of the first kind.

Likewise, the cross-correlation in (11.44) becomes

$$\phi_{\tilde{h}_I \tilde{h}_Q}(\tau) = \frac{\Omega_p}{2\pi}\int_{-\pi}^{\pi} \sin\left(2\pi\nu_m\tau\cos(\theta)\right)\mathrm{d}\theta = 0. \tag{11.47}$$

The cross-correlation $\phi_{\tilde{h}_I\tilde{h}_Q}(\tau) = 0$ implies that $\tilde{h}_I(t)$ and $\tilde{h}_Q(t)$ are uncorrelated and, since they are Gaussian, independent random processes. In this case, the envelope $\alpha = |\tilde{h}|$ is Rayleigh distributed as discussed in Sect. 11.2.1. The cross-correlation $\phi_{\tilde{h}_I\tilde{h}_Q}(\tau)$ in (11.47 is generally nonzero for non-isotropic scattering environments. However, any $p(\theta)$ that is symmetric about $\theta = \pi/2$ (a line perpendicular to the direction of motion) will have $\phi_{\tilde{h}_I\tilde{h}_Q}(\tau) = 0$ and the envelope α will be Rayleigh distributed.

For 2-D isotropic scattering, the channel correlation function is

$$\phi_{\tilde{h}\tilde{h}}(\tau) = 2\phi_{\tilde{h}_I\tilde{h}_I}(\tau) + j2\phi_{\tilde{h}_I\tilde{h}_Q}(\tau) = \Omega_p J_0(2\pi\nu_m\tau), \tag{11.48}$$

The normalized channel correlation function $\phi_{\tilde{h}\tilde{h}}(\tau)/\Omega_p$ is plotted against the normalized time delay $\nu_m\tau$ in Fig. 11.7. The channel correction function is zero at certain time lags τ, the first zero occurs at $\nu_m\tau = 0.3875$. This means that samples of the faded envelope $\tilde{h}(t)$ taken at intervals of $\tau = 0.3875/\nu_m$ seconds will be mutually uncorrelated and, since they are complex Gaussian, independent.

For Ricean fading channels, the power angle distribution $p(\theta)$ may be modeled by the mixture distribution

$$p(\theta) = \frac{1}{K+1}\hat{p}(\theta) + \frac{K}{K+1}\delta(\theta - \theta_0), \tag{11.49}$$

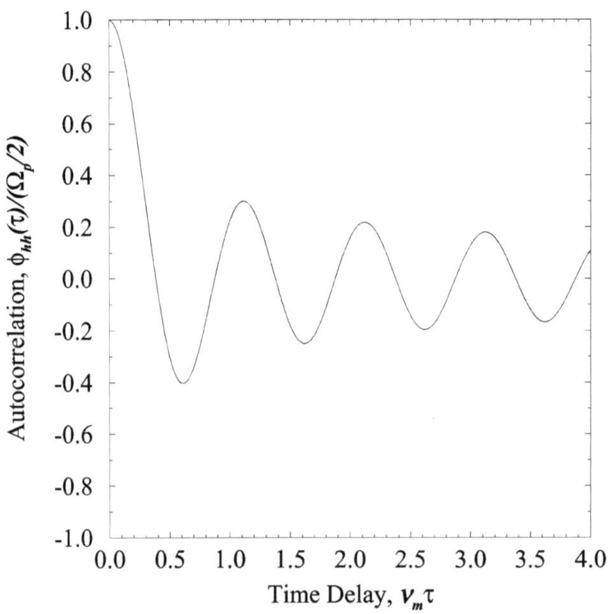

Fig. 11.7 Normalized auto-correlation function of the quadrature components of the received complex envelope with 2-D isotropic scattering

where $\hat{p}(\theta)$ is the power angle distribution of the diffuse or scatter component, θ_0 is the angle-of-arrival (AoA) of the specular or LoS component, and K is the Rice factor. Fig. 11.8 shows a plot of $p(\theta)$ for one such scattering environment, where the there is a line of sight component arriving at angle $\theta_0 = \pi/2$ while the scatter component has 2-D isotropic scattering. The correlation functions $\phi_{g_I g_I}(\tau)$ and $\phi_{g_I g_Q}(\tau)$ in this case are

$$\phi_{g_I g_I}(\tau) = \frac{1}{K+1}\frac{\Omega_p}{2} J_0(2\pi \nu_m \tau) + \frac{K}{K+1}\frac{\Omega_p}{2}\cos(2\pi \nu_m \tau \cos\theta_0) \tag{11.50a}$$

$$\phi_{g_I g_Q}(\tau) = \frac{K}{K+1}\frac{\Omega_p}{2}\sin(2\pi \nu_m \tau \cos\theta_0). \tag{11.50b}$$

Spatial Correlation

For 2-D isotropic scattering, the moving direction of the MS has no effect on the channel correlation function. Suppose the MS is moving along a straight line trajectory with velocity v. Two samples of the faded envelope taken τ seconds apart correspond to a two-element virtual antenna array with an element spacing of $d = v\tau$. Since $\nu_m = v/\lambda_c$, it follows

$$\nu_m \tau = d/\lambda_c \tag{11.51}$$

Hence, for 2-D isotropic scattering, the spatial correlation as a function of the spatial distance d is

$$\phi_{\tilde{h}\tilde{h}}(d) = \Omega_p J_0(2\pi d/\lambda_c). \tag{11.52}$$

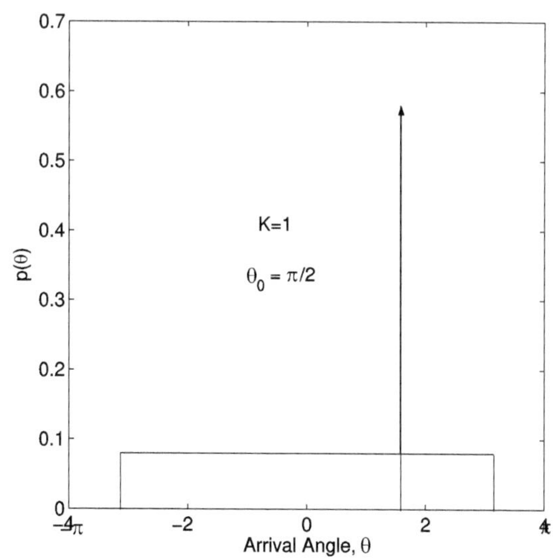

Fig. 11.8 Plot of $p(\theta)$ vs. θ with 2-D isotropic scattering mixed with an LoS or specular component arriving at angle $\theta_0 = \pi/2$

The spatial correlation is zero when $d/\lambda_c = 0.3875$, i.e., when the antenna elements are spaced $0.3875\lambda_c \approx \lambda_c/2$ (a half-wavelength) apart.

11.3.1 Doppler Power Spectrum

The Doppler power spectrum and channel auto-correlation function are a Fourier transform pair

$$S_{\tilde{h}\tilde{h}}(\nu) = \int_{-\infty}^{\infty} \phi_{\tilde{h}\tilde{h}}(\tau) \mathrm{e}^{-j2\pi\nu\tau} \mathrm{d}\tau \tag{11.53a}$$

$$\phi_{\tilde{h}\tilde{h}}(\tau) = \int_{-\infty}^{\infty} S_{\tilde{h}\tilde{h}}(\nu) \mathrm{e}^{j2\pi\nu\tau} \mathrm{d}\nu, \tag{11.53b}$$

From the channel correlation function in (11.38)

$$S_{\tilde{h}\tilde{h}}(\nu) = 2S_{\tilde{h}_I\tilde{h}_I}(\nu) + j2S_{\tilde{h}_I\tilde{h}_Q}(\nu), \tag{11.54}$$

is the Doppler power spectrum.

From (11.54), it may appear at first glance that the Doppler spectrum is complex-valued. However, the auto-correlation $\phi_{\tilde{h}_I\tilde{h}_I}(\tau)$, as with any auto-correlation, is an even function such that $\phi_{\tilde{h}_I\tilde{h}_I}(\tau) = \phi_{\tilde{h}_I\tilde{h}_I}(-\tau)$. Hence, its Fourier transform $S_{\tilde{h}_I\tilde{h}_I}(\nu)$ is real-valued. Moreover, the cross-correlation $\phi_{\tilde{h}_I\tilde{h}_Q}(\tau)$ in (11.38), as with any cross-correlation, satisfies the property

$$\phi_{\tilde{h}_I\tilde{h}_Q}(\tau) = \phi_{\tilde{h}_Q\tilde{h}_I}(-\tau). \tag{11.55}$$

Combining (11.55) with (4.25b), yields the property

$$\phi_{\tilde{h}_I\tilde{h}_Q}(\tau) = -\phi_{\tilde{h}_I\tilde{h}_Q}(-\tau). \tag{11.56}$$

Hence, the cross-correlation $\phi_{\tilde{h}_I\tilde{h}_Q}(\tau)$ is an odd function and its Fourier transform $S_{\tilde{h}_I\tilde{h}_Q}(\nu)$ is purely imaginary-valued. Therefore, the Doppler power spectrum in (11.54) is always real-valued, but not necessarily an even function.

For the case of 2-D isotropic scattering, $S_{\tilde{h}_Q\tilde{h}_I}(\nu) = 0$ so that $S_{\tilde{h}\tilde{h}}(\nu) = 2S_{\tilde{h}_I\tilde{h}_I}(\nu)$. For the auto-correlation in (11.45), the Doppler power spectrum is [12, 6.671.7]

$$\begin{aligned} S_{\tilde{h}\tilde{h}}(\nu) &= 2S_{\tilde{h}_I\tilde{h}_I}(\nu) = \mathcal{F}[2\phi_{\tilde{h}_I\tilde{h}_I}(\tau)] \\ &= \begin{cases} \frac{\Omega_p}{\pi\nu_m}\frac{1}{\sqrt{1-(\nu/\nu_m)^2}} \;, & 0 \le |\nu| \le \nu_m \\ 0 \;, & \text{otherwise} \end{cases} . \end{aligned} \tag{11.57}$$

Sometimes, this is called the *classical* Doppler power spectrum. The normalized Doppler power spectrum $S_{hh}(\nu)/(\Omega_p/\pi\nu_m)$ from (11.57) is plotted against the normalized Doppler frequency ν/ν_m in Fig. 11.9. Notice that $S_{\tilde{h}\tilde{h}}(\nu)$ is limited to the range of frequencies $0 \leq |\nu| \leq \nu_m$ and $S_{\tilde{h}\tilde{h}}(\nu) = \infty$ at $\nu = \pm\nu_m$. In reality, the Doppler power spectrum is bounded, and the singular behavior at $\nu = \pm\nu_m$ is due to the assumption of 2-D propagation.

Finally, the Doppler power spectrum of the complex channel envelope $\tilde{h}(t)$ can be related to the Doppler power spectrum of the received band-pass waveform $r(t)$ in (11.32). From (11.39),

$$\phi_{rr}(\tau) = \tfrac{1}{2}\mathrm{Re}\left\{\phi_{\tilde{h}\tilde{h}}(\tau)\mathrm{e}^{j2\pi f_c\tau}\right\}. \tag{11.58}$$

By using the identity

$$\mathrm{Re}\{z\} = \frac{z + z^*}{2} \tag{11.59}$$

and the property $\phi_{\tilde{h}\tilde{h}}(\tau) = \phi^*_{\tilde{h}\tilde{h}}(-\tau)$, the bandpass Doppler power spectrum is

$$S_{rr}(\nu) = \tfrac{1}{4}\Big(S_{\tilde{h}\tilde{h}}(\nu - f_c) + S_{\tilde{h}\tilde{h}}(-\nu - f_c)\Big). \tag{11.60}$$

The bandpass Doppler power spectrum $S_{rr}(\nu)$ is always real-valued and even.

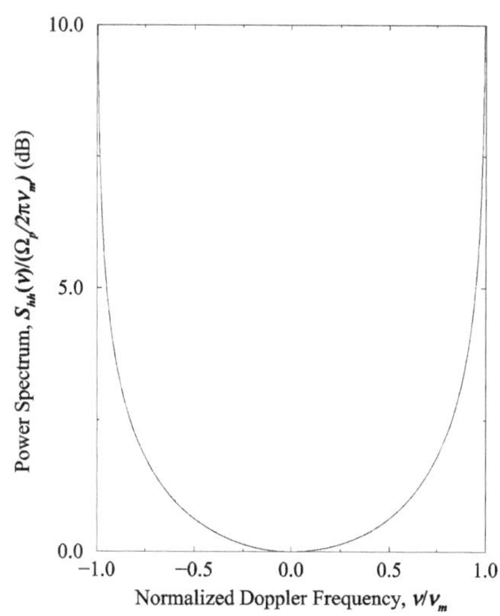

Fig. 11.9 Normalized Doppler spectrum of the received complex envelope with 2-D isotropic scattering

11.3.1.1 Doppler Power Spectrum Alternate Approach

The Doppler power spectrum can be derived by using a different approach that is sometimes useful because it can avoid the need to evaluate integrals. As mentioned earlier, the incident power on the receiver antenna is a function of the azimuth angle θ, such that the fraction of the total received power arriving between angles θ and $\theta + \mathrm{d}\theta$ is $p(\theta)\mathrm{d}\theta$.

From (11.1), the Doppler frequency $\nu = \nu_m \cos(\theta)$ is an even function of the arrival angle θ. Therefore, the Doppler spectrum may be expressed as [14]

$$S_{\tilde{h}\tilde{h}}(\nu)|\mathrm{d}\nu| = \Omega_p\Big(p(\theta) + p(-\theta)\Big)|\mathrm{d}\theta| , \quad 0 \le \theta \le \pi. \tag{11.61}$$

However,

$$|\mathrm{d}\nu| = \nu_m| - \sin(\theta)\mathrm{d}\theta| = \sqrt{\nu_m^2 - \nu^2}\, |\mathrm{d}\theta|. \tag{11.62}$$

The second equality in the above equation can be obtained by considering Fig. 11.10 from which $\sin(\theta) = \sqrt{\nu_m^2 - \nu^2}/\nu_m$.

From the above, it follows

$$S_{\tilde{h}\tilde{h}}(\nu) = \frac{\Omega_p}{\sqrt{\nu_m^2 - \nu^2}}\Big(p(\theta) + p(-\theta)\Big) , \quad 0 \le \theta \le \pi, \tag{11.63}$$

where

$$\theta = \cos^{-1}(\nu/\nu_m) . \tag{11.64}$$

Hence, if $p(\theta)$ is known, the Doppler power spectrum can be easily calculated without resorting to integrals. Instead, it reduces to a problem of substituting $\theta = \cos^{-1}(/\nu/\nu_m)$ into $p(\theta)$ and $p(-\theta)$ in (11.63) and simplifying.

For 2-D isotropic scattering, $p(\theta) = 1/(2\pi)$, so that (11.63) becomes

$$S_{\tilde{h}\tilde{h}}(\nu) = \begin{cases} \frac{\Omega_p}{\pi \nu_m} \frac{1}{\sqrt{1-(\nu/\nu_m)^2}} , & 0 \le |\nu| \le \nu_m \\ 0 , & \text{otherwise} \end{cases} . \tag{11.65}$$

The same result was obtained in (11.45).

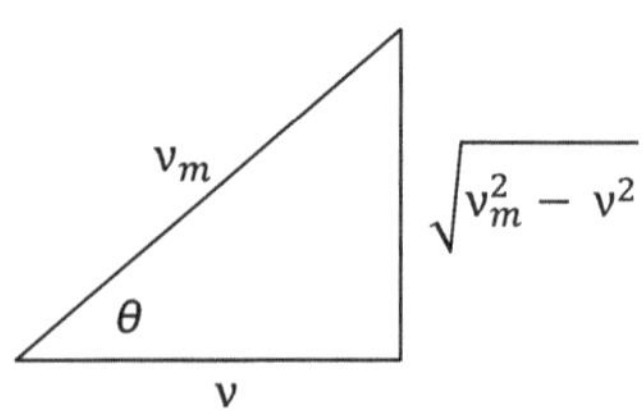

Fig. 11.10 Geometry relating angle of arrival θ and Doppler frequency ν

For a Rician fading channel with the power angle distribution $p(\theta)$ in (11.49) it follows readily from (11.63) that

$$S_{\tilde{h}\tilde{h}}(\nu) = \begin{cases} \frac{1}{K+1} \cdot \frac{\Omega_p}{\pi \nu_m} \frac{1}{\sqrt{1-(\nu/\nu_m)^2}} + \frac{K}{K+1}\Omega_p \delta(\nu - \nu_m \cos(\theta_0)) \ , & 0 \leq |\nu| \leq \nu_m \\ 0 & , \quad \text{otherwise} \end{cases} . \tag{11.66}$$

Note that the Doppler power spectrum in (11.66) has the same shape as (11.65), except for the discrete spectral tone at frequency $\nu = \nu_m \cos(\theta_0)$. The same result may be obtained by taking the Fourier transform of the components in (11.50a) and (11.50b) and substituting into (11.54).

11.4 Simulation of Multipath-Fading Channels

Software simulation of mobile radio channels is necessary for the development and testing of mobile radio systems. It is desirable to generate complex fading envelopes that match the statistical characteristics of a mathematical reference model while at the same time having low computational complexity.

Two fundamentally different approaches to channel simulation are the filtered white Gaussian noise models and the sum-of-sinusoids models. The filtered white noise approach shapes the power spectrum of a white Gaussian noise source by using a pair of filters whose magnitude response is the square root of the desired Doppler power spectrum. The sum-of-sinusoids approach synthesizes the complex fading envelope as the superposition of a finite number of sinusoids with properly chosen amplitudes, frequencies and phases to yield the desired Doppler power spectrum, very often the classical Doppler power spectrum.

11.4.1 Filtered White Gaussian Noise Models

Recall that the complex fading envelope $\tilde{h}(t) = \tilde{h}_I(t) + j\tilde{h}_Q(t)$ is a complex Gaussian random process. A straightforward method for constructing a fading simulator is to low-pass filter two independent white Gaussian noise sources, as shown in Fig. 11.11. Since the input processes to the low-pass filters are independent, the random processes $\tilde{h}_I(t)$ and $\tilde{h}_Q(t)$ will also be independent and have power spectral densities that are shaped by the squared magnitude response of the low-pass filters. If the white Gaussian noise sources have power spectral densities equal to $\Omega_p/2$ watts/Hz, then

$$S_{\tilde{h}_I\tilde{h}_I}(\nu) = S_{\tilde{h}_Q\tilde{h}_Q}(\nu) = \frac{\Omega_p}{2}|H(\nu)|^2 \tag{11.67a}$$

$$S_{\tilde{h}_I\tilde{h}_Q}(\nu) = 0, \tag{11.67b}$$

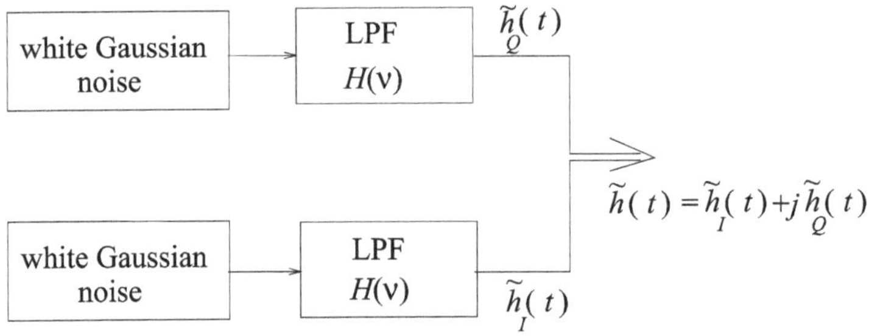

Fig. 11.11 Fading simulator that uses low-pass filtered white Gaussian noise

where the normalization $\int_{-\infty}^{\infty} |H(\nu)|^2 df = 1$ is assumed so that the envelope power is equal to Ω_p. Since the processes $\tilde{h}_I(t)$ and $\tilde{h}_Q(t)$ are independent, zero mean and Gaussian, the envelope $\alpha \equiv |\tilde{h}(t_1)|$ is Rayleigh distributed at any time t_1. The problem is to construct a low-pass filter $H(\nu)$ that will closely approximate the Doppler spectrum of the reference model. For the classical Doppler power spectrum with its asymptotes at $\pm\, \nu_m$, the filter design is possible but challenging.

11.4.2 Sum-of-Sinusoids Methods

Sum-of-sinusoids (SoS) channel models simulate the complex fading envelope by the sum of multiple complex exponentials having appropriately selected frequencies, amplitudes, and phases. SoS models are broadly categorized as being either deterministic or statistical. Deterministic SoS models use fixed frequencies, amplitudes, and phases for the complex exponential components. These parameters are chosen so that the statistical properties of the faded envelope, particularly the temporal auto-correlation, match those of the reference model. Deterministic SoS models are the simplest to use.

Statistical SoS models require multiple simulation trials, where one or more of the parameter sets (frequencies and/or phases) are randomly selected for each trial. As a result, the simulated channels have statistical properties that vary for each trial, but they converge to the desired statistical properties when averaged over a sufficiently large number of trials. An ergodic statistical SoS model is one whose statistical properties converge to those of the reference model in a single trial. The ergodic statistical model differs from the deterministic model, in that one of the parameter sets, typically the phases, is randomly generated prior to the simulation trial.

11.4.2.1 Clarke's model

Clarke's model begins with (11.11) and the phases in (11.6) and assumes equal strength multipath components ($C_n = \sqrt{1/N}$).[1] The received complex envelope is modeled as

$$\tilde{h}(t) = \sqrt{\frac{1}{N}} \sum_{n=1}^{N} \mathrm{e}^{j(2\pi \nu_m \cos(\theta_n)t + \phi_n)} , \tag{11.68}$$

where N is the number of sinusoids and the phases ϕ_n are i.i.d. uniform random variables on $[-\pi, \pi)$. For a 2-D isotropic scattering reference model, the θ_n are also i.i.d. uniform random variables on $[-\pi, \pi)$, and are independent of the ϕ_n.

Clarke's model is a statistical simulation model where the C_n are fixed, while ϕ_n and θ_n are generated anew with each simulation run. The ensemble averaged correlation functions of Clarke's model in (11.68) for a *finite* N are [14]

$$\phi_{\tilde{h}_I \tilde{h}_I}(\tau) = \phi_{\tilde{h}_Q \tilde{h}_Q}(\tau) = \tfrac{1}{2} J_0(2\pi \nu_m \tau) \tag{11.69}$$

$$\phi_{\tilde{h}_I \tilde{h}_Q}(\tau) = \phi_{\tilde{h}_Q \tilde{h}_I}(-\tau) = 0 \tag{11.70}$$

$$\phi_{\tilde{h}\tilde{h}}(\tau) = J_0(2\pi \nu_m \tau) \tag{11.71}$$

$$\phi_{\alpha^2\alpha^2}(\tau) = 1 + \frac{N-1}{N} J_0^2(2\pi \nu_m \tau). \tag{11.72}$$

For finite N, the auto- and cross-correlation of $\tilde{h}_I(t)$ and $\tilde{h}_Q(t)$ match those of the reference model in (11.45) and (11.47).

11.4.2.2 Jakes' Model

Jakes [14] derived a deterministic SoS fading simulation model that is widely cited in literature. Jakes' model begins by choosing the N multipath components to be uniformly distributed in angle of arrival, i.e.,

$$\theta_n = \frac{2\pi n}{N} , \quad n = 1, 2, \ldots, N. \tag{11.73}$$

[1] Note that $\Omega_p = \sum_{n=1}^{N} C_n^2 = 1$ in this case; other values of Ω_p can be obtained by straightforward scaling.

By choosing $N/2$ to be an odd integer, the sum in (11.68) can be rearranged into the form

$$\tilde{h}(t) = \sqrt{\frac{1}{N}} \left\{ \sum_{n=1}^{N/2-1} \left[e^{-j(2\pi \nu_m \cos(\theta_n)t - \phi_{N/2+n})} + e^{j(2\pi \nu_m \cos(\theta_n)t + \phi_n)} \right] \right.$$
$$\left. + e^{-j(2\pi \nu_m t - \phi_{N/2})} + e^{j(2\pi \nu_m t + \phi_N)} \right\}. \tag{11.74}$$

As depicted in Fig. 11.12 for $N = 10$, as n progresses from 1 to $N/2 - 1$ the Doppler shifts progress from $-2\pi \nu_m \cos(2\pi/N)$ to $+2\pi \nu_m \cos(2\pi/N)$ in the first sum, while they progress from $+2\pi \nu_m \cos(2\pi/N)$ to $-2\pi \nu_m \cos(2\pi/N)$ in the second sum in (11.74). Due to symmetry, the Doppler shifts in these terms overlap. To reduce the number of complex exponentials used in the model, the Jakes method removes the overlapping Doppler shifts to write $\tilde{h}(t)$ as

$$\tilde{h}(t) = \sqrt{\frac{1}{N}} \left\{ \sqrt{2} \sum_{n=1}^{M} \left[e^{-j(2\pi \nu_m t \cos(\theta_n) - \phi_{N/2+n})} + e^{j(2\pi \nu_m t \cos(\theta_n) + \phi_n)} \right] \right.$$
$$\left. + e^{-j(2\pi \nu_m t - \phi_{N/2})} + e^{j(2\pi \nu_m t + \phi_N)} \right\} \tag{11.75}$$

where

$$M = \frac{1}{2} \left(\frac{N}{2} - 1 \right) \tag{11.76}$$

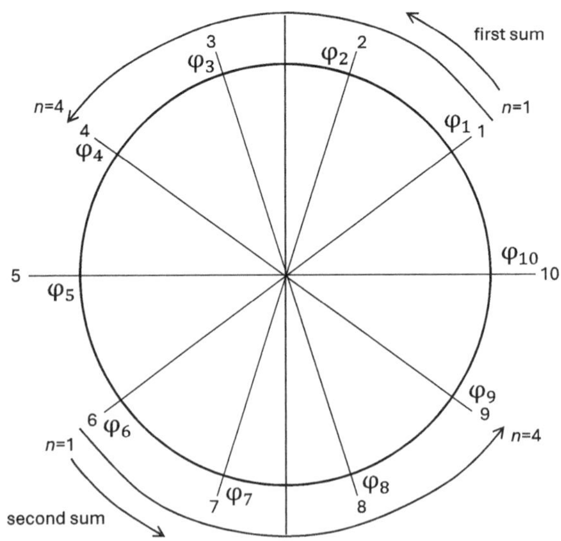

Fig. 11.12 Depiction of the frequencies and phases of the complex exponential components in Jakes' method with $N = 10$

and the factor $\sqrt{2}$ is included so that the envelope power remains at unity, since the sums now contain half the number of complex exponentials. Although not immediately obvious, (11.74) and (11.75) are not exactly the same. In (11.74) all phases are independent. However, the model reduction in (11.75) implies that $\phi_{N-n} = \phi_n$ and $\phi_{N/2+n} = \phi_{N/2-n}$ for $n = 1, \ldots, M$.

The Jakes method further constrains the phases by setting $\phi_{N/2+n} = \phi_n, n = 1, \ldots, M$ and $\phi_{N/2} = \phi_N$. By doing so, $\tilde{h}(t) = \tilde{h}_I(t) + j\tilde{h}_Q(t)$ in (11.75) may be reduced further such that its real and imaginary components have the form

$$\tilde{h}_I(t) = \sqrt{\frac{1}{N}} \left(\sum_{n=0}^{M} a_n \cos(2\pi \nu_m \cos(\theta_n) t) \right) \tag{11.77a}$$

$$\tilde{h}_Q(t) = \sqrt{\frac{1}{N}} \left(\sum_{n=0}^{M} b_n \cos(2\pi \nu_m \cos(\theta_n) t) \right), \tag{11.77b}$$

where

$$a_n = \begin{cases} 2\sqrt{2}\cos(\beta_n), & n = 1, \ldots, M \\ 2\cos(\beta_n), & n = 0 \end{cases} \tag{11.78a}$$

$$b_n = \begin{cases} 2\sqrt{2}\sin(\beta_n), & n = 1, \ldots, M \\ 2\sin(\beta_n), & n = 0 \end{cases} \tag{11.78b}$$

and $\beta_0 = \phi_N$ and $\beta_n = \phi_n$.

The Jakes method next chooses the phases β_n, $n = 0, 1, \ldots, M$ with the objective of making $\langle \tilde{h}_I^2(t) \rangle = \langle \tilde{h}_Q^2(t) \rangle = 1/2$ and $\langle \tilde{h}_I(t)\tilde{h}_Q(t) \rangle = 0$, where $\langle \cdot \rangle$ denotes time averaging. From (11.77a) and (11.77b),

$$\begin{aligned} \langle \tilde{h}_I^2(t) \rangle &= \frac{2}{N} \left[2\sum_{n=1}^{M} \cos^2(\beta_n) + \cos^2(\beta_0) \right] \\ &= \frac{2}{N} \left[M + \cos^2(\beta_0) + \sum_{n=1}^{M} \cos(2\beta_n) \right], \end{aligned} \tag{11.79}$$

$$\begin{aligned} \langle \tilde{h}_Q^2(t) \rangle &= \frac{2}{N} \left[2\sum_{n=1}^{M} \sin^2(\beta_n) + \sin^2(\beta_0) \right] \\ &= \frac{2}{N} \left[M + \sin^2(\beta_0) - \sum_{n=1}^{M} \cos(2\beta_n) \right], \end{aligned} \tag{11.80}$$

and

$$\langle \tilde{h}_I(t)\tilde{h}_Q(t)\rangle = \frac{2}{N}\left[2\sum_{n=1}^{M}\sin(\beta_n)\cos(\beta_n) + \sin(\beta_0)\cos(\beta_0)\right]. \tag{11.81}$$

Finally, the Jakes method chooses the phases as

$$\beta_n = \frac{\pi n}{M}, \quad n = 0, 1, \ldots, M \tag{11.82}$$

to yield

$$\begin{aligned}\langle \tilde{h}_Q^2(t)\rangle &= M/(2M+1)\\ \langle \tilde{h}_I^2(t)\rangle &= (M+1)/(2M+1)\\ \langle \tilde{h}_I(t)\tilde{h}_Q(t)\rangle &= 0\end{aligned}$$

Note that there is a slight imbalance in the values of $\langle \tilde{h}_I^2(t)\rangle$ and $\langle \tilde{h}_Q^2(t)\rangle$ that diminishes with M and $\langle \tilde{h}_I^2(t)\rangle + \langle \tilde{h}_Q^2(t)\rangle = 1$ as desired. Finally, the envelope power can be set to any desired value Ω_p by simply scaling the $\tilde{h}(t)$ obtained above, viz., $\sqrt{\Omega_p}\tilde{h}(t)$ will have envelope power Ω_p.

A typical faded envelope obtained by using Jakes' method with $N = 34$ ($M = 8$) and $\Omega_p = 1$ is shown in Fig. 11.13, where the envelope level in decibel units $20\log_{10}|\tilde{h}(t)|$ is plotted against time t/T where T is the discrete-time step size. Note that the plot in Fig. 11.13 does not start at $t = 0$. The value $\nu_m T = 0.1$ is chosen to illustrate the faded envelope, and the normalized Doppler $\nu_m T$ is typically much smaller.

The accuracy of Jakes' simulator can be evaluated by comparing the auto-correlation of the complex fading envelope with that of the 2-D isotropic scattering reference model. To do so, suppose Clarke's statistical method is modified by imposing the additional restrictions of even N and $\theta_n = 2\pi n/N$, $n = 1, \ldots, N$. The resulting simulation model is a statistical model where the phases ϕ_n are chosen as i.i.d. uniform random variables on $[-\pi, \pi)$ for each simulation trial. The resulting ensemble auto-correlation is

$$\phi_{\tilde{h}\tilde{h}}(\tau) = \frac{1}{N}\sum_{n=1}^{N}\cos\left(2\pi\nu_m\tau\cos\left(\frac{2\pi n}{N}\right)\right), \tag{11.83}$$

where ensemble averaging is performed over the random phases $\hat{\phi}_n$. Note that the auto-correlation in (11.83) tends to the Bessel function $J_0(2\pi\nu_m\tau)$ as $N \to \infty$. The normalized auto-correlations in (11.71) and (11.83) with $N = 8$ are plotted

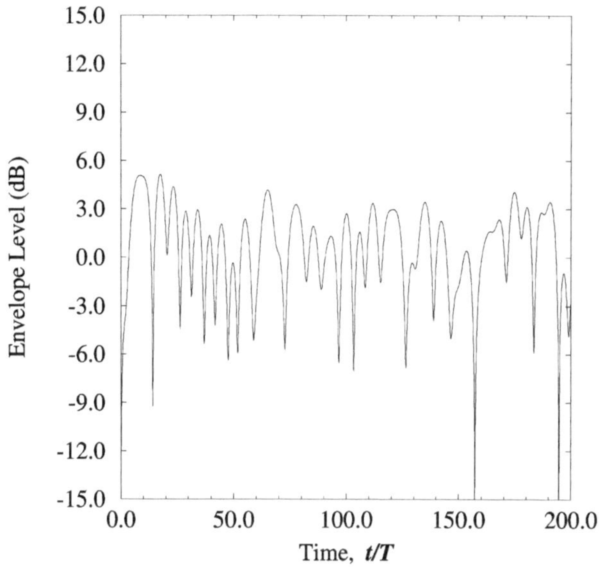

Fig. 11.13 Faded envelope generated by using Jakes' method with $M = 8$ oscillators; $\nu_m T = 0.1$

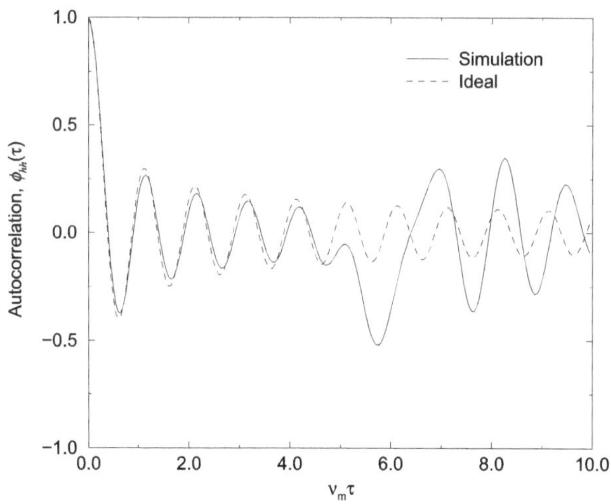

Fig. 11.14 Auto-correlation of in-phase and quadrature components obtained by using Clarke's method with $\theta_n = 2\pi n/N$ and $N = 8$

against the normalized time delay $\nu_m \tau$ in Fig. 11.14. Observe that Clark's model with $\theta_n = 2\pi n/N$, $n = 1, \ldots, N$ yields an auto-correlation that deviates from the reference model at large lags. This can be improved upon by increasing N. For example, Fig. 11.15 shows the auto-correlation when $N = 16$.

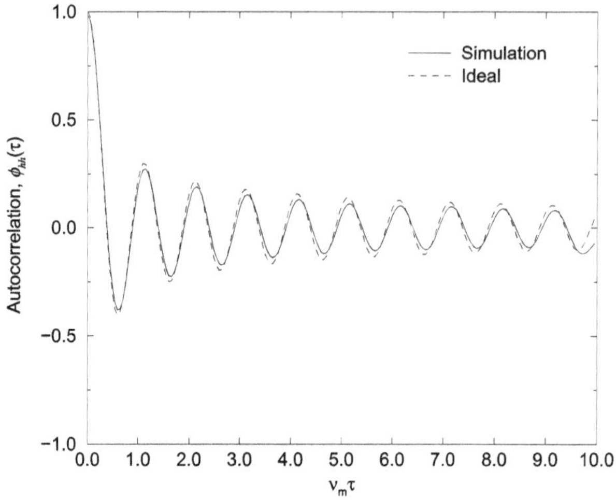

Fig. 11.15 Auto-correlation of in-phase and quadrature components obtained by using Clarke's method with $\theta_n = 2\pi n/N$ and $N = 16$

Returning to the reduced model in (11.77b) and (11.78b), the ensemble auto-correlation $\phi_{\tilde{h}\tilde{h}}(t, t+\tau)$ may be obtained by treating the phases $\beta_n,\ n = 0, \ldots, M$ as i.i.d. unform random variables on $[-\pi, \pi)$ and averaging to give [24]

$$\hat{\phi}_{\tilde{h}\tilde{h}}(t, t+\tau) = \frac{1}{N}\left(\cos(2\pi\nu_m\tau) + \cos(2\pi\nu_m(2t+\tau))\right) \\ + \frac{2}{N}\sum_{n=1}^{M}\left(\cos(2\pi\nu_m\cos(\theta_n)\tau) + \cos(2\pi\nu_m\cos(\theta_n)(2t+\tau))\right). \tag{11.84}$$

From (11.84), it is apparent that the ensemble averaged auto-correlation depends on the time variable t. Hence, the reduced model from which the Jakes simulation model is derived yields a faded envelope that is not stationary or even WSS. The cause of this non-stationary behavior is the correlation introduced into the phases by setting $\phi_{N-n} = \phi_n$ and $\phi_{N/2+n} = \phi_{N/2-n}$ for $n = 1, \ldots, M$.

Due to the non-stationary behavior of the original Jakes simulator, and also because the Jakes' method yields only a single faded envelope, a number of improved sum-of-sinusoids models have been proposed in the literature that also generate multiple faded envelopes. One such model is presented here.

11.4.2.3 Multi-envelope Deterministic Model

A variety of enhanced Jakes' models have been suggested in the literature to improve on the statistical characteristics and to produce multiple uncorrelated fading envelopes. One such deterministic model generates P faded envelopes, where $1 \leq P \leq M$, with M is the number of sinusoids. The kth faded envelope $\tilde{h}_k(t) = \tilde{h}_{I,k}(t) + j\tilde{h}_{Q,k}(t)$ is generated according to [36]

$$\tilde{h}_{I,k}(t) = \sqrt{\frac{1}{N}} \sum_{n=0}^{M} a_n \cos(2\pi \nu_m t \cos(\theta_{n,k}) + \phi_{n,k}) \tag{11.85a}$$

$$\tilde{h}_{Q,k}(t) = \sqrt{\frac{1}{N}} \sum_{n=0}^{M} b_n \sin(2\pi \nu_m t \cos(\theta_{n,k}) + \phi_{n,k}), \tag{11.85b}$$

where

$$M = \frac{1}{2}\left(\frac{N}{2} - 1\right), \tag{11.86}$$

and

$$a_n = \begin{cases} 2\sqrt{2}\cos(\beta_n), & n = 1, \ldots, M \\ 2\cos(\beta_n), & n = 0 \end{cases} \tag{11.87a}$$

$$b_n = \begin{cases} 2\sqrt{2}\sin(\beta_n), & n = 1, \ldots, M \\ 2\sin(\beta_n), & n = 0 \end{cases} \tag{11.87b}$$

$$\beta_n = \frac{\pi n}{M}, \quad n = 0, \ldots, M \tag{11.87c}$$

$$\theta_{n,k} = \frac{2\pi n}{N} + \frac{2\pi k}{PN} + \theta_{0,0}, \quad n = 0, \ldots, M, \quad k = 0, \ldots, P-1. \tag{11.87d}$$

The angle $\theta_{n,k}$ is chosen so that the arrival angles associated with the kth faded envelope are obtained by rotating the arrival angles of the $(k-1)$th faded envelope by $2\pi/PN$. This ensures that the Doppler shifts for each faded envelope are all distinct, which minimizes the correlation between the multiple faded envelopes. The initial arrival angle, $\theta_{0,0}$, can be numerically optimized to minimize the cross-correlation between the quadrature components of each faded envelope, resulting in the choice $\theta_{0,0} = 0.2\pi/PN$. Finally, the $M+1$ phases $\phi_{n,k}$ are chosen as i.i.d. uniform random variables on the interval $[0, 2\pi)$, a process that only needs to be done once. A completely deterministic model can be obtained by approximating the uniform random phases by equally spaced phases that are all distinct, such that

$$\phi_{n,k} = \frac{2\pi n}{M} + \frac{2\pi k}{PM}, \quad n = 0, \ldots, M, \quad k = 0, \ldots, P-1. \tag{11.88}$$

In the case where the $\phi_{n,k}$ are chosen as i.i.d. uniform random variables on the interval $[0, 2\pi)$, the auto- and cross-correlation of the quadrature components, the auto- and cross-correlation of the multiple faded envelopes, and the squared envelope auto-correlation, respectively, have been derived in [36] as

$$\lim_{N\to\infty} \phi_{\tilde{h}_{I,k}\tilde{h}_{I,k}}(\tau) = \lim_{N\to\infty} \frac{2}{N}\sum_{n=0}^{M} \frac{a_n^2}{2}\cos(2\pi\nu_m\tau\cos(\theta_{n,k}))$$
$$= \tfrac{1}{2}J_0(2\pi\nu_m\tau) + \tfrac{1}{2}J_4(2\pi\nu_m\tau), \tag{11.89a}$$

$$\lim_{N\to\infty} \phi_{\tilde{h}_{Q,k}\tilde{h}_{Q,k}}(\tau) = \lim_{N\to\infty} \frac{2}{N}\sum_{n=0}^{M} \frac{b_n^2}{2}\cos(2\pi\nu_m\tau\sin(\theta_{n,k}))$$
$$= \tfrac{1}{2}J_0(2\pi\nu_m\tau) - \tfrac{1}{2}J_4(2\pi\nu_m\tau), \tag{11.89b}$$

$$\phi_{\tilde{h}_{I,k}\tilde{h}_{Q,k}}(\tau) = \phi_{\tilde{h}_{Q,k}\tilde{h}_{I,k}}(\tau) = \phi_{\tilde{h}_k\tilde{h}_{l\neq k}}(\tau) = 0, \tag{11.89c}$$

$$\lim_{N\to\infty} \phi_{\tilde{h}_k\tilde{h}_k}(\tau) = \lim_{N\to\infty} \frac{2}{N}\sum_{n=0}^{M} \frac{a_n^2}{2}\cos(2\pi\nu_m\tau\cos(\theta_{n,k}))$$
$$+ \lim_{N\to\infty} \frac{2}{N}\sum_{n=0}^{M} \frac{b_n^2}{2}\cos(2\pi\nu_m\tau\sin(\theta_{n,k}))$$
$$= J_0(2\pi\nu_m\tau), \tag{11.89d}$$

$$\phi_{\alpha_k^2\alpha_k^2}(\tau) = \frac{2}{N^2}\sum_{n=0}^{M} a_n^4 + \frac{2}{N^2}\sum_{n=0}^{M} b_n^4 + \tfrac{1}{2}\phi^2_{\tilde{h}_{I,k}\tilde{h}_{I,k}}(\tau)$$
$$+ \tfrac{1}{2}\phi^2_{\tilde{h}_{Q,k}\tilde{h}_{Q,k}}(\tau) + \phi^2_{\tilde{h}_{I,k}\tilde{h}_{Q,k}}(\tau), \tag{11.89e}$$

where $J_0(\cdot)$ is the zero-order Bessel function of the first kind and $J_4(\cdot)$ is the fourth-order Bessel function of the first kind.

Figures 11.16 and 11.17 confirm that, for $M = 8$, the auto- and cross-correlations of the quadrature components and the auto- and cross-correlation of the multiple faded envelopes approach the values given in (11.89a). The model satisfies (11.70) and (11.71) of the reference model. However, the auto-correlations of the quadrature components and the auto-correlation of the squared envelope are close to, but do not perfectly match, (11.69) and (11.72), respectively.

11.5 Wide-band Channel Models

A simple geometric model for frequency-selective fading channels is depicted in Fig. 11.18, where the transmitter and receiver are located at the foci of concentric ellipses. Considering only single bounce reflections between the transmitter and receiver, all paths that are associated with scatterers on the nth ellipse will have the same propagation delay τ_n. If the delay spread $\max_{i,j}|\tau_i - \tau_j|$ exceeds roughly 20% of the modulation symbol period T, then the fading channel magnitude response

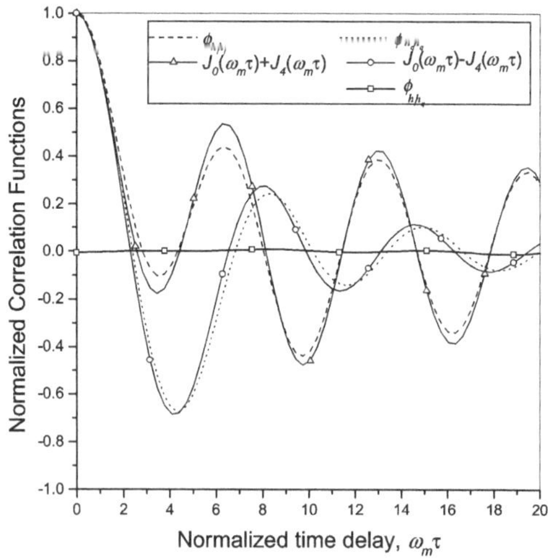

Fig. 11.16 Theoretical and simulated auto-correlation functions and the cross-correlation function of the in-phase and quadrature components of the Multi-envelope Deterministic Model

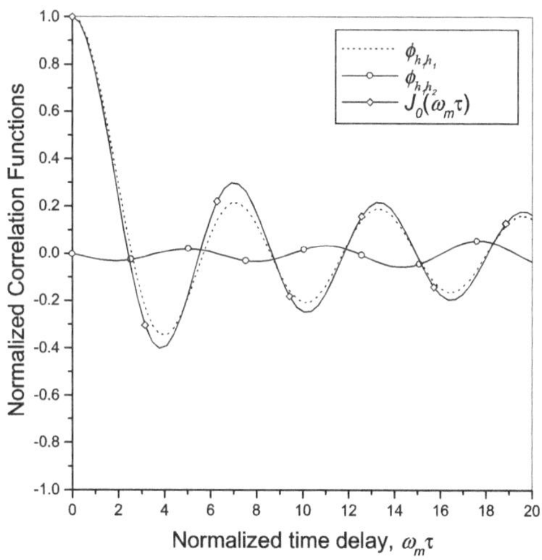

Fig. 11.17 Theoretical and simulated auto-correlation functions and the cross-correlation function of the first and the second complex envelope of the Multi-envelope Deterministic Model

$|\tilde{H}(t, f)|$ will no longer be flat with respect to the frequency variable f and the channel will exhibit frequency-selective fading. Sometimes frequency-selective fading channels are called wide-band channels because they arise with high baud rates $R = 1/T$ that have high bandwidth and small values of T.

Wide-band channels can be modeled by a tapped delay line with irregularly spaced tap delays. Each channel tap is the superposition of a large number of component plane waves that arrive with approximately the same delay, but with random phases and, therefore, the channel taps will undergo fading. The wide-band channel has the time-variant impulse response

$$\tilde{h}(t, \tau) = \sum_{i=1}^{\ell} \tilde{h}_i(t)\delta(\tau - \tau_i), \tag{11.90}$$

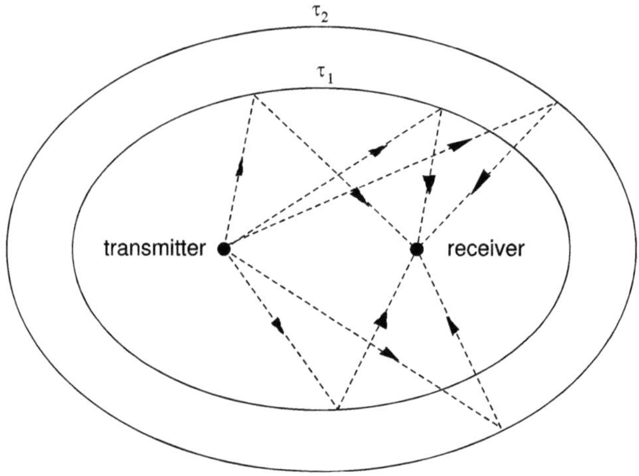

Fig. 11.18 Concentric ellipses model for frequency-selective fading channels

where ℓ is the number of channel taps or rays, and the $\{\tilde{h}_i(t)\}$ and $\{\tau_i\}$ are the complex gains and path delays associated with the channel taps. The corresponding time-variant transfer function is

$$\tilde{H}(t, f) = \sum_{i=1}^{\ell} \tilde{h}_i(t) e^{j2\pi f \tau_i} . \tag{11.91}$$

Usually, the tap delays are multiples of some very small delay τ, such that $\tau_i = K_i \tau$, $i = 1, \ldots, \ell$, which yields the sparse tapped delay line channel model shown in Fig. 11.19, here called the "τ-spaced" model. Many of the tap coefficients in the sparse tapped delay line are zero, reflecting the fact that no energy is received at these delays. Also, the K_i should be chosen to be relatively prime so as to prevent any periodicity in the channel.

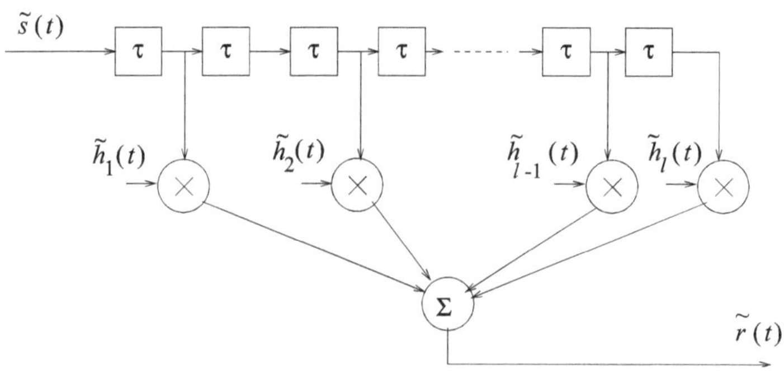

Fig. 11.19 Sparse tapped delay line model for wide-band multipath-fading channels

Example 2.2 (Two-Ray Multipath Channel) Suppose that the waveform $\tilde{s}(t)$ is transmitted over the time-invariant channel

$$\tilde{h}(t) = \alpha\delta(t) + \beta\delta(t - \tau) .$$

such that the received waveform is $\tilde{r}(t) = \alpha\tilde{s}(t) + \beta\tilde{s}(t - \tau_1)$. The channel transfer function $\tilde{H}(f)$ is

$$\tilde{H}(f) = \alpha + \beta e^{-j2\pi f\tau} .$$

Using $|\tilde{H}(f)|^2 = \tilde{H}(f)\tilde{H}^*(f)$ yields

$$|\tilde{H}(f)| = \sqrt{\alpha^2 + \beta^2 + 2\alpha\beta\cos(2\pi f\tau)}$$

Using the Euler identity, $e^{j\theta} = \cos(\theta) + j\sin(\theta)$ in $\tilde{H}(f)$ above,

$$\angle(\tilde{H}(f) = -\mathrm{Tan}^{-1}\frac{\beta\sin(2\pi f\tau)}{\alpha + \beta\cos(2\pi f\tau)}$$

Suppose $\alpha = \beta = 1$. Then

$$|\tilde{H}(f)| = \sqrt{2 + 2\cos(2\pi f\tau)}$$

$$\angle\tilde{H}(f) = -\mathrm{Tan}^{-1}\frac{\sin(2\pi f\tau)}{1 + \cos(2\pi f\tau)}$$

The magnitude response $|\tilde{H}(f)|$ and phase response $\angle\tilde{H}(f)$ are plotted in Figs. 11.20 and 11.21, respectively. Observe that the multipath channel is *frequency selective.*

Land mobile radio channels are often modeled as locally wide-sense stationary with uncorrelated scattering (WSSUS). For WSSUS channels, there is uncorrelated scattering in both the time-delay and Doppler-shift domains. This implies that multipath components arriving at different time delays or from different angles-of-arrival are uncorrelated, which is reasonable since the set of scatterers will vary with distance and angle from the receiver. For a WSSUS channel, the auto-correlation of the time-variant channel impulse response in (11.90) may be derived as

$$\phi_{\tilde{h}\tilde{h}}(t, s; \tau, \eta) = \mathrm{E}[\tilde{h}^*(t, \tau)\tilde{h}(s, \eta)] \tag{11.92a}$$

$$= \sum_{i=1}^{\ell} \phi_{\tilde{h}_i\tilde{h}_i}(\Delta t)\delta(\tau - \tau_i)\delta(\eta - \tau) \tag{11.92b}$$

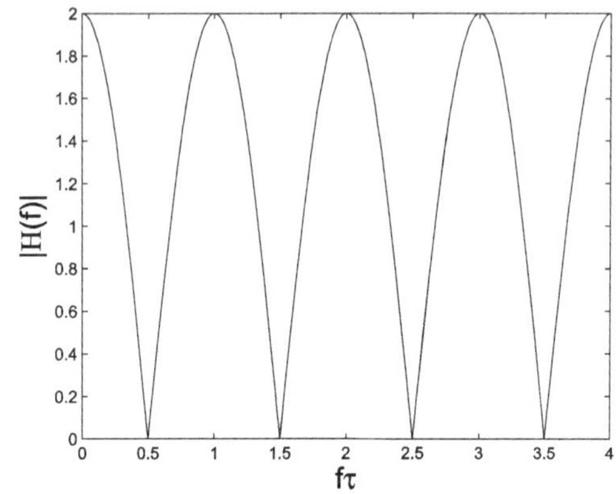

Fig. 11.20 Two-ray multipath channel magnitude response $|\tilde{H}(f)|$

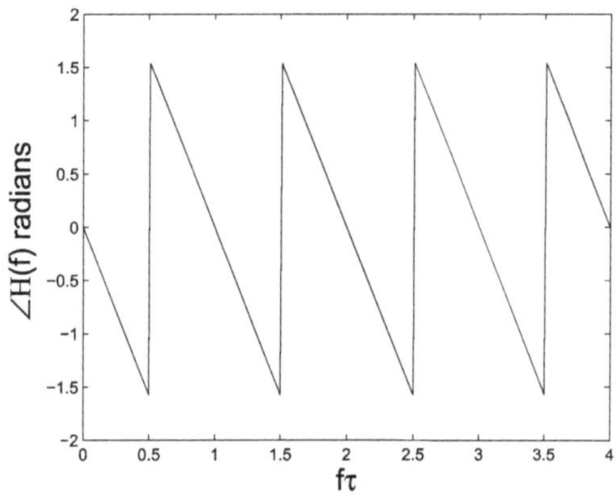

Fig. 11.21 Two-ray multipath channel phase response $\angle\tilde{H}(f)$

where the dependency on the time difference $\Delta_t = s - t$ implies that the channel is wide-sense stationary and the delta functions imply uncorrelated scattering at different delays.

The *wide-band channel correlation function* is defined as

$$\psi_{\tilde{h}\tilde{h}}(\Delta t; \tau) \triangleq \sum_{i=1}^{\ell} \phi_{\tilde{h}_i\tilde{h}_i}(\Delta t)\delta(\tau - \tau_i). \tag{11.93}$$

If each channel tap exhibits 2-D isotropic scattering then, from (11.48),

$$\phi_{\tilde{h}_i\tilde{h}_i}(\Delta t) = \Omega_i J_0(2\pi \nu_m \Delta_t). \tag{11.94}$$

where $\Omega_i = \phi_{\tilde{h}_i\tilde{h}_i}(0)$ is the envelope power of the ith channel tap, and the total envelope power is

$$\Omega_p - \sum_{k=1}^{\ell} \Omega_k. \tag{11.95}$$

The *power-delay profile* is

$$\psi_{\tilde{h}\tilde{h}}(\tau) \triangleq \psi_{\tilde{h}\tilde{h}}(0;\tau) = \sum_{i=1}^{\ell} \Omega_i \delta(\tau - \tau_i), \tag{11.96}$$

which characterizes a wide-band WSSUS fading channel by its power profile

$$\boldsymbol{\Omega} = (\Omega_0,\ \Omega_1, \ldots, \Omega_\ell). \tag{11.97}$$

and its delay profile

$$\boldsymbol{\tau} = (\tau_1,\ \tau_2, \ldots, \tau_\ell). \tag{11.98}$$

Wide-band WSSUS fading channels are often described by their *mean delay*[2]

$$\mu_\tau \triangleq \frac{\int_0^\infty \tau \psi_{\tilde{h}\tilde{h}}(\tau)\mathrm{d}\tau}{\int_0^\infty \psi_{\tilde{h}\tilde{h}}(\tau)\mathrm{d}\tau} \tag{11.99}$$

and their *rms delay spread*,

$$\sigma_\tau \triangleq \sqrt{\frac{\int_0^\infty (\tau - \mu_\tau)^2 \psi_{\tilde{h}\tilde{h}}(\tau)\mathrm{d}\tau}{\int_0^\infty \psi_{\tilde{h}\tilde{h}}(\tau)\mathrm{d}\tau}}. \tag{11.100}$$

The inverse of μ_τ and σ_τ gives a measure of the *coherence bandwidth* of a multipath channel, since a larger delay spread from the ideal channel indicates increased frequency selectivity. The *mean Doppler* μ_ν and *rms Doppler spread* σ_ν may be similarly defined, by replacing the power delay profile $\psi_{\tilde{h}\tilde{h}}(\tau)$ with the Doppler power spectrum $S_{\tilde{h}\tilde{h}}(\nu)$ and the delay variable τ with the Doppler variable ν in (11.99) and (11.100), respectively. The inverse of μ_ν and σ_ν gives a measure of the *coherence time* of a fading channel, since a larger Doppler spread from the static channel indicates increase time selectivity.

[2] Note that the normalization $\int_0^\infty \psi_h(\tau)\mathrm{d}\tau$ is applied because $\psi_{\tilde{h}\tilde{h}}(\tau)$ integrates to Ω_p instead of unity like a probability density function.

11.5.1 COST 207 Models

The COST 207 models were developed for the GSM cellular system. Four different Doppler spectra, $S_{\tilde{h}\tilde{h}}(\nu)$ are specified in the COST 207 models [7]. First, define the function

$$G(A, \nu_1, \nu_2) = A \exp\left\{-\frac{(\nu - \nu_1)^2}{2\nu_2^2}\right\} \tag{11.101}$$

which has the shape of a Gaussian pdf with mean ν_1 and variance ν_2^2. COST 207 specifies the following Doppler spectra:

(a) CLASS is the classical Doppler spectrum and is used for excess path delays not exceeding 500 ns ($\tau_i \leq 500$ ns);

$$\text{(CLASS)} \qquad S_{\tilde{h}\tilde{h}}(\nu) = \frac{A}{\sqrt{1 - (\nu/\nu_m)^2}}, \qquad 0 \leq |\nu| \leq \nu_m \tag{11.102}$$

Multipath components arriving with short path delays are those that experience local scattering around the MS. In the COST 207 models, the local scattering is modeled as 2-D isotropic scattering, and the MS is assumed to have an isotropic antenna.

(b) GAUS1 is the sum of two Gaussian functions and is used for excess path delays ranging from 500 ns to 2 μs; (500 ns $\leq \tau_i \leq 2$ μs)

$$\text{(GAUS1)} \qquad S_{\tilde{h}\tilde{h}}(\nu) = G(A, -0.8\nu_m, 0.05\nu_m) + G(A_1, 0.4\nu_m, 0.1\nu_m), \tag{11.103}$$

where A_1 is 10 dB below A. The GAUS1 Doppler spectra implies two clusters of scattering objects. Considering that the MS is moving along the positive x-axis, the larger cluster is located behind the MS such that the mean angle-of-arrival is $\theta = \cos^{-1}(-0.8) = \pm 143^{\circ}$. A smaller cluster is located in front of the MS such that the mean angle-of-arrival is $\theta = \cos^{-1}(0.4) = \pm 66^{\circ}$.

(c) GAUS2 is the sum of two Gaussian functions, used for excess path delays exceeding 2 μs; ($\tau_i > 500$ ns)

$$\text{(GAUS2)} \qquad S_{\tilde{h}\tilde{h}}(\nu) = G(B, 0.7\nu_m, 0.1\nu_m) + G(B_1, -0.4\nu_m, 0.15\nu_m), \tag{11.104}$$

where B_1 is 15 dB below B. Like GAUS1, GAUS2 implies two clusters of scattering objects such that the mean angles-of-arrival are $\theta = \cos^{-1}(0.7) = \pm 45^{\circ}$ and $\theta = \cos^{-1}(-0.4) = \pm 113^{\circ}$.

(d) RICE is a combination of 2-D isotropic scattering and a line-of-sight component;

$$\text{(RICE)} \qquad S_{\tilde{h}\tilde{h}}(\nu) = \frac{0.5}{\pi \nu_m \sqrt{1-(\nu/\nu_m)^2}} + 0.5\delta(\nu - 0.7\nu_m)\,, \qquad 0 \leq |\nu| \leq \nu_m \tag{11.105}$$

The RICE scattering environment is depicted in Fig. 11.8, where $\theta_0 = \cos^{-1}(0.7) = 45^\circ$. The above RICE reference model has equal power contribution of the line-of-sight and scattered power, i.e., $K = 1$ and has an envelope power of unity.

Several power-delay profiles have been defined by COST 207 for different propagation environments [7], as described in Appendix 2A. Typical urban (TU) (non-hilly) and bad urban (BU) (hilly) discrete power-delay profiles in Table 11.2 of Appendix 2A are shown in Fig. 11.22. In these figures and tables, the fractional power sums to unity, i.e., the total envelope power Ω_p is normalized to unity.

11.6 Shadowing

It was shown earlier in (11.43) that the received envelope power is

$$\Omega_p = \mathrm{E}[|\tilde{h}(t)|^2] = \sum_{n=1}^{N} C_n^2, \tag{11.106}$$

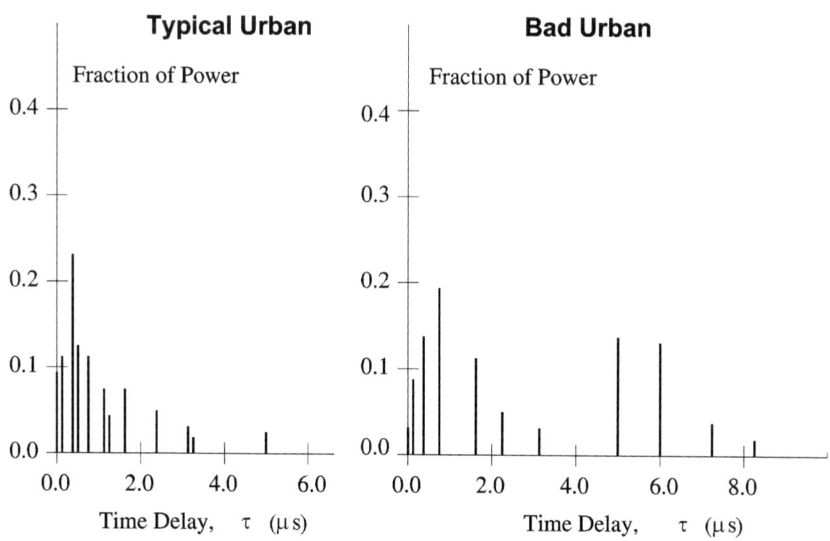

Fig. 11.22 COST 207 typical urban (TU) and bad urban (BU) power-delay profiles

where C_n depends on the cross-sectional area of the nth reflecting surface or length of the nth diffracting edge. The local mean may be estimated by computing the time average

$$\hat{\Omega}_p = \frac{1}{T}\int_T |\tilde{h}(t)|^2 \mathrm{d}t, \tag{11.107}$$

where the time interval T is chosen to correspond to a spatial averaging interval that is large enough to average over the envelope fades but not the shadows. In practice, this corresponds to a spatial averaging distance of about 20 wavelengths. Note that the required averaging interval T will depend on the MS velocity.

The averaging interval must be small enough so that the $\{C_n\}$ do not change over the averaging interval. A *local area* is defined as the largest volume of space where this condition holds true. Sometimes, Ω_p is called the local mean because it is computed within a local area. If the receiver moves outside a local area, the $\{C_n\}$ will change due to the continuous appearance and disappearance of large terrain features like hills, valleys, and buildings. The local mean Ω_p (or its estimate $\hat{\Omega}_p$) changes with time and location in a process known as shadowing.

Empirical studies have shown that Ω_p has the log-normal pdf

$$p_{\Omega_p}(x) = \frac{1}{x\sigma_\Omega \xi\sqrt{2\pi}} \exp\left\{-\frac{\left(10\log_{10}\{x\} - \mu_{\Omega_p\,(\mathrm{dBm})}\right)^2}{2\sigma_\Omega^2}\right\}, \tag{11.108}$$

where

$$\mu_{\Omega_p\,(\mathrm{dBm})} = \mathrm{E}[10\log_{10}\{\Omega_p\}], \tag{11.109}$$

is the mean value of Ω_p in decibel units, σ_Ω is the shadow standard deviation in decibel units, and $\xi = \ln(10)/10$. Here, the local mean is expressed in dBm (decibels with respect to 1 milliwatt), so Ω_p in (11.108) and (11.109) are in units of milliwatts (mW). The shadow standard deviation σ_Ω ranges from 5 to 12 dB with 8 dB being a typical value for macrocellular applications.

The mean value $\mu_{\Omega_p\,(\mathrm{dBm})}$ is sometimes called the area mean because it corresponds to an extended local area that can average over shadowing. The mean envelope power depends on the propagation path loss between the transmitter and receiver, such that the path loss is constant within the extended local area.

By using a transformation of random variables, Problem 11.22 shows that the logarithm $\Omega_{p\,(\mathrm{dBm})} = 10\log_{10}\{\Omega_p\}$ of the log-normal random variable in (11.108) has the normal pdf

$$p_{\Omega_{p\,(\mathrm{dBm})}}(x) = \frac{1}{\sqrt{2\pi}\sigma_\Omega} \exp\left\{-\frac{(x - \mu_{\Omega_p\,(\mathrm{dBm})})^2}{2\sigma_\Omega^2}\right\}. \tag{11.110}$$

11.7 Path Loss Models

Path loss is the largest and most important quantity in a communication link budget. Path loss depends on frequency, antenna heights, distance and topography. A variety of theoretical and empirical path loss models exist in the literature. Our discussion starts with a discussion of theoretical models, followed by empirical models.

11.7.1 *Free Space Path Loss (FSPL)*

Free-space path loss (FSPL) is proportional to the square of the distance between the transmitter and receiver and also proportional to the square of the frequency f_c of the radio signal. The FSPL equation is

$$L_{\mathrm{FS}} = \left(\frac{4\pi d}{\lambda_c}\right)^2 = \left(\frac{4\pi d f_c}{c}\right)^2. \tag{11.111}$$

Free-space path loss (FSPL) is a combination of two effects: First, the intensity of an electromagnetic wave in free space decays with the square of the radio path length, d, such that the received power per unit area or power spatial density (watts/m^2) at distance d is

$$\Omega_r(d) = \Omega_t \frac{1}{4\pi d^2}, \tag{11.112}$$

where Ω_t is the transmitted power in watts. Note that this term is not frequency-dependent.

The second effect is due to aperture, which determines how well an antenna picks up power from an incoming electromagnetic wave. For an isotropic antenna (equal antenna gain in all directions),

$$\Omega_p(d) = \Omega_r(d) \frac{\lambda_c^2}{4\pi}, \tag{11.113}$$

where $\Omega_p(d)$ is the received power at distance d. Note that the aperture is entirely dependent on the wavelength λ_c, which is how the frequency-dependent behavior arises.

Using (11.112) and (11.113) gives the free space propagation path loss as

$$L_{\mathrm{FS\ (dB)}} = 10\log_{10}\left\{\frac{\Omega_t}{\Omega_p(d)}\right\} = 10\log_{10}\left\{\left(\frac{4\pi d}{\lambda_c}\right)^2\right\}$$

$$= 10\log_{10}\left\{\left(\frac{4\pi d f_c}{c}\right)^2\right\}$$

$$= 20\log_{10}\{f_c\} + 20\log_{10}\{d\} - 147.55 \text{ dB}. \tag{11.114}$$

11.7.2 Flat Earth Path Loss (FEPL)

The signals in land mobile radio environments do not experience free space propagation. A more realistic model assumes propagation over a flat reflecting surface, the Earth, as shown in Fig. 11.23. The length of the direct path is

$$d_1 = \sqrt{d^2 + (h_b - h_m)^2} \tag{11.115}$$

and the length of the reflected path is

$$d_2 = \sqrt{d^2 + (h_b + h_m)^2} \tag{11.116}$$

Given that $d \gg h_b h_m$, it follows that $d_1 \approx d$ and $d_2 \approx d$. However, since the wavelength is small, the direct and reflected paths may add constructively or destructively over relatively small distances. The carrier phase difference between the direct and reflected paths is

$$\phi_2 - \phi_1 = \frac{2\pi}{\lambda_c}(d_2 - d_1) \tag{11.117}$$

Taking into account the phase difference, the received envelope power is

$$\mu_{\Omega_p} = \Omega_t \left(\frac{\lambda_c}{4\pi d}\right)^2 \left|1 + a\mathrm{e}^{-jb}\mathrm{e}^{j(\phi_2-\phi_1)}\right|^2, \tag{11.118}$$

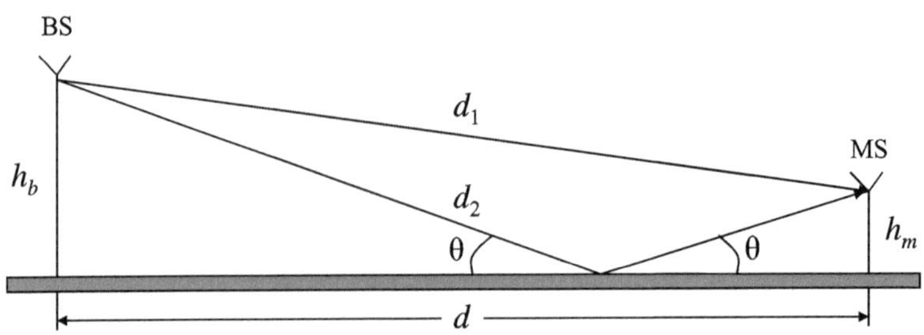

Fig. 11.23 Radio propagation over a flat reflecting surface

where a and b are the amplitude attenuation and phase change introduced by the flat reflecting surface. If a perfect specular reflection is assumed, then $a = 1$ and $b = \pi$ and

$$\mu_{\Omega_p} = \Omega_t \left(\frac{\lambda_c}{4\pi d}\right)^2 \left|1 - \mathrm{e}^{j(\frac{2\pi}{\lambda_c}\Delta_d)}\right|^2$$
$$= 4\Omega_t \left(\frac{\lambda_c}{4\pi d}\right)^2 \sin^2\left(\frac{\pi}{\lambda_c}\Delta_d\right), \quad (11.119)$$

where $\Delta_d = (d_2 - d_1)$. Given that $d \gg h_b$ and $d \gg h_m$, and applying the approximation $\sqrt{1+x} \approx 1 + x/2$ for small x to (11.115) and (11.116),

$$\Delta_d \approx \frac{2h_b h_m}{d}. \quad (11.120)$$

Finally, the received envelope power is

$$\mu_{\Omega_p} \approx 4\Omega_t \left(\frac{\lambda_c}{4\pi d}\right)^2 \sin^2\left(\frac{2\pi h_b h_m}{\lambda_c d}\right), \quad (11.121)$$

where h_b and h_m are the heights of the BS and MS antennas, respectively. Under the condition that $d \gg h_b h_m$, (11.121) reduces to

$$\mu_{\Omega_p} \approx \Omega_t \left(\frac{h_b h_m}{d^2}\right)^2, \quad (11.122)$$

where the small angle approximation $\sin(x) \approx x$ for small x has been invoked. When $d \gg h_b h_m$, propagation over a flat reflecting surface differs from free space propagation in two respects. First, the path loss is not frequency-dependent, and second, the received power is inversely proportional to the fourth power of the distance rather than the square of the distance. Finally, the model in (11.122) shows that the received power is proportional to the product of the square of the BS and MS antenna heights.

Figure 11.24 plots the flat Earth path loss (FEPL)

$$L_{\mathrm{FE\ (dB)}} = 10\log_{10}\left\{\frac{\Omega_t}{\mu_{\Omega_p}}\right\}$$
$$= -10\log_{10}\left\{4\left(\frac{\lambda_c}{4\pi d}\right)^2 \sin^2\left(\frac{2\pi h_b h_m}{\lambda_c d}\right)\right\} \ \mathrm{dB} \quad (11.123)$$

against the distance d. The path loss and, hence, the received power have alternate minima and maxima when the path length is small. This phenomenon has been

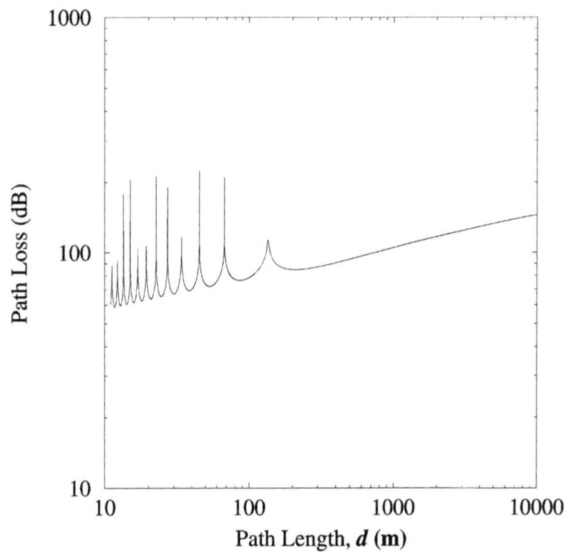

Fig. 11.24 Propagation path loss with distance over a flat reflecting surface; $h_b = 7.5\,\text{m}$, $h_m = 1.5\,\text{m}$, $f_c = 1800\,\text{MHz}$

noted in experiments [22]. The last local maxima in the path loss occurs when

$$\frac{2\pi h_b h_m}{\lambda_c d} = \frac{\pi}{2},$$

giving the breakpoint distance

$$d_{BP} = \frac{4h_b h_m}{\lambda_c d}. \tag{11.124}$$

11.7.3 Empirical Path Loss Models

Several empirical path loss models have been developed by curve fitting experimental measurement data. Two of the more widely cited historical models for 900 MHz cellular systems are Hata's model [23] based on Okumura's prediction method [13] and Lee's model [15].

11.7.3.1 Okumura–Hata and CCIR Models

Hata's empirical model [13] is simple to use and distinguishes between various degrees of urbanization. The empirical data for this model was collected by Okumura [23] in the city of Tokyo. The Okumura–Hata model is expressed in terms of the carrier frequency $150 \leq f_c \leq 1000\,(\text{MHz})$, BS antenna height $30 \leq h_b \leq 200\,(\text{m})$, the MS antenna height $1 \leq h_m \leq 10\,(\text{m})$, and the distance $1 \leq d \leq 20\,(\text{km})$ between the BS and MS. The units in the parentheses are those used in the model. The Okumura–Hata model is known to match the experimental data to within 1 dB for range of parameters indicated. The path loss between a BS and an MS, each with an isotropic antenna, is

$$L_{p\ (\text{dB})} = \begin{cases} A + B\log_{10}\{d\} & \text{for urban area} \\ A + B\log_{10}\{d\} - C & \text{for suburban area} \\ A + B\log_{10}\{d\} - D & \text{for open area} \end{cases} \tag{11.125}$$

where

$$\begin{aligned} A &= 69.55 + 26.16\log_{10}\{f_c\} - 13.82\log_{10}\{h_b\} - a(h_m) \\ B &= 44.9 - 6.55\log_{10}\{h_b\} \\ C &= 5.4 + 2\left(\log_{10}\{f_c/28\}\right)^2 \\ D &= 40.94 + 4.78\left(\log_{10}\{f_c\}\right)^2 - 18.33\log_{10}\{f_c\} \end{aligned} \tag{11.126}$$

and

$$a(h_m) = \begin{cases} \left(1.1\log_{10}\{f_c\} - 0.7\right)h_m - \left(1.56\log_{10}\{f_c\} - 0.8\right) & \\ & \text{for medium or small city} \\ \begin{cases} 8.28\left(\log_{10}\{1.54h_m\}\right)^2 - 1.1 & \text{for } f_c \leq 200\text{ MHz} \\ 3.2\left(\log_{10}\{11.75h_m\}\right)^2 - 4.97 & \text{for } f_c \geq 400\text{ MHz} \end{cases} & \\ & \text{for large city} \end{cases} \tag{11.127}$$

Typical values from the Okumura–Hata "large city" model are plotted in Fig. 11.25, for a BS height of 70 m, an MS antenna height of 1.5 m, and a carrier frequency of 900 MHz. The reader is cautioned that, due to lower urbanization or building density, the path losses for Japanese suburban areas do not match North American suburban areas very well. The latter are more like the quasi-open areas in Japan. Also, the North American urban areas have path losses more like Japanese suburban areas.

To account for varying degrees of urbanization, the CCIR (Comité International des Radio-Communication, now ITU-R) developed the empirical path loss model:

$$L_{p\ (\text{dB})} = A + B\log_{10}\{d\} - E, \tag{11.128}$$

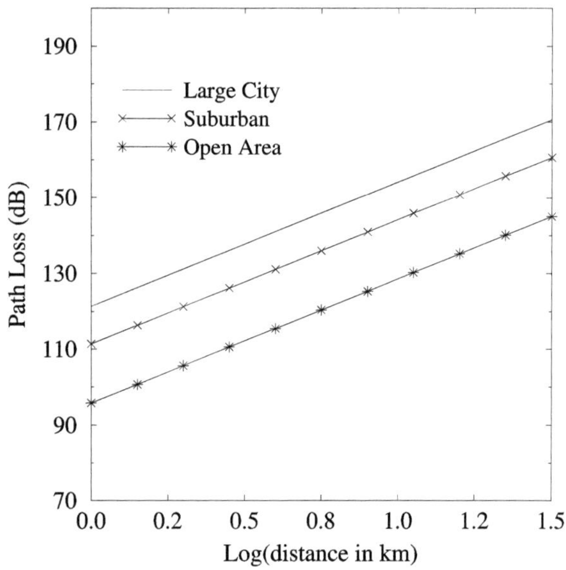

Fig. 11.25 Path loss obtained from the Okumura–Hata model; $h_b = 70\,\text{m}$, $h_m = 1.5\,\text{m}$, $f_c = 900\,\text{MHz}$

where A and B are defined in (11.126) with $a(h_m)$ being the medium or small city value in (11.127). The parameter E accounts for the degree of urbanization and is given by

$$E = 30 - 25\log_{10}\{\% \text{ of area covered by buildings}\}, \tag{11.129}$$

where $E = 0$ when the area is covered by approximately 16% buildings.

11.7.3.2 Lee's Area-to-Area Model

W. C. Y. Lee's area-to-area model [15] can be used to predict a path loss over flat terrain. If the actual terrain is not flat, e.g., hilly, there will be large prediction errors, and other site-specific path loss prediction models should be used [15]. Two parameters are required for Lee's area-to-area model; the power from a 10 watt transmitter that is received at a 1 mile (1.6 km) point of interception, $\mu_P(d_o)$, and the path-loss exponent, β. The received power at distance d can be expressed as

$$\mu_P(d) = \mu_P(d_o)\left(\frac{d}{d_o}\right)^{-\beta}\left(\frac{f}{f_o}\right)^{-n}\alpha_0 \tag{11.130}$$

or in decibel units

$$\mu_{P\ (\mathrm{dBm})}(d) = \mu_{P\ (\mathrm{dBm})}(d_o) - 10\beta\log_{10}\left\{\frac{d}{d_o}\right\} - 10n\log_{10}\left\{\frac{f}{f_o}\right\} + 10\log_{10}\{\alpha_0\}, \tag{11.131}$$

where d is in units of kilometers and $d_o = 1.6\,\mathrm{km}$. The parameter α_0 is a correction factor used to account for different BS and MS antenna heights, transmit powers, and antenna gains from a set of nominal values. The nominal values in Lee's area-to-area model are

- frequency $f_o = 900\,\mathrm{MHz}$
- BS antenna height = 30.48 m
- BS transmit power = 10 watts
- BS antenna gain = 6 dB above dipole gain
- MS antenna height = 3 m
- MS antenna gain = 0 dB above dipole gain

If the actual conditions are different from those listed above, then the following parameters are computed:

$$\begin{aligned}
\alpha_1 &= \left(\frac{\text{BS antenna height (m)}}{30.48\ \text{m}}\right)^2 \\
\alpha_2 &= \left(\frac{\text{MS antenna height (m)}}{3\ \text{m}}\right)^\kappa \\
\alpha_3 &= \frac{\text{transmitter power}}{10\ \text{watts}} \\
\alpha_4 &= \frac{\text{BS antenna gain with respect to } \lambda_c/2 \text{ dipole}}{4} \\
\alpha_5 &= \text{different antenna-gain correction factor at the MS}
\end{aligned} \tag{11.132}$$

From these parameters, the correction factor α_0 is

$$\alpha_0 = \alpha_1 \cdot \alpha_2 \cdot \alpha_3 \cdot \alpha_4 \cdot \alpha_5. \tag{11.133}$$

The parameters β and $\mu_{P\ (\mathrm{dBm})}(d_o)$ have been determined from empirical measurements and are listed in Table 11.1.

Table 11.1 Parameters for Lee's area-to-area model in various propagation environments, from [15]

Terrain	$\mu_{P\ (\mathrm{dBm})}(d_o)$	β
Free space	– 45	2
Open area	– 49	4.35
North American Suburban	– 61.7	3.84
North American Urban (Philadelphia)	– 70	3.68
North American Urban (Newark)	– 64	4.31
Japanese Urban (Tokyo)	– 84	3.05

Experimental data suggest that n in (11.131) ranges between 2 and 3 with the exact value depending upon the carrier frequency and the geographic area. For $f_c <$ 450 MHz in a suburban or open area, $n = 2$ is recommended. In an urban area with $f_c > 450$ MHz, $n = 3$ is recommended. The value of κ in (11.132), also optimized using empirical data, is

$$\kappa = \begin{cases} 2 \text{ for an MS antenna height } > 10 \text{ m} \\ 3 \text{ for an MS antenna height } < 3 \text{ m} \end{cases} . \tag{11.134}$$

The path loss $L_{p\ (\mathrm{dB})}$ is the difference between the transmitted and received power, $L_{p\ (\mathrm{dB})} = \mu_{P\ (\mathrm{dBm})}(d) - \mu_{P_T\ (\mathrm{dBm})}(d)$. To compare directly with the Okumura-Hata model in Fig. 11.25, assume an isotropic BS antenna with 0 dB gain, such that $\alpha_4 = -6$ dB. Then by using the same parameters as in Fig. 11.25, $h_b = 70$ m, $h_m = 1.5$ m, $f_c = 900$ MHz, a nominal BS transmitter power of 40 dBm (10 watts), and the parameters in Table 11.1 for $\mu_{P\ (\mathrm{dBm})}(d_o)$ and β, the following path losses are obtained:

$$L_{p\ (\mathrm{dB})} = \begin{cases} 85.74 + 20.0 \log_{10}\{d\} & \text{Free Space} \\ 84.94 + 43.5 \log_{10}\{d\} & \text{Open Area} \\ 98.68 + 38.4 \log_{10}\{d\} & \text{Suburban} \\ 107.31 + 36.8 \log_{10}\{d\} & \text{Philadelphia} \\ 100.02 + 43.1 \log_{10}\{d\} & \text{Newark} \\ 122.59 + 30.5 \log_{10}\{d\} & \text{Tokyo} \end{cases} \tag{11.135}$$

These typical values from Lee's area-to-area model are plotted in Fig. 11.26.

Appendix 2A: COST 207 Channel Models

The COST 207 study has specified typical realizations for the power-delay profile in the following environments; Typical Urban (TU), Bad Urban (BA), Reduced Typical Urban (RTU), Reduced Bad Urban (RBU), Rural Area (RA), Hilly Terrain (HT), and Reduced Hilly Terrain (RHT) [7]. The models below are identical to the COST 207 models, except that fractional powers have been normalized so as to sum to unity, i.e., the envelope power $\Omega_p = 1$ (Tables 11.2, 11.3, 11.4, 11.5, and 11.6).

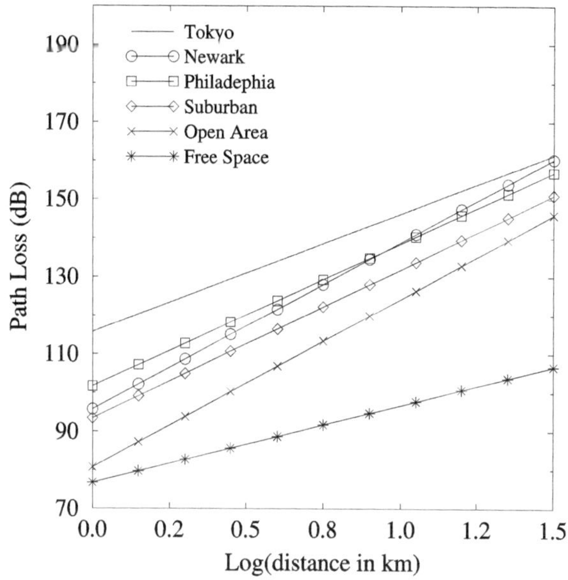

Fig. 11.26 Path loss obtained by using Lee's method; $h_b = 70$ m, $h_m = 1.5$ m, $f_c = 900$ MHz, and an isotropic BS antenna

Table 11.2 COST 207 Typical Urban (TU) ($\sigma_\tau = 1.0\,\mu$s) and Bad Urban (BU) ($\sigma_\tau = 2.5\,\mu$s) power-delay profiles, from [7]

Typical Urban (TU)			Bad Urban (BU)		
Delay µs	Fractional power	Doppler category	Delay µs	Fractional power	Doppler category
0.0	0.092	CLASS	0.0	0.033	CLASS
0.1	0.115	CLASS	0.1	0.089	CLASS
0.3	0.231	CLASS	0.3	0.141	CLASS
0.5	0.127	CLASS	0.7	0.194	GAUS1
0.8	0.115	GAUS1	1.6	0.114	GAUS1
1.1	0.074	GAUS1	2.2	0.052	GAUS2
1.3	0.046	GAUS1	3.1	0.035	GAUS2
1.7	0.074	GAUS1	5.0	0.140	GAUS2
2.3	0.051	GAUS2	6.0	0.136	GAUS2
3.1	0.032	GAUS2	7.2	0.041	GAUS2
3.2	0.018	GAUS2	8.1	0.019	GAUS2
5.0	0.025	GAUS2	10.0	0.006	GAUS2

Table 11.3 COST 207 Reduced Typical Urban (RTU) ($\sigma_\tau = 1.0\,\mu s$) and Reduced Bad Urban (RBU) ($\sigma_\tau = 2.5\,\mu s$) power-delay profiles, from [7]

Typical Urban (TU)			Bad Urban (BU)		
Delay µs	Fractional power	Doppler category	Delay µs	Fractional power	Doppler category
0.0	0.189	CLASS	0.0	0.164	CLASS
0.2	0.379	CLASS	0.3	0.293	CLASS
0.5	0.239	CLASS	1.0	0.147	GAUS1
1.6	0.095	GAUS1	1.6	0.094	GAUS1
2.3	0.061	GAUS2	5.0	0.185	GAUS2
5.0	0.037	GAUS2	6.6	0.117	GAUS2

Table 11.4 COST 207 Typical Rural (non-hilly) Area (RA) power-delay profile ($\sigma_\tau = 0.1\,\mu s$), from [7]

Delay µs	Fractional power	Doppler category
0.0	0.602	RICE
0.1	0.241	CLASS
0.2	0.096	CLASS
0.3	0.036	CLASS
0.4	0.018	CLASS
0.5	0.006	CLASS

Table 11.5 COST 207 Typical Hilly Terrain (HT) power-delay profile ($\sigma_\tau = 5.0\,\mu s$), from [7]

Delay µs	Fractional power	Doppler category
0.0	0.026	CLASS
0.1	0.042	CLASS
0.3	0.066	CLASS
0.5	0.105	CLASS
0.7	0.263	GAUS1
1.0	0.263	GAUS1
1.3	0.105	GAUS1
15.0	0.042	GAUS2
15.2	0.034	GAUS2
15.7	0.026	GAUS2
17.2	0.016	GAUS2
20.0	0.011	GAUS2

Table 11.6 COST 207 Reduced Hilly Terrain (RHT) power-delay profile ($\sigma_\tau = 5.0\,\mu$s), from [7]

Delay μs	Fractional power	Doppler category
0.0	0.413	CLASS
0.1	0.293	CLASS
0.3	0.145	CLASS
0.5	0.074	CLASS
15.0	0.066	GAUS2
17.2	0.008	GAUS2

Problems

11.1 Consider Rayleigh fading, with the fading amplitude α pdf

$$p_\alpha(x) = \frac{2x}{\Omega_p} \exp\left\{-\frac{x^2}{\Omega_p}\right\}, \qquad x \geq 0,$$

where $\Omega_p = \mathrm{E}[\alpha^2]$.

(a) What is the probability that the fading amplitude α is 10 dB or more below the rms fading amplitude $\alpha_{\text{rms}} = \sqrt{\Omega_p}$?

(b) What is the probability that the squared fading amplitude α^2 is 10 dB or more below the mean-squared fading amplitude Ω_p?

11.2 Consider a Ricean fading channel with Rice factor $K = s^2/(2b_0)$ and envelope power $\Omega_p = s^2 + 2b_0$. Assume that the means $m_I(t)$ and $m_Q(t)$ of the in phase and quadrature components are given by, respectively,

$$m_I(t) = s \cdot \cos(2\pi f_m \cos(\theta_0)t + \phi_0)$$
$$m_Q(t) = s \cdot \sin(2\pi f_m \cos(\theta_0)t + \phi_0),$$

and where $f_m \cos(\theta_0)$ and ϕ_0 are the Doppler shift and random phase associated with the LoS or specular component arriving at the receiver with azimuth angle θ_0, respectively.

(a) Derive an integral expression for the probability density function of the envelope phase $\angle \tilde{h}(t) = \mathrm{Tan}^{-1}(\tilde{h}_Q(t)/\tilde{h}_I(t))$ at any time instant t in terms of K and Ω_p.

(b) Find a closed-form expression for the probability density function of the envelope phase in part (a) when $K = 0$, i.e., the channel exhibits Rayleigh fading.

11.3 When a message vector $\tilde{\mathbf{s}}$ is transmitted, the received vector is $\tilde{\mathbf{r}} = \alpha\tilde{\mathbf{s}} + \tilde{\mathbf{w}}$, where the channel gain, α, has the discrete probability density function

$$p_\alpha(x) = 0.2\delta(x) + 0.5\delta(x-1) + 0.3\delta(x-2).$$

and $\tilde{\mathbf{w}}$ is AWGN.

(a) Determine the average probability of bit error for binary BPSK signaling in terms of the *average* received bit energy-to-noise ratio $\bar{\gamma}_b = \mathrm{E}[\alpha^2]E_b/N_o$.
(b) What value does the probability of bit error approach as $\bar{\gamma}_b$ gets large?

11.4 Consider a discrete-time multipath channel with input $x[n]$ and output $y[n]$ that is modeled as the following FIR filter:

$$y[n] = x[n] - 1.5x[n-1] + 4.2x[n-2] - 2.1x[n-3] + 1.5x[n-4]$$

(a) Construct a block diagram description of the channel.
(b) Find the transfer function $H(e^{j\hat{\omega}})$ of the channel, where $\hat{\omega} = 2\pi f/f_s$ is the normalized angular frequency with f_s being the sample frequency.
(c) Plot the magnitude response $|H(e^{j\hat{\omega}})|$ and phase response $\angle H(e^{j\hat{\omega}})$ of the channel.

11.5 Consider a linear time-invariant channel having the impulse response

$$\tilde{h}(t,\tau) = \delta(\tau) + 2\delta(\tau - \tau_1) + \delta(\tau - 2\tau_1).$$

(a) Derive a closed-form expression for magnitude response of the channel $|\tilde{H}(t, f)|$ and sketch it showing all important points.
(b) Repeat part (a) for the phase response of the channel $\angle\tilde{H}(t, f)$.

11.6 Consider a linear time-invariant channel having the impulse response

$$\tilde{h}(t,\tau) = \frac{1}{\sqrt{L}}\sum_{k=1}^{L}\delta(\tau - (k-1)\Delta_\tau)$$

(a) Derive a closed-form expression for magnitude response of the channel $|\tilde{H}(t, f)|$ and sketch it showing all important points.
(b) Suppose that a root raised cosine pulse having roll-off factor $\beta = 0.5$ is used to transmit data with baud rate $R = 3/(L\Delta_\tau)$ on this channel. What frequencies will not be present in the received complex envelope at the output of the channel?

11.7 A cellular subscriber is sitting on a high-speed train traveling at a speed of $v = 220$ km/h.

(a) What is the maximum positive Doppler shift, ν_m, that will occur on the "downlink" channel (935 to 960 MHz RF band), and under what conditions will it occur?
(b) What is the maximum negative Doppler shift, ν_m, that will occur on the "uplink" channel (890 to 915 MHz RF band), and under what conditions will it occur?

(c) What is the minimum possible Doppler shifting that will occur, and under what conditions will it occur?

11.8 A mobile station, moving at a known speed of 80 km/h, has a directional antenna that is tuned to a carrier-frequency of 1 GHz. Plane waves are received from angles that happen to produce Doppler frequencies ranging from 10 Hz to 50 Hz. What is the beam width of the directional antenna, assuming an ideal directional antenna having a uniform gain over its beam width and zero gain elsewhere?

11.9 Determine and plot the Doppler power spectrum $S_{\tilde{h}\tilde{h}}(f)$ for the following power angle distributions:

(a) $p(\theta) = \frac{1}{\pi}\sin^2(\theta)\,, \quad -\pi \leq \theta \leq \pi$
(b) $p(\theta) = \frac{1}{\pi}\cos^2(\theta)\,, \quad -\pi \leq \theta \leq \pi.$

11.10 Consider a multipath channel where the power angle distribution is

$$p(\theta) = \begin{cases} \frac{1}{2}\sin(\theta) & 0 \leq \theta \leq \pi \\ 0 & \text{elsewhere} \end{cases}$$

The mobile station is traveling at 60 mph and the carrier frequency is 1900 MHz. An isotropic antenna is used and the received signal has envelope power of Ω_p.

(a) What is the maximum Doppler frequency, ν_m?
(b) Find the Doppler power spectrum, $S_{\tilde{h}\tilde{h}}(\nu)$, and sketch it.

11.11 Consider channel with a 700 MHz carrier frequency. The Doppler spectrum has

$$S_{\tilde{h}_I\tilde{h}_I}(\nu) = \begin{cases} \text{rect}\left(\frac{\nu}{200}\right), & |\nu| \leq 100\ Hz \\ 0, & \text{elsewhere} \end{cases}$$

$$S_{\tilde{h}_I\tilde{h}_Q}(\nu) = 0.$$

(a) What is the minimum possible speed of the mobile station?
(b) What is the cross-correlation function $\phi_{\tilde{h}_I\tilde{h}_Q}(\tau)$ of the I and Q components of the faded envelope?
(c) Write down an expression for $\phi_{\tilde{h}_Q\tilde{h}_Q}(\tau)$.

11.12 Suppose the Doppler spectrum is

$$S_{\tilde{h}\tilde{h}}(\nu) = A \wedge (\nu - \nu_1) + A \wedge (\nu + \nu_1)$$

where

$$\wedge(\nu) = \begin{cases} 1 - \frac{|\nu|}{W} & , \quad |\nu| < W \\ 0 & , \quad |\nu| \geq W \end{cases}$$

and A, ν_1, and W are constants.

(a) Sketch the Doppler spectrum.
(b) Find the envelope correlation function $\phi_{\tilde{h}\tilde{h}}(\tau)$.
(c) For which values of τ are $\tilde{h}(t)$ and $\tilde{h}(t+\tau)$ uncorrelated?
(d) What is the envelope power Ω_p?
(e) If the mobile station happens to be moving directly toward the base station that is transmitting a 1 GHz unmodulated carrier, and $W = 40$ Hz and $\nu_1 = 44$ Hz, what is the speed of the mobile station?

11.13 Suppose the Doppler power spectrum is

$$S_{\tilde{h}\tilde{h}}(\nu) = A\cos\left(\frac{\pi\nu}{2\nu_m}\right)\operatorname{rect}\left(\frac{\nu}{2\nu_m}\right)$$

(a) Find the corresponding envelope auto-correlation function $\phi_{\tilde{h}\tilde{h}}(\tau)$.
(b) For what values of τ are $\tilde{h}_I(t)$ and $\tilde{h}_Q(t+\tau)$ uncorrelated?
(c) Given that

$$\Omega_p = \phi_{\tilde{h}\tilde{h}}(0) = \int_{-\nu_m}^{\nu_m} S_{\tilde{h}\tilde{h}}(\nu)d\nu$$

find the value of A in terms of Ω_p.

11.14 Suppose the Doppler power spectrum is

$$S_{\tilde{h}\tilde{h}}(\nu) = \begin{cases} \frac{\Omega_p}{2}\left(1 - \frac{|\nu|}{\nu_m}\right), & |\nu| < \nu_m \\ 0, & |\nu| \geq \nu_m \end{cases}$$

(a) Find and sketch the envelope correlation function $\phi_{\tilde{h}\tilde{h}}(\tau)$.
(b) What is the minimum τ such that $\tilde{h}(t)$ and $\tilde{h}(t+\tau)$ are uncorrelated?

11.15 The Doppler power spectrum is

$$S_{\tilde{h}\tilde{h}}(\nu) = \frac{\Omega_p}{2\nu_m}\operatorname{rect}\left(\frac{\nu}{2\nu_m}\right)$$

(a) What is the channel time correlation function $\phi_{\tilde{h}\tilde{h}}(\tau)$?
(b) What is the correlation coefficient ρ between the pair of real-valued Gaussian random variables $X = \tilde{h}_I(t)$ and $Y = \tilde{h}_I\left(t + \frac{1}{4f_m}\right)$?

11.16 The Doppler power spectrum is

$$S_{\tilde{h}\tilde{h}}(\nu) = G(A, -0.8\nu_m, 0.05\nu_m)$$

where

$$G(A, \nu_1, \nu_2) = A \exp\left\{-\frac{(\nu - \nu_1)^2}{2\nu_2^2}\right\}$$

(a) Sketch the Doppler spectrum.
(b) Find the envelope correlation function $\phi_{\tilde{h}\tilde{h}}(\tau)$.
(c) For which values of τ are $\tilde{h}_I(t)$ and $\tilde{h}_Q(t+\tau)$ uncorrelated?
(d) Find the envelope power Ω_p.

11.17 Consider the Jakes' sum-of-sinusoids simulation model. Jakes chooses $\alpha = 0$ and $\beta_n = \pi n/M$. With these choices show that

$$\begin{aligned} < \tilde{h}_I(t)\tilde{h}_Q(t) > &= 0 \\ < \tilde{h}_Q^2(t) > &= M/(2M+1) \\ < \tilde{h}_I^2(t) > &= (M+1)/(2M+1) \end{aligned}$$

11.18 (Computer Exercise) Implement a software fading simulator that uses Jakes' method and plot a typical sample function of the faded envelope $\alpha(t) = |\tilde{h}(t)|$ versus time, where $\tilde{h}(t) = \tilde{h}_I(t) + j\tilde{h}_Q(t)$. By scaling $\tilde{h}_I(t)$ and $\tilde{h}_Q(t)$ appropriately, generate a Rayleigh faded envelope having the mean-squared envelope $\Omega_p = 1$. Plot a sample function of your faded envelope $\alpha(t) = |\tilde{h}(t)|$ assuming a maximum normalized Doppler frequency of $\nu_m T = 0.1$, where T is the simulation step size. The timescale should be indexed in terms of the discrete-time index n. Plot the faded envelope in decibel units, i.e., plot $20\log_{10}\alpha(nT)$.

11.19 (Computer Exercise) This problem uses the fading simulator from Problem 11.18. Estimates of the mean-square envelope Ω_p are used in resource management algorithms such as handoff algorithms. An estimate of the mean-square envelope Ω_p can obtained by forming the empirical average

$$\hat{\Omega}_p = \frac{1}{N}\sum_{i=1}^{N}\left(\tilde{h}_I^2(iT) + \tilde{h}_Q^2(iT)\right),$$

where T is the simulation step size. Assume that $f_m T = 0.1$

(a) Generate 1000 estimates $\hat{\Omega}_p$ by using 1000 non-overlapping averaging windows of length $N = 50, 100, 150, 200, 250, 300$. Construct a plot with the *sample variance* of the 1000 estimates on the ordinate and the average window length N on the abscissa.
(b) Draw qualitative conclusions from part (a)?

11.20 Consider the statistical fading simulator in Fig. 11.27, where independent white Gaussian noise waveforms $w_I(t)$ and $w_Q(t)$ each having a flat power density

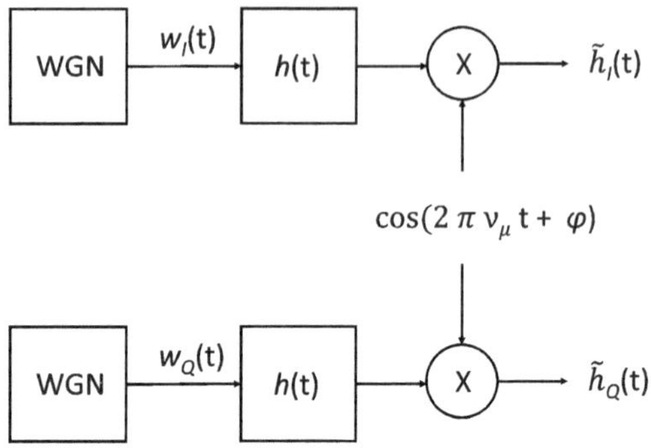

Fig. 11.27 Filtered white noise statistical simulator for Problem 11.20

spectrum of 2 milliwatts/Hz are input to a pair of filters each having impulse response $h(t)$, followed by multiplication by a cosine waveform having frequency ν_μ and a random phase π that is chosen independently and uniformly over the interval $[-\pi, \pi)$ for each simulation run.

The desired Doppler power spectrum is

$$S_{\tilde{h}\tilde{h}}(\nu) = \frac{1}{16 + (2\pi(\nu + 100))^2} + \frac{1}{16 + (2\pi(\nu - 100))^2} \quad \text{milliwatts/Hz}$$

(a) What is the channel time auto-correlation function $\phi_{\tilde{h}\tilde{h}}(\tau)$?
(b) What is the filter impulse response $h(t)$?
(c) What is the value of ν_μ?
(d) Why is the random phase ϕ necessary?

11.21 Consider a Rayleigh faded signal corrupted by Rayleigh faded co-channel interference, and neglect the effects of AWGN. The received signal power, s_0, and interference power, s_1, have the exponential distributions

$$f_{s_0}(x) = \frac{1}{\Omega_0} e^{-x/\Omega_0}$$

$$f_{s_1}(y) = \frac{1}{\Omega_1} e^{-y/\Omega_1},$$

where Ω_0 and Ω_1 are the average received signal power and average received interference power, respectively.

(a) Assuming that s_0 and s_1 are independent random variables, derive the pdf of the carrier-to-interference ratio

$$\lambda = \frac{s_0}{s_1}.$$

as a function of $\bar{\lambda} = \Omega_0/\Omega_1$.

(b) What is the probability $\lambda < \lambda_{\text{th}}$?
(c) What is the mean of λ?

11.22 With log-normal shadowing, the envelope power expressed in dBm units, $\Omega_{p\ (\text{dBm})}$, is a Gaussian random variable having the pdf

$$f_{\Omega_{p\ (\text{dBm})}}(x) = \frac{1}{\sqrt{2\pi}\sigma_\Omega} \exp\left\{ -\frac{(x - \mu_{\Omega_{p\ (\text{dBm})}})^2}{2\sigma_\Omega^2} \right\}.$$

where $\mu_{\Omega_{p\ (\text{dBm})}}$ is the mean envelope power as described by a path loss model. The envelope power in milliwatt units, Ω_p is

$$\Omega_p = 10^{\Omega_{p\ (\text{dBm})}/10}$$

The random variable Ω_p is said to be a log-normal random variable because its logarithm $\Omega_{p\ (\text{dBm})}$ is a normal or Gaussian random variable. Derive the probability density function of the log-normal random variable Ω_p.

Chapter 12
Multi-antenna Techniques

Rayleigh fading converts an exponential dependency of the bit error probability on the average received bit energy-to-noise ratio into an inverse linear one, yielding a very large performance loss. Diversity is a remedy that provides the receiver with multiple independently faded replicas of the same information bearing signal. Sometimes, these replicas are called diversity branches. To understand how diversity works, let p be the probability that the instantaneous received bit energy-to-noise ratio for any one diversity branch falls below a critical threshold γ_{th} such that the bit error probability for that diversity branch exceeds a target value. With independent diversity branches, the probability that the instantaneous bit energy-to-noise ratio simultaneously falls below the same critical threshold γ_{th} for all L diversity branches is p^L, and $p^L \ll p$.

The methods by which diversity can be achieved fall into several categories: (i) space, (ii) angle, (iii) polarization, (iv) frequency, (v) multipath, and (vi) time. Spatial diversity is achieved by using multiple transmit or receiver antennas. The spatial separation between the antenna elements at the transmitter and/or receiver is chosen so that the envelope fades on the various diversity branches are mutually uncorrelated. Sect. 11.3 showed that with 2-D isotropic scattering, antennas spaced roughly a half-wavelength ($\lambda_c/2$) apart will have uncorrelated fading. Angle (or direction) diversity requires a number of directional antennas. Each antenna receives multipath components arriving from a different angle-of-arrival range, so the branches are uncorrelated.

Polarization diversity exploits the depolarization of a radio wave propagating in a scattering environment. Dual-polarized or tri-polarized antennas can be used at the receiver to obtain uncorrelated diversity branches on their orthogonal polarizations.

Frequency diversity exploits frequency selectivity to send the same message on multiple channels that are separated in frequency by at least the channel coherence bandwidth. This can be done efficiently by using frequency-hop spread spectrum techniques. Multipath diversity exploits an uncorrelated scattering channel

G. Stuber, *Introductory Digital Wireless Communications*,
https://doi.org/10.1007/978-3-032-01559-4_12

by resolving uncorrelated multipaths that are received at different delays. This can be done efficiently by using direct-sequence spread spectrum techniques.

Time diversity is obtained by transmitting the same modulation symbols or messages at multiple time instants that are separated by at least the coherence time of the channel. Error correction coding and interleaving techniques are an efficient method for implementing time diversity. Unfortunately, for time diversity, the coherence time of the channel depends on the Doppler spread, and a small Doppler spread implies a large coherence time. Under this condition, it may not be possible to obtain time diversity without introducing an unacceptably large interleaving delay.

Once the diversity branches are available, they must be combined together. The remainder of this chapter begins with an introduction to diversity combining techniques, including selective combining and maximal ratio combining. Multi-input multiple-output (MIMO) channels are considered in Sect. 12.2, while various types of MIMO systems are presented in Sect. 12.3. This chapter concludes with a more detailed treatment of transmitter diversity in Sect. 12.4, namely the Alomouti transmit diversity scheme, and basic spatial multiplexing in Sect. 12.5.

12.1 Diversity Combining

There are many methods for combining the signals that are received on the different diversity branches, and several ways of categorizing them. Diversity combining that weights and combines the continuous-time complex envelopes received on the different branches before matched filtering or correlation detection is called predetection combining. Diversity combining that takes place in the digital baseband after A/D conversion and matched filtering or correlation detection is called postdetection combining. There is no theoretical difference in the performance of these two approaches, but their implementation is quite different. This textbook concentrates on postdetection combining.

Consider the L-branch postdetection receive diversity system shown in Fig. 12.1. The received complex envelope $\tilde{r}_k(t)$ on each diversity branch is passed through a bank of correlators or matched filters shown in Figs. 6.2 and 6.4, respectively, to yield the corresponding received vector $\tilde{\mathbf{r}}_k$. Afterward, the vectors $\tilde{\mathbf{r}}_k$ are input to a diversity combiner that outputs the combined vector $\tilde{\mathbf{r}}$.

The received complex envelopes for the L diversity branches are

$$\tilde{r}_k(t) = \tilde{h}_k \tilde{s}_m(t) + \tilde{w}_k(t), \qquad k = 1, \ldots, L, \tag{12.1}$$

where $\tilde{s}_m(t)$ is the transmitted message and $\tilde{h}_k = \alpha_k e^{j\phi_k}$ is the complex channel fading gain associated with the kth branch. The AWGN processes $\tilde{w}_k(t)$, $k = 1, \ldots, L$ are reasonably assumed to be independent from branch to branch. After correlation or matched filter detection, the L received signal vectors are

$$\tilde{\mathbf{r}}_k = \tilde{h}_k \tilde{\mathbf{s}}_m + \tilde{\mathbf{w}}_k, \qquad k = 1, \ldots, L, \tag{12.2}$$

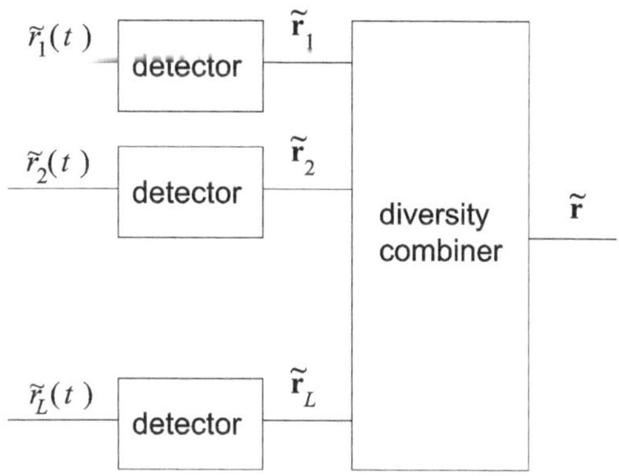

Fig. 12.1 Postdetection diversity combiner

where

$$\tilde{r}_{k,i} = \tilde{h}_k \tilde{s}_{m,i} + \tilde{w}_{k,i}, \qquad i = 1, \ldots, N. \tag{12.3}$$

The fading gains of the various diversity branches will have some degree of correlation that will depend on the type of diversity being used and the propagation environment. To simplify any theoretical analysis, the diversity branches are often assumed to be uncorrelated. However, this is a best-case condition and any branch correlation will more or less reduce the achievable gain from using diversity. For our purpose, the performance of the various diversity combining techniques will be evaluated under the best-case condition of uncorrelated branches.

The envelope fade distribution will also affect the diversity gain. In general, the advantage of diversity is greater for Rayleigh fading than for Ricean fading, because as the Rice factor K increases, there is less difference between the instantaneous received bit energy-to-noise ratios on the various diversity branches. However, the performance will always be better with Ricean fading than with Rayleigh fading, for a given average received branch bit energy-to-noise ratio and diversity order L. For our purpose, the performance with slow flat Rayleigh fading will be considered, although the methodology will apply to slow flat Ricean fading as well.

12.1.1 Selective Combining

With selective combining (SC), the diversity branch yielding the highest bit energy-to-noise ratio is always selected. In this case, the diversity combiner in Fig. 12.1 performs the operation

$$\tilde{\mathbf{r}} = \max_{|\tilde{h}_k|} \tilde{\mathbf{r}}_k. \tag{12.4}$$

For communication links that use continuous transmission, SC is impractical because it requires continuous monitoring of all diversity branches to estimate and track the time-varying complex gains $\tilde{h}_k$. If such channel estimation is required, then it is better to use maximal ratio combining, as discussed in the next section. For communication systems that transmit messages in frames, a form of SC can be implemented where the complex channel gains $\tilde{h}_k$ are estimated on a frame-by-frame basis, by using a synchronization word or training sequence that is inserted into each frame. The diversity branch is then selected and used for each received frame under the assumption that the $\tilde{h}_k$ change slowly enough to remain constant over the duration of a frame. Fortunately, this is often the case for cellular land mobile radio applications.

With Rayleigh fading, the instantaneous received symbol energy-to-noise ratio on the kth diversity branch has the exponential pdf

$$p_{\gamma_k}(x) = \frac{1}{\bar{\gamma}_c} \mathrm{e}^{-x/\bar{\gamma}_c} \ , \quad x \geq 0, \tag{12.5}$$

where $\bar{\gamma}_c$ is the average received branch symbol energy-to-noise ratio, assumed here to be the same for all diversity branches. For non-binary modulation $\bar{\gamma}_c = (\log_2 M)\bar{\gamma}_b$.

With ideal SC, the branch having the largest symbol energy-to-noise ratio is always selected so that the instantaneous symbol energy-to-noise ratio at the selective combiner output is

$$\gamma_s^{\mathrm{s}} = \max\{\gamma_1, \ \gamma_2, \ \cdots, \ \gamma_L\}, \tag{12.6}$$

where L is the number of branches. If the branches are independently faded, then order statistics gives the cumulative distribution function (cdf) of γ_s^{s} as

$$F_{\gamma_s^{\mathrm{s}}}(x) = \mathrm{P}[\gamma_1 \leq x, \gamma_2 \leq x, \ \cdots, \gamma_L \leq x] = \left(1 - \mathrm{e}^{-x/\bar{\gamma}_c}\right)^L \ , \quad x \geq 0. \tag{12.7}$$

Differentiating the above expression gives the pdf of the selective combiner output symbol energy-to-noise ratio as

$$p_{\gamma_s^{\mathrm{s}}}(x) = \frac{L}{\bar{\gamma}_c}\left(1 - \mathrm{e}^{-x/\bar{\gamma}_c}\right)^{L-1} \mathrm{e}^{-x/\bar{\gamma}_c} \ , \quad x \geq 0. \tag{12.8}$$

The average output symbol energy-to-noise ratio with SC is

$$\begin{aligned}
\bar{\gamma}_s^{\mathrm{s}} &= \int_0^\infty x p_{\gamma_s^{\mathrm{s}}}(x)\mathrm{d}x \\
&= \int_0^\infty \frac{Lx}{\bar{\gamma}_c}\left(1 - \mathrm{e}^{-x/\bar{\gamma}_c}\right)^{L-1} \mathrm{e}^{-x/\bar{\gamma}_c}\mathrm{d}x \\
&= \bar{\gamma}_c \sum_{k=1}^{L} \frac{1}{k}.
\end{aligned} \tag{12.9}$$

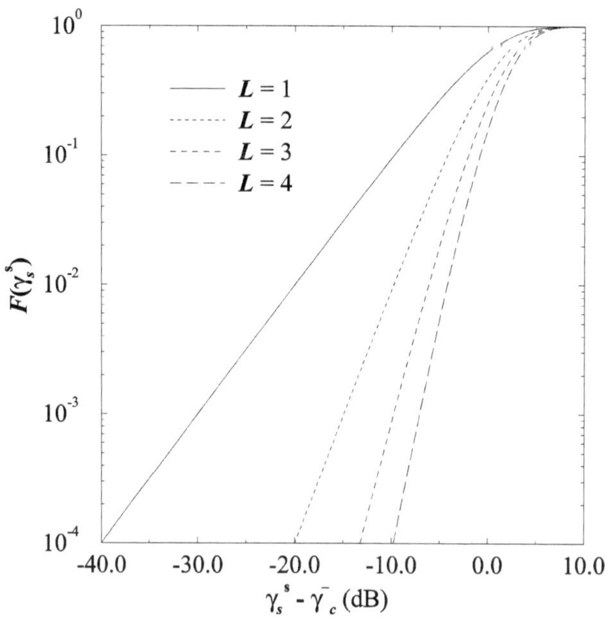

Fig. 12.2 Cdf of γ_b^{s} for selective combining; $\bar{\gamma}_c$ is the average branch symbol energy-to-noise ratio

Figure 12.2 plots the cdf $F_{\gamma_s^{\mathrm{s}}}(x)$ in (12.7) against the normalized symbol energy-to-noise ratio $x/\bar{\gamma}_c$. Note that the largest diversity gain is obtained in going from $L = 1$ to $L = 2$, and diminishing returns are obtained with increasing L. This behavior is typical for all diversity techniques.

The bit error probability is obtained by averaging the bit error probability, as a function of the symbol energy-to-noise ratio, over the pdf of γ_s^{s}. The bit error probability for BPSK on an AWGN channel is

$$P_b(\gamma_b^{\mathrm{s}}) = Q\left(\sqrt{2\gamma_b^{s}}\right),$$

where γ_b^{s} is the instantaneous bit energy-to-noise ratio at the output of the selective combiner. Hence, with SC

$$\begin{aligned} P_b &= \int_0^\infty P_b(x) p_{\gamma_b^{\mathrm{s}}}(x) \mathrm{d}x \\ &= \int_0^\infty Q\left(\sqrt{2x}\right) \frac{L}{\bar{\gamma}_c} \left[1 - e^{-x/\bar{\gamma}_c}\right]^{L-1} e^{-x/\bar{\gamma}_c} \mathrm{d}x \\ &= \frac{1}{2} \sum_{k=0}^{L} (-1)^k \binom{L}{k} \left(1 + \frac{k}{\bar{\gamma}_c}\right)^{-1/2} \end{aligned} \tag{12.10}$$

where $\bar{\gamma}_c$ is the average branch bit energy-to-noise ratio.

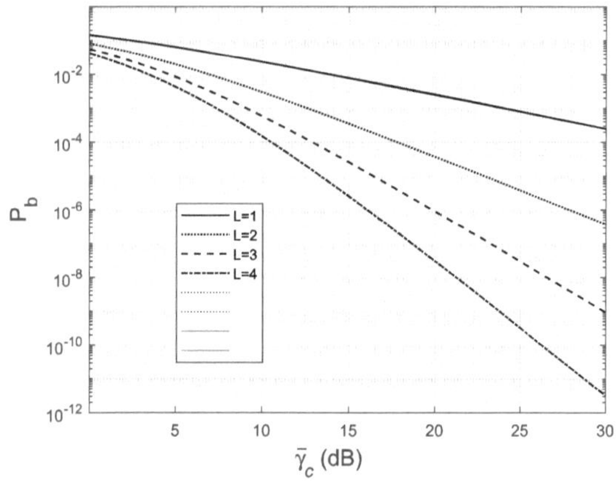

Fig. 12.3 Bit error probability for BPSK and L-branch selective diversity combining

The bit error probability is plotted in Fig. 12.3, where SC is seen to give a substantial improvement in bit error rate performance. Again, the greatest benefit of using diversity is achieved in going from $L = 1$ to $L = 2$, and diminishing returns are obtained by further increasing L.

12.1.2 Maximal Ratio Combining

Maximal ratio combining (MRC) realizes a maximum likelihood receiver for fading channels that are corrupted with AWGN. Referring to (12.2), the vector

$$\tilde{\mathbf{r}} = \mathrm{vec}\left(\tilde{\mathbf{r}}_1,\ \tilde{\mathbf{r}}_2,\ \cdots,\ \tilde{\mathbf{r}}_L\right) \tag{12.11}$$

has the conditional multivariate complex Gaussian distribution

$$\begin{aligned} p(\tilde{\mathbf{r}}|\tilde{\mathbf{h}}, \tilde{\mathbf{s}}_m) &= \prod_{k=1}^{L}\prod_{i=1}^{N} \frac{1}{2\pi N_o} \exp\left\{-\frac{1}{2N_o}|\tilde{r}_{k,i} - \tilde{h}_k \tilde{s}_{m,i}|^2\right\} \\ &= \frac{1}{(2\pi N_o)^{LN}} \exp\left\{-\frac{1}{2N_o}\sum_{k=1}^{L} \|\tilde{\mathbf{r}}_k - \tilde{h}_k \tilde{\mathbf{s}}_m\|^2\right\} \end{aligned} \tag{12.12}$$

where $\tilde{\mathbf{h}} = (\tilde{h}_1, \tilde{h}_2, \ldots, \tilde{h}_L)$ is the channel vector. The maximum likelihood receiver chooses the message vector $\tilde{\mathbf{s}}_m$ that maximizes the likelihood function $p(\tilde{\mathbf{r}}|\tilde{\mathbf{h}}, \tilde{\mathbf{s}}_m)$. This is equivalent to choosing the message vector $\tilde{\mathbf{s}}_m$ that minimizes the metric

$$\begin{aligned} \mu(\tilde{\mathbf{s}}_m) &= \sum_{k=1}^{L} \|\tilde{\mathbf{r}}_k - \tilde{h}_k \tilde{\mathbf{s}}_m\|^2 \\ &= \sum_{k=1}^{L} \left(\|\tilde{\mathbf{r}}_k\|^2 - 2\mathrm{Re}\{\tilde{\mathbf{r}}_k \cdot \tilde{h}_k^* \tilde{\mathbf{s}}_m^*\} + |\tilde{h}_k|^2 \|\tilde{\mathbf{s}}_m\|^2\right). \end{aligned} \tag{12.13}$$

Since $\sum_{k=1}^{L} \|\tilde{\mathbf{r}}_k\|^2$ is independent of the hypothesis as to which $\tilde{\mathbf{s}}_m$ was sent and $\|\tilde{\mathbf{s}}_m\|^2 = 2E_m$, the receiver just needs to maximize the metric

$$\mu_2(\tilde{\mathbf{s}}_m) = \sum_{k=1}^{L} \mathrm{Re}\{\tilde{\mathbf{r}}_k \cdot \tilde{h}_k^* \tilde{\mathbf{s}}_m^*\} - E_m \sum_{k=1}^{L} |\tilde{h}_k|^2$$

$$= \mathrm{Re}\left\{\sum_{k=1}^{L} \tilde{h}_k^* \tilde{\mathbf{r}}_k \cdot \tilde{\mathbf{s}}_m^*\right\} - E_m \sum_{k=1}^{L} |\tilde{h}_k|^2 \qquad (12.14)$$

If the messages have equal energy, $E_m = E$, then the second term is irrelevant since it is the same for all messages. Hence, the message vector $\tilde{\mathbf{s}}_m$ may be chosen to maximize the metric

$$\mu_3(\tilde{\mathbf{s}}_m) = \mathrm{Re}\left\{\sum_{k=1}^{L} \tilde{h}_k^* \tilde{\mathbf{r}}_k \cdot \tilde{\mathbf{s}}_m^*\right\}. \qquad (12.15)$$

From (12.14) or (12.15) as appropriate, the ML receiver can be constructed. The diversity combiner in Fig. 12.1 generates the sum

$$\tilde{\mathbf{r}} = \sum_{k=1}^{L} \tilde{h}_k^* \tilde{\mathbf{r}}_k \qquad (12.16)$$

which is then applied to the metric computer shown in Fig. 12.4.

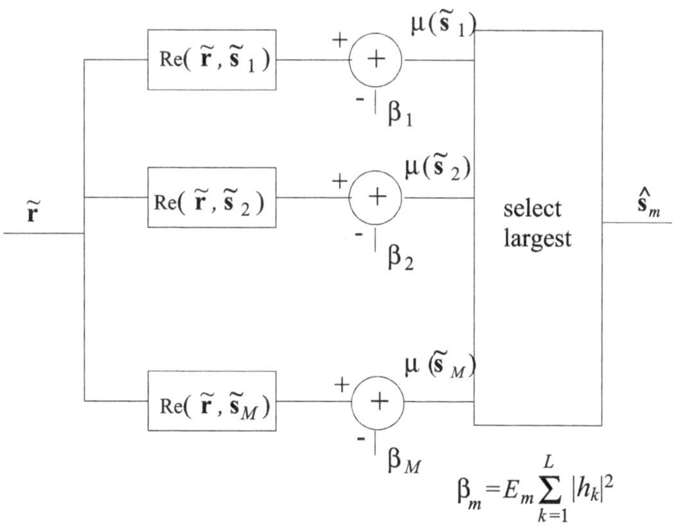

Fig. 12.4 Metric computer for maximal ratio combining

To evaluate the performance gain with MRC, the received branch vectors in (12.2) are substituted into (12.16). This yields

$$\begin{aligned}\tilde{\mathbf{r}} &= \sum_{k=1}^{L} \tilde{h}_k^* \left(\tilde{h}_k \tilde{\mathbf{s}}_m + \tilde{\mathbf{w}}_k\right) \\ &= \left(\sum_{k=1}^{L} \alpha_k^2\right) \tilde{\mathbf{s}}_m + \sum_{k=1}^{L} \tilde{h}_k^* \tilde{\mathbf{w}}_k \\ &\equiv \alpha_{\mathrm{M}}^2 \tilde{\mathbf{s}}_m + \tilde{\mathbf{w}}_{\mathrm{M}},\end{aligned} \tag{12.17}$$

where $\alpha_{\mathrm{M}}^2 \triangleq \sum_{k=1}^{L} \alpha_k^2$ with $\alpha_k^2 = |\tilde{h}_k|^2$ and $\tilde{\mathbf{w}}_{\mathrm{M}} \triangleq \sum_{k=1}^{L} \tilde{h}_k^* \tilde{\mathbf{w}}_k$. The first term in (12.17) is the message vector $\tilde{\mathbf{s}}_m$ scaled by α_{M}^2. Thus, the energy in the received message is

$$\tfrac{1}{2}\mathrm{E}[\alpha_{\mathrm{M}}^4 \|\tilde{\mathbf{s}}_m\|^2] = \alpha_{\mathrm{M}}^4 E, \tag{12.18}$$

where E is the symbol energy in the signal constellation. The second term in (12.17) is the complex Gaussian noise vector at the combiner output, having variance per dimension

$$\sigma_{\tilde{n}_{\mathrm{M}}}^2 = \frac{1}{2N}\mathrm{E}[\|\tilde{\mathbf{w}}_{\mathrm{M}}\|^2] = N_o \sum_{k=1}^{L} \alpha_k^2 = N_o \alpha_{\mathrm{M}}^2. \tag{12.19}$$

The ratio of the above two gives the symbol energy-to-noise ratio at the MRC combiner output

$$\gamma_s^{\mathrm{mr}} = \frac{\frac{1}{2}\mathrm{E}[\alpha_{\mathrm{M}}^4 \|\tilde{\mathbf{s}}_m\|^2]}{\sigma_{\tilde{n}_{\mathrm{M}}}^2} = \frac{\alpha_{\mathrm{M}}^2 E}{N_o} = \sum_{k=1}^{L} \frac{\alpha_k^2 E}{N_o} = \sum_{k=1}^{L} \gamma_k, \tag{12.20}$$

where $\gamma_k = \alpha_k^2 E/N_o$. Hence, for MRC γ_s^{mr} is the sum of the symbol energy-to-noise ratios, γ_k, of the L diversity branches.

If the branches are balanced (which is a reasonable assumption for antenna diversity) and uncorrelated, then γ_s^{mr} is the sum of L i.i.d exponential random variables each with mean value $\bar{\gamma}_c$. The pdf of γ_s^{mr} has the following Erlang distribution, named after the Danish mathematician and teletraffic engineer A.K. Erlang,

$$p_{\gamma_s^{\mathrm{mr}}}(x) = \frac{1}{(L-1)!(\bar{\gamma}_c)^L} x^{L-1} \mathrm{e}^{-x/\bar{\gamma}_c}, \quad x \geq 0, \tag{12.21}$$

where

$$\bar{\gamma}_c = \mathrm{E}[\gamma_k] \qquad k = 1, \ \ldots, \ L. \tag{12.22}$$

The cdf of γ_s^{mr} is

$$F_{\gamma_s^{\mathrm{mr}}}(x) = 1 - \mathrm{e}^{-x/\bar{\gamma}_c} \sum_{k=0}^{L-1} \frac{1}{k!} \left(\frac{x}{\bar{\gamma}_c} \right)^k , \quad x \geq 0. \tag{12.23}$$

It follows from (12.20) that the average symbol energy-to-noise ratio with MRC and balanced branches is

$$\bar{\gamma}_s^{\mathrm{mr}} = \sum_{k=1}^{L} \bar{\gamma}_k = \sum_{k=1}^{L} \bar{\gamma}_c = L\bar{\gamma}_c. \tag{12.24}$$

Figure 12.5 plots the cdf $F_{\gamma_b^{\mathrm{mr}}}(x)$ against the normalized symbol energy-to-noise ratio $x/\bar{\gamma}_c$. Plots of the cumulative distribution function can be used to compare the various diversity combining schemes, independent of the modulation scheme being employed. For example, with $L = 2$ and SC, the cdf in Fig. 12.2 shows that $F_{\gamma_s^{\mathrm{s}}}(x) = 10^{-4}$ at $x/\bar{\gamma}_c = 0.01$ ($\gamma_s^{\mathrm{s}} - \bar{\gamma}_c = -20\,\mathrm{dB}$). However, with $L = 2$

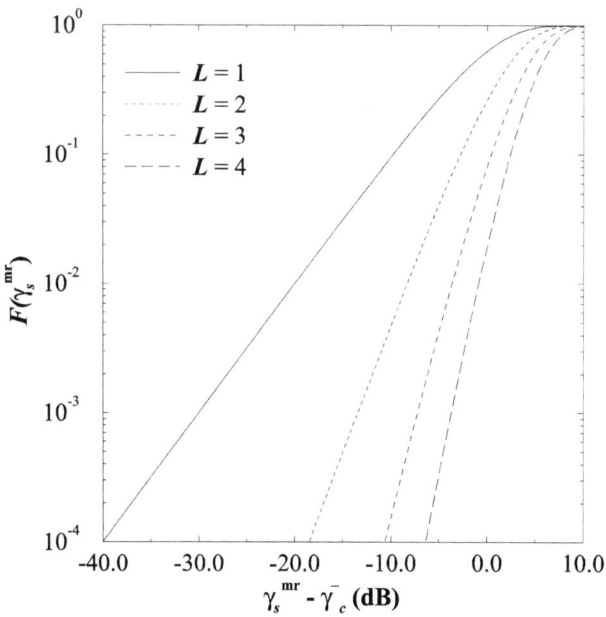

Fig. 12.5 Cdf of γ_s^{mr} for maximal ratio combining; $\bar{\gamma}_c$ is the average branch symbol energy-to-noise ratio

and MRC, Fig. 12.5 shows that $F_{\gamma_s^{\rm mr}}(x) = 10^{-4}$ at $x/\bar{\gamma}_c = 0.01585$ ($\gamma_s^{\rm mr} - \bar{\gamma}_c = -18$ dB). The implication is that with $L = 2$ MRC is 2 dB more power efficient than SC with all other things being equal.

When computing the probability of bit error, it must be kept in mind that MRC is a coherent detection technique. The bit error probability with coherently detected BPSK is

$$P_b(\gamma_s^{\rm mr}) = Q(\sqrt{2\gamma_s^{\rm mr}}), \tag{12.25}$$

where $\gamma_s^{\rm mr}$ is the instantaneous *bit* energy-to-noise ratio. Hence, the bit error probability is

$$\begin{aligned} P_b &= \int_0^\infty P_b(x) p_{\gamma_s^{\rm mr}}(x) {\rm d}x \\ &= \int_0^\infty Q\left(\sqrt{2x}\right) \frac{1}{(L-1)!(\bar{\gamma}_c)^L} x^{L-1} {\rm e}^{-x/\bar{\gamma}_c} {\rm d}x \\ &= \left(\frac{1-\mu}{2}\right)^L \sum_{k=0}^{L-1} \binom{L-1+k}{k} \left(\frac{1+\mu}{2}\right)^k \end{aligned} \tag{12.26}$$

where

$$\mu = \sqrt{\frac{\bar{\gamma}_c}{1+\bar{\gamma}_c}}. \tag{12.27}$$

The last step follows after considerable algebra. The bit error probability in (12.26) is plotted in Fig. 12.6 against the normalized symbol energy-to-noise ratio $x/\bar{\gamma}_c$ per diversity branch. Once again, diversity significantly improves the performance. The largest diversity gain is obtained in going from $L = 1$ to $L = 2$ branches, and diminishing returns are obtained as L increases further. Finally, by comparing the bit error rate performance in Figs. 12.3 and 12.6 with $L = 2$ and $P_b = 10^{-5}$, maximal ratio combining is about 2 dB more power efficient than selective combining. This agrees with the same result obtained earlier by comparing the cdfs with $L = 2$ in Figs. 12.2 and 12.5.

12.2 MIMO Channels

A multi-input multiple-output (MIMO) channel is one characterized by multiple inputs and multiple outputs. For example, multiple transmit and receive antennas will create an MIMO channel as shown in Fig 12.7. For a system consisting of L_T

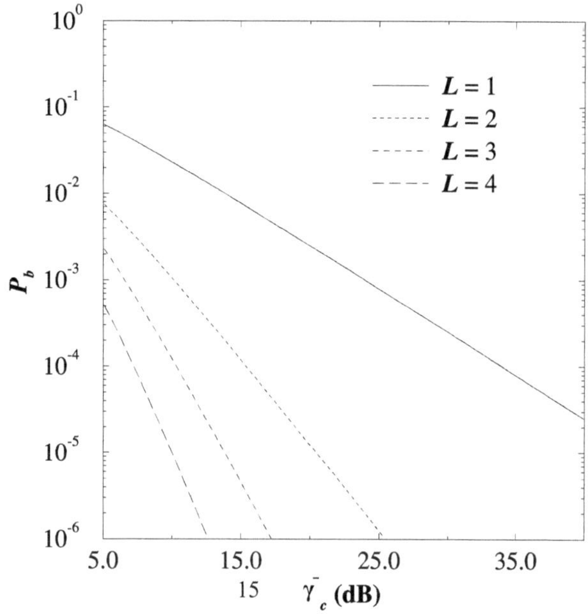

Fig. 12.6 Bit error probability for DPSK with maximal ratio combining

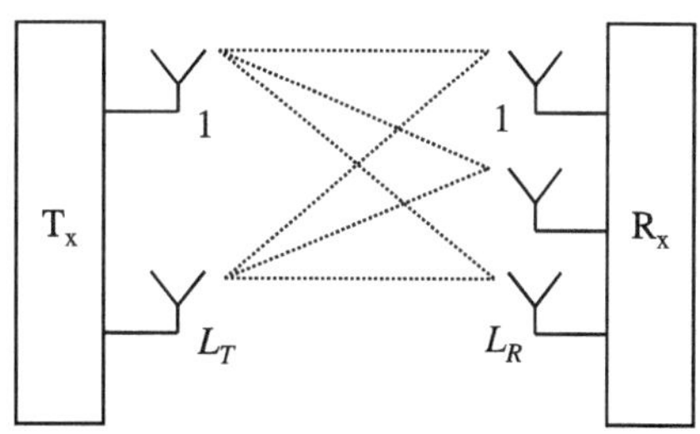

Fig. 12.7 MIMO system with multiple transmit and multiple receive antennas

transmit and L_R receive antennas, the MIMO channel is described by the $L_R \times L_T$ matrix

$$\tilde{\mathbf{H}}(t,\tau) = \begin{bmatrix} \tilde{h}_{1,1}(t,\tau) & \tilde{h}_{1,2}(t,\tau) & \cdots & \tilde{h}_{1,L_T}(t,\tau) \\ \tilde{h}_{2,1}(t,\tau) & \tilde{h}_{2,2}(t,\tau) & \cdots & \tilde{h}_{2,L_T}(t,\tau) \\ \vdots & \vdots & & \vdots \\ \tilde{h}_{L_R,1}(t,\tau) & \tilde{h}_{L_R,2}(t,\tau) & \cdots & \tilde{h}_{L_R,L_T}(t,\tau) \end{bmatrix}, \tag{12.28}$$

where $\tilde{h}_{qp}(t,\tau)$ denotes the time-varying subchannel impulse response between the pth transmit antenna and qth receive antenna.

Suppose that the complex envelopes of the signals transmitted from the L_T transmit antennas are described by the column vector

$$\tilde{\mathbf{s}}(t) = (\tilde{s}_1(t), \tilde{s}_2(t), \ldots, \tilde{s}_{L_T}(t))^T, \tag{12.29}$$

where $\tilde{s}_p(t)$ is the signal transmitted from the pth transmit antenna. Likewise, let

$$\tilde{\mathbf{r}}(t) = (\tilde{r}_1(t), \tilde{r}_2(t), \ldots, \tilde{r}_{L_R}(t))^T, \tag{12.30}$$

denote the column vector of received complex envelopes, where $\tilde{r}_q(t)$ is the signal received at the qth receiver antenna. Then $\tilde{\mathbf{s}}(t)$ and $\tilde{\mathbf{r}}(t)$ are related though by convolution integral

$$\tilde{\mathbf{r}}(t) = \int_0^t \tilde{\mathbf{H}}(t, \tau)\tilde{\mathbf{s}}(t - \tau)\mathrm{d}\tau \tag{12.31}$$

If each MIMO subchannel is flat faded, then

$$\tilde{\mathbf{H}}(t, \tau) = \tilde{\mathbf{H}}(t)\delta(\tau - \bar{\tau}), \tag{12.32}$$

where $\bar{\tau}$ is the MIMO channel delay and

$$\tilde{\mathbf{r}}(t) = \tilde{\mathbf{H}}(t)\tilde{\mathbf{s}}(t - \hat{\tau}). \tag{12.33}$$

If the MIMO channel is characterized by slow fading, then

$$\tilde{\mathbf{r}}(t) = \int_0^t \tilde{\mathbf{H}}(\tau)\tilde{\mathbf{s}}(t - \tau)\mathrm{d}\tau. \tag{12.34}$$

In this case, the MIMO channel impulse response $\tilde{\mathbf{H}}(t)$ is modeled to remain constant over the shaping pulse duration of the transmitted message vector $\tilde{\mathbf{s}}(t)$, but it changes from one channel use to the next. MIMO systems often transmit information using a space-time code, where the space-time codewords are matrices of message symbols that have dimension $L_T \times n$, where n is the space-time codeword length. In this case, the MIMO channel is sometimes modeled as being quasi-static, where the MIMO channel impulse response $\tilde{\mathbf{H}}(t)$ remains constant over the duration of a space-time codeword, but it varies from one space-time codeword to the next.

Finally, if the MIMO channel is characterized by slow flat quasi-static fading, then

$$\tilde{\mathbf{r}}(t) = \tilde{\mathbf{H}}\tilde{\mathbf{s}}(t). \tag{12.35}$$

A variety of models have been suggested for the MIMO channel $\tilde{\mathbf{H}}$. The simplest model assumes that the entries of the matrix $\tilde{\mathbf{H}}$ are i.i.d. complex Gaussian random variables, meaning that the MIMO subchannels have independent and identical Rayleigh fading. This model implies a rich scattering or spatially white environment. The i.i.d model greatly simplifies the performance analysis of MIMO systems. However, in reality, the subchannels will be correlated and, therefore, the i.i.d. MIMO model will lead to best-case results.

12.3 MIMO Systems

The MIMO channel can be exploited in a number of ways. Figure 12.8 illustrates the concepts of *spatial multiplexing*, *transmit diversity*, and *spatial modulation*, for the case of $L_T = 2$ transmit antennas and a modulation alphabet size $M = 2$. With spatial multiplexing, two messages $\mathbf{s}_1$ and $\mathbf{s}_2$ are sent simultaneously from each of two transmit antennas. For an arbitrary number of transmit antennas L_T and modulation alphabet size M, the modulation efficiency is $L_T \log_2 M$ bits/channel use.

With transmit diversity, two messages are sent in two channel uses. In Fig. 12.8, the messages happen to be sent using an Alamouti space-time block code. In general, for L_T transmit antennas and modulation alphabet size M, the space-time block code transmits N_M modulation symbols in N_c channel uses. Hence, the overall modulation efficiency is $(N_M/N_c)\log_2 M$ bits/channel use.

Finally, with spatial modulation, one of the two modulation symbols $\mathbf{s}_1$ is explicitly transmitted, while the other modulation symbol $\mathbf{s}_2$ is implicitly transmitted by the transmitter antenna selection. For L_T transmit antennas and a modulation alphabet size M, the modulation efficiency is $\log_2 L_T + \log_2 M$ bits/channel use. For the special case when $M = 1$, information is only transmitted in the activated antenna index. This scheme is called space-shift keying (SSK).

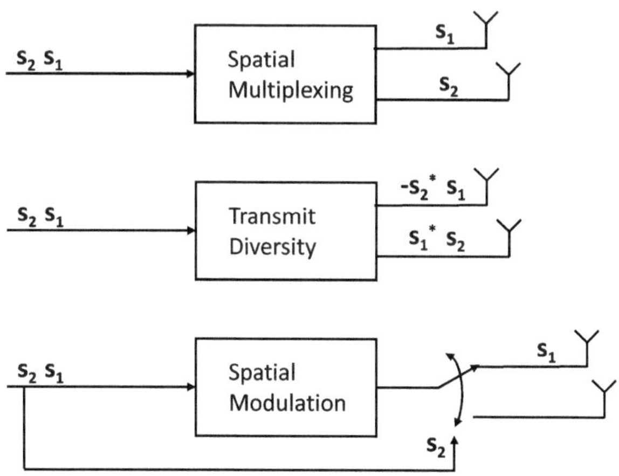

Fig. 12.8 Spatial multiplexing, spatial diversity, and spatial modulation

12.4 Transmitter Diversity

Transmitter diversity uses multiple transmit antennas to provide the receiver with multiple uncorrelated replicas of the same message. Transmitter diversity places the complexity of having multiple antennas onto the transmitter side of the radio link, such as cellular base station or other access point, that is shared by many receivers. The end user terminals may use just a single antenna and still benefit from a diversity gain. Transmitter diversity can take on many variations, distinguished by the method of using the multiple transmit antennas.

Transmitter diversity is straightforward for systems that use time division duplexing (TDD), where alternating time slots on the same RF carrier are used for the forward link and reverse link burst transmissions. For TDD systems, the channel impulse response is reciprocal. At the base station or access point, the signals received on all antennas can be processed during every received burst, and used to estimate the corresponding channel impulse responses. The antenna that provides the largest received bit or symbol energy-to-noise ratio on the reverse link is selected and used for the next forward burst transmission. This is a form of selective transmit diversity (STD). This scheme requires that the channel coherence time be larger than the burst duration.

For frequency division duplexed (FDD) systems, transmit diversity is more complicated, because the forward and reverse links are not reciprocal. Time division transmit diversity (TDTD) can be used for FDD by switching the transmitted signal between two or more transmit antennas. Alternate bursts are transmitted through two or more separate antennas, a technique known as time switched transmit diversity (TSTD). Delay transmit diversity is another method, where copies of the same symbol are transmitted through multiple antennas along a symbol-spaced tapped delay line. This creates artificial delay spread so that the resulting channel looks like a fading ISI channel. A time- or frequency-domain equalizer can then be used to recover the signal and provide a diversity gain.

More elaborate forms of transmitter diversity use space-time, space-frequency, or space-time-frequency encoding of the transmitted information. In general, these schemes require three functions: (i) a method for encoding and transmitting the information sequence at the transmitter, (ii) a combining scheme at the receiver, (iii) a rule for making decisions.

12.4.1 Alamouti's Transmit Diversity Scheme

Alamouti [1] introduced a simple repetition transmit diversity scheme with maximum likelihood combining at the receiver. By using two transmit antennas and one receiver antenna, the scheme provides the same diversity order as maximal ratio combining with one transmit antenna and two receiver antennas. This scheme requires no feedback from the receiver to the transmitter and requires little

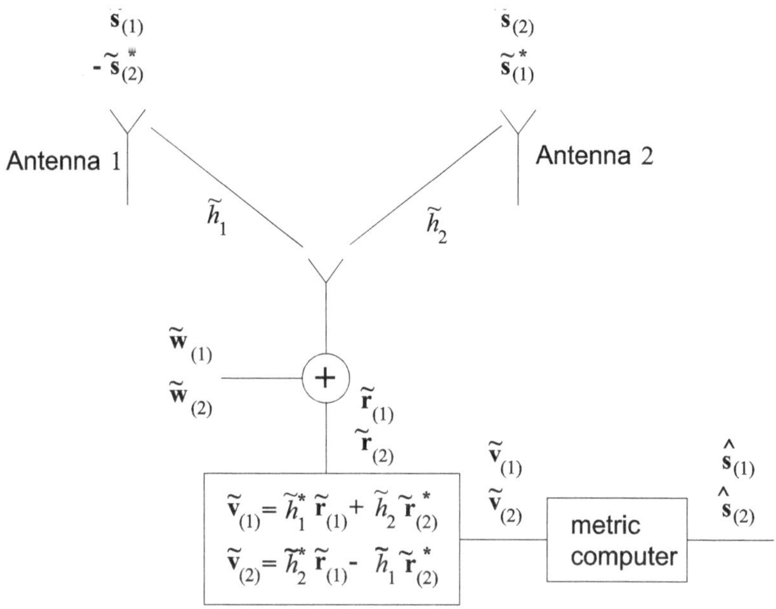

Fig. 12.9 Space-time diversity receiver for 2 × 1 diversity

bandwidth expansion. To estimate the channel, the scheme may insert distinct pilot sequences into the bursts that are transmitted from each of the two transmit antennas.

Alamouti's transmit diversity scheme [1] uses two transmit antennas and one receiver antenna (2 × 1) diversity as shown in Fig. 12.9. Two messages are transmitted over two successive channel uses by using two transmit antennas. During the first channel use, the messages transmitted from Antennas 1 and 2 are $\tilde{\mathbf{s}}_{(1)}$ and $\tilde{\mathbf{s}}_{(2)}$, respectively. During the second channel use, the same messages are transmitted from Antennas 1 and 2 as $-\tilde{\mathbf{s}}^*_{(2)}$ and $\tilde{\mathbf{s}}^*_{(1)}$, respectively. Assuming a quasi-static flat fading channel over the two channel uses, the complex channel gains associated with transmit Antennas 1 and 2 are $\tilde{h}_1$ and $\tilde{h}_2$, respectively, The complex-valued received vectors for the first and second channel uses are, respectively,

$$\tilde{\mathbf{r}}_{(1)} = \tilde{h}_1 \tilde{\mathbf{s}}_{(1)} + \tilde{h}_2 \tilde{\mathbf{s}}_{(2)} + \tilde{\mathbf{w}}_{(1)} \tag{12.36a}$$

$$\tilde{\mathbf{r}}_{(2)} = -\tilde{h}_1 \tilde{\mathbf{s}}^*_{(2)} + \tilde{h}_2 \tilde{\mathbf{s}}^*_{(1)} + \tilde{\mathbf{w}}_{(2)}, \tag{12.36b}$$

and $\tilde{\mathbf{w}}_{(1)}$ and $\tilde{\mathbf{w}}_{(2)}$ are the corresponding complex AWGN noise vectors.

The diversity combiner in Fig. 12.9 outputs the following two vectors

$$\tilde{\mathbf{v}}_{(1)} = \tilde{h}^*_1 \tilde{\mathbf{r}}_{(1)} + \tilde{h}_2 \tilde{\mathbf{r}}^*_{(2)} \tag{12.37a}$$

$$\tilde{\mathbf{v}}_{(2)} = \tilde{h}^*_2 \tilde{\mathbf{r}}_{(1)} - \tilde{h}_1 \tilde{\mathbf{r}}^*_{(2)}. \tag{12.37b}$$

Afterwards, the vectors $\tilde{\mathbf{v}}_{(1)}$ and $\tilde{\mathbf{v}}_{(2)}$ are input to the metric computer in Fig 12.4, to make decisions by maximizing the decision metric

$$\mu(\tilde{\mathbf{s}}_{(1),m}) = \text{Re}\left(\tilde{\mathbf{v}}_{(1)}, \tilde{\mathbf{s}}_{(1),\text{m}}\right) - E_m(|\tilde{h}_1|^2 + |\tilde{h}_2|^2) \tag{12.38a}$$

$$\mu(\tilde{\mathbf{s}}_{(2),m}) = \text{Re}\left(\tilde{\mathbf{v}}_{(2)}, \tilde{\mathbf{s}}_{(2),\text{m}}\right) - E_m(|\tilde{h}_1|^2 + |\tilde{h}_2|^2). \tag{12.38b}$$

where E_m is the energy in message $\tilde{\mathbf{s}}_{(1/2),m}$ and the terms involving E_m can be neglected for equal energy messages.

Substituting (12.36b) in (12.37b) gives

$$\tilde{\mathbf{v}}_{(1)} = (\alpha_1^2 + \alpha_2^2)\tilde{\mathbf{s}}_{(1)} + \tilde{h}_1^*\tilde{\mathbf{w}}_{(1)} + \tilde{h}_2\tilde{\mathbf{w}}_{(2)}^* \tag{12.39a}$$

$$\tilde{\mathbf{v}}_{(2)} = (\alpha_1^2 + \alpha_2^2)\tilde{\mathbf{s}}_{(2)} - \tilde{h}_1\tilde{\mathbf{w}}_{(2)}^* + \tilde{h}_2^*\tilde{\mathbf{w}}_{(1)}. \tag{12.39b}$$

This is to be compared with the output of the MRC metric computer in Fig. 12.4. With $L = 2$,

$$\begin{aligned}\tilde{\mathbf{r}} &= \tilde{h}_1^*\tilde{\mathbf{r}}_1 + \tilde{h}_2^*\tilde{\mathbf{r}}_2 \\ &= (\alpha_1^2 + \alpha_2^2)\tilde{\mathbf{s}}_m + \tilde{h}_1^*\tilde{\mathbf{w}}_1 + \tilde{h}_2^*\tilde{\mathbf{w}}_2.\end{aligned} \tag{12.40}$$

Comparison of (12.39b) and (12.40) shows that the combiner output is the same in each case. The only difference is the phase rotations of the complex Gaussian noise vectors, which will not matter statistically due to their circularly symmetric components. The bit error probability of 2×1 Alamouti transmit diversity is the same as 1×2 receive diversity with MRC. The only difference is a 3 dB loss due to power splitting at the transmitter between the two transmit antennas.

12.4.1.1 2 x 2 Diversity

The 2×2 Alamouti transmit diversity scheme is shown in Fig. 12.10. The following notation is introduced

$\tilde{h}_{i,j}$ = complex channel gain between transmit antenna i and receiver antenna j.

$\tilde{\mathbf{r}}_{(1),j}$ = complex received vector at antenna j during the first channel use.

$\tilde{\mathbf{r}}_{(2),j}$ = complex received vector at antenna j during the second channel use.

The encoding scheme remains the same as before: two messages $\tilde{\mathbf{s}}_{(1)}$ and $\tilde{\mathbf{s}}_{(2)}$ are sent from Antennas 1 and 2, respectively, during the first channel use, and the same

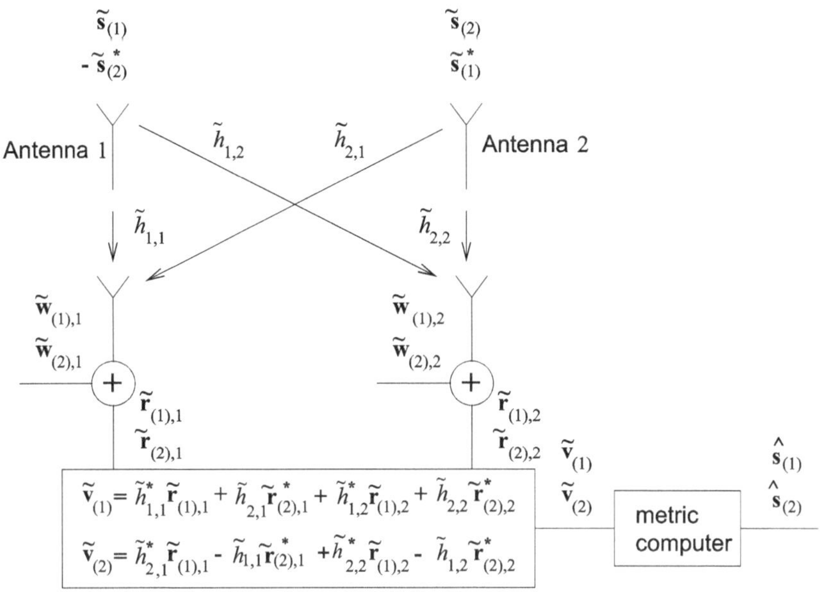

Fig. 12.10 Space-time diversity receiver for 2 × 2 diversity

messages are sent as $-\tilde{\mathbf{s}}^*_{(2)}$ and $\tilde{\mathbf{s}}^*_{(1)}$ from Antennas 1 and 2, respectively during the second channel use. The four complex received signal vectors are

$$\tilde{\mathbf{r}}_{(1),1} = \tilde{h}_{1,1}\tilde{\mathbf{s}}_{(1)} + \tilde{h}_{2,1}\tilde{\mathbf{s}}_{(2)} + \tilde{\mathbf{w}}_{(1),1} \tag{12.41a}$$

$$\tilde{\mathbf{r}}_{(2),1} = -\tilde{h}_{1,1}\tilde{\mathbf{s}}^*_{(2)} + \tilde{h}_{2,1}\tilde{\mathbf{s}}^*_{(1)} + \tilde{\mathbf{w}}_{(2),1} \tag{12.41b}$$

$$\tilde{\mathbf{r}}_{(1),2} = \tilde{h}_{1,2}\tilde{\mathbf{s}}_{(1)} + \tilde{h}_{2,2}\tilde{\mathbf{s}}_{(2)} + \tilde{\mathbf{w}}_{(1),2} \tag{12.41c}$$

$$\tilde{\mathbf{r}}_{(2),2} = -\tilde{h}_{1,2}\tilde{\mathbf{s}}^*_{(2)} + \tilde{h}_{2,2}\tilde{\mathbf{s}}^*_{(1)} + \tilde{\mathbf{w}}_{(2),2}. \tag{12.41d}$$

The combiner in Fig. 12.10 outputs the following two vectors

$$\tilde{\mathbf{v}}_{(1)} = \tilde{h}^*_{1,1}\tilde{\mathbf{r}}_{(1),1} + \tilde{h}_{2,1}\tilde{\mathbf{r}}^*_{(2),1} + \tilde{h}^*_{1,2}\tilde{\mathbf{r}}_{(1),2} + \tilde{h}_{2,2}\tilde{\mathbf{r}}^*_{(2),2} \tag{12.42a}$$

$$\tilde{\mathbf{v}}_{(2)} = \tilde{h}^*_{2,1}\tilde{\mathbf{r}}_{(1),1} - \tilde{h}_{1,1}\tilde{\mathbf{r}}^*_{(2),1} + \tilde{h}^*_{2,2}\tilde{\mathbf{r}}_{(1),2} - \tilde{h}_{1,2}\tilde{\mathbf{r}}^*_{(2),2}. \tag{12.42b}$$

Again, the receiver applies the vectors $\tilde{\mathbf{v}}_{(1)}$ and $\tilde{\mathbf{v}}_{(2)}$ to the metric computer in Fig 12.4 and decisions are made that maximize the metric in (12.38b).

To compare the 2 × 2 transmit diversity scheme to 1 × 4 receive diversity with MRC, (12.41d) may be substituted into (12.42a) and (12.42b) to obtain

$$\begin{aligned}\tilde{\mathbf{v}}_{(1)} = &(\alpha^2_{1,1} + \alpha^2_{1,2} + \alpha^2_{2,1} + \alpha^2_{2,2})\tilde{\mathbf{s}}_{(1)} \\ &+\tilde{h}^*_{1,1}\tilde{\mathbf{w}}_{(1),1} + \tilde{h}_{2,1}\tilde{\mathbf{w}}^*_{(2),1} + \tilde{h}^*_{1,2}\tilde{\mathbf{w}}_{(1),2} + \tilde{h}_{2,2}\tilde{\mathbf{w}}^*_{(2),2}\end{aligned} \tag{12.43a}$$

$$\tilde{\mathbf{v}}_{(2)} = (\alpha_{1,1}^2 + \alpha_{1,2}^2 + \alpha_{2,1}^2 + \alpha_{2,2}^2)\tilde{\mathbf{s}}_{(2)}$$
$$+\tilde{h}_{2,1}^*\tilde{\mathbf{w}}_{(1),1} - \tilde{h}_{1,1}\tilde{\mathbf{w}}_{(2),1}^* + \tilde{h}_{2,2}^*\tilde{\mathbf{w}}_{(1),2} - \tilde{h}_{1,2}\tilde{\mathbf{w}}_{(2),2}^*. \quad (12.43\text{b})$$

This is to be compared with the output of the MRC in Fig. 12.4. With $L = 4$,

$$\tilde{\mathbf{r}} = \tilde{h}_1^*\tilde{\mathbf{r}}_1 + \tilde{h}_2^*\tilde{\mathbf{r}}_2 + \tilde{h}_3^*\tilde{\mathbf{r}}_3 + \tilde{h}_4^*\tilde{\mathbf{r}}_4$$
$$= (\alpha_1^2 + \alpha_2^2 + \alpha_3^2 + \alpha_4^2)\tilde{\mathbf{s}}_m + \tilde{h}_1^*\tilde{\mathbf{w}}_1 + \tilde{h}_2^*\tilde{\mathbf{w}}_2 + \tilde{h}_3^*\tilde{\mathbf{w}}_3 + \tilde{h}_4^*\tilde{\mathbf{w}}_4. \quad (12.44)$$

Again, Alamouti's 2×2 transmit diversity scheme is equivalent in bit error rate performance to a 1×4 receive diversity scheme with MRC except for a 3 dB loss due to power splitting between two transmit antennas. The extension of Alamouti's 2×2 transmit diversity scheme to $1 \times 2L$ transmit diversity is left as an exercise.

12.4.1.2 Implementation Issues

There are several key implementation issues with Alamouti's transmit diversity scheme, including the following:

- Since there are two transmit antennas, the power per antenna must be halved to maintain a constant transmit power. Hence, the $2 \times L$ transmit diversity scheme suffers a 3 dB performance loss when compared to a $1 \times 2L$ diversity scheme with MRC.
- With two transmit antennas, twice as many pilot symbols are needed compared to the case of one transmit antenna. The pilots must alternate between the antennas. Alternatively, orthogonal pilot sequences can be transmitted simultaneously from the two transmit antennas over two channel uses.
- The transmit antennas must be spaced sufficiently far apart in order to achieve sufficient spatial decorrelation between the subchannels $h_{1,i}$ and $h_{2,i}$, $i = 1, \ldots, L$. Chap. 11 showed that the required antenna separation can be on the order of a half-wavelength with 2-D isotropic scattering. However, due to the narrow angle-of-arrival spread of multipath components, the required separation can be several tens of wavelengths at a cellular base station.

12.5 Spatial Multiplexing

In wireless systems that use time division duplexing (TDD), knowledge of the channel $\tilde{\mathbf{H}}$ is available at both the transmitter and receiver due to channel reciprocity. In this case, spatial multiplexing schemes may be devised based on singular value decomposition (SVD) of the channel matrix $\tilde{\mathbf{H}}$. Suppose that $\tilde{\mathbf{H}}$ has rank r, which is at most $\min\{L_T, L_R\}$. Then $\tilde{\mathbf{H}}$ may be decomposed as

$$\tilde{\mathbf{H}} = \mathbf{U}\boldsymbol{\Lambda}\mathbf{V}^H \tag{12.45}$$

where $\mathbf{U}$ is an $L_R \times r$ matrix, $\mathbf{V}$ is an $L_T \times r$ matrix, and $\boldsymbol{\Lambda}$ is an $r \times r$ diagonal matrix whose diagonal elements $\lambda_1, \lambda_2, \ldots, \lambda_r$ are the singular values of the channel matrix $\tilde{\mathbf{H}}$. The matrices $\mathbf{U}$ and $\mathbf{V}$ are unitary matrices, meaning that $\mathbf{U}\mathbf{U}^H = \mathbf{I}_{r\times r}$ and $\mathbf{V}\mathbf{V}^H = \mathbf{I}_{r\times r}$ where $\mathbf{I}_{r\times r}$ is the $r \times r$ identity matrix.

Given knowledge of the channel matrix $\tilde{\mathbf{H}}$ having rank r at the transmitter, r messages are sent for each channel use. To do so, each $r \times 1$ message vector $\tilde{\mathbf{s}}$ is precoded at the transmitter by using the linear transformation

$$\tilde{\mathbf{s}}_p = \mathbf{V}\tilde{\mathbf{s}} \tag{12.46}$$

and transmitted from the L_T transmit antennas. The corresponding received signal vector $\tilde{\mathbf{r}}$ observed across the L_R receiver antennas is

$$\tilde{\mathbf{r}} = \tilde{\mathbf{H}}\tilde{\mathbf{s}}_p + \tilde{\mathbf{w}} = \tilde{\mathbf{H}}\mathbf{V}\tilde{\mathbf{s}} + \tilde{\mathbf{w}}. \tag{12.47}$$

At the receiver, the vector $\tilde{\mathbf{r}}$ is processed by the linear transformation $\mathbf{U}^H$ as

$$\begin{aligned}\hat{\mathbf{s}} &= \mathbf{U}^H\tilde{\mathbf{r}} \\ &= \mathbf{U}^H\tilde{\mathbf{H}}\mathbf{V}\tilde{\mathbf{s}} + \mathbf{U}^H\tilde{\mathbf{w}} \\ &= \mathbf{U}^H\mathbf{U}\boldsymbol{\Lambda}\mathbf{V}^H\mathbf{V}\tilde{\mathbf{s}} + \mathbf{U}^H\tilde{\mathbf{w}} \\ &= \boldsymbol{\Lambda}\tilde{\mathbf{s}} + \mathbf{U}^H\tilde{\mathbf{w}}.\end{aligned} \tag{12.48}$$

Multiplication of the complex Gaussian noise vector $\tilde{\mathbf{w}}$ by the unitary matrix $\mathbf{U}^H$ will not alter the statistics of the noise vector $\tilde{\mathbf{w}}$. Due to the multiplication of each message $\tilde{s}_k$ in the vector $\tilde{\mathbf{s}}$ by the corresponding singular value λ_k of the diagonal matrix $\boldsymbol{\Lambda}$, the r parallel data streams have different received bit energy-to-noise ratios $\lambda_k^2 E_b/N_o$ that will depend on the channel matrix $\tilde{\mathbf{H}}$. Adaptive bit loading techniques may be used where each of the r streams is sent using a modulation/coding scheme whose constellation size M varies with $\tilde{\mathbf{H}}$ to maintain a target bit error probability.

Problems

12.1 Consider BPSK modulation with simple binary repetition coding on a Rayleigh fading channel. Each data bit is assumed to be repeated L times, and each copy has energy E_b/L, where E_b is the energy per data bit. Each copy is affected by independent Rayleigh fading and independent noise. The receiver is assumed to have a single antenna.

(a) At the receiver, a decision is first made on each received copy as to whether a data bit "0" or data bit "1" was sent. Afterward, a final majority logic decision is made such that if a majority of the L copies are decided to be a "0" than a "1," then the message bit is decided to be a "0" and vice versa. Derive an expression for the probability of bit error, P_b, as a function of the average received bit energy-to-noise ratio $\bar{\gamma}_b$.

(b) Evaluate P_b for $L = 3$ and $\bar{\gamma}_b = 20\,\text{dB}$.

12.2 Consider again the system outlined in Problem 12.1.

(a) At the receiver, maximal ratio combining is used to combine together the L copies of each data bit that are received. Derive an expression for the probability of bit error, P_b, as a function of the average received bit energy to noise ratio $\bar{\gamma}_b$.

(b) Evaluate P_b for $L = 3$ and $\bar{\gamma}_b = 20\,\text{dB}$.

12.3 Consider BPSK signaling on an AWGN channel where the channel fading attenuation, α, has the pdf

$$p_\alpha(x) = 0.2\delta(x) + 0.5\delta(x-1) + 0.3\delta(x-2).$$

Suppose that two-branch antenna diversity is used with selective combining. Assume that the diversity branches experience independent and identical fading.

(a) Determine the average probability of error in terms of the average received bit energy-to-noise ratio per diversity branch $\bar{\gamma}_c$.

(b) What value does the probability of error approach as $\bar{\gamma}_c$ gets large?

12.4 Similar to Problem 12.3, again consider BPSK signaling on an AWGN channel where the channel fading attenuation, α, has the pdf

$$p_\alpha(x) = 0.2\delta(x) + 0.5\delta(x-1) + 0.3\delta(x-2).$$

However, this time two-branch antenna diversity is used with maximal ratio combining. Assume that the diversity branches experience independent and identical fading.

(a) Determine the pdf of the bit energy-to-noise ratio at the output of the maximal ratio combiner γ_b^{mr}. Express your answer in terms of the average received bit energy-to-noise ratio per diversity branch $\bar{\gamma}_c$.

(b) Determine the average probability of error in terms of the average received bit energy-to-noise ratio per diversity branch $\bar{\gamma}_c$. What value does the probability of error approach as $\bar{\gamma}_c$ gets large?

12.5 A strange communication receiver for BPSK signals on a Rayleigh fading channel employs two-branch selection diversity, but where each selection diversity

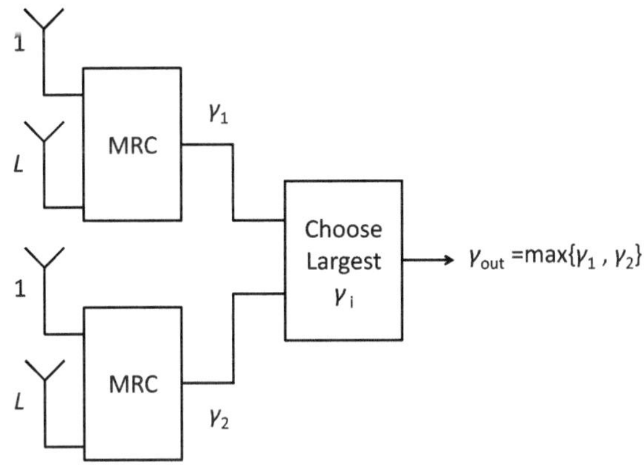

Fig. 12.11 Diversity arrangement for Problem 12.5

branch consists of L antennas with maximal ratio combining as shown in Fig. 12.11. Assume balanced diversity branches at the input to the two MRC combiners, i.e., $\bar{\gamma}_{ij} = \bar{\gamma}_c$, $i = 1, 2$, $j = 1, \ldots, L$, where

$$\bar{\gamma}_{ij} = \frac{\mathrm{E}[\alpha_{ij}^2]E_b}{N_o} = \bar{\gamma}_c$$

Derive an expression for the cdf and pdf of the symbol energy-to-noise ratio, γ_{out}, at the output of the selective combiner.

12.6 Consider binary orthogonal FSK (BFSK) modulation with simple repetition coding on a Rayleigh fading channel. Each data bit is repeated L times, and each copy is transmitted with energy E_b/L, where E_b is the energy per data bit. Each of the L copies is affected by independent Rayleigh fading and independent AWGN.

- **(a)** At the receiver, selective combining is used to combine together the L copies of each data bit that are transmitted. What is the probability of bit error, P_b, as a function of the average received bit energy-to-noise ratio $\bar{\gamma}_b$.
- **(b)** Repeat part a) if maximal ratio combining is used.
- **(c)** What is the probability of bit error in parts a) and b) if all L copies experience identical fading?

12.7 A mobile communication system uses orthogonal binary frequency shift keying (BFSK) with non-coherent detection. The probability of bit error on an additive white Gaussian noise (AWGN) channel is

$$P_b(\gamma_b) = \frac{1}{2}\mathrm{e}^{-\gamma_b/2} ,$$

where $\gamma_b = \alpha^2 E_b/N_o$ is the received bit energy-to-noise ratio.

(a) Derive a closed form expression for the probability of bit error if the channel exhibits Rayleigh fading.
(b) Evaluate the probability of bit error in part a) if the average received bit energy-to-noise ratio is $\bar{\gamma}_b = 20$ dB.
(c) Now suppose that the receiver employs L-branch antenna diversity with selective combining. Assume that the L diversity branches experience independent and identical (balanced) Rayleigh fading. Find a closed-form expression for the probability of bit error in terms of the average received bit energy-to-noise ratio per diversity branch $\bar{\gamma}_c$.
(d) Evaluate the probability of bit error in part c) for $L = 2$ and $\bar{\gamma}_c = 20$ dB.

12.8 Consider Alamouti's transmit diversity scheme. Extend the concept and construct the combiner for the case of $2 \times L$ diversity.

Chapter 13
Cellular Systems

A cellular land mobile radio network is a collection of base stations (BSs) that serve defined geographical areas called cells. By integrating the coverage over a plurality of cells, radio coverage is provided over a large geographic area. A cellular land mobile radio system has two basic functions; it must locate and track both active and idle mobile stations (MSs), and it must always attempt to connect the MSs to the best available set of BSs. The former task is the subject of mobility management and requires a continuous location updating procedure as the MS roams. The latter task is the subject of radio resource management and requires the continuous evaluation of the radio link quality with the serving BS(s) and the radio link qualities of alternate BSs to allow for handoffs and call continuity.

Cellular systems partition the available spectrum among the cells, and the radio resources are reused in geographically separated cells. Such frequency reuse causes co-channel interference (CCI), which is the primary impediment to high-capacity cellular systems. Radio resources are reused at the closest possible geographical distance that the cellular link budget will allow. Since the radio links in high-capacity cellular systems will interfere with each other due to frequency reuse, it is always desirable to maintain each radio link at a target quality of service (QoS) while using the lowest possible transmit power. This means that radio links should not significantly exceed their target QoS since doing so will cause unnecessary interference to other radio links.

Section 13.1 begins the chapter with the concept of frequency reuse in cellular systems. Section 13.2 considers cell splitting to accommodate traffic growth in a cellular deployment. Dramatic improvement in spectral efficiency is the main reason to deploy small cells known as microcells, picocells, and femtocells. However, the small cell radio propagation environment can be highly erratic, and the radio links are more difficult to control. Section 13.3 considers frequency planning for OFDM/OFDMA air interfaces used in 4G and 5G cellular systems. Section 13.4 introduces the cellular link budget that is critical for cellular deployment and

G. Stuber, *Introductory Digital Wireless Communications*,
https://doi.org/10.1007/978-3-032-01559-4_13

includes margins for CCI, shadowing and handoff gain. Finally, the chapter concludes with the coverage and spectral efficiency of cellular systems in Sects. 13.5 and 13.6, respectively.

13.1 Frequency Reuse and the Cellular Concept

Cellular mobile radio systems rely upon frequency reuse, where users in geographically separated cells simultaneously use the same radio resource. A conventional cellular deployment is accomplished using a grid of hexagonal cells. The hexagon is the preferred choice for representing cellular coverage areas, because it approximates a circle, and it offers a wide range of tessellating frequency reuse cluster sizes meaning that the reuse clusters will fit together without leaving any gaps.

Consider the deployment of uniform hexagonal cells in Fig. 13.1 showing two adjacent clusters of seven cells with points A_1 and A_2 at their centers. The individual cells with A_1 and A_2 at their center reuse the same radio resources. The cells have radius R, where R is the distance from the center to a corner of a hexagonal cell also equal to the length of a cell edge. Applying basic trigonometry, the distance from the center to the edge of a hexagon cell is $\sqrt{3}R/2$.

The integer-valued $u - v$ axis in Fig. 13.1 is used to calculate the distance D between the cell centers A_1 and A_2 having the integer labels (u_1, v_1) and (u_2, v_2), respectively. The distance D can be calculated by applying the cosine formula to the triangle $\Delta A_1 A_2 B$ as follows:

$$\begin{aligned} D^2 &= \left((u_2 - u_1)\sqrt{3}R\right)^2 + \left((v_2 - v_1)\sqrt{3}R\right)^2 \\ &\quad -2(u_2 - u_1)\sqrt{3}R(v_2 - v_1)\sqrt{3}R\cos(2\pi/3) \\ &= 3R^2\left((u_2 - u_1)^2 + (v_2 - v_1)^2 + (u_2 - u_1)(v_2 - v_1)\right) \end{aligned} \tag{13.1}$$

Fig. 13.1 Two adjacent clusters of seven cells

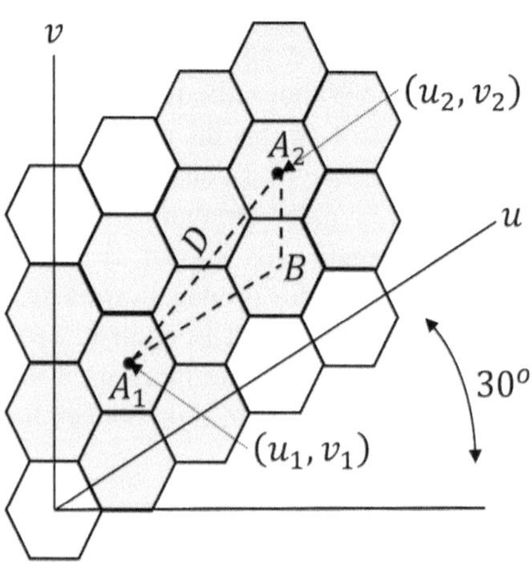

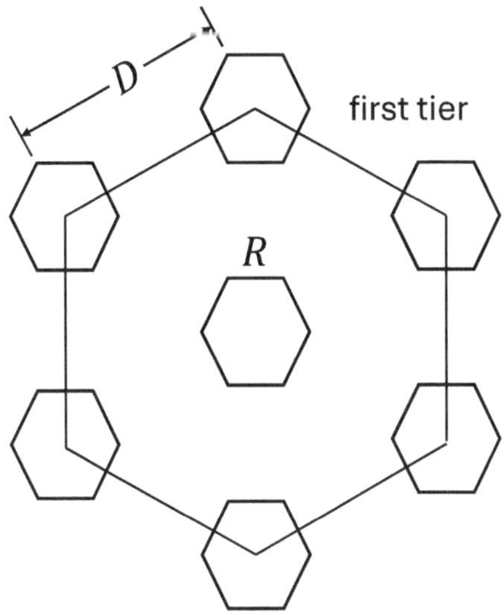

Fig. 13.2 Deployment of co-channel cells

Now let $(u_1, v_1) = (0, 0)$ and $(u_2, v_2) = (i, j)$ where i and j are integers. Then

$$D^2 = 3R^2(i^2 + j^2 + ij) \tag{13.2}$$

Now consider Fig. 13.2, where the radius of the large hexagon D, or the co-channel reuse distance, satisfies $D^2 = 3R^2(i^2 + j^2 + ij)$. The area of the large hexagon in Fig. 13.2 is equal to

$$A_{\text{large}} = kD^2 = k3R^2(i^2 + j^2 + ij)$$

and the area of the small hexagon is equal to

$$A_{\text{small}} = kR^2$$

Hence,

$$\frac{A_{\text{large}}}{A_{\text{small}}} = \left(\frac{D}{R}\right)^2 = 3(i^2 + j^2 + ij)$$

From symmetry, the large hexagon completely encloses the center cluster of N cells plus one third of the cluster area associated with each of the six peripheral co-channel cells. The area enclosed by the large hexagon in units of the small hexagon area is

$$N + 6(N/3) = 3N$$

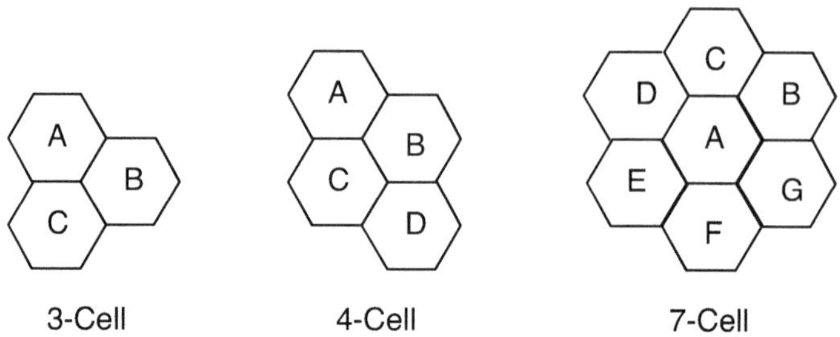

Fig. 13.3 Commonly used cellular reuse clusters

Hence,

$$\frac{A_{\text{large}}}{A_{\text{small}}} = 3N = \left(\frac{D}{R}\right)^2 = 3(i^2 + j^2 + ij) \tag{13.3}$$

From (13.3), the cluster size N satisfies the integer equation

$$N = i^2 + ij + j^2, \tag{13.4}$$

where i and j are non-negative integers, and $i \geq j$. It follows that the allowable hexagonal cluster sizes are $N = 1, 3, 4, 7, 9, 12,$, etc. Examples of 3-, 4-, and 7-cell tessellating reuse clusters are shown in Fig. 13.3.

The co-channel reuse factor D/R, is defined as the ratio of the co-channel reuse distance D between cells using the same set of radio resources and the cell radii R. From (13.3)

$$D/R = \sqrt{3N}. \tag{13.5}$$

The reuse clusters are tessellated to form a frequency plan. A simplified seven-cell frequency reuse plan is shown in Fig. 13.4, where cells marked with the same letter labels use identical sets of radio resources. A correct reuse cluster shape will result in a deployment with six equally spaced first-tier co-channel cells having centers lying on a circle of radius D from the center of the reference cell. For example, in Fig. 13.4 with $N = 7$, there are six first-tier co-channel cells with label "A" at distance $D = \sqrt{21}R$ from the cell with label "A" in the center of the circle. The second-tier co-channel cells may have centers that lay on multiple circles having radius approximately $2D$.

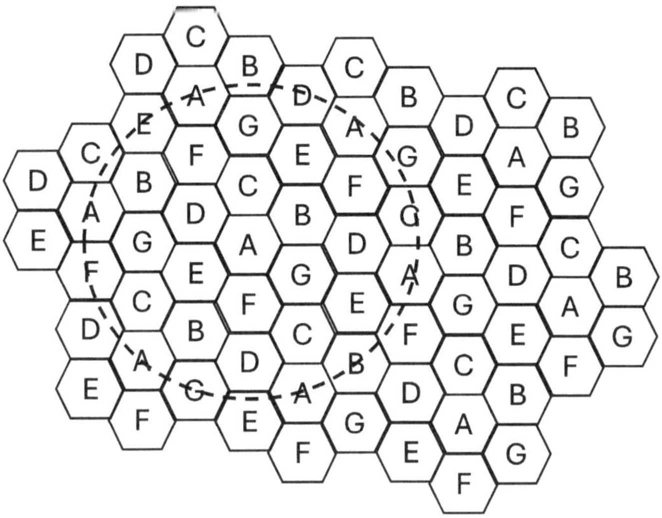

Fig. 13.4 Cellular deployment using seven-cell reuse pattern

13.1.1 Co-channel Interference

Consider a uniform deployment of hexagonal cells, where the BSs employ omnidirectional antennas. Figure 13.5 depicts the forward (BS to MS) link with isotropic BS antennas, where six first-tier co-channel BSs are present. The MS is located at distance d_0 from its serving BS and at distances $d_k, k = 1, 2, \cdots, N_I$ from the first tier of $N_I = 6$ co-channel BSs. If $\mathbf{d} = (d_0,\ d_1, \cdots,\ d_{N_I})$ denotes the vector of distances at a particular MS location, then the forward link signal-to-interference power ratio as a function of the distance vector $\mathbf{d}$ is

$$\Lambda(\mathbf{d}) = \frac{P(d_0)}{\sum_{k=1}^{N_I} P(d_k)}\ , \tag{13.6}$$

or in decibel units

$$\begin{aligned}\Lambda_{(\mathrm{dB})}(\mathbf{d}) &= P_{(\mathrm{dBm})}(d_0) - 10\log_{10}\left\{\sum_{k=1}^{N_I} P(d_k)\right\} \\ &= P_{(\mathrm{dBm})}(d_0) - 10\log_{10}\left\{\sum_{k=1}^{N_I} 10^{P_{(\mathrm{dBm})}(d_k)/10}\right\}.\end{aligned} \tag{13.7}$$

The worst-case forward channel CCI situation occurs when the MS is located at the corner of a cell at distance R from its serving BS, as depicted in Fig. 13.6. Under this condition and with $N = 7$, the six first-tier co-channel BSs are

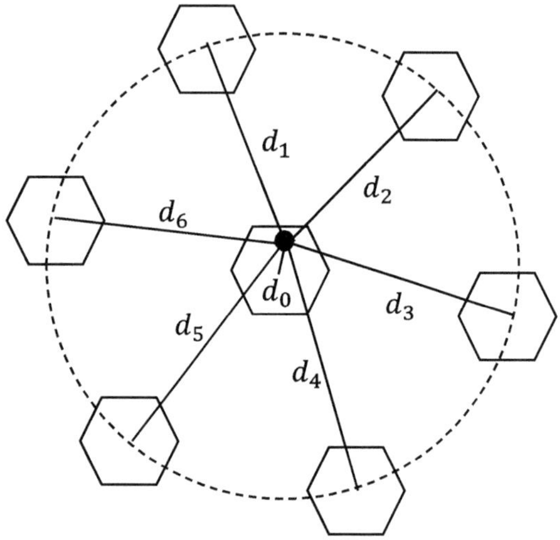

Fig. 13.5 Co-channel interference on the forward link at a MS receiver. There are six first-tier co-channel BSs

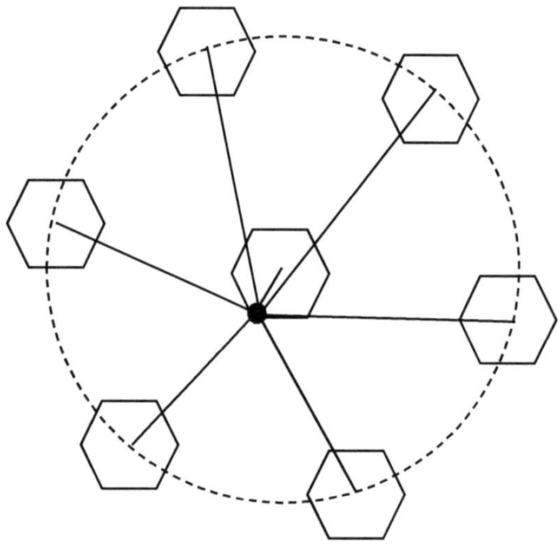

Fig. 13.6 Worst-case CCI situation on the forward link

located at distances $\{\sqrt{13}R, 4R, \sqrt{19}R, 5R, \sqrt{28}R, \sqrt{31}R\}$ from the MS. These distances can be obtained by putting seven-cell clusters on hex-paper and using basic trigonometry.

Going further requires a path loss model. Based on Lee's Area-to-Area Model in Sect. 11.7.3.2, the following simplified path loss model is used here

$$P(d) = \frac{P(d_o)}{(d/d_o)^{\beta}} \tag{13.8}$$

or in decibels with respect to 1 milliwatt

$$P_{(\text{dBm})}(d) = P_{(\text{dBm})}(d_o) - 10\beta \log_{10}\{d/d_o\} \ \ (\text{dBm}). \tag{13.9}$$

Applying the above path loss model to each radio link, the value $P(d_o)$ becomes irrelevant and the worst-case forward link signal-to-interference ratio, Λ, is

$$\Lambda = \frac{R^{-\beta}}{(\sqrt{13}R)^{-\beta} + (4R)^{-\beta} + (\sqrt{19}R)^{-\beta} + (5R)^{-\beta} + (\sqrt{28}R)^{-\beta} + (\sqrt{31}R)^{-\beta}}$$
$$= \frac{1}{(\sqrt{13})^{-\beta} + (4)^{-\beta} + (\sqrt{19})^{-\beta} + (5)^{-\beta} + (\sqrt{28})^{-\beta} + (\sqrt{31})^{-\beta}}. \quad (13.10)$$

For the case $N = 3$, the six first-tier co-channel interferers are at distances $\{2R, \sqrt{7}R, \sqrt{7}R, \sqrt{13}R, \sqrt{13}R, 4R\}$. For the case $N = 4$, the six first-tier co-channel interferers are at distances $\{\sqrt{7}R, \sqrt{7}R, \sqrt{13}R, \sqrt{13}R, \sqrt{19}R, \sqrt{19}R\}$. With a path loss exponent $\beta = 3.5$, the worst case Λ in each case is

$$\Lambda_{(\text{dB})} = \begin{cases} 14.56 \text{ dB} & \text{for } N = 7 \\ 9.98 \text{ dB} & \text{for } N = 4 \\ 7.33 \text{ dB} & \text{for } N = 3 \end{cases}.$$

The minimum allowable cluster size is determined by the threshold signal-to-interference, Λ_{th}, of the radio receiver needed to achieve a specified QoS. As long as the worst case $\Lambda_{(\text{dB})}$ exceeds Λ_{th} the QoS will always be met.

Figure 13.7 depicts CCI on the reverse link at the serving BS, considering first-tier co-channel interferers. Although the MS is in the same location of the center cell, the CCI conditions are not the same on the forward and reverse links, because the distance vector **d** is different in each direction. This phenomenon creates *link imbalance*. The worst-case reverse channel CCI situation occurs when the MS is located at the corner of a cell at distance R from its serving BS, as depicted in

Fig. 13.7 Co-channel interference on the reverse link at a BS. There are six first-tier co-channel MSs

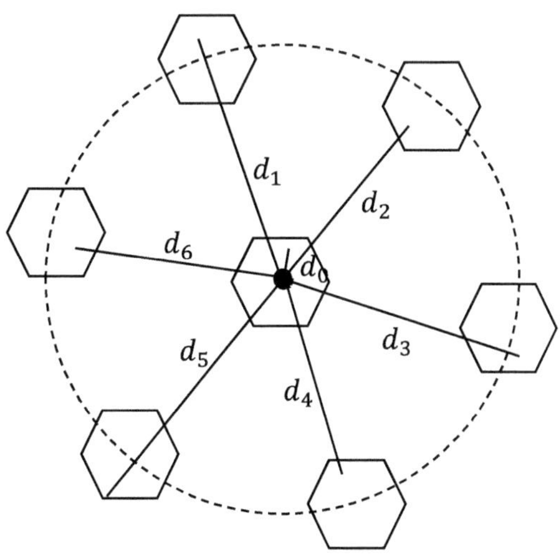

Fig. 13.8. The co-channel MSs are located at the closest possible locations in their cells to the BS located in the center cell.

Finally, the worst-case signal-to-interference ratio condition occurs when all six first-tier co-channel interferers are actively transmitting. However, the number of active co-channel interferers will be random, and the average number that are active will increase or decrease with the system traffic loading. As will be explained in Sect. 13.6.3, it is not possible that all radio resources be continuously utilized.

13.1.2 Cell Sectoring

A simple method to reduce CCI uses wide-beam directional antennas at the BSs to partition the cells into radial sectors. On the reverse link, CCI arriving from directions outside the directional antenna beam can be neglected due to a high front-to-back ratio. On the forward link, the wide-beam directional BS antennas reduce CCI by transmitting the signals to the MSs with a narrower angle-of-departure (AoD) spread than omnidirectional antennas.

Cellular systems are often deployed with 120° or 60° cell sectoring. An N-cell reuse cluster with 120° sectors yields an $N/3N$ reuse plan (N cells and $3N$ sectors). With cell sectoring, the radio resources assigned to each cell are further partitioned into disjoint sets, and each cell sector uses a disjoint or orthogonal set of radio resources. This is illustrated in Fig. 13.9 for the case of 120° cell sectoring, where the radio resources for cell A are further partitioned into sets $A1$, $A2$, and $A3$. With cell sectoring, two different sector orientations are possible as illustrated in Fig. 13.9. The finer partitioning of radio resources with cell sectoring, however, results in a loss in trunking efficiency, as discussed in Sect. 13.6.3.

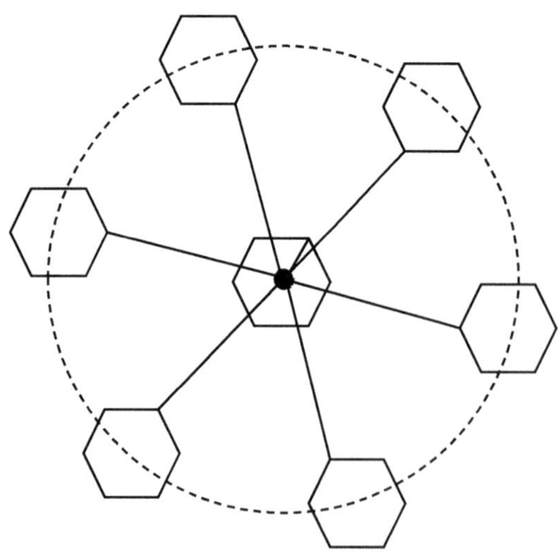

Fig. 13.8 Worst-case CCI situation on the reverse link

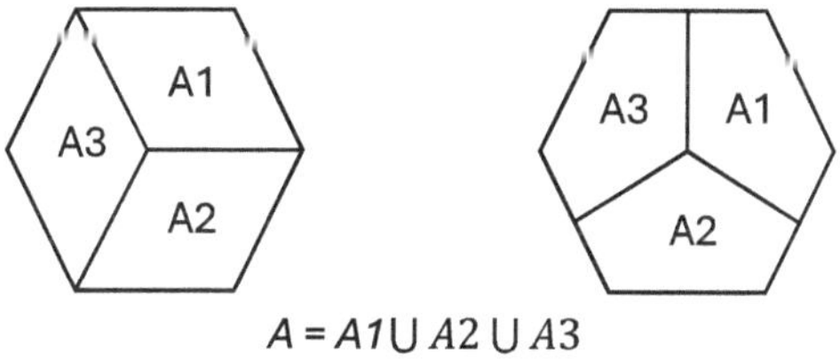

Fig. 13.9 120° cell sectoring with two different sector orientations

For each cluster size N, the sector orientation that reduces the number of first-tier interferers from 6 to 2 is selected, as depicted in Fig. 13.10. The same concept applies to 60° cell sectoring, where two different sector orientations are possible, and the sector orientation that reduces the number of first-tier co-channel interferers from 6 to 1 is selected.

For $N = 7$, careful layout on a hexagonal grid and considering the forward link, there are two first-tier co-channel BSs located at distances $\sqrt{19}R$, $\sqrt{28}R$ from the worst-case MS location, in this case, in the corner of a sector at distance R from its serving BS. The worst-case forward link signal-to-interference ratio, Λ, is

$$\begin{aligned}\Lambda &= \frac{R^{-\beta}}{(\sqrt{19}R)^{-\beta} + (\sqrt{28}R)^{-\beta}} \\ &= \frac{1}{(\sqrt{19})^{-\beta} + (\sqrt{28})^{-\beta}}.\end{aligned} \tag{13.11}$$

For $N = 3$, the two co-channel interferers are at distances $\{\sqrt{7}R, \sqrt{13}R\}$, and for $N = 4$ they are at distances $\{\sqrt{13}R, \sqrt{19}R\}$. Hence,

$$\Lambda_{(\mathrm{dB})} = \begin{cases} 20.60 \text{ dB} & \text{for } N = 7 \\ 17.69 \text{ dB} & \text{for } N = 4 \\ 13.52 \text{ dB} & \text{for } N = 3 \end{cases}.$$

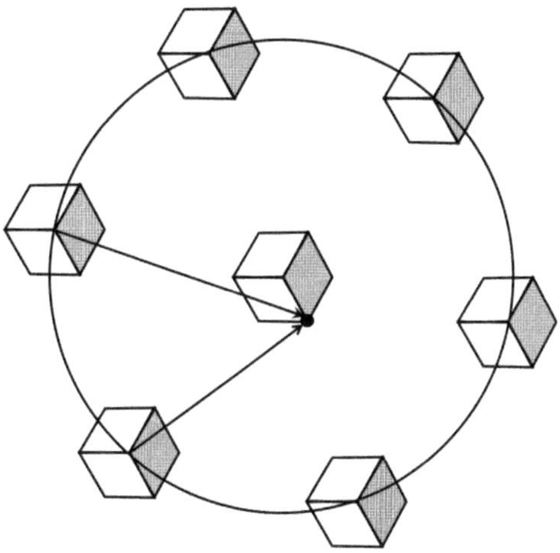

Fig. 13.10 Worst-case CCI situation on the forward channel with 120° cell sectoring

For $N = 7$, 120° sectoring yields a 6.04 dB gain in the worst-case signal-to-interference ratio, Λ, over the case when omnidirectional antennas are used.

13.2 Cell Splitting

Conventional cell splitting is a means to introduce new, smaller, cells into an existing cellular deployment to accommodate traffic growth. Consider the uniform grid of hexagonal cells shown in Fig. 13.11. If heavy traffic loading is experienced at the midpoint between two the cells labeled A, then a split cell labeled A1' is introduced at that location. The area of the split cell is 1/4 of the area of the parent cells. Additional split cells can be introduced to accommodate traffic loading in other locations throughout the system area. For example, the split cell labeled B' can be located at the midpoint between the two cells labeled B.

Because the split cells are smaller, the transmit power can be reduced. To estimate the transmit power requirements in the split cells, consider the the forward link with the simple path loss model in (13.8). The signal power received at an MS located at the corner of a parent cell has the form

$$P(R_o) = \frac{kP_o}{(R_o/d_o)^{\beta}}, \tag{13.12}$$

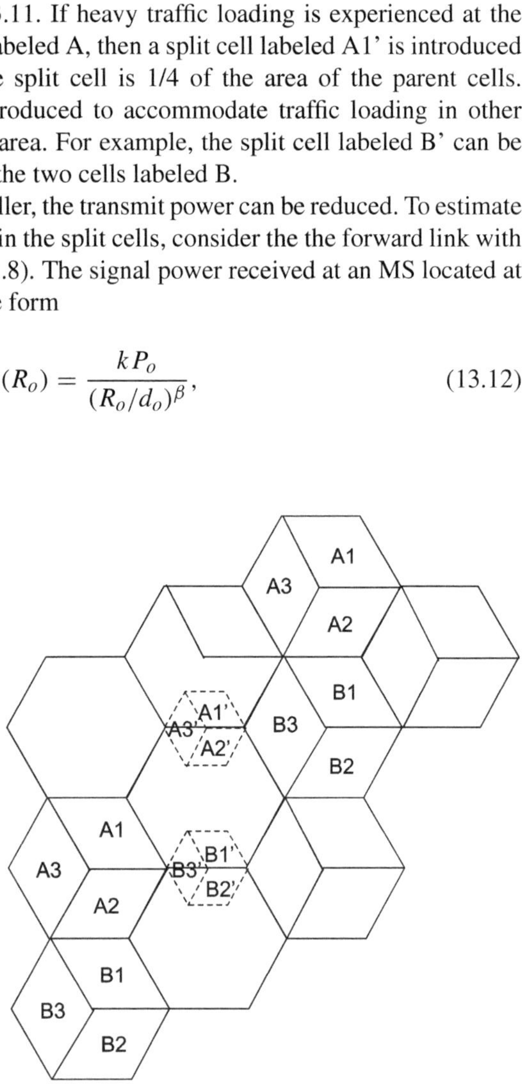

Fig. 13.11 Conventional cell splitting is used to accommodate an increased traffic load by introducing smaller cells

while the signal power received at the corner of a split cell is

$$P(R_s)\frac{kP_s}{(R_o/d_o)^\beta}. \tag{13.13}$$

where P_o and P_s, and R_o and R_s, are the transmit power and cell radius associated with the parent cells and split cells, respectively, and k is a constant of proportionality. To keep the signal power received by a MS located in the corner of a parent or split cell the same, the transmit power of the split cell and parent cell BSs are related as

$$P_s = P_o \left(\frac{R_s}{R_o}\right)^{-\beta}. \tag{13.14}$$

Since $R_s = R_o/2$, if $\beta = 4$, then $P_s = P_o/16$. Hence, the split cell BSs can reduce their transmit power levels by 12 dB.

After introducing the split cells, changes in the frequency plan are required to avoid violations of the frequency reuse constraint. One possible approach is shown in Fig. 13.12, where overlaid inner cells are introduced into the parent cells. Once again, the radio resources are further partitioned into two disjoint sets. MSs located within the overlaid inner cells and the split cells use one set of radio resources, while MSs located within the outer cells use the other set of radio resources. Whenever an MS moves between the inner and outer areas of a parent cell, an intra-cell handoff must be executed.

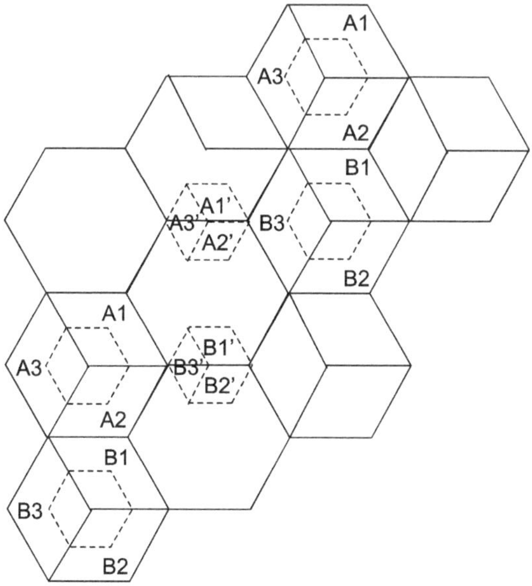

Fig. 13.12 Overlaid inner cells may be used to maintain the frequency reuse constraint when cell splitting is used

13.3 OFDMA Frequency Planning

OFDMA allocates radio resources according to the OFDM subcarrier frequency and time slot that may be combined together to create radio resource blocks. For the OFDMA forward channel, intra-cell interference is essentially eliminated due to the time orthogonality property of waveforms transmitted on different OFDM subcarriers. However, to achieve high spectral efficiency, aggressive frequency planning is necessary that will increase the level of inter-cell interference. For OFDMA, adjacent channel interference is mitigated by the orthogonal nature of the subcarriers. However, with frequency reuse, co-channel interference will be present. A typical OFDMA forward link may employ 120° sectoring and a sector reuse of 1 or 3.

With a sector reuse of 1, also known as universal frequency reuse, all sectors use the same subcarrier frequencies, as shown in Fig. 13.13. With universal frequency reuse, two different schemes can be used to mitigate the co-channel interference between the cells and along the sector boundaries, namely interference avoidance and interference randomization. Interference avoidance assigns sets of subcarriers dynamically so as to avoid interference. Randomization is achieved by scrambling the set of subcarriers that are used within the same band or by frequency hopping over a larger band. In either case, the interference experienced by the users is randomized, and a strong interferer is shared by all the users.

With a sector reuse of 3, the three reuse sectors in each cell are assigned a disjoint set of subcarrier frequencies as shown in Fig. 13.14. Reuse 3 will eliminate the co-channel interference that occurs along the sector boundaries within the same cell. Furthermore, it also increases the co-channel reuse distance. This will reduce the level of co-channel interference. However, a sector reuse of 3 requires that the set of radio resources be partitioned into three groups, which will tend to decrease the Erlang capacity as discussed in Sect. 13.6.3.

A third alternative employs the concept of reuse partitioning. With reuse partitioning, a cell is divided into multiple regions, for example, an inner cell and three outer cell sectors as shown in Fig. 13.15. MSs that are located in the inner cells can

Fig. 13.13 Deployment with a sector reuse factor of 1

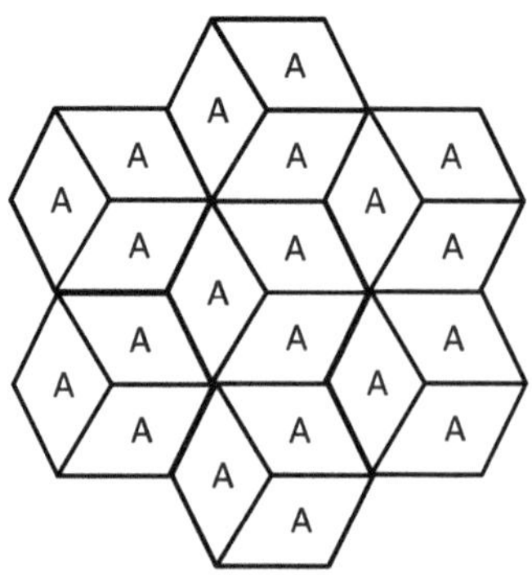

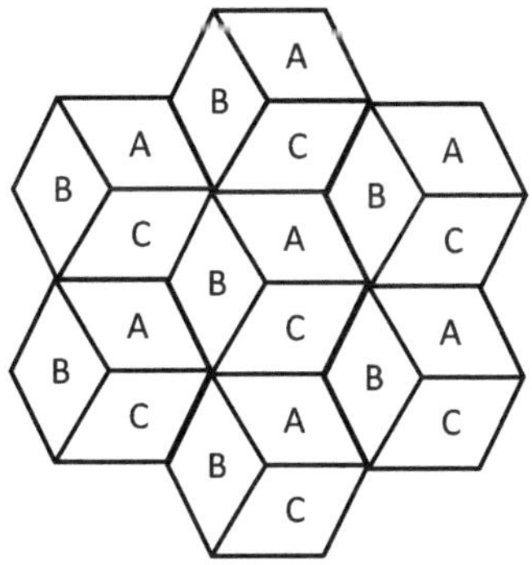

Fig. 13.14 Deployment with a sector reuse factor of 3

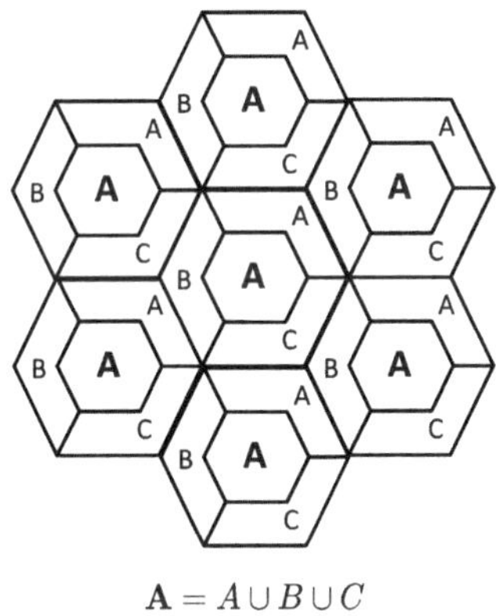

Fig. 13.15 Deployment with reuse partitioning

reuse all the available subcarrier frequencies. MSs that are located in the outer cell areas use a fraction of the available subcarriers according to a 3 sector reuse. This arrangement will yield high spectral efficiency by using all the available subcarrier frequencies in the inner cells, while at the same time provide robustness against co-channel interference by using a three-sector reuse of subcarrier frequencies in the outer cell areas. The spectral efficiency will lay somewhere between that of universal frequency reuse and three-sector reuse.

Reuse partitioning requires that the locations of the users be tracked and the set of subcarriers allocated to the users be dynamically updated according to a policy that promotes a uniform subcarrier signal-to-interference ratio while maximizing a network level performance metric such as the sum capacity. For cellular downlink applications, power and subcarrier allocation schemes can be implemented under the assumption that the BS has mobile-assisted channel state information for all forward channels it serves, and the total BS transmit power is constrained.

13.4 Link Budget

A link budget is a detailed accounting of all the power gains and losses that a communication signal experiences, from the transmitter through the communication channel and input to the detector in the receiver, as well as the noise power that is input to the detector. The end-to-end performance of the communication link will be determined by the ratio of the received signal power, P, and noise power, N, in the samples at the output of the receiver matched filter that are input to the detector, otherwise known as the signal-to-noise power ratio $\Gamma = P/N$.

A simplified link budget for most communication systems can be expressed in terms of the following parameters:

$$\begin{aligned}
P_T &= \text{transmit signal power} \\
G_\text{T} &= \text{transmit antenna gain} \\
L_p &= \text{propagation path loss} \\
G_\text{R} &= \text{receiver antenna gain} \\
P &= \Omega_p/2 = \text{received signal power} \\
E &= \text{received energy per modulated symbol} \\
T_o &= \text{receiving system noise temperature in degrees Kelvin} \\
B_w &= \text{receiver noise equivalent bandwidth in Hz} \\
N_o/2 &= \text{two-sided thermal noise power spectral density} \\
R &= \text{modulation symbol rate or baud rate} \\
k &= 1.38 \times 10^{-23}\, Ws/K = \text{Boltzmann's constant} \\
F &= \text{receiver noise figure} \\
L_\text{Rx} &= \text{receiver implementation losses} \\
P_\text{th} &= \text{receiver sensitivity} \\
E/N_{o\,\text{th(dB)}} &= \text{threshold symbol energy-to-noise ratio}
\end{aligned}$$

The received signal power input to the detector can be expressed as

$$P = \frac{P_T G_\text{T} G_\text{R}}{L_\text{Rx} L_p}. \tag{13.15}$$

The receiver implementation loss, $L_\text{Rx} \geq 1$, is included in the denominator of (13.15). Receiver timing synchronization errors are manifested as a loss in the effective received signal power and an increase in the effective received noise power that is input to the detector due to ISI, as discussed in Sect. 5.4.

The thermal noise power that is input to the detector is [9]

$$N = kT_o B_w F. \tag{13.16}$$

The value of kT_o at a room temperature of 17 °C(290 °K) is $kT_o = -174$ dBm/Hz. The noise figure F is a degradation caused by elements in a radio frequency signal chain that introduce noise. The noise figure of any device or system, F, is defined as the ratio of the input signal-to-noise ratio to the output signal-to-noise ratio. The noise figure is always greater than or equal to unity, and a lower noise figure indicates better performance. The noise figure of a typical receiver is in the range of 4–6 dB.

The received signal-to-noise ratio, Γ, is the power ratio

$$\Gamma = \frac{P}{N} = \frac{P_T G_\mathrm{T} G_\mathrm{R}}{kT_o B_w F L_{\mathrm{Rx}} L_p}. \tag{13.17}$$

The signal-to-noise ratio, Γ, and modulation symbol energy-to-noise ratio, E/N_o, are related as follows:

$$\frac{E}{N_o} = \frac{PT}{N/B_w} = \Gamma \frac{B_w}{R}. \tag{13.18}$$

where $E = PT$ and the one-sided complex white noise power spectral density is $N_o = N/B_w$ with B_w being the noise equivalent bandwidth of the receiver matched filter. The modulation symbol energy-to-noise ratio is related to the bit energy-to-noise ratio as $E/N_o = (\log_2 M) E_b/N_o$ and determines the link QoS such as the bit error rate or frame error rate.

Using (13.18), the link budget may be rewritten as

$$\frac{E}{N_o} = \frac{P_T G_\mathrm{T} G_\mathrm{R}}{kT_o R F L_{\mathrm{Rx}} L_p}. \tag{13.19}$$

Converting to decibel units gives

$$\begin{aligned} E/N_{o(\mathrm{dB})} = {} & P_{T(\mathrm{dBm})} + G_{\mathrm{T}\ (\mathrm{dB})} + G_{\mathrm{R}\ (\mathrm{dB})} - kT_{o(\mathrm{dBm})/\mathrm{Hz}} - R_{(\mathrm{dBHz})} \\ & - F_{(\mathrm{dB})} - L_{\mathrm{Rx}\ (\mathrm{dB})} - L_{p\ (\mathrm{dB})}. \end{aligned} \tag{13.20}$$

The receiver sensitivity is defined as

$$P_{\mathrm{th}} = L_{\mathrm{Rx}} kT_o F (E/N_o)_{\mathrm{th}} R \tag{13.21}$$

or converting to decibel units

$$P_{\mathrm{th}\ (\mathrm{dBm})} = L_{\mathrm{Rx}\ (\mathrm{dB})} + kT_{o(\mathrm{dBm})/\mathrm{Hz}} + F_{(\mathrm{dB})} + E/N_{o\mathrm{th}(\mathrm{dB})} + R_{(\mathrm{dBHz})}. \tag{13.22}$$

The receiver sensitivity can be thought of as the smallest received signal power that will give an acceptable QoS. In (13.22), all parameters are typically fixed by the communication standard and vendor equipment except for the symbol energy-to-noise ratio threshold, $E/N_{o\text{th(dB)}}$, which varies depending on the required QoS, e.g., the required bit error probability.

Substituting the receiver sensitivity $P_{\text{th (dBm)}}$ in (13.22) into (13.20), gives the maximum allowable propagation path loss $L_{p\text{ (dB)}}$ that a link can tolerate as

$$L_{\text{max (dB)}} = P_{T\text{(dBm)}} + G_{\text{T (dB)}} + G_{\text{R (dB)}} - P_{\text{th (dBm)}}. \tag{13.23}$$

Minus the transmitter and receiver antenna gains, the maximum path loss is the difference between the transmit power and the receiver sensitivity. Once $L_{\text{max (dB)}}$ is known, a path loss model as discussed in Sect. 11.7 can be applied to determine the maximum radio path length.

13.4.1 Cellular Link Budget

For cellular radio systems, there are three other major link budget parameters;

$$\begin{aligned} L_{\text{I}} &= \text{interference margin} \\ M_{\text{shad}} &= \text{shadow margin} \\ G_{\text{HO}} &= \text{handoff gain} \end{aligned}$$

The two margins reduce the maximum allowable path loss, while the handoff gain increases it. The maximum allowable path loss with the above three parameters is

$$\begin{aligned} L_{\text{max (dB)}} = {} & P_{T\text{(dBm)}} + G_{\text{T (dB)}} + G_{\text{R (dB)}} - P_{\text{th (dBm)}} - M_{\text{shad (dB)}} \\ & - L_{I\text{ (dB)}} + G_{\text{HO (dB)}}. \end{aligned} \tag{13.24}$$

13.4.2 Co-channel Interference

Cellular radio links are not only affected by noise but also CCI. They exhibit a link performance threshold according to a defined QoS criterion, such that QoS is acceptable if the received signal-to-interference plus noise ratio $P/(I+N) = (P/N)/(1+I/N)$ exceeds a certain threshold value. Two extreme cases are often of interest. The first case is the noise limited case where $I = 0$ and $I/N = 0$. In the noise limited case, there is no CCI, and the link performance is limited by thermal noise in the receiver itself according to the received signal-to-noise ratio $\Gamma = P/N$. Such conditions may be typical of mm-wave fixed wireless access where high-gain narrow-beam directional antenna is used to mitigate CCI. The second case is the interference limited case where $N = 0$ and $I/N = \infty$. In the interference limited case, the link performance is determined by the received signal-to-interference ratio $\Lambda = P/I$. Such conditions are typical of high-capacity cellular land mobile radio systems with dense frequency reuse.

The signal-to-interference plus noise ratio $P/(I + N)$ depends on the sum impairment power $I + N$. To a first approximation, interference may be treated as AWGN, although CCI of equal power is more deleterious to receiver performance than AWGN. If there are multiple interferers of the same type, the receiver performance is insensitive to the distribution of the individual interfering powers, P_n, in the power sum $\sum_{n \in \mathcal{P}} P_n = P$, and largely depends on the total interfering power.

The level of CCI will depend on the traffic loading on the cellular network. As the traffic load increases, the level of CCI will also increase. The radio resources are used with a probability that increases with traffic load. The example in Fig. 13.6 is a worst-case scenario where all six first tier co-channel interferers are actually present. However, at normal system loading, only a random fraction of the first- and higher-tier co-channel interferers will be active.

MSs that are located near the edges of the cells will often experience an unacceptably low QoS due to excessive CCI. The links associated with MSs located near the cell edges will be dropped, effectively shrinking the cell radii. However, as these links are dropped, the overall level of CCI decreases. Eventually, MSs that are located near the edges of the cells may again be able to establish links with an acceptable QoS. However, as more MSs establish links, the CCI will once again shrink the effective cell radii and links will be dropped. This phenomenon of shrinking and expanding cell boundaries due to the statistical nature of CCI is called cell breathing.

To mitigate the effects of cell breathing, an interference margin is included in the link budget. The received signal-to-interference-plus-noise ratio is

$$\Lambda_{IN} = \frac{C}{I + N} \tag{13.25}$$

where I is the total interference power. An interference margin, L_I, is inserted as

$$\Lambda_{IN} = \frac{C}{I + L_I + N} \tag{13.26}$$

The required interference margin depends on the particular cellular system under consideration and the traffic load and can be quite difficult to derive since it depends not only on the parameters of the radio link but also on the detailed radio resource management algorithms being used for power control and handoff. With universal frequency reuse, every cell and cell sector in the network use the entire bandwidth available, and every co-channel emitter will act as a source of interference to every other receiver, so the required interference margin is typically higher.

13.4.3 Shadow Margin

With log-normal shadowing, the power received at distance d in decibel units has a Gaussian distribution $P_{(\text{dBm})}(d) \sim \mathcal{N}(\mu_{(\text{dBm})}(d), \sigma_\Omega^2)$. where $\mu_{(\text{dBm})}(d)$ is the mean received signal power in dBm at distance d as predicted by the pass loss model and σ_Ω is the shadow standard deviation.

An outage occurs when $P_{(\text{dBm})}(d)$ drops below the receiver sensitivity, i.e., $P_{(\text{dBm})}(d) < P_{\text{th (dBm)}}$. The outage probability at a random location on the cell edge at distance $d = R$ is

$$\begin{aligned} O_N(R) &= \mathrm{P}[P_{(\text{dBm})}(R) < P_{\text{th (dBm)}}] \\ &= \int_{-\infty}^{P_{\text{th(dBm)}}} \frac{1}{\sqrt{2\pi}\sigma_\Omega} \exp\left\{-\frac{\left(x - \mu_{P_{(\text{dBm})}}(R)\right)^2}{2\sigma_\Omega^2}\right\} \mathrm{d}x \\ &= Q\left(\frac{M_{\text{shad}}}{\sigma_\Omega}\right), \end{aligned} \tag{13.27}$$

where $Q(x)$ is the Gaussian Q function in (2.63) and

$$M_{\text{shad}} \triangleq \mu_{P_{(\text{dBm})}}(R) - P_{\text{th (dBm)}}. \tag{13.28}$$

is the shadow margin. The cell edge outage probability, $O_N(R)$ is plotted against M_{shad} in Fig. 13.16 for various shadow standard deviations.

Example 13.1 Suppose it is desired to have $O_N(R) = 0.1$. The shadow margin M_{shad} is chosen such that the shaded area under the Gaussian density function in Fig. 13.17 is equal to 0.1. Thus,

$$0.1 = Q\left(\frac{M_{\text{shad}}}{\sigma_\Omega}\right). \tag{13.29}$$

Hence,

$$\frac{M_{\text{shad}}}{\sigma_\Omega} = Q^{-1}(0.1) = 1.28. \tag{13.30}$$

For $\sigma_\Omega = 8$ dB, the required shadow margin is

$$M_{\text{shad}} = 1.28 \times 8 = 10.24\,\text{dB}. \tag{13.31}$$

The area averaged outage probability is defined as the probability that $P_{(\text{dBm})}(d) < P_{\text{th (dBm)}}$ when averaged over the random location of an MS in

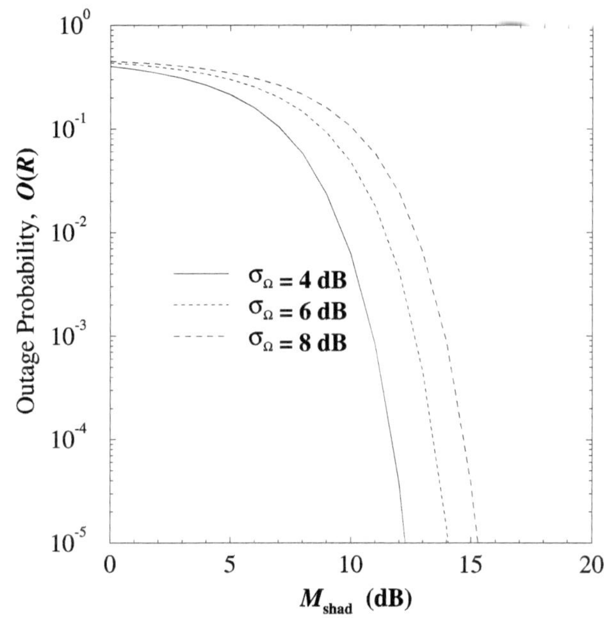

Fig. 13.16 Outage probability on the cell edge, $O_N(R)$, against the shadow margin M_{shad}

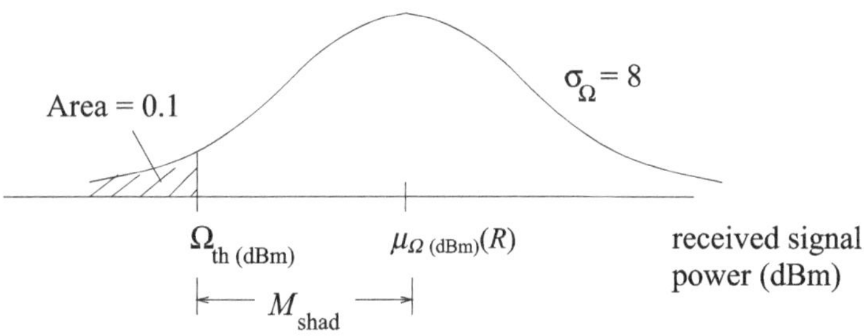

Fig. 13.17 Determining the required shadow margin

the entire cell area. To obtain the area averaged outage probability, models are needed for the propagation path loss and spatial distribution of MSs. One useful model assumes that the MSs are uniformly distributed throughout the cell area. The propagation path loss may assume Lee's model in Sect. 11.7.3.2.

$$\mu_{(\text{dBm})}(d) = \mu_{(\text{dBm})}(d_o) - 10\beta \log_{10}\{d/d_o\} \text{ (dBm)}. \tag{13.32}$$

The above yields an area averaged outage probability [8]

$$\begin{aligned} O_N &= \frac{1}{\pi R^2} \int_0^R O_N(r) 2\pi r \mathrm{d}r \\ &= Q(X) - \mathrm{e}^{XY + Y^2/2} Q(X + Y) \end{aligned} \tag{13.33}$$

where

$$X = \frac{M_{\text{shad}}}{\sigma_\Omega}, \qquad Y = \frac{2\sigma_\Omega \xi}{\beta} \tag{13.34}$$

where $\xi = \ln(10)/10$. The first term in (13.33) happens to be the cell edge outage probability, $O_N(R)$, while the second term accounts for the spatial averaging.

13.4.4 Handoff Gain

As MSs move from one cell to the next, handoffs or handovers are executed to maintain service continuity. To understand how handoffs improve the link budget, suppose that an MS is located in the boundary area between two cells. Although the link to the serving BS may be in an outage condition due to shadowing, the link to an alternate BS may provide an acceptable link QoS due to uncorrelated shadowing on the two links. In fact, in the boundary area, it may be possible to establish links with multiple BSs at once and handoff seamlessly. A macrodiversity gain is obtained by using handoffs equal to the handoff gain, G_{HO}.

There are a variety of handoff algorithms used in cellular systems. Cellular systems with universal frequency reuse employ soft handoff where the MS may connect to multiple BSs at once. With soft handoff, the MS makes the link to the new BS before breaking the link to the serving BS (make before break). If a frequency plan is used on the other hand, a hard handoff is necessary where the MS breaks the link with the serving BS before making the link with the new BS (break before make). This is necessary since the local oscillator in the MS must be tuned to a new carrier frequency when crossing cell boundaries due to the frequency plan.

To compare hard and soft handoffs, consider a cluster of seven cells; the target cell is in the center and surrounded by six adjacent cells as shown in Fig. 13.18, where the cells are modeled as overlapping circles rather than hexagons. Although the MS is located in the center cell as shown, it is possible that the MS could be connected to any one of the seven BSs due to the handoff mechanism. Of course, this is more likely in the boundary area of the target cell. The area averaged outage probability may be obtained by assuming a uniform distribution of the MS location over the circular target cell area.

The links to the center serving BS and the six neighboring BSs have the path loss in (13.32) and correlated log-normal shadowing. For each random MS location, seven path loss values are calculated to each BS. The log-normal shadowing with BS_i, in dB units, is obtained as

$$\epsilon_i = a\zeta + b\zeta_i, \tag{13.35}$$

where

$$a^2 + b^2 = 1,$$

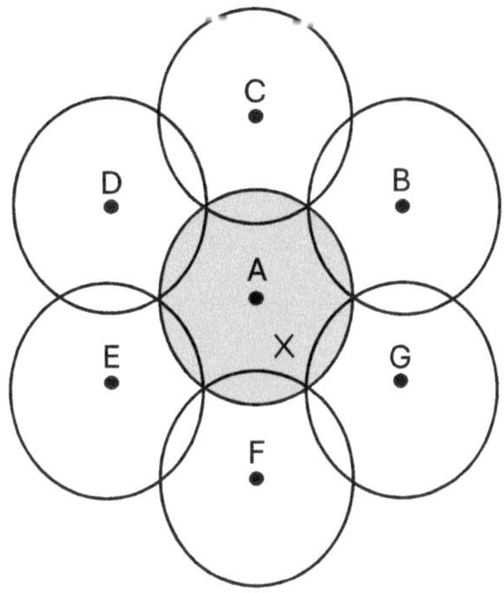

Fig. 13.18 Computing hard and soft handoff gain

and ζ and ζ_i, $1 = 1, \ldots, 6$ are i.i.d. Gaussian random variables with zero mean and variance σ_Ω^2. The log-normal shadows for the above model have covariance

$$\mathrm{E}[\epsilon_i \epsilon_j] = a^2 \sigma_\Omega^2 = \rho \sigma_\Omega^2 \ , \quad i \neq j. \tag{13.36}$$

where $\rho = a^2$ is the correlation coefficient. Once ρ is specified, the required values of a and b can be obtained.

Let $P_{k\ (\mathrm{dBm})}$, $k = 0, \ldots, 6$ denote the received signal strength associated with the target BS ($k = 0$) and each of the six neighboring BSs ($k = 1, \ldots, 6$). Three cases are considered; a single isolated cell to compare with earlier results, soft handoffs and hard handoffs. For the case of a single isolated cell, no handoffs are used and the area averaged outage probability is identical to that obtained in (13.33).

With a soft handoff algorithm, the BS that provides the largest received power is always selected as the serving BS. Hence, if any of the seven BSs has an associated received power at the MS that is above the receiver sensitivity, $P_{\mathrm{th}\ (\mathrm{dBm})}$, then the link quality is acceptable. Only if all have associated received power levels below the receiver sensitivity does an outage occur.

With a hard handoff algorithm, the received power at the MS from the serving BS is denoted as $P_{0\ (\mathrm{dBm})}(d_0)$. If this value exceeds the receiver sensitivity, $P_{\mathrm{th}\ (\mathrm{dBm})}$, then the link quality is acceptable. Otherwise, the six surrounding BSs are considered for handoff candidacy by using a mobile-assisted handoff (MAHO) algorithm that continuously monitors the received power from all seven BSs. BS k is included in the handoff candidate set if $P_{k\ (\mathrm{dBm})}(d_k) - P_{0\ (\mathrm{dBm})}(d_0) \geq H_{(\mathrm{dB})}$, where $H_{(\mathrm{dB})}$ is the *handoff hysteresis*, as depicted in Fig. 13.19. The purpose of the hysteresis is to discourage the MS from executing hard handoffs back and forth, or ping ponging, between BSs due to shadowing in the boundary area. If the received signal power for any member of the handoff candidate set is above the receiver sensitivity, $P_{\mathrm{th}\ (\mathrm{dBm})}$, then a handoff is executed to keep the link quality is acceptable. If the handoff candidate set is empty, then an outage occurs.

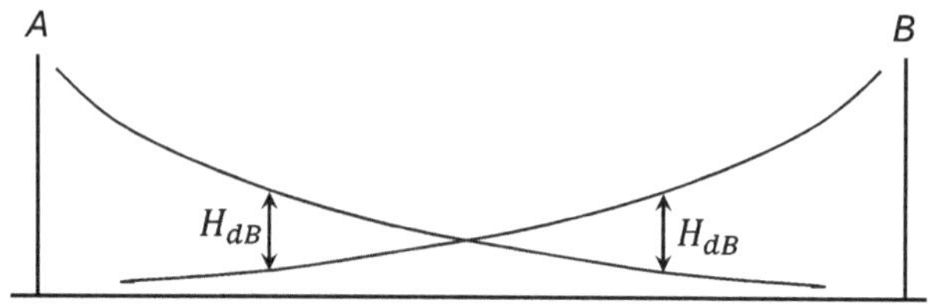

Fig. 13.19 Illustration of hard handoff hysteresis

Fig. 13.20 Required shadow margin with hard and soft handoffs and 95% coverage. Shadowing parameters are $\sigma_{\Omega}^2 = 6$ dB and $\rho = 0.5$

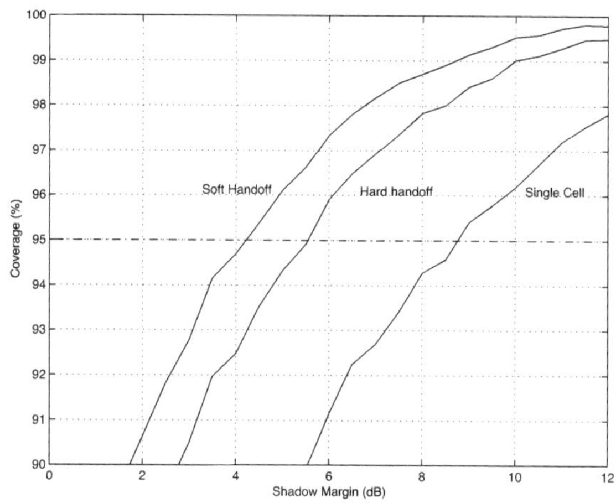

Figure 13.20 compares hard and soft handoff performance in terms of coverage versus shadow margin. A 10% area averaged outage probability (90% coverage) requires a shadow margin of 5.6 dB for a single isolated cell (no handoff). This is to be compared with the 10.24 dB required for a 10% edge outage probability. With soft handoffs, the required shadow margin is 1.8 dB. The difference of 3.8 dB between these two values is the soft handoff gain. The corresponding hard handoff gain is roughly 2.8 dB. The soft handoff gain is always greater than the hard handoff gain depending on the hysteresis $H_{(\mathrm{dB})}$. When $H_{(\mathrm{dB})} = 0$, hard handoff gain will theoretically be the same as the soft handoff gain.

13.5 Coverage

Coverage refers to the number of base stations or cell sites that are required to cover or provide service to a given geographical area with an acceptable QoS. This is an important consideration for a greenfield cellular network. A cellular network option that requires a smaller number of cell sites to cover a given geographic area has an infrastructure cost advantage. However, in order to accommodate future traffic loading, it is important to include an interference margin into the link budget, and this complicates the comparison of different options. The traffic load must be kept the same with the different cellular network options to allow for a fair coverage comparison. However, the dependency of the required interference margin, $L_{I\ (\text{dB})}$ on the traffic loading is complicated, especially for cellular systems with universal frequency reuse.

The number of cell sites that are required to cover a service area is determined by the maximum allowable path loss from the link budget and the path loss model. To compare the coverage of different cellular network options, the maximum allowable path loss, $L_{\max\ (\text{dB})}$, is first determined for the different cellular network choices at the same QoS. Ignoring shadowing and from Lee's Area-to-Area model (11.7.3.2),

$$L_{\max\ (\text{dB})} = B + 10\beta\log_{10}\{d_{\max}\} \quad (\text{dB}) \tag{13.37}$$

where $d_{\max}$ is equal to the radius of the cell, and B is a constant. Naturally, it is desirable that $d_{\max}$ be as large as possible.

Once the $L_{\max}$ have been determined for the various cellular network options, their coverage can be compared. For example, suppose that Option 1 has $L_{\max\ (\text{dB})} = L_1$ and Option 2 has $L_{\max\ (\text{dB})} = L_2$, with corresponding maximum radio path lengths of d_1 and d_2, respectively. The difference in the maximum allowable path loss is related to the cell radii through the following relationship:

$$\begin{aligned} L_1 - L_2 &= 10\beta\left(\log_{10}\{d_1\} - \log_{10}\{d_2\}\right) \\ &= 10\beta\log_{10}\left\{\frac{d_1}{d_2}\right\}. \end{aligned} \tag{13.38}$$

Inverting the above

$$\frac{d_1}{d_2} = 10^{(L_1-L_2)/(10\beta)}. \tag{13.39}$$

Since the area of a cell is equal to $A = \pi d^2$ (assuming a circular cell), the ratio of the cell areas is

$$\frac{A_1}{A_2} = \frac{d_1^2}{d_2^2} = \left(\frac{d_1}{d_2}\right)^2 \tag{13.40}$$

and, hence,

$$\frac{A_1}{A_2} = 10^{2(L_1-L_2)/(10\beta)}. \tag{13.41}$$

Let A_{tot} be the total geographical area to be covered. The ratio of the required number of cell sites for Options 1 and 2 is

$$\frac{N_1}{N_2} = \frac{A_{\text{tot}}/A_1}{A_{\text{tot}}/A_2} = \frac{A_2}{A_1} = 10^{-2(L_1-L_2)/(10\beta)}. \tag{13.42}$$

As an example, suppose that $\beta = 3.5$ and $L_1 - L_2 = 2\,\text{dB}$. Then $N_2/N_1 = 1.30$. Hence, Option 2 requires 30% more base stations to cover the same geographical area as Option 1.

In conclusion, a modest difference in link budget translates into a substantial difference in infrastructure cost. Efforts to reduce the receiver threshold $E/N_{o\text{th}}$ for a given QoS, reduce the required interference margin L_{I}, and improve the handoff gain G_{HO} will all have a substantial impact.

13.6 Spectral Efficiency and Cell Capacity

Spectral efficiency and cell capacity are of primary concern to cellular system operators. Spectral efficiency may be expressed in terms of the following parameters:

$$\begin{aligned} U &= \text{resource usage efficiency (Erlangs/resource)} \\ N_{\text{slot}} &= \text{number of radio resources per RF carrier or subcarrier (resources/carrier)} \\ N_c &= \text{number of RF carriers or subcarriers per cell (carriers/cell)} \\ W_{\text{sys}} &= \text{system bandwidth (Hz/reuse cluster)} \\ W_c &= \text{bandwidth per radio resource (Hz/resource)} \\ A &= \text{area per cell } (\text{km}^2/\text{cell}) \\ N &= \text{reuse cluster size} \end{aligned}$$

One Erlang is the traffic intensity in a radio resource that is continuously occupied, such that if a radio resource is occupied for $x\%$ of the time it carries $x/100$ Erlangs of traffic. The spectral efficiency may defined as

$$\eta_{\text{S}} \stackrel{\Delta}{=} \frac{N_c N N_{\text{slot}} U}{W_{\text{sys}} A} \text{ Erlangs/Hz/km}^2. \tag{13.43}$$

Since

$$W_c = W_{\text{sys}}/(N N_c N_{\text{slot}}),$$

the spectral efficiency can be written as the product of three efficiencies, viz.,

$$\begin{aligned} \eta_{\text{S}} &= \frac{1}{W_c} \cdot \frac{1}{A} \cdot U && (13.44\text{a}) \\ &= \eta_{\text{M}} \cdot \eta_{\text{C}} \cdot \eta_{\text{T}}, && (13.44\text{b}) \end{aligned}$$

where

$$\eta_{\mathrm{M}} = \text{modulation efficiency}$$
$$\eta_{\mathrm{C}} = \text{spatial efficiency}$$
$$\eta_{\mathrm{T}} = \text{trunking efficiency}$$

For cellular systems, the cell capacity N_{cell}, defined as number of resources per cell, is sometimes used instead of the spectral efficiency. The cell capacity is

$$N_{\text{cell}} = N_c N_{\text{slot}} = \frac{W_{\text{sys}}}{W_c \cdot N}. \tag{13.45}$$

Note that the cell capacity does not account for trunking efficiency.

13.6.1 Modulation Efficiency

Bandwidth efficiency is measured in bits per second per unit bandwidth (b/s/Hz). High bandwidth efficiency can be achieved by using MIMO techniques, bandwidth efficient modulation, and coding. Bandwidth efficiency (bits/s/Hz) is equal to modulation efficiency (resources/Hz) defined above multiplied by the bit rate per resource (bits/s/resource).

The modulation efficiency increases with a reduction in the reuse cluster size N. For interference limited cellular systems, small N is achieved by i) minimizing the production of co-channel interference, and ii) mitigating the effect of co-channel interference. Production of co-channel interference can be controlled by using techniques such adaptive power control and handoff algorithms. Co-channel interference can be mitigated by using techniques such as adaptive antenna beam forming and single antenna interference cancelation.

13.6.2 Spatial Efficiency

High spatial efficiency can be achieved by minimizing the area per cell. This is done using microcell, picocell, and femtocell technologies, where progressively smaller cells are used to cover high-traffic areas.

13.6.3 Trunking Efficiency

High trunking efficiency is achieved by using resource allocation schemes that maximize resource utilization. There is a trade-off between trunking efficiency (or offered traffic per channel) and grade of service such as new call and handoff call blocking probabilities and the link QoS. Various fundamental formula were developed by Erlang almost a century ago that laid the foundation of modern teletraffic theory. One of Erlang's most famous results is the Erlang-B formula, first derived in 1917, that gives the probability that a newly arriving customer will not find any free resource in a shared pool of m resources and is denied service. Sometimes, this policy is called a blocked calls cleared queueing discipline, because the denied calls are not buffered or queued, and if no free resource is immediately available, they are denied service. The Erlang-B formula is not entirely applicable to cellular systems, because it does not distinguish between new call and handoff traffic. Normally, handoff calls are given priority over new call requests.

The Erlang-B blocking formula is

$$B(\rho_T, m) = \frac{\rho_T^m}{m! \sum_{k=0}^{m} \frac{\rho_T^k}{k!}}, \tag{13.46}$$

where $B(\rho_T, m)$ is the blocking probability, m is the number of resources and $\rho_T = \lambda_a \mu_d$ is the total offered traffic (λ_a is the call arrival rate (calls/second), and μ_d is the mean call duration (seconds/call)). The Erlang-B formula is derived under standard Markovian assumptions, meaning that there is an infinite population of users, call arrivals are Poisson distributed with an average rate λ_a calls/s, and call durations exponentially distributed with a mean call duration μ_d s/call. Note that the total offered traffic ρ_T is a dimensionless quantity as the units cancel, but is expressed in Erlangs.

Figure 13.21 plots the blocking probability $B(\rho_T, m)$ as a function of the offered traffic per resource $G_c = \rho_T/m$. The benefit from trunking is obvious, since the offered traffic per resource, G_c, increases as the number m of trunked resources increases, at any blocking probability. However, diminishing returns are obtained as m becomes larger. Note that doubling the number of resources in a cell or cell sector will double the cell or sector capacity. However, the Erlang capacity will more than double due to trunking efficiency. The resource usage efficiency is given by

$$U = \rho_T(1 - B(\rho_T, m))/m , \tag{13.47}$$

where the term $(1 - B(\rho_T, m))$ excludes the blocked calls from the carried traffic. For small blocking probabilities, the resource usage efficiency approaches the offered traffic per resource G_c.

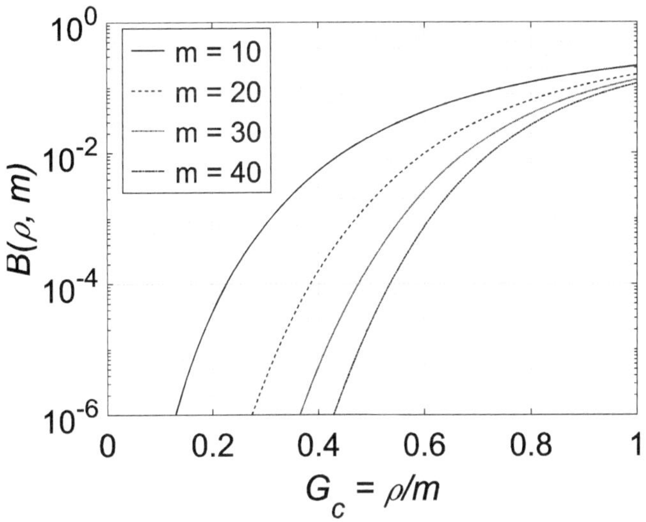

Fig. 13.21 Erlang-B blocking probability $B(\rho_T, m)$ vs. offered traffic per resource $G_c = \rho_T/m$. Trunking is shown to improve the spectral efficiency

Problems

13.1 Consider a hexagonal cellular deployment with omnidirectional BS antennas and reuse cluster size $N = 4$. Assume a path loss exponent $\beta = 3.2$.

(a) Depict the worst-case co-channel interference situation on the reverse link, considering the first tier of co-channel MSs.
(b) Determine the set of exact distances from the first-tier co-channel MSs to the center (serving) BS in part (a).
(c) Calculate the worst-case signal-to-interference ratio, Λ.
(d) Repeat parts (a) to (c) for the case when 120° cell sectoring is used.

13.2 Consider a hexagonal cellular deployment with omnidirectional BS antennas and reuse cluster size $N = 4$. Assume a path loss exponent $\beta = 3.2$.

(a) Depict the worst-case co-channel interference situation on the forward link, considering the first two tiers of co-channel BSs.
(b) Determine the set of exact distances from the first two tiers of co-channel BSs to the target MS.
(c) Calculate the worst-case signal-to-interference ratio, Λ considering the first two tiers of co-channel BSs and compare with the case when only the first-tier co-channel BSs is considered. What is the difference in decibel (dB) units between the two cases?

(d) Considering the first tier of co-channel interfrerers, compare the worst-case signal-to-interference ratio, Λ, on the forward and reverse links using results from Problem 13.1.

13.3 Consider a hexagonal cellular deployment with 120° cell sectoring and a reuse cluster size $N = 4$.

(a) Depict the worst-case co-channel interference situation on the forward link, considering the first tier of co-channel BSs.
(b) Determine the set of exact distances from the first-tier co-channel BSs to the target MS.
(c) Calculate the worst-case signal-to-interference ratio, Λ, assuming a path loss exponent $\beta = 3.5$. Express your result in dB units.

13.4 Consider a hexagonal cellular deployment with 60° *cell sectoring* and a reuse cluster size $N = 3$.

(a) Depict the worst-case co-channel interference situation on the *forward link*, considering the first tier of co-channel interferers.
(b) Calculate the worst-case *forward link* signal-to-interference ratio, Λ, assuming a path loss exponent $\beta = 3$. Express Λ in dB units, Λ_{dB}. Assume that the base stations transmit with the same power and all network entities have isotropic antennas.
(c) Repeat parts (a) and (b) for the *reverse link.*

13.5 Suppose that a mobile station is situated in the corner of a cell and is equidistant from three base stations, BS_1, BS_2, and BS_3 as shown in Fig. 13.22. The received signal power at the mobile station is described by the path loss model

$$\mu_{\Omega_p \text{ (dBm)}}(d) = \mu_{\Omega_p \text{ (dBm)}}(d_o) - 10\beta \log_{10}(d/d_o) \text{ (dBm)}$$

where $\mu_{\Omega_p \text{ (dBm)}}(d)$ the received signal power at distance d (km). Assume that $\beta = 3.5$ and $\mu_{\Omega_p \text{ (dBm)}}(d_o) = -82.36$ dBm, where $d_o = 1.6$ km, and assume that the receiver sensitivity threshold is $-$ 110 dBm.

Fig. 13.22 Deployment for Problem 13.5

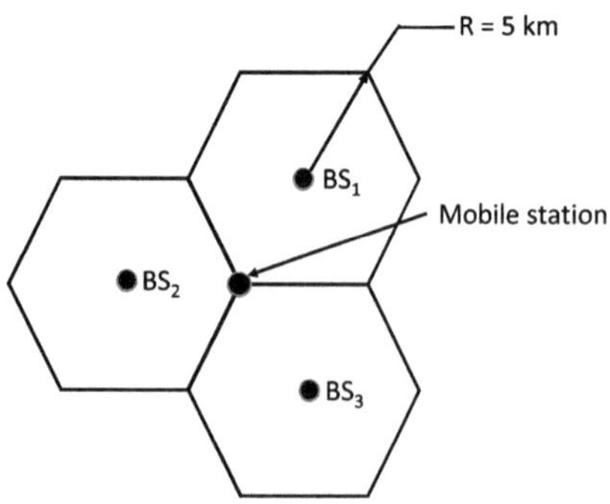

Each of the three links experiences independent log-normal shadowing such that the received signal power in dBm units at distance d, denoted by $\Omega_{p\ (\text{dBm})}(d)$, is a Gaussian random variable with mean $\mu_{\Omega_{p\ (\text{dBm})}}(d)$ and variance σ_Ω^2, where $\sigma_\Omega =$ 8 dB.

(a) If the mobile station is attached at random to one of the three base stations what is the probability of outage?

(b) If the mobile station is attached to the base station providing the largest received signal power, what is the probability of outage?

13.6 An MS receives a desired signal from its serving BS in the presence of a single co-channel BS. The serving BS is located 1 km from the MS, and the co-channel BS is located 5 km from the MS.

The received power in dBm units is described by the path loss model

$$\Omega_{p\ (\text{dBm})}(d) = \mu_{\Omega_{p\ (\text{dBm})}}(d_o) - 10\beta \log_{10}\{d/d_o\} + \epsilon_{(\text{dB})},$$

where $\mu_{\Omega_{p\ (\text{dBm})}}(d_o) = -70$ dBm at $d_o = 1$ km and $\beta = 3.5$. Each link is affected by independent log-normal shadowing, with a shadow standard deviation $\sigma_\Omega = 8$ dB.

(a) What is the probability density function of the signal-to-interference ratio P/I in dB units at the MS?

(b) What is the probability that the received P/I is less than 10 dB?

13.7 Consider that the traffic intensity for one user A_u is 0.5 Erlangs and the number of resources are $m = 2$ and 20. How many users can be supported for 1% blocking probability.

13.8 Suppose that the number of resources per cell is $m = 10$, the total number of users per cell is 500, and the busy hour call blocking probability is 0.1%. What is the average length of time that each subscriber is being served during a busy hour?

13.9 Suppose that an urban area has three competing trunked mobile networks (systems A, B, and C) to provide cellular service. System A has 400 cells with 15 radio resources/cell, System B has 50 cells with 100 radio resources/cell, and System C has 100 cells with 60 radio resources/cell. Ignore handoff traffic and assume uniform cell traffic loading.

(a) Plot the (Erlang-B) blocking probability, $B(A, m)$, for each system versus A, the total offered traffic per cell.

(b) Find the number of users that can be accommodated by each system for a blocking probability of 2%, if the traffic loading offered by each user is 0.1 Erlangs.

13.10 An MS is located midway between two BSs that are 20 km apart. The pass loss is known to follow Lee's suburban model

$$\mu_{\Omega_{p\ (\text{dBm})}}(d) = -61.7 - 38.4 \log_{10}(d/d_o) \ \ (\text{dBm})$$

where $d_o = 1.6$ km. The link to each BS is affected by independent log-normal shadowing having a shadow standard deviation of $\sigma_\Omega = 6$ dB. Assuming soft handoff, the MS is always connected to the BS that provides the largest received power. If that largest received power happens to fall below the receiver sensitivity threshold, $P_{\text{th (dBm)}}$, there is a service outage.

(a) What shadow margin M_{shad} is required on each link to maintain a target service outage probability of 1%?
(b) What receiver sensitivity threshold $P_{\text{th (dBm)}}$ is a least required to meet the 1% target service outage probability?

13.11 Cell splitting is a process whereby smaller cells are introduced into an existing cellular deployment as shown in Fig. 13.11. Assume the following path loss model:

$$\mu_{\Omega_p}(d) = \mu_{\Omega_p}(d_o, h_{b_o}, h_{m_o}) \frac{[(h_b/h_{b_o})(h_m/h_{m_o})]^2}{(d/d_o)^\beta}$$

where $\mu_{\Omega_p\ (dBm)} = -70$ dBm at $d_o = 1\ km$, $\beta = 3.5$, $h_{b_o} = 70\ m$, $h_{m_o} = 1.5\ m$. Set $h_b = h_{b_o} = 70$ m and $h_m = h_{m_o} = 1.5$ m unless otherwise specified.

(a) Convert the path loss model to dBm units, involving the variables, d, h_b and h_m.
(b) If the receiver sensitivity is − 115 dBm, and a combined "shadow margin + handoff gain + interference margin" of 6 dB is used on each link, and the transmit and receive antenna gains are unity (0 dB), what is the radius of the original cells?
(c) Following part (a), what is the radius of the split cells?
(d) If the power received by an MS in the corner of a split cell is kept equal to the power received by an MS in the corner of a parent cell, by what factor must the split cell base station adjust its transmit power? Express your result in dB units.
(e) Repeat part d) if the split cell uses an antenna height of 30 m.

13.12 Suppose the propagation path loss can be modeled by the following equation

$$L_p = 25 \text{ dB} + 10\log_{10}\{d^{2.8}\},$$

where d is in units of meters, i.e., the path loss exponent is $\beta = 2.8$, and there is a 25 dB fixed loss.

(a) Assuming a receiver sensitivity of −95 dBm, what transmitter power is required to service a circular area of radius 10 km?
(b) Suppose the measurements were optimistic and $\beta = 3.1$ is more appropriate. What is the corresponding increase in transmit power (in decibels) that would be required?
(c) If log-normal shadowing is present with $\sigma_\Omega = 8$ dB, how much additional transmit power is required in each case to ensure 10% thermal noise outage at a distance of 10 km?

Correction to: Introductory Digital Wireless Communications

Correction to:
G. Stuber, *Introductory Digital Wireless Communications*, https://doi.org/10.1007/978-3-032-01559-4

The book was published recently, and the author wants to update the supplementary information for all chapters in the book as follows:

The solution manual available in individual chapters needs to be removed, and it will be made available only upon request, not to students.

The updated version of this book can be found at
https://doi.org/10.1007/978-3-032-01559-4

G. Stuber, *Introductory Digital Wireless Communications*,
https://doi.org/10.1007/978-3-032-01559-4_14

Appendix A
Mathematical Tables and Identities

See Table A.1.

Table A.1 Fourier transform pairs

Pair Number	$x(t)$	$X(f)$
1.	$\text{rect}\left(\frac{t}{T}\right)$	$T\text{sinc}(fT)$
2.	$TW\text{sinc}(2Wt)$	$\text{rect}\left(\frac{f}{2W}\right)$
3.	$\wedge\left(\frac{t}{T}\right)$	$T\text{sinc}^2(fT)$
4.	$\exp(-\alpha t)u(t),\ \alpha > 0$	$\frac{1}{\alpha+j2\pi f}$
5.	$t\exp(-\alpha t)u(t),\ \alpha > 0$	$\frac{1}{(\alpha+j2\pi f)^2}$
6.	$\exp(-\alpha\|t\|),\ \alpha > 0$	$\frac{2\alpha}{\alpha^2+(2\pi f)^2}$
7.	$\exp(-\alpha t^2)$	$\sqrt{\frac{\pi}{\alpha}}\exp\left(-\frac{\pi^2 f^2}{\alpha}\right)$
8.	$\delta(t)$	1
9.	1	$\delta(f)$
10.	$\delta(t-t_o)$	$\exp(j2\pi f t_o)$
11.	$\exp(j2\pi f_o t)$	$\delta(f-f_o)$
12.	$\cos(2\pi f_o t)$	$\frac{1}{2}\delta(f-f_o)+\frac{1}{2}\delta(f+f_o)$
13.	$\sin(2\pi f_o t)$	$\frac{1}{j2}\delta(f-f_o)-\frac{1}{j2}\delta(f+f_o)$
14.	$u(t)$	$\frac{1}{j2\pi f}+\frac{1}{2}\delta(f)$
15.	$\text{sgn}(t)$	$\frac{1}{j2\pi f}$
16.	$\frac{1}{\pi t}$	$-j\text{sgn}(f)$
17.	$\sum_{k=-\infty}^{\infty}\delta(t-kT)$	$\frac{1}{T}\sum_{k=-\infty}^{\infty}\delta\left(f-\frac{k}{T}\right)$

$$\text{sinc}(x) = \frac{\sin(\pi x)}{\pi x} \qquad \text{sgn}(x) = \begin{cases} 1 , & x > 0 \\ -1 , & x < 0 \end{cases}$$

$$\text{rect}\left(\frac{t}{T}\right) = \begin{cases} 1 , & |t| < T/2 \\ 0 , & |t| \geq T/2 \end{cases} \qquad \wedge\left(\frac{t}{T}\right) = \begin{cases} 1 - \frac{|t|}{T} , & |t| < T \\ 0 , & |t| \geq T \end{cases}$$

G. Stuber, *Introductory Digital Wireless Communications*,
https://doi.org/10.1007/978-3-032-01559-4

See Table A.2.

Table A.2 Fourier transform properties

Property Number	Time Domain	Frequency Domain
1. Linearity	$a_1x_1(t) + a_2x_2(t)$	$a_1X_1(f) + a_2X_2(f)$
2. Time delay	$x(t - t_o)$	$X(f)\exp(j2\pi f t_o)$
3. Time scaling	$x(at)$	$\frac{1}{\lvert a\rvert} X\left(\frac{f}{a}\right)$
4. Time reversal	$x(-t)$	$X(-f)$
5. Conjugation	$x^*(t)$	$X^*(-f)$
6. Duality	$X(t)$	$x(-f)$
7. Frequency shifting	$x(t)\exp(j2\pi f_o t)$	$X(f - f_o)$
8. Modulation	$x(t)\cos(2\pi f_o t)$	$\frac{1}{2}X(f - f_o) + \frac{1}{2}X(f + f_o)$
9. Area under $x(t)$	$\int_{-\infty}^{\infty} x(t)\mathrm{d}t$	$X(0)$
10. Area under $X(f)$	$x(0)$	$\int_{-\infty}^{\infty} X(f)\mathrm{d}f$
11. Differentiation	$\frac{d^n x(t)}{dt^n}$	$(j2\pi f)^n X(f)$
12. Integration	$\int_{-\infty}^{t} x(t')\mathrm{d}t'$	$\frac{1}{j2\pi f} X(f) + \frac{1}{2}X(0)\delta(f)$
13. Convolution	$\int_{-\infty}^{\infty} x_1(\tau)x_2(t - \tau)\mathrm{d}\tau$	$X_1(f)X_2(f)$
14. Multiplication	$x_1(t)x_2(t)$	$\int_{-\infty}^{\infty} X_1(\nu)X_2(f - \nu)\mathrm{d}\nu$

Some Useful Trigonometric Identities

Angle sum and angle difference relations

$$\sin(\alpha+\beta) = \sin\alpha\cos\beta + \cos\alpha\sin\beta$$
$$\sin(\alpha-\beta) = \sin\alpha\cos\beta - \cos\alpha\sin\beta$$
$$\cos(\alpha+\beta) = \cos\alpha\cos\beta - \sin\alpha\sin\beta$$
$$\cos(\alpha-\beta) = \cos\alpha\cos\beta + \sin\alpha\sin\beta$$

Function product relations

$$\sin\alpha\sin\beta = \tfrac{1}{2}\cos(\alpha-\beta) - \tfrac{1}{2}\cos(\alpha+\beta)$$
$$\cos\alpha\cos\beta = \tfrac{1}{2}\cos(\alpha-\beta) + \tfrac{1}{2}\cos(\alpha+\beta)$$
$$\sin\alpha\cos\beta = \tfrac{1}{2}\sin(\alpha+\beta) + \tfrac{1}{2}\sin(\alpha-\beta)$$
$$\cos\alpha\sin\beta = \tfrac{1}{2}\sin(\alpha+\beta) - \tfrac{1}{2}\sin(\alpha-\beta)$$

Function sum and difference relations

$$\sin\alpha + \sin\beta = 2\sin\tfrac{1}{2}(\alpha+\beta)\cos\tfrac{1}{2}(\alpha-\beta)$$
$$\sin\alpha - \sin\beta = 2\cos\tfrac{1}{2}(\alpha+\beta)\sin\tfrac{1}{2}(\alpha-\beta)$$
$$\cos\alpha + \cos\beta = 2\cos\tfrac{1}{2}(\alpha+\beta)\cos\tfrac{1}{2}(\alpha-\beta)$$
$$\cos\alpha - \cos\beta = -2\sin\tfrac{1}{2}(\alpha+\beta)\sin\tfrac{1}{2}(\alpha-\beta)$$

Power relations

$$\sin^2\alpha = \tfrac{1}{2}(1-\cos 2\alpha)$$
$$\cos^2\alpha = \tfrac{1}{2}(1+\cos 2\alpha)$$
$$\sin^2\alpha + \cos^2\alpha = 1$$

Double angle relations

$$\sin 2\alpha = 2\sin\alpha\cos\alpha$$
$$\cos 2\alpha = \cos^2\alpha - \sin^2\alpha = 2\cos^2\alpha - 1 = 1 - 2\sin^2\alpha$$

References

1. S. M. Alamouti, "A Simple Transmit Diversity Technique for Wireless Communications," *IEEE J. Selec. Areas Commun.*, Vol. 16, pp. 1451–1458, October 1998.
2. T. Aulin, "A modified model for the fading signal at a mobile radio channel," *IEEE Trans. Veh. Technol.*, Vol. 28, pp. 182–203, August 1979.
3. E. R. Berlekamp, *Algebraic Coding Theory*. New York: McGraw-Hill, 1968.
4. E. Biglieri, D. Divsalar, P. McLane, and M. Simon, *Introduction to Trellis-Coded Modulation with Applications*. New York, NY: McMillan, 1991.
5. G. C. Clark, Jr. and J. B. Cain, *Error-Correction Coding for Digital Communications*. New York, NY: Plenum, 1981.
6. R. Clarke, "A statistical theory of mobile radio reception," *Bell System Tech. J.*, Vol. 47, pp. 957–1000, 1968.
7. COST 207 TD(86)51-REV 3 (WG1):, "Proposal on channel transfer functions to be used in GSM tests late 1986," September 1986.
8. K. Feher, *Advanced Digital Communications*. Englewood Cliffs, NJ: Prentice-Hall, 1987.
9. K. Feher, *Wireless Digital Communications*. Englewood Cliffs, NJ: Prentice-Hall, 1995.
10. G. D. Forney, Jr., "Burst-correcting codes for the classic bursty channel," *IEEE Trans. Commun. Technol.*, Vol. 19, pp. 772–781, October 1971.
11. R. Gallager, *Information Theory and Reliable Communications*. New York, NY: Wiley, 1968.
12. I. Gradshteyn and I. Ryzhik, *Tables of Integrals, Series, and Products*. San Diego, CA: Academic Press, 1980.
13. M. Hata and T. Nagatsu, "Mobile location using signal strength measurements in cellular systems," *IEEE Trans. Veh. Technol.*, Vol. 29, pp. 245–251, 1980.
14. W. C. Jakes, *Microwave Mobile Communication*. New York, NY: IEEE Press, 1993.
15. W. C. Y. Lee, *Mobile Communications Design Fundamentals*. Indianapolis, IN: Sams, 1986.
16. S. Lin and D. J. Costello, Jr., *Error Control Coding: Fundamentals and Applications*. Englewood Cliffs, NJ: Prentice-Hall, 1983.
17. R. W. Lucky, "Automatic equalization for digital communication," *Bell System Tech. J.*, Vol. 44, pp. 547–588, April 1965.
18. R. W. Lucky, "Techniques for adaptive equalization of digital communication systems," *Bell System Tech. J.*, Vol. 45, pp. 255–286, February 1966.
19. F. J. MacWilliams, "A Theorem on the Distribution of Weights in a Systematic Code," *Bell System Tech. J.*, Vol. 42, pp. 79–94, 1963.
20. S. J. Mason, "Feedback theory: further properties of signal flow graphs," *IRE*, Vol. 44, pp. 920–926, July 1956.

G. Stuber, *Introductory Digital Wireless Communications*,
https://doi.org/10.1007/978-3-032-01559-4

21. J. L. Massey, "Coding and modulation in digital communications," in *Int. Zurich Sem. Digital Communications*, Zurich, Switzerland, pp. E2(1)–E2(4), March 1974.
22. L. B. Milstein, D. L. Schilling, R. L. Pickholtz, V. Erceg, M. Kullback, E. G. Kanterakis, D. Fishman, W. H. Biederman, and D. C. Salerno, "On the feasibility of a CDMA overlay for personal communications networks," *IEEE J. Selec. Areas Commun.*, Vol. 10, pp. 655–668, May 1992.
23. Y. Okumura, E. Ohmuri, T. Kawano, and K. Fukuda, "Field strength and its variability in VHF and UHF land mobile radio service," *Rev. of the ECL*, Vol. 16, pp. 825–873, 1968.
24. M. F. Pop and N. C. Beaulieu, "Limitations of Sum-of-Sinusoids Fading Channel Simulators," *IEEE Trans. Commun.*, Vol. 49, pp. 699–708, April 2001.
25. J. G. Proakis and M. Salehi, *Digital Communications, 5th ed.* New York, NY: McGraw-Hill, 2007.
26. J. L. Ramsey, "Realization of optimum interleavers," *IEEE Trans. Inform. Theory*, Vol. 16, pp. 338–345, May 1970.
27. C. E. Shannon, "A Mathematical Theory of Communication," *Bell System Tech. J.*, Vol. 27, pp. 379–423 and 623–656, 1948.
28. G. L. Stüber, *Principles of Mobile Communication*. Cham, Switzerland: Springer International Publishing, AG, 2017.
29. G. Ungerboeck, "Channel coding with multilevel phase signals," *IEEE Trans. Inform. Theory*, Vol. 28, pp. 55–67, January 1982.
30. G. Ungerboeck, "Trellis coded modulation with redundant signal sets – part I: introduction," *IEEE Commun. Mag.*, Vol. 25, pp. 5 – 11, February 1987.
31. G. M. Vachula and J. F. S. Hill, "On optimal detection of band-limited PAM signals with excess bandwidth," *IEEE Trans. Commun.*, Vol. 29, pp. 886–890, June 1981.
32. A. J. Viterbi, "Error bounds for convolutional codes and an asymptotically optimum decoding algorithm," *IEEE Trans. Inform. Theory*, Vol. 13, pp. 260–269, April 1967.
33. A. J. Viterbi, "Convolutional codes and their performance in communication systems," *IEEE Trans. Commun.*, Vol. 19, pp. 751–772, October 1971.
34. S. B. Weinstein and P. M. Ebert, "Data transmission by frequency division multiplexing using the discrete Fourier transform," *IEEE Trans. Commun.*, Vol. 19, pp. 628–634, October 1971.
35. J. M. Wozencraft and I. M. Jacobs, *Principles of Communication Engineering*. Prospect Heights, IL: Waveland Press, 1990.
36. A. G. Zajić and G. L. Stüber, "Efficient Simulation of Rayleigh Fading with Enhanced De-Correlation Properties," *IEEE Trans on Wireless Commun.*, Vol. 5, pp. 1866–1875, July 2006.

Index

G. Stuber, *Introductory Digital Wireless Communications*,
https://doi.org/10.1007/978-3-032-01559-4

MIX
Papier aus verantwortungsvollen Quellen
Paper from responsible sources
FSC® C105338

If you have any concerns about our products,
you can contact us on
ProductSafety@springernature.com

In case Publisher is established outside the EU,
the EU authorized representative is:
Springer Nature Customer Service Center GmbH
Europaplatz 3, 69115 Heidelberg, Germany

Printed by Libri Plureos GmbH
in Hamburg, Germany